Iterative Learning Control Algorithms and Experimental Benchmarking

Eric Rogers, Bing Chu, Christopher Freeman and Paul Lewin
University of Southampton, UK

The right of Eric Rogers, Bing Chu, Christopher Freeman and Paul Lewin to be identified as the authors of this work has been asserted in accordance with law.

Registered Office
John Wiley & Sons, Inc., 111 River Street, Hoboken, NJ 07030, USA
John Wiley & Sons Ltd, The Atrium, Southern Gate, Chichester, West Sussex, PO19 8SQ, UK

For details of our global editorial offices, customer services, and more information about Wiley products visit us at www.wiley.com.

Wiley also publishes its books in a variety of electronic formats and by print-on-demand. Some content that appears in standard print versions of this book may not be available in other formats.

Library of Congress Cataloging-in-Publication Data Applied for:

ISBN 9780470745045 (hardback)

Cover Design: Wiley
Cover Image: © agsandrew/Getty Images

Set in 9.5/12.5pt STIXTwoText by Straive, Chennai, India
Printed and bound by CPI Group (UK) Ltd, Croydon, CR0 4YY

C9780470745045_040123

Contents

Preface

Humans and other species seeking to complete many finite duration tasks proceed by an attempt at it. If not successful, then aim to learn from the outcome to complete the task after many repeated attempts if necessary. This approach led Japanese robotics researcher Suguru Arimoto to say, "It is human to make mistakes, but it also human to learn from such experience. Is it possible to think of a way to implement such a learning ability in the automatic operation of dynamic systems?" This statement is widely seen as the starting point for iterative learning control (ILC), which is applicable, in its basic form, to systems that complete the same finite duration task over and over again. The sequence of operations is that the task is executed, and then the system resets to the starting position, ready for the next in the sequence. As in many areas, the terminology in ILC varies between terming an execution a trial or an iteration, or a pass. In this book, a trial is the default terminology, and the associated finite duration of the operation is termed the trial length.

The standard ILC design specification assumes that a reference trajectory is available. The difference between the output on any trial and the reference trajectory is the error on this trial. Then the task is to design a control law that forces the sequence of errors to converge from trial-to-trial, ideally to zero or to within an acceptable tolerance. The convergence of ILC designs is a very well-studied problem for various forms of dynamics. It has led to many design algorithms and, in many cases, at least laboratory-level supporting experimental evaluation.

Robotics is a source of many problems to which ILC can be applied. A particular example is "pick and place" operations, e.g. a robot required to collect an object from a given location, transfer it over a finite time, place it on a conveyor under synchronization, and then return to the starting location for the next one, and so on. The control input for the subsequent trial can be computed during the reset time, with all of the previous trial error data available if required since the data are generated on the previous trial. In this application, error convergence in the trial variable must be supported by control of the dynamics generated along the trials, e.g. to prevent unacceptable behavior, such as oscillations that would be of particular concern should the object be an open-necked container containing a liquid.

This book focuses on ILC analysis and design with experimental benchmarking. The analysis and design methods coverage include linear and nonlinear dynamics, time-varying, and stochastic dynamics. Both linear and nonlinear designs are considered, where one design for the latter case is for examples where actuation saturation is a possibility.

Two-dimensional (2D) systems, i.e. information propagation in two independent directions, are a well-established area of research where image- and signal-processing applications are not the case for control. Repetitive processes are a distinct class of 2D systems where information propagation occurs only over a finite duration in one of the independent directions. Moreover, this is a feature of the dynamics and not an assumption introduced for analysis purposes. These processes are a natural fit for ILC dynamics. This book develops a range of repetitive process-based designs. These designs complement/improve those based, e.g. on the lifting approach where the finite trial length allows the trial-to-trial error dynamics to be represented by a standard linear systems difference equation. Hence, results for this latter case can be applied.

A particular feature of this book is algorithm, or control law, development supported by at least laboratory experimental benchmarking. The latter aspect contains results from experimental testbeds, such as a laboratory-based system replicating the pick-and-place task for robotics, and also from implementation on physical systems such as a free-electron laser system. In some cases, a laboratory testbed has been specially designed for ILC experiments, e.g. a servomechanism and a rack feeder system.

Another distinctive feature is the application of ILC in healthcare. This area is represented by robotic-assisted upper-limb stroke rehabilitation. In physiotherapy, the recommended approach to rehabilitation after a stroke is based on repeated practice of a task where experience from previous attempts is used to update the effort applied on the next attempt. Still, the difficulty is that the patient cannot move the affected limb and therefore does not get feedback. In such cases, functional electrical stimulation, appropriately designed, can be applied to assist movement.

In a program of research stretching back to the mid-2000s, collaborative research at the University of Southampton, between physiotherapists and control engineers, has used ILC to regulate the level of assistive stimulation applied. This research has followed through with small-scale clinical trials with patients. The critical feature is that if the patient improves with each successive attempt, the level of applied simulation goes down. Therefore, the patient's voluntary effort is increasing. This feature is essential in rehabilitation, where the ultimate aim is to restore independent living. A particular feature of this book is the use of the rehabilitation application to highlight the application of ILC designs, ranging from the tuning of simple structure controllers to nonlinear model-based designs.

The first two chapters give background to the subject area, including the basics of the settings used for analysis and design and the models for some physical examples used to validate the designs, respectively, experimentally. Chapter 3 gives an overview of design for performance. Chapter 4 covers the design of control laws with simple structures and supporting experimental results. Chapters 5–7 cover design based on linear models using various settings, including the repetitive processes/2D systems setting and robust design, with supporting experimental results. Chapter 8 begins the coverage of nonlinear effects, focusing on input constraints. Two designs are developed, one of which uses the repetitive process setting for the case of a saturating input and the other constrained design for linear time-varying dynamics, again with experimental results. Chapter 9 then addresses ILC for distributed parameter systems, with particular emphasis on building approximate dynamics models, and one of the designs is supported by experimental results.

This book has taken a long time to complete and has benefited hugely from collaboration with others. The first and second authors have been working with David Owens (University of Sheffield, UK) since their PhD studies, and results from this collaboration have contributed significantly to several chapters. A very long-standing collaboration with Krzysztof Galkowski, Wojciech Paszke, and Lukasz Hladowski (University of Zielona Gora, Poland) and Slawek Mandra (Nicolaus Copernicus University in Toruń, Poland) forms the basis for the results on repetitive process-based design and experimental validation for linear systems. The extension to design based on nonlinear repetitive process stability theory has results from a collaboration with Pavel Pakshin and Julia Emelianova (R.E. Alekseev Nizhny Novgorod State Technical University, Russia). The experimental results for constrained design for linear time-varying systems in Chapter 8 and the those in Chapter 9 for distributed parameter systems were obtained from experimental testbeds designed and commissioned by Harald Aschemann and Andreas Rauh (University of Rostock, Germany). The data-driven model example in Chapter 13 arose from collaboration with Simon Vestergaard Johansen and Jan Dimon Bendtsen (Aalborg University, Denmark) and the quantum control application from collaboration with Re-Bing Wu (Tsinghua University, PR China).

In the University of Southampton, our completed PhD students Dr. James Ratcliffe, Dr. Jack Cai, and Dr. Weronika Nowicka contributed to the design and commissioning of the gantry robot test facility, the reference shift algorithm of Chapter 4, and the wind turbine application of Chapter 9, respectively. The contributions of Professor Owen Tutty to the flow modeling for the wind turbine application are also acknowledged. A chance conversation at a university event with Professor Jane Burridge led to the start of a long and ongoing research program on using ILC in robotic-assisted stroke rehabilitation with her, Dr. Ann-Marie Hughes, and Dr. Katie Meadmore.

University of Southampton, UK

December 2002 *Eric Rogers, Bing Chu, Christopher Freeman, and Paul Lewin*

1

Iterative Learning Control: Origins and General Overview

A commonly encountered requirement in some industries is for a machine to repeat the same finite duration operation over and over again. The exact sequence is that the procedure is completed and then the system or process involved resets to the starting location and the next one begins. A typical scenario is a gantry robot, such as the one shown in Figure 1.1, undertaking a "pick and place" operation encountered in many industries where the following steps must be conducted in synchronization with a conveyor system: (i) collect an object from a fixed location, (ii) transfer it over a finite duration, (iii) place it on the moving conveyor, (iv) return to the original location for the next object and then, (v) repeat the previous four steps for as many objects as required or can be transferred before it is necessary to stop for maintenance or other reasons. Stopping these robots for such reasons in high-throughput applications means down time and lost production.

Figure 1.2 shows a 3D reference trajectory for the gantry robot and Figure 1.3 its X-axis component. On each execution a variable, say y, is defined over the finite duration taken to move from the pick to place locations, e.g. $y(t)$, $0 \leq t \leq \alpha < \infty$, but it is also required to distinguish variables according to which execution is under consideration. One option, used except where stated otherwise in this book, is to write $y_k(t)$, $0 \leq t \leq \alpha$, where k is a nonnegative integer termed the trial number with α denoting the trial duration or length.

Let $r(t)$ be a prespecified 3D path or trajectory that the robot is required to follow between the "pick" and "place" locations (and back to the "pick" location), such as Figure 1.2 or, to focus on one axis, Figure 1.3. Then on trial k, the error is $e_k(t) = r(t) - y_k(t)$ and if $e_k(t) \neq 0$, the question is: how should the control input signal be adjusted to reduce or remove this error?

In applications such as the one considered, the system is performing the same operation repeatedly under the same operating conditions. One approach to control design is to copy human behavior and aim to learn from experience, i.e. the errors generated on previous trials are rich in information. This information is not exploited by a standard controller that would produce the same error on each trial. Iterative learning control (ILC) aims to improve performance by using information from previous trials to update the control law to be applied on the next one. As the above example illustrates, a significant application area for ILC is robotics.

As with other areas within control systems, there has been a debate, see the next section, on the origins of ILC. Since the 1980s, when concentrated research began, applications for ILC have spread beyond robotics in the industrial domain and outside engineering into healthcare. An example from the latter area, in the form of robotic-assisted upper-limb stroke rehabilitation, is introduced in Section 2.4 and considered in depth in Sections 4.2, 10.2, 10.3, and 11.4 of this book.

Iterative Learning Control Algorithms and Experimental Benchmarking, First Edition.
Eric Rogers, Bing Chu, Christopher Freeman and Paul Lewin.

Figure 1.1 A gantry robot for a pick-and-place operation with the axes marked.

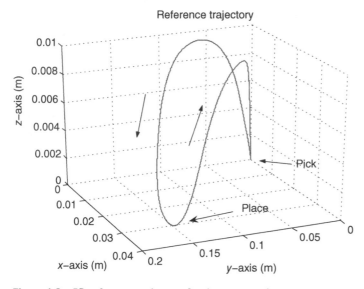

Figure 1.2 3D reference trajectory for the gantry robot.

1.1 The Origins of ILC

According to [2, 33] and others, the basic idea of ILC was first proposed in a 1971 patent [89] and the journal article [256] published in Japanese. The first concerted volume of work that initiated widespread interest was, in particular, the journal paper [12], which considers a simple first-order linear servomechanism system for speed control of a voltage-controlled DC-servomotor. In this section, this system is used to highlight the essence of ILC, and it is appropriate to start by quoting parts of the opening two paragraphs in this paper.

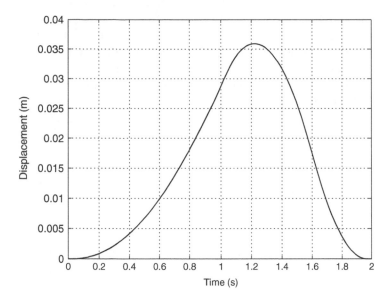

Figure 1.3 *X*-axis component of the gantry robot reference trajectory.

"It is human to make mistakes, but it also human to learn from such experience. Athletes have improved their form of body motion by learning through repeated training and skilled hands have mastered the operation of machines or plants by acquiring skill in practice and gaining knowledge. Is it possible to think of a way to implement such a learning ability in the automatic operation of dynamic systems?"

Motivated by this consideration, Arimoto et al. [12] proposed "a practical approach to the problem of bettering the present operation of mechanical robots by using the data of previous operations." This work constructed an "iterative betterment process for the dynamics of robots so that the trajectory of their motion approaches asymptotically a given desired trajectory as the number of operation trials increases." The example in [12] is next used to illustrate the construction of ILC laws and the behavior that can arise. A critical feature is "the direct use of the underlying dynamics of the objective systems."

The form of the control law developed in [12] applies to systems that are required to track a desired reference trajectory **of a fixed length** α and **specified a priori**. After each trial, **resetting of the system states** occurs, during which time the **measured output** is used in the construction of the control input for application on the next trial. In [12], the system dynamics were assumed **trial-invariant** and **invertible**. These six distinguishing features of ILC highlighted in bold provided the basis for a major area of research in the control systems community internationally, both in terms of theory and an ever-broadening list of applications, many with supporting experimental verification or actual implementation.

As a motivating example, Arimoto et al. [12] considered speed control of a voltage-controlled DC servomotor where if the armature inductance is sufficiently small and mechanical friction is ignored, the resulting controlled dynamics are described by

$$\dot{y}(t) + ay(t) = bv(t) \tag{1.1}$$

where $\dot{y}(t)$ denotes the angular velocity of the motor, $v(t)$ is the input voltage, and a and b are constants.

Suppose that a reference signal, or trajectory, $r(t)$, for the angular velocity $y(t)$ is given over the fixed finite duration $0 \leq t \leq \alpha$ and also that the system dynamics are unknown in the sense that the exact values of a and b in (1.1) may not be available. Also, it is assumed that $r(t)$ is continuously differentiable, which is a commonly used assumption in differential ILC design.

The design problem is to construct an input voltage $v(t)$ that coincides with $r(t)$ over $0 \leq t \leq \alpha$. If an arbitrary input $v_0(t)$ is applied, the error between the desired output $r(t)$ and the first response $y_0(t)$ is

$$e_0(t) = r(t) - y_0(t) \tag{1.2}$$

where

$$\dot{y}_0(t) + ay_0(t) = bv_0(t) \tag{1.3}$$

Also, construct the voltage

$$v_1(t) = av_0(t) + b^{-1}\dot{e}_0(t) \tag{1.4}$$

and apply this as the input on the next trial. Storing the resulting error $e_1(t)$, constructing $e_2(t)$, and continuing this sequence of operations give on trial $k + 1$

$$v_{k+1}(t) = v_k(t) + b^{-1}\dot{e}_k(t)$$
$$e_k(t) = r(t) - y_k(t)$$
$$y_k(t) = e^{-at}y_k(0) + b \int_0^t e^{-a(t-\tau)}v_k(\tau)d\tau \tag{1.5}$$

Suppose that $y_0(0) = r(0) - y_k(0)$ for all k. Then since $r(t)$ is continuously differentiable and using (1.5) gives

$$\dot{e}_k(t) = a \int_0^t e^{-a(t-\tau)}\dot{e}_{k-1}(\tau)d\tau \tag{1.6}$$

Hence,

$$|\dot{e}_k(t)| \leq |a|^2 \int_0^t dt_1 \int_0^{t_1} e^{-a(t-t_2)}|\dot{e}_{k-2}(t_2)|dt_2$$
$$\leq |a|^k \int_0^t \int_0^{t_1} \cdots \int_0^{t_{k-1}} |\dot{e}_0(t_k)|dt_k dt_{k-1}\cdots dt_1$$
$$\leq \frac{\mathcal{H}|a|^k \alpha^k}{k!} \to 0, \quad k \to \infty \tag{1.7}$$

where

$$\mathcal{H} = \max_{0 \leq t \leq \alpha}|\dot{e}_0(t)|$$

Supporting simulation studies are in [12]. Since this first work, various definitions of ILC have been given in the literature, including the following quoted in [2]:

1. The learning control concept stands for the repeatability of operating a given objective system and the possibility of improving the control input on the basis of previous actual operation data [12].
2. It is a recursive online control method that relies on less calculation and less a priori knowledge about the system dynamics. The idea is to apply a simple algorithm repetitively to an unknown system, until perfect tracking is achieved [26].
3. ILC is an approach to improve the transient response of the system that operates repetitively over a fixed time interval [166].

4. ILC considers systems that repetitively perform the same task with a view to sequentially improving accuracy [4].
5. ILC is to utilize the system repetitions as an experience to improve the system control performance even under incomplete knowledge of the system to be controlled [50].
6. The controller learns to produce zero-tracking error during repetitions of a command or learns to eliminate the effects of a repeating disturbance on a control system [210].
7. The main idea behind ILC is to iteratively find an input sequence such that the output of the system is as close as possible to a desired output. Although ILC is directly associated with control, it is important to note that the end result is that the system has been inverted.
8. We learned that ILC is about enhancing a system's performance by means of repetition, but we did not learn how it is done. This brings us to the core activity in ILC research, which is the construction and subsequent analysis of algorithms [266].

Each of these definitions has their focus, but the underlying question is the same: how to improve performance using information from previous trials to update the control law applied on the current one? In some applications, ILC will form only one possible way of designing the controller to be employed. Therefore, it is essential to understand how ILC differs from other forms of control. One place to start is with other forms of control that can be categorized as using some form of learning.

Learning-type control algorithms include adaptive control, neural networks, and repetitive control (RC), where an adaptive control solution for a problem aims to modify the controller, whereas ILC adjusts the control input. Hence, these approaches differ since the controller is a system, and the control input is a signal. In most cases, adaptive controllers do not take advantage of the information contained in repetitive reference or command signals.

In a healthcare application of ILC [84], see Section 2.4 for a general introduction with a more detailed treatment in Chapters 4, 10, and 11, to robotic-assisted stroke rehabilitation, the control signal is electrical stimulation. If a patient is improving, then the level of electrical stimulation applied must decrease, i.e. control of the input signal is of paramount importance. Neural network learning in a control system problem involves modifying the parameters in the controller and not the control signal.

In many respects, ILC is closest to RC, where in the latter [103] there is continuous operation, and the initial conditions at the start of any trial are those at the end of the previous trial, and there is no resetting. An RC counterpart problem to the robotic pick-and-place operation of the last section, for which ILC is suitable, is the control of the read/write head of a hard disk drive, where each trial is a full rotation of the disk, and the next one immediately follows the completion of the current one. The critical difference between ILC and RC is the resetting of the initial conditions for each trial. Still, it can be shown that a duality exists between them under certain conditions, see, e.g. [78].

1.2 A Synopsis of the Literature

Research in ILC can, as in almost all other areas of control systems research, be broadly broken down into developments in the underlying systems theory and also applications with some work combining both in the form of theory followed by experimental validation. In the theoretical/algorithm development research, several of the six postulates given in Section 1.1 have been relaxed in recent years, but the concept of learning from experience gained over repetitions of a task remains unchanged. The mature algorithmic development, especially for linear systems, has, in turn, led to increased effort on extending the boundaries of achievable performance, together with new analysis and design methods often inspired by practical applications.

In the case of nonlinear systems, the primary focus of applying ILC designed using a linear model approximation of the dynamics has shifted toward practical control problems and associated theoretical challenges. Also, interest in the development of nonlinear model-based ILC has significantly increased, again driven by practical applications. These trends can be seen in the survey papers [2, 33] and papers published in learned journals, conferences, and research monographs since the publication of these two surveys.

Since the publication of [12], ILC has broadened in breadth and depth, fusing with established fields, such as robust, adaptive, and optimal control, together with neural networks, fuzzy logic, and many others. Application areas have also expanded, most heavily in robotics, rotary systems, process control, bioapplications, power systems, and semiconductors and, more recently, a transfer to other areas including healthcare. While ILC is still a relatively young field, with a combined body of research work amassing around 1/30th of more established areas, there is a strong linear trend of increasing year-on-year publications on both theory and applications.

The relative merits of ILC and feedback and feedforward design have been considered in the literature. For example, in [33] it has been asserted that ILC has advantages for some problem classes but cannot be applied to all control problems, and the question is how does ILC compare with well-designed feedback and feedforward control? This question is wide-ranging, but the following general remarks [33] can be made.

First, the goal of ILC in the basic case is to generate a feedforward control that tracks a specified reference or rejects a repeating disturbance. In contrast, a feedback controller reacts to inputs and disturbances, and hence, there is always a lag in transient tracking. Feedforward control can eliminate this lag for known or measurable signals but typically not for disturbances. Also, ILC is anticipatory and can compensate for exogenous signals, such as repeating disturbances, in advance by learning and using information from previous trials and does not require that the exogenous signals, either reference or disturbances, be known or measured. Such signals are, however, required to repeat from trial to trial. Moreover, a feedback controller can deal with variations or uncertainties in the system model, i.e. robust control, but the performance of a feedforward controller depends critically on how accurately the system is known, and unmodeled nonlinear behavior and disturbances can restrict the effectiveness of feedforward control.

Noise and nonrepeating disturbances degrade tracking performance, and in common with feedback control, sensitivity to noise can be mitigated by an observer. Again, however, performance depends on to what extent the system dynamics are known. The trial-to-trial structure of ILC and, in particular, the reset between trials allows for advanced filtering and signal processing. In particular, zero-phase noncausal filtering can be used for high-frequency attenuation without introducing phase lag. To reject nonrepeating disturbances, a feedback controller in combination with ILC is advantageous.

The relative merits of ILC against alternatives will be discussed at many points in this book, and in the next section, representations or models widely used in ILC research are introduced, together with some commonly used control law structures.

1.3 Linear Models and Control Structures

As noted in the introduction to this chapter, there is no uniformity of notation in the ILC literature. The notation for continuous-time variables is in the previous section. For discrete variables, the notation is $z_k(p)$, $0 \leq p \leq N - 1$, $k \geq 0$, where z denotes the vector or scalar valued variable under consideration, N denotes the number of samples along a trial and, hence, multiplying this number by the sampling period gives the trial length. This notation is used throughout this book except where stated otherwise.

Analysis of ILC schemes can begin in either the continuous or discrete-time settings, but a large number of actual implementations will be in the latter. This section considers the continuous-time, or differential, dynamics next.

1.3.1 Differential Linear Dynamics

If the system considered is represented by a linear time-invariant model, the state-space model in the ILC setting is

$$\dot{x}_{k+1}(t) = Ax_{k+1}(t) + Bu_{k+1}(t)$$
$$y_{k+1}(t) = Cx_{k+1}(t), \quad 0 \leq t \leq \alpha, \quad k \geq 0 \tag{1.8}$$

where on trial k, $x_k(t) \in \mathbb{R}^n$ is the state vector, $y_k(t) \in \mathbb{R}^m$ is the output vector, and $u_k(t) \in \mathbb{R}^l$ is the control input vector. The control task is to force $y_k(t)$ to track a supplied reference signal $r(t)$ over the fixed trial length $0 \leq t \leq \alpha$, and it is assumed that (i) trial length is the same for all trials, (ii) the state initial vector is of the form $x_{k+1}(0) = d_{k+1}$, $k \geq 0$, where the entries in d_{k+1} are known constants, and (iii) the system dynamics are time-invariant and deterministic, i.e. noise-free. As discussed at the relevant places in this book, these assumptions can be relaxed. On trial k, $e_k(t) = r(t) - y_k(t)$, $0 \leq t \leq \alpha$, and the control law used in [12] for differential linear time-invariant systems, given in this chapter as the first equation in (1.5), is a particular case of

$$u_{k+1}(t) = u_k(t) + K_d \dot{e}_k(t) \tag{1.9}$$

Consequently, the current trial input is the sum of that used on the previous trial plus a correction term. In this particular case, the correction is based on the derivative of the previous trial error, and it is assumed that this term can be computed for measured signals without introducing unacceptable error. This law has the proportional plus derivative (PD) structure.

Error convergence in k, termed trial-to-trial error convergence, i.e.

$$\lim_{k \to \infty} \|e_k(t)\| = 0 \tag{1.10}$$

where $\| \cdot \|$ is any appropriate norm on the underlying function space, is a fundamental requirement in ILC. One condition for this property, see also Chapter 4, under the ILC law (1.9) applied to (1.8) is that

$$\|I - CBK_d\| < 1 \tag{1.11}$$

and $\| \cdot \|$ also denotes the induced operator norm. The immediate observations on this condition are that it does not depend on the system state matrix A, and it cannot be satisfied if $CB = 0$. Hence, this ILC law guarantees trial-to-trial error convergence for single-input single-output (SISO) systems of relative degree one. (Design for linear systems with relative degree greater than unity is considered in Section 7.3.) As in the standard linear systems case, it is possible to write down ILC equivalents of commonly used control laws. For example, a proportional plus integral plus derivative (PID) ILC law for (1.8) is

$$u_{k+1}(t) = u_k(t) + K_p e_k(t) + K_i \int e_k(\tau)d\tau + K_d \dot{e}_k(t) \tag{1.12}$$

A proportional, or P-type, ILC law is obtained by setting $K_i = 0$ and $K_d = 0$ in (1.12), a D-type law by setting $K_p = 0$ and $K_i = 0$, and a PD-type law by setting $K_i = 0$. The ILC law in [12] is D-type, and the following observations [270] on this law are relevant.

1. The right-hand side of (1.12) for D-type ILC has the property that the input $u_{k+1}(t)$ is used at time instance t and its directly produced result $\dot{x}_{k+1}(t)$ is then transmitted to the output derivative $\dot{y}_{k+1}(t) = C\dot{x}_{k+1}(t)$, also termed a causal pair [270]. These operations are algebraic and occur at the same instance t when the control input is applied, i.e. $u_{k+1}(t)$ and $\dot{y}_{k+1}(t)$ are algebraically related.

2. To implement D-type ILC, the highest-order derivative signals of the system are required and, hence, potential difficulties if these cannot be measured or are very noisy due to numerical differentiation. For example, many robots are equipped with position, but not velocity or acceleration, sensors. Therefore, accelerations have to be obtained by differentiating the position measurements twice, and the result can be estimated with severe noise corruption. Hence, high-level noise on measurements can severely reduce the effectiveness of D-type ILC. (The design of stochastic ILC laws is considered in Chapter 12.)

Consider P-type ILC, i.e.

$$u_{k+1}(t) = u_k(t) + K_p e_k(t) \tag{1.13}$$

or

$$u_{k+1}(t) = u_k(t) + K_p(r(t) - y_k(t)) \tag{1.14}$$

Then $(y_k(t), u_k(t))$ on the right-hand side is not a causal pair [270] since the result of the input action applied on trial k is not produced on this trial. In particular, when $u_k(t)$ is applied to the system state-space model, its effects cannot be seen in the state $x_k(t)$ because of the system updating structure. Equivalently, the state $x_k(t)$ is the result of an input applied before t and not $u_k(t)$. Hence, this ILC law does not capture direction or trend information in the errors produced on previous trials. For example, if $r(t) - y_k(t) = 0$ the control law (1.14) stops learning but $\dot{r}(t) - \dot{y}_k(t)$ could be of any value. However, this form of control law does not require higher-order derivatives.

The combination of P- and D-type ILC, resulting in PD ILC, is known to be effective, and the considerations summarized above led [270], in the case of nonlinear system dynamics, to propose anticipatory ILC with the following features:

- Use a causal pair of the action taken on the previous trial and use the result of this pair to compute the next trial input.
- Capture trend or direction information from the recorded previous trial errors but avoid using the highest-order derivatives present in the model.
- Keep the noise levels to a minimum to ease implementation.

Using the state-space model (1.8)

$$
\begin{aligned}
x_{k+1}(t + \Delta) &= x_{k+1}(t) + \int_t^{t+\Delta} \dot{x}_{k+1}(\tau)d\tau \\
&= x_{k+1}(t) + \int_t^{t+\Delta} (Ax_{k+1}(\tau) + Bu_{k+1}(\tau))d\tau
\end{aligned} \tag{1.15}
$$

and then $y_{k+1}(t + \Delta) = Cx_{k+1}(t + \Delta)$. Hence, $u_{k+1}(t)$ and $y_k(t + \Delta)$ form a causal pair in the sense of [270] with anticipatory action since $y_{k+1}(t + \Delta)$ is comparable to $\dot{y}_k(t)$ in capturing trend/directional information. This structure leads to the following form of ILC law:

$$u_{k+1}(t) = u_k(t) + L(r(t + \Delta) - y_k(t + \Delta)) \tag{1.16}$$

The design parameters are L and $\Delta > 0$, and this law is also known as phase-lead ILC.

Phase-lead ILC is not causal in the standard systems sense, where the general concept of time imposes a natural ordering into past, present, and future. Under this definition, $t + \Delta$ is in the future. The critical feature in ILC is that once a trial is complete, all information from it is available for use on the next trial at the cost of data storage. Still, this last aspect is application-specific, and the following is the definition of causality for ILC.

Definition 1.1 An ILC law is causal if and only if the input at time τ on trial $k + 1$ is computed only from data that are available from trial $k + 1$ in the time interval $0 \leq t \leq \tau$ (also written $t \in [0, \alpha]$) and from previous trials over the complete trial length $[0, \alpha]$.

In ILC, information from $t' > t$ can be used but only from previous trials. Control laws that use information from more than one previous trials to compute the current trial input are also possible, sometimes termed higher-order in trial [2] or simply higher-order iterative learning control (HOILC). Such control laws are considered again in Section 1.5 (and in Section 7.4).

The ILC laws discussed above do not necessarily require an accurate model of the system dynamics for implementation and design by tuning is quite common. Model-based control must be used for some applications, and this general area has been very heavily researched. One approach is to design the control law to minimize a suitably chosen cost function. The most common choice is the sum of quadratic terms in the state and control or, particular to ILC, the difference between the control signals applied on successive trials. Such cost functions, in an area known as norm optimal iterative learning control (NOILC), have been the subject of much research to derive algorithms that, in some cases, have been experimentally tested in the laboratory and then applied to physical systems. Optimal ILC is considered in Chapter 5.

1.4 ILC for Time-Varying Linear Systems

Consider linear time-varying (LTV) systems with input–output description

$$y(t) = g(t) + \int_0^t H(t, \tau)u(\tau)d\tau \tag{1.17}$$

where $u(t) \in \mathbb{R}^m$ and $y(t) \in \mathbb{R}^m$ are the input and output vectors, respectively, and the vector $g \in \mathbb{R}^m$ represents initial conditions and disturbances. Suppose that the reference trajectory vector $r(t)$ is given and also that the entries in $r(t), u(t), g(t)$, and $H(t, t')$ are continuously differentiable in t and t'. Then one ILC problem is described by

$$y_k(t) = g(t) + \int_0^t H(t, t')u_k(t')dt'$$
$$e_k(t) = r(t) - y_k(t)$$
$$u_{k+1}(t) = u_k(t) + \Gamma(t)\dot{e}_k(t) \tag{1.18}$$

where $\Gamma(t)$ is an $m \times m$ matrix in the PD ILC law to be designed. Also introduce the so-called λ norm

$$\|e_k\|_\lambda = \sup_{0 \leq t \leq \alpha} e^{-\lambda t} \max_{1 \leq i \leq m} |e_i(t)|, \quad \lambda > 0 \tag{1.19}$$

Then the following result is Theorem 1 in [12], where $\| \cdot \|_\infty$ is the matrix norm induced by the vector norm $\|f\| = \max_{1 \leq i \leq m} |f_i|$ and, hence, for an $m \times m$ matrix F

$$\|F\|_\infty = \max_{1 \leq i \leq m} \sum_{j=1}^m |f_{ij}| \tag{1.20}$$

Theorem 1.1 *Suppose that $r(0) = g(0)$ and*

$$\|I - H(t,t)\Gamma(t)\|_\infty < 1 \tag{1.21}$$

for all $0 \leq t \leq \alpha$ and a given initial input $u_0(t)$ is continuous on $0 \leq t \leq \alpha$. Then there exists positive constants λ and ρ_0 such that

$$\|\dot{e}_{k+1}\|_\lambda \leq \rho_0 \|\dot{e}_k\|_\lambda \quad \text{and} \quad 0 \leq \rho_0 < 1 \tag{1.22}$$

Moreover,

$$\|\dot{e}_k(t)\|_\lambda \to 0, \quad k \to \infty \tag{1.23}$$

and hence,

$$\dot{y}_k(t) \to \dot{r}(t), \quad k \to \infty \tag{1.24}$$

By the definition of the norm $\|\cdot\|_\lambda$, the convergence in both cases is uniform for $0 \leq t \leq \alpha$. Moreover, since $y_k(0) = g(0) = r(0)$, (1.24) implies that

$$y_k(t) \to r(t), \quad k \to \infty$$

uniformly for $0 \leq t \leq \alpha$.

Consider systems described by the state-space model (1.8), where the matrices are time-varying, i.e. A, B, and C are, respectively, replaced by $A(t)$, $B(t)$, and $C(t)$. Then

$$y_{k+1}(t) = C(t)X(t)x_{k+1}(0) + \int_0^t C(t)X(t)X^{-1}(\tau)B(\tau)u_{k+1}(\tau)d\tau \tag{1.25}$$

where $X(t)$ is the unique matrix solution of the homogenous matrix differential equation:

$$\dot{X}(t) = A(t)X(t), \quad X(0) = I \tag{1.26}$$

Also, the time-invariant case is recovered when

$$g(t) = C(t)X(t)x(0) \quad \text{and} \quad H(t,\tau) = C(t)X(t)X^{-1}(\tau)B(\tau)$$

Hence, if $B(t) = B$, $C(t) = C$ and $\det(CB) \neq 0$, there exists a constant matrix such that

$$\|I - H(t,t)\Gamma\|_\infty = \|I - CB\Gamma\|_\infty < 1 \tag{1.27}$$

and if this condition holds application of the ILC law ensures that $y_k(t) \to r(t)$ uniformly in $t \in [0, \alpha]$ as $k \to \infty$. More recent results on time-varying ILC are given in Chapter 8, with supporting experimental results.

The λ-norm (1.19) is extensively used in trial-to-trial error convergence analysis, especially for nonlinear systems. However, it is a sufficient but not a necessary condition for convergence in the norm, and consequently, it can be very conservative. This aspect is considered again for linear systems in Chapter 3.

With the emphasis on digital implementation, discrete-time ILC is considered next. The outcome will be another representation for the system dynamics that provides the setting for a substantial volume of research, with follow-through to implementations and an extension to nonlinear discrete dynamics.

1.5 Discrete Linear Dynamics

The counterpart to (1.8), with N denoting the number of samples along a trial, for discrete-time linear systems is

$$x_k(p+1) = Ax_k(p) + Bu_k(p)$$

$$y_k(p) = Cx_k(p), \quad x_k(0) = x_0, \quad 0 \leq p \leq N-1, \quad k \geq 0 \tag{1.28}$$

where the state, input, and output vectors are as for the differential case (1.8) with t replaced by p, and it is assumed that the state initial conditions are the same on each trial (where this assumption can be relaxed). Also, let q denote the forward shift-time operator defined by $qx(p) = x(p+1)$. Then an operator representation for the dynamics is

$$y_k(p) = G(q)u_k(p) + d(p) \tag{1.29}$$

and $d(p)$ is an exogenous signal that repeats on each trial. (Equivalently, the initial conditions are the same on each trial and there are no external disturbances.) Also,

$$y_k(p) = C(qI - A)^{-1}Bu_k(p) + d(p)$$

$$d(p) = CA^p x(0) \tag{1.30}$$

The system $G(q)$ in (1.29) is a proper rational function of q and, in general, has a delay, or equivalently, a relative degree of h and is sufficiently general to capture infinite impulse response (IIR) and finite impulse response (FIR) systems. Assume that the system is stable, i.e. all eigenvalues of A have modulus strictly less than unity. The case when this property does not hold is considered again later in this section. Repeating disturbances, repeated nonzero initial conditions, and systems augmented with feedback and feedforward control can be included in the term $d(p)$ in (1.29).

Figure 1.4 illustrates the two-dimensional, or 2D, systems nature of ILC, information propagation from trial-to-trial (k) and along a trial (p). Control systems analysis for 2D systems is well developed in theory, and ILC provides an application area discussed in the next section. Let $r(p)$ denote the reference trajectory, and then $e_k(p) = r(p) - y_k(p)$ is the error on trial k. An extensively used ILC

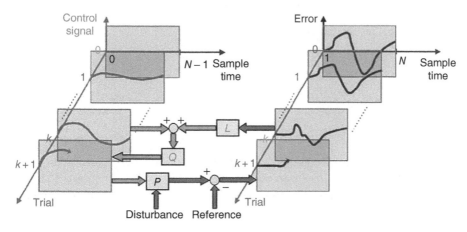

Figure 1.4 Illustrating the 2D structure of ILC.

law, stated for SISO systems with an immediate generalization to multiple-input multiple-output (MIMO) systems, is

$$u_{k+1}(p) = Q(q)(u_k(p) + L(q)e_k(p+1)) \tag{1.31}$$

where $Q(p)$ is termed the Q-filter and $L(p)$ the learning function. There are many variations of (1.31), including time-varying, nonlinear functions, and trial-varying functions. An application where trial-varying ILC has a role is in smart rotors for wind turbines considered in Section 9.1.2. Current trial feedback is a method of applying feedback action in conjunction with ILC and in this case (1.31) is extended to

$$u_{k+1}(p) = Q(q)(u_k(p) + L(q)e_k(p+1)) + C(q)e_{k+1}(p) \tag{1.32}$$

The current error term $C(q)e_{k+1}(p)$ is feedback action on the current trial.

Write (1.32) as

$$u_{k+1}(p) = w_{k+1}(p) + C(q)e_{k+1}(p) \tag{1.33}$$

where, by routine algebraic manipulations,

$$w_{k+1}(p) = Q(q)\left[w_k(p) + (L(q) + q^{-1}C(q))e_k(p+1)\right] \tag{1.34}$$

and the feedforward part of current trial ILC is identical to (1.31) with learning function $L(q) + q^{-1}C(q)$. Consequently, the ILC law (1.31) with learning function $L(q) + q^{-1}C(q)$ combined with a feedback controller in a parallel architecture is equivalent to the complete current trial ILC.

Expanding $G(q)$ in (1.29), with $h = 1$ for simplicity, as an infinite power series gives

$$G(q) = g_1 q^{-1} + g_2 q^{-2} + g_3 q^{-3} + \cdots \tag{1.35}$$

where the g_i are the Markov parameters. Moreover, the g_i forms the impulse response and since $CB \neq 0$ is assumed, $g_1 \neq 0$ and from (1.28), $g_j = CA^{j-1}B$. Introduce the vectors

$$Y_k = \begin{bmatrix} y_k(1) \\ y_k(2) \\ \vdots \\ y_k(N) \end{bmatrix}, \quad U_k = \begin{bmatrix} u_k(0) \\ u_k(1) \\ \vdots \\ u_k(N-1) \end{bmatrix}, \quad d = \begin{bmatrix} d(1) \\ d(2) \\ \vdots \\ d(N) \end{bmatrix} \tag{1.36}$$

and, hence, the dynamics can be written as follows:

$$Y_k = GU_k + d \tag{1.37}$$

$$G = \begin{bmatrix} g_1 & 0 & \cdots & 0 \\ g_2 & g_1 & \cdots & 0 \\ \vdots & \vdots & \ddots & \vdots \\ g_N & g_{N-1} & \cdots & g_1 \end{bmatrix} \tag{1.38}$$

The entries in Y_k and d are shifted by one-time step. Moreover, the relative degree is unity, to account for the one-step delay in the system, which ensures that G is invertible. If there is an $h > 1$ step delay or, equivalently, the first nonzero Markov parameter is $CA^{h-1}B$, the lifted system representation is

$$\begin{bmatrix} y_k(h) \\ y_k(h+1) \\ \vdots \\ y_k(h+N-1) \end{bmatrix} = G_h \begin{bmatrix} u_k(0) \\ u_k(1) \\ \vdots \\ u_k(N-1) \end{bmatrix} + \begin{bmatrix} d(h) \\ d(h+1) \\ \vdots \\ d(h+N-1) \end{bmatrix} \tag{1.39}$$

$$G_h = \begin{bmatrix} g_h & 0 & \cdots & 0 \\ g_{h+1} & g_h & \cdots & 0 \\ \vdots & \vdots & \ddots & \vdots \\ g_{h+N-1} & g_{h+N-2} & \cdots & g_h \end{bmatrix} \qquad (1.40)$$

In the lifted form, the time and trial domain dynamics are replaced by an algebraic updating in the trial index. Consequently, the along the trial dynamics are absorbed into the super-vector, and simultaneous consideration of along the trial dynamics and trial-to-trial error convergence is not possible. In the 2D approach considered in the next section, such simultaneous consideration is possible.

The ILC law (1.31) can also be written in the lifted form, and the Q-filter and learning function L can be noncausal functions of the impulse response

$$Q(q) = \cdots + q_{-2}q^2 + q_{-1}q + q_0 + q_1 q^{-1} + q_2 q^{-2} + \cdots$$
$$L(q) = \cdots + l_{-2}q^2 + l_{-1}q + l_0 + l_1 q^{-1} + l_2 q^{-2} + \cdots \qquad (1.41)$$

and hence,

$$U_{k+1} = QU_k + LE_k \qquad (1.42)$$

$$E_k = R - Y_k \qquad (1.43)$$

$$R = \begin{bmatrix} r(1) & r(2) & \cdots & r(N) \end{bmatrix}^T \qquad (1.44)$$

$$Q = \begin{bmatrix} q_0 & q_{-1} & \cdots & q_{-(N-1)} \\ q_1 & q_0 & \cdots & q_{-(N-2)} \\ \vdots & \vdots & \ddots & \vdots \\ q_{N-1} & q_{N-2} & \cdots & q_0 \end{bmatrix} \qquad (1.45)$$

$$L = \begin{bmatrix} l_0 & l_{-1} & \cdots & l_{-(N-1)} \\ l_1 & l_0 & \cdots & l_{-(N-2)} \\ \vdots & \vdots & \ddots & \vdots \\ l_{N-1} & l_{N-2} & \cdots & l_0 \end{bmatrix} \qquad (1.46)$$

If $Q(q)$ and $L(q)$ are causal functions, then

$$q_{-1} = q_{-2} = \cdots = 0, \quad l_{-1} = l_{-2} = \cdots = 0 \qquad (1.47)$$

and the matrices Q and L are lower triangular. The matrices G, Q, and L are also Toeplitz, i.e. all entries along each diagonal are equal. This setting also extends to LTV systems, but the corresponding matrices do not have the Toeplitz structure. Next, the z transform description is introduced.

The one-sided z-transform of a signal $x(j)$, $j = 0,1,\ldots$, is

$$x(z) = \sum_{j=0}^{\infty} x(j)z^{-j} \qquad (1.48)$$

and the frequency response is defined by $z = e^{j\theta}$, $\theta \in [-\pi, \pi]$. To apply the z transform, it is necessary to assume $N = \infty$, even though the trial length is finite, and the implications of this difference will arise at various places throughout this book (next in Section 3.1).

In z-transform terms, the system and control law dynamics are described by

$$y_k(z) = G(z)u_k(z) + D(z) \qquad (1.49)$$

$$u_{k+1}(z) = Q(z)(u_k(z) + zL(z)e_k(z)) \tag{1.50}$$

$$e_k(z) = r(z) - y_k(z) \tag{1.51}$$

The z term in this last equation emphasizes the forward time shift, and for an h time-step delay, z^h replaces z and the definition of causality given in Definition 1.1 for differential dynamics has an obvious discrete counterpart not formally stated.

Unlike the standard concept of causality, a noncausal ILC law is implementable in practice because the entire time sequence of data is available from all previous trials. Consider the noncausal ILC law

$$u_{k+1}(p) = u_k(p) + k_p e_k(p+1) \tag{1.52}$$

and the causal ILC law

$$u_{k+1}(p) = u_k(p) + k_p e_k(p) \tag{1.53}$$

Then a disturbance $d(p)$ contributes to the error as

$$e_k(p) = r(p) - G(q)u_k(p) - d(p) \tag{1.54}$$

Hence, noncausal ILC anticipates the disturbance $d(p+1)$ and compensates with the control action $u_{k+1}(p)$. The causal ILC law has no anticipation since $u_{k+1}(p)$ compensates for the disturbance $d(p)$ with the same time index p. Causality also has consequences for feedback equivalence where the final, or converged, control, denoted u_∞, can instead be obtained by a feedback controller. It can be shown [92] that there is a feedback equivalence for causal ILC laws, and the equivalent controller can be obtained directly from the ILC law.

The assertion in [92] is "causal ILC laws are of limited (or no) use since the same control action can be obtained by applying the equivalent feedback controller without the learning process." There are, however, critical limitations [33, 193] to this equivalence, and these are as follows:

- A noise-free requirement must be assumed.
- As the ILC performance increases, the gain of the equivalent feedback controller also increases. In the presence of noise, a high gain can lead to performance degradation and equipment damage. Hence, casual ILC laws are still of interest, and this equivalence was already known in the repetitive process/2D systems literature. See the next section.
- The equivalent feedback controller may not be stable.
- There is no equivalence for noncausal ILC as a feedback controller reacts to errors.

In [176] the system of Figure 1.5 with four inputs is considered, i.e. the reference signal $r(p)$, an externally generated control signal $u(p)$ and $w(p)$ and $v(p)$, respectively, denote load and measurement disturbances. The measured output is $y(p)$ and the controlled variable is $z(p)$. This system can also have an internal feedback and hence, the blocks G_u, G_r, G_w, and G_v contain the system to be controlled in addition to the controlled dynamics.

The equations describing the dynamics of this general representation in the SISO case, with an immediate generalization to MIMO systems, are given in lifted form by a system of the form

$$\begin{aligned} Z_k &= G_{r,k}R + G_{u,k}U_k + G_{w,k}W_k \\ Y_k &= Z_k + G_{v,k}V_k \end{aligned} \tag{1.55}$$

where the notation is as above, and W_k, V_k are the lifted vectors corresponding to $v_k(p)$, $w_k(p)$ and

$$Z_k = \begin{bmatrix} z_k(0) & \cdots & z_k(N-1) \end{bmatrix}^T \tag{1.56}$$

Figure 1.5 ILC in general form.

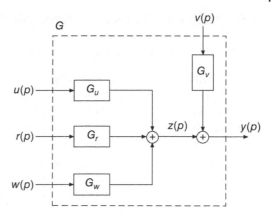

This representation includes trial-variant and time-variant systems. If the system is both time- and trial-invariant, the dynamics can be described in the operator setting as

$$z_k(p) = G_r(q)r(p) + G_u(q)u_k(p) + G_w(q)w_k(p)$$

$$y_k(p) = z_k(p) + G_v v_k(p) \tag{1.57}$$

and this representation can be used for frequency-domain analysis. If time-varying dynamics is written in lifted form, some of the structural properties of the corresponding matrices in the time-invariant case are not preserved.

A lifted model can also be constructed for HOILC where, as one example, consider the following law from [176], which uses data from the previous $M > 1$ trials:

$$u_{k+1}(p) = \sum_{j=k-M+1}^{k} \left(Q_{k-j+1}(u_j(p) + L_{k-j+1}e_j(p)) \right) \tag{1.58}$$

Then

$$u_{k+1}(p) = \sum_{j=k-M+1}^{k} Q_{k-j+1}(I - L_{k-j+1}G_u)u_j(p) + \sum_{j=1}^{M} Q_j L_j (I - G_r)r(p) \tag{1.59}$$

or, on defining,

$$U_k = \left[\, u_k^T(p) \ \ u_{k-1}^T(p) \ \ \cdots \ \ u_{k-M+1}^T(p) \, \right]^T \tag{1.60}$$

(1.59) can be written as follows:

$$u_{k+1} = \begin{bmatrix} H_1 & H_2 & \cdots & H_M \\ I & 0 & \cdots & 0 \\ \vdots & \ddots & \ddots & \vdots \\ 0 & \cdots & I & 0 \end{bmatrix} u_k + \begin{bmatrix} F \\ 0 \\ \vdots \\ 0 \end{bmatrix} R \tag{1.61}$$

with $H_j = Q_j(I - L_j G_u)$ and $F = \sum_{j=1}^{M} Q_j L_j (I - G_r)$.

A HOILC law for differential dynamics can be found in [50]. These forms of HOILC laws can also include anticipatory action. An obvious question for investigation is when does HOILC bring advantages? Further coverage of HOILC is given in Chapters 3 and 7.

1.6 ILC in a 2D Linear Systems/Repetitive Processes Setting

1.6.1 2D Discrete Linear Systems and ILC

Multidimensional, or nD, systems originally arose from the need to have a mathematical setting to address problems whose formulation and solution required the use of functions and polynomials in more than one complex or real variable. Recently emerging areas where nD systems theory has been used include grid sensor networks and evidence filtering. In particular, wireless sensor networks consist of large numbers of resource-constrained, embedded sensor nodes and are an emerging candidate for many distributed applications. Some of these applications require regularly placed nodes in a spatial grid, often sampling the sensors periodically over time. Agriculture and environmental monitoring applications often favor a grid or mesh topology. Spatially distributed sensor lattices are also essential in surveillance, target location, and tracking applications. In [229, 230], several case studies on the application of nD control systems design are given, where for ILC, the case of interest is when $n = 2$..

Many models are used to describe nD linear systems, and one of the two heavily studied from a control systems standpoint is the Fornasini–Marchesini model [71]. An alternative model that describes how a dynamic process evolves over the 2D plane is the Roesser state-space model [227] where a state vector is defined for each direction of information propagation. If these vectors are denoted by $x^h(n_1, n_2)$, and $x^v(n_1, n_2)$, respectively, the state-space model is

$$\begin{bmatrix} x^h(n_1 + 1, n_2) \\ x^v(n_1, n_2 + 1) \end{bmatrix} = \begin{bmatrix} A_1 & A_2 \\ A_3 & A_4 \end{bmatrix} \begin{bmatrix} x^h(n_1, n_2) \\ x^v(n_1, n_2) \end{bmatrix} + \begin{bmatrix} B_1 \\ B_2 \end{bmatrix} u(n_1, n_2)$$

$$y(n_1, n_2) = \begin{bmatrix} C_1 & C_2 \end{bmatrix} \begin{bmatrix} x^h(n_1, n_2) \\ x^v(n_1, n_2) \end{bmatrix} \tag{1.62}$$

The links between ILC and 2D linear systems are evident, information propagation from trial-to-trial (k) and along the trial (p). Moreover, the first reported work [136] on the use of the 2D systems setting for ILC analysis and design used the Roesser model. Consider the state-space model (1.28) and

$$u_{k+1}(p) = u_k(p) + \Delta u_k(p) \tag{1.63}$$

where the control applied on the next trial is the sum of that used on the previous trial plus a correction denoted by $\Delta u_k(p)$. Using $e_k(p) = r(p) - y_k(p)$ it follows that

$$e_{k+1}(p) - e_k(p) = -CA\eta_k(p) - CB\Delta u_k(p - 1) \tag{1.64}$$

where

$$\eta_k(p) = x_{k+1}(p - 1) - x_k(p - 1)$$

or, using (1.63) and the state equation in (1.28),

$$\eta_k(p + 1) = A\eta_k(p) + B\Delta u_k(p - 1) \tag{1.65}$$

Selecting

$$\Delta u_k(p) = Ke_k(p + 1) \tag{1.66}$$

results in controlled systems dynamics described by the following 2D Roesser state-space model

$$\begin{bmatrix} \eta_k(p + 1) \\ e_{k+1}(p) \end{bmatrix} = \begin{bmatrix} A & BK \\ -CA & I - CBK \end{bmatrix} \begin{bmatrix} \eta_k(p) \\ e_k(p) \end{bmatrix} \tag{1.67}$$

Hence, 2D systems theory for the Roesser state-space model, or that for the Fornasini–Marchesini state-space model, can be applied to ILC analysis and design.

One immediate conclusion from (1.67) is that the system-state matrix A and, therefore, the dynamics along any trial are independent of this control law. Hence, if A is an unstable matrix or along the trial dynamics have unacceptable transients, then one option is to design a preliminary feedback loop to remove these difficulties, resulting in a two-stage control law design.

The general 2D systems approach to ILC analysis and design, see Chapter 7, provide a means of avoiding this resulting two-stage design. For example, consider the ILC law (1.63) with

$$\Delta u_k(p) = -K_1 \eta_k(p+1) + K_2 e_k(p+1) \tag{1.68}$$

where K_1 and K_2 are compatibly dimensioned matrices to be selected.

Then the controlled dynamics are described by the following Roesser state-space model:

$$\left[\begin{array}{c} \eta_k(p+1) \\ e_{k+1}(p) \end{array} \right] = \left[\begin{array}{cc} A - BK_1 & BK_2 \\ -CA + CBK_1 & I - CBK_2 \end{array} \right] \left[\begin{array}{c} \eta_k(p) \\ e_k(p) \end{array} \right] \tag{1.69}$$

and it is possible to design K_1 and K_2 simultaneously in this setting to control trial-to-trial error convergence and the along the trial dynamics.

1.6.2 ILC in a Repetitive Process Setting

Many processes make repeated completions of the same task or operation over a finite interval. Upon completing each, the process is reset to the initial location, ready for the next execution. Call each completion a pass, and the output produced the pass profile. In [70] the term "multi-pass process," subsequently changed to repetitive process, was introduced to describe the operation of longwall coal cutting. The novel feature is that the pass profile produced on the previous pass contributes to the next pass profile and hence has the ILC structure. Thus, as in most of the ILC literature, pass and pass length, respectively, are replaced by trial and trial length from this point onward.

Variables in a repetitive process model have to be described by a variable denoting the trial number and another for the dynamics produced on any trial. Moreover, any trial dynamics can be functions of differential or discrete variables, where the latter can arise from sampling of the former. In this book, the notation used for differential dynamics is of the form $h_k(t)$, $0 \le t \le \alpha$, $k \ge 0$, where h is the vector or scalar-valued variable under consideration, $\alpha < \infty$ denotes the trial length and k denotes the trial number. For discrete dynamics, the notation is of the form $h_k(p)$, $0 \le p \le N - 1 < \infty$, $k \ge 0$, where N denotes the number of samples along the trial (N times the sampling period gives the trial length α).

Differential linear repetitive processes are described by the state-space model [228]

$$\begin{aligned} \dot{x}_{k+1}(t) &= Ax_{k+1}(t) + Bu_{k+1}(t) + B_0 y_k(t) \\ y_{k+1}(t) &= Cx_{k+1}(t) + Du_{k+1}(t) + D_0 y_k(t) \end{aligned} \tag{1.70}$$

where on trial k, $x_k(t) \in \mathbb{R}^n$ is the state vector, $y_k(t) \in \mathbb{R}^m$ is the trial profile vector, and $u_k(t) \in \mathbb{R}^l$ is the vector of control inputs. The simplest possible form of boundary conditions for these processes is

$$\begin{aligned} x_{k+1}(0) &= d_{k+1}, \quad k \ge 0 \\ y_0(t) &= f(t), \quad 0 \le t \le \alpha \end{aligned} \tag{1.71}$$

where the vector d_{k+1} has constant entries and the entries in $f(t)$ are known functions of t over the trial length.

Discrete linear repetitive processes are described by the state-space model

$$
\begin{aligned}
x_{k+1}(p+1) &= Ax_{k+1}(p) + Bu_{k+1}(p) + B_0y_k(p) \\
y_{k+1}(p) &= Cx_{k+1}(p) + Du_{k+1}(p) + D_0y_k(p)
\end{aligned}
\tag{1.72}
$$

where on trial k, $x_k(p), y_k(p)$, and $u_k(p)$ are of the form of (1.70) with t replaced by p. Also, the boundary conditions are the discrete equivalent of (1.71).

The essential difference between discrete linear repetitive processes and 2D linear systems described by the Roesser and Fornasini–Marchesini state-space models is that information propagation in one of the two separate directions, along the trial, only occurs over the fixed-finite duration of the trial length. Therefore, as shown in Chapter 7, the repetitive process setting has more advantages than the Roesser and Fornasini–Marchesini state-space models for ILC analysis. Moreover, repetitive process-based ILC designs have seen experimental application, unlike those in the Roesser or Fornasini–Marchesini state-space model setting.

Remark 1.1   Other forms of boundary conditions for repetitive processes can be defined, see, e.g. [228] but are not required in this book. No explicit mention of the boundary conditions for linear repetitive processes will be made from this point onward.

1.7   ILC for Nonlinear Dynamics

The original work on ILC for nonlinear dynamics was for robotics where [12] considered an n-degree of freedom serial link robot whose dynamics, adopting the notation in [12], are described by the general form

$$
R(\theta(t))\ddot{\theta}(t) + f(\theta(t), \dot{\theta}(t)) + g(\theta(t)) = \tau(t)
\tag{1.73}
$$

where

$$
\theta(t) = \begin{bmatrix} \theta_1(t) & \theta_2(t) & \dots & \theta_n(t) \end{bmatrix}^T
\tag{1.74}
$$

is the joint angles coordinates vector and

$$
\tau(t) = \begin{bmatrix} \tau_1(t) & \tau_2(t) & \dots & \tau_n(t) \end{bmatrix}^T
\tag{1.75}
$$

is the generalized force vector, $R(\theta(t))$ is the inertia matrix and is usually positive-definite, $f(\theta(t), \dot{\theta}(t))$ consists of centrifugal and Coriolis forces and other nonlinear characteristics such as frictional forces and $g(\theta(t))$ is the potential energy.

Often in robotics, the task, or reference signal in ILC terminology, is described in Cartesian or other task-specific coordinates. In the current case, it is assumed, for simplicity, that a desired motion of the manipulator is determined from a description of the task as a time function $r(t)$, $0 \leq t \leq \alpha$, for the joint coordinates $\theta(t)$ and an extension to task-oriented coordinates is straightforward.

In [12], a local linear model approximation in the form of a time-varying state-space was used to design a PD ILC law. Of course, this analysis has limitations, and nonlinear ILC model-based analysis and design has received a substantial amount of attention in the literature, particularly trial-to-trial error convergence proofs using the λ-norm defined by (1.19). The role of this norm in ILC analysis will be considered in Chapter 3.

Nonlinear systems can, in general terms, be split into two groups, those that are affine in the control and those that are not. Differential systems that are affine in the control are assumed to be described by

$$\dot{x}(t) = f(x(t)) + B(x(t))u(t)$$
$$y(t) = g(x(t)) \tag{1.76}$$

where x is the state vector, u is the input, and y is the output. A special case is

$$M_r(x)\ddot{x} - C_r(x, \dot{x})\dot{x} - g_r(x) - d_r(x, \dot{x}) = \tau \tag{1.77}$$

where for robotics the vectors $x, \dot{x}, \ddot{x}$ are the positions, velocities, and accelerations of the link, τ is the torque input, $M_r(x)$ is the symmetric positive-definite matrix of link inertias, $C_r(x, \dot{x})$ is the Coriolis and centripetal acceleration matrix, $g_r(x)$ is the gravitational force vector and $d_r(x, \dot{x})$ is the friction torque vector.

The analysis of ILC for affine nonlinear systems uses a wide variety of algorithms, but a critical common assumption is smoothness, which is often expressed as a global Lipshitz assumption on each of the functions in (1.76) of the form

$$|f(x_1) - f(x_2)| \le f_0|x_1 - x_2|$$
$$|B(x_1) - B(x_2)| \le b_0|x_1 - x_2|$$
$$|g(x_1) - g(x_2)| \le g_0|x_1 - x_2| \tag{1.78}$$

The constants f_0, b_0, and g_0 are used in a contraction mapping setting to obtain (sufficient) conditions for trial-to-trial error convergence and ILC law design. Nonaffine systems have the form

$$\dot{x}(t) = f(x(t)) + B(x(t), u(t))$$
$$y(t) = g(x(t)) \tag{1.79}$$

The analysis and design of ILC laws have also been extended to nonlinear discrete-time systems. Obtaining discrete-time models for nonlinear systems is sometimes nontrivial, but, for example, trial-to-trial error convergence proofs are more straightforward, and the final design is directly compatible with digital implementation. The case of systems with saturating actuators is considered in Chapter 8, and nonlinear ILC design is also considered in Chapters 10 and 11 with measured results from the stroke rehabilitation application.

1.8 Robust, Stochastic, and Adaptive ILC

Robustness is also an issue in ILC. As for other linear systems classes, one approach assumes that the uncertainty belongs to a particular class, e.g. polytopic or norm bounded. For discrete dynamics where the uncertainty is assumed to be additive (norm bounded), the starting point is the state-space model

$$\dot{x}_{k+1}(p + 1) = (A + \Delta A)x_{k+1}(p) + (B + \Delta B)u_{k+1}(p)$$
$$y_{k+1}(p) = (C + \Delta C)x_{k+1}(p), \quad 0 \le p \le \alpha - 1, \quad k \ge 0 \tag{1.80}$$

where ΔA, ΔB, and ΔC represent the model uncertainties. Suppose also that the ILC law applied is

$$u_{k+1}(p) = u_k(p) + Ke_k(p + 1) \tag{1.81}$$

Then the error dynamics are described by

$$e_{k+1}(p) = Qe_k(p) \tag{1.82}$$

where

$$Q = \begin{bmatrix} I - KCB & 0 & \cdots & 0 \\ -KCAB & I - KCB & \cdots & 0 \\ \vdots & \vdots & \ddots & \vdots \\ -KCA^{\alpha-1}B & -KCA^{\alpha-2}B & \cdots & I - KCB \end{bmatrix} \tag{1.83}$$

Hence, when the state-space model matrices A, B, and C have uncertainty associated with them that belong to a convex set, the operator Q does not belong to this set. Consequently, only a bound is possible that increases the conservativeness of the results. Robust ILC with experimental validation is considered in Chapter 6.

In some applications, assuming noise-free signals is not feasible for even initial control-related studies. Instead, a stochastic setting must be used, and for discrete linear systems in the ILC setting, one starting point is the state-space model

$$x_k(p + 1) = Ax_k(p) + B_k u(p) + \omega_k(p)$$
$$y_k(p) = Cx_k(p) + v_k(p), \quad 0 \le t \le \alpha \tag{1.84}$$

where $\omega_k(p)$ is the state disturbance vector and $v_k(p)$ is the measurement error or noise vector. In [232–234], a range of ILC laws were considered. The results in these papers are considered in Chapter 12 and compared against a new design developed using repetitive process stability theory.

An alternative setting [32] for analysis (in the SISO case for ease of presentation and specializing to control law parameters that are independent of the trial number) starts from the system model

$$e_k(p) = -G(q)u_k(p) + d(p) + w_k(p) \tag{1.85}$$

where $w_k(p)$ is a stationary random disturbance, G is a stable system, and the deterministic signal d represents disturbances and initial conditions. The ILC law is

$$u_{k+1}(p) = Q(q)(u_k(p) + L(q)\hat{e}_k(p)) \tag{1.86}$$

where $\hat{e}_k(p)$ is the noise-corrupted error measurement modeled as follows:

$$\hat{e}_k(p) = e_k(p) + v_k(p) \tag{1.87}$$

and $v_k(p)$ is stationary random noise, see Figure 1.6.

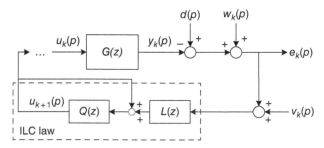

Figure 1.6 Stochastic ILC – a block diagram representation.

Adaptive ILC has also seen productive research. For example, [191] has considered the following control law in the case of differential linear time-invariant SISO dynamics:

$$u_{k+1}(t) = u_k(t) + \text{sgn}(CB)((K_{k+1}e_{k+1})(t) + (F_{k+1}e_k)(t)) \tag{1.88}$$

where sign denotes the signum function, K_{k+1} is a causal feedback learning operator feeding back the current trial error, and F_{k+1} is a feedforward learning operator feeding forward the previous trial error. Some results on adaptive ILC are given in Section 5.4.1.

1.9 Other ILC Problem Formulations

This chapter has introduced the main settings for analysis and design used in ILC research and application. Others related to specific application areas will be introduced as required in subsequent chapters. In many applications, the reference trajectory in ILC design does not change after the trials begin. It is this area that has seen the vast majority of ILC analysis, design, and implementation. However, applications exist where a more general objective than tracking a predefined static reference signal over a fixed trial length is required.

Early applications in this last area considered include gas metal arc welding [170], underwater robotics [129], and liquid slosh in a packaging machine [95]. These deal with specific applications, but many applications, e.g. production line automation and crane positioning, require a motion profile to be followed repeatedly. Also, the error is only critical at specific points.

An approach to such problems is point-to-point motion control. The objective is to ensure that the output equals a corresponding set of desired values at prescribed time instants. Point-to-point control algorithms often involve generating a suitable motion profile in advance and then designing a tracking controller. One approach is input shaping, see, e.g. [1].

Applying ILC to point-to-point motion control offers the potential benefits of learning from experience gained over previous trials. A possible approach is to apply a standard ILC law using any reference trajectory connecting the desired points. In Chapter 4, one design in this area is developed and supported by experimental testing results from application to a nonminimum phase electromechanical system, whose design and construction is detailed in Section 2.2.1.

A particular case of point-to-point ILC is to specify the start and finish locations of a trial, known as terminal ILC. This refinement of ILC has specific relevance in stroke rehabilitation, a specification application area within this book. The stroke application is longer-term, and in Chapter 13, a very recent application in food production is described with experimental results from an application study.

Returning to the six postulates of Section 1.2, one relaxation is to consider the case when the trial length varies with the trial number, i.e. trial-varying ILC, which has seen many papers published. Application areas for this theory to date are very limited, and the only work that reports a physical example is [239], where the application area is human lower-limb rehabilitation. No further consideration of varying trial length ILC is given in this book.

Other areas that have been considered include varying the reference trajectory, e.g. the system operates under one reference trajectory for a fixed number of trials, and then the reference trajectory is changed. This action can occur multiple times and is one form of switching in ILC, where the other is switching dynamics along the trials. Switching in ILC systems has been the subject of some research, see, e.g. [200].

In some cases, a mathematical model developed from first principles is not available, but data linking the inputs and outputs of a system is available. One option is to use system identification to build a model as a basis for control law design. Also, it is possible to couple ILC with empirical or semiempirical models of the dynamics, where an example with supporting evidence from an industrial process in [122], which is also considered again in Chapters 5 and 13, a form of data-driven control.

For some implementations, noise may arise in the system and thereby corrupt the previous trial error. If this is sufficiently strong, then stochastic ILC laws will be required, and (as noted in the last section) this topic is the subject of Chapter 12. However, in some applications, the noise present may not need a stochastic design and can be dealt with by a simple structure filter, e.g. low-pass, high-pass. One extra option in ILC is zero-phase filtering, which is possible due to the finite trial length. At the end of each trial, the error generated can be zero-phase filtered prior between trials before being used to compute the new trial input. See Chapters 7 and 12 for particular application studies.

1.10 Concluding Remarks

This chapter has introduced the basics of ILC and given a short overview of the literature and the main settings for ILC analysis and design. The next chapter describes the experimental testbeds used to obtain most of the experimental results given in this book.

2

Iterative Learning Control: Experimental Benchmarking

As discussed in Chapter 1, the survey papers [2, 33] form a comprehensive overview up to their time of publication of experimental benchmarking and actual implementation of iterative learning control (ILC). Since the publication of these papers, there have been significant developments on the applications front within process control and especially run-to-run control. One starting point for the literature in this area, not covered in this book, is the survey paper [276]. One area that has seen significant applications is high-resolution printing see, e.g. the PhD thesis [29] and the follow on journal and conference papers. The remainder of this chapter details the systems from engineering and healthcare, which forms the basis for most of the given experimental benchmarking results.

2.1 Robotic Systems

Robotics is a central application area for ILC, and this section gives the required details of two robotic systems used to obtain experimental results throughout this book.

2.1.1 Gantry Robot

The gantry robot [218] of Figure 1 is constructed from two types of linear motion device where the lowest horizontal axis, X, consists of one brushless linear DC motor and a free-running slide parallel to the motor. The next horizontal axis, Y, is perpendicular to the X-axis and has one end attached to the linear motor and the other to the slide. The Y-axis is a single brushless linear DC motor, and the X and Y-axes, respectively, are 1.02 and 0.91 m long. Also, the vertical axis, Z, is a short 0.10 m travel linear ball-screw and a rotary brushless DC motor drive stage. All axes are powered by matched brushless motor dc amplifiers, and axis motion is detected and recorded with appropriate optical encoder systems.

Each axis of the gantry robot has been modeled individually in the velocity control mode of operation. The dynamics of each axis were determined in [218] by performing a series of open-loop frequency response tests. The experimental Bode plots and their related linear approximations, respectively, are shown in Figures 2.1–2.3.

These Bode plots were then used to construct reduced transfer-function models for each axis, which was then sampled at 0.01 seconds to develop a discrete linear state-space model of the

Iterative Learning Control Algorithms and Experimental Benchmarking, First Edition.
Eric Rogers, Bing Chu, Christopher Freeman and Paul Lewin.
© 2023 John Wiley & Sons Ltd. Published 2023 by John Wiley & Sons Ltd.

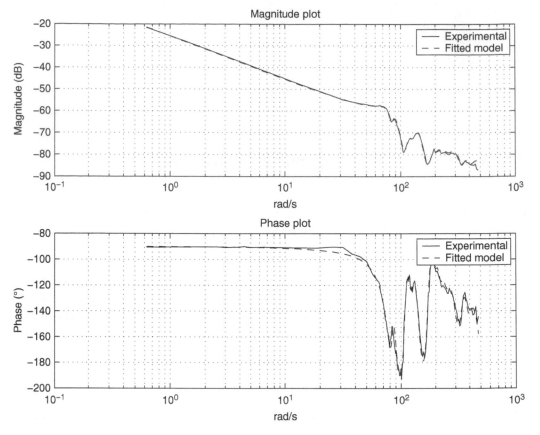

Figure 2.1 X-axis Bode plot, experimental data, and fitted model.

form (1.28) for the X-axis with

$$A_X = \begin{bmatrix} 0.3879 & 1.0000 & 0.2138 & 0 & 0.1041 & 0 & 0.0832 \\ -0.3898 & 0.3879 & 0.1744 & 0 & 0.0849 & 0 & 0.0678 \\ 0 & 0 & -0.1575 & 0.2500 & -0.2006 & 0 & -0.1603 \\ 0 & 0 & -0.3103 & -0.1575 & -0.0555 & 0 & -0.0444 \\ 0 & 0 & 0 & 0 & 0.0353 & 0.5000 & 0.2809 \\ 0 & 0 & 0 & 0 & -0.0164 & 0.0353 & -0.2757 \\ 0 & 0 & 0 & 0 & 0 & 0 & 1.0000 \end{bmatrix}$$

$$B_X = \begin{bmatrix} 0 & 0 & 0 & 0 & 0 & 0 & 0.0910 \end{bmatrix}^T$$

$$C_X = \begin{bmatrix} 0.0391 & 0 & 0.0146 & 0 & 0.0071 & 0 & 0.0057 \end{bmatrix} \qquad (2.1)$$

The state-space model matrices for the Y and Z axes are

$$A_Y = \begin{bmatrix} -0.1067 & 0.1250 & 0.0777 \\ -0.0211 & -0.1067 & 0.1016 \\ 0 & 0 & 1.0000 \end{bmatrix}, \quad B_Y = \begin{bmatrix} 0 \\ 0 \\ 0.0286 \end{bmatrix}, \quad C_Y = \begin{bmatrix} 0.0360 & 0 & 0.0286 \end{bmatrix}$$

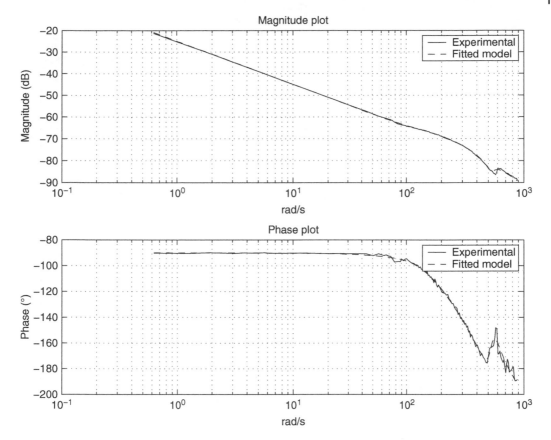

Figure 2.2 *Y*-axis Bode plot, experimental data, and fitted model.

and

$$A_Z = \begin{bmatrix} -0.0030 & 0.0625 & 0.0758 \\ -0.0134 & -0.0030 & 0.0637 \\ 0 & 0 & 1.0000 \end{bmatrix}, \quad B_Z = \begin{bmatrix} 0 \\ 0 \\ 0.0191 \end{bmatrix}, \quad C_Z = \begin{bmatrix} 0.0232 & 0 & 0.0191 \end{bmatrix}$$

respectively. The reference signals for the *Y* and *Z* axes corresponding to Figure 1.3 for the *X* axis are given in Figures 2.4 and 2.5, respectively.

2.1.2 Anthromorphic Robot Arm

The anthropomorphic robotic arm considered is modeled as a two-input and two-output model constructed from frequency domain tests. Moreover, the robot arm has been configured to perform "pick and place" tasks in a horizontal plane using two joints and is shown in Figure 2.6. Its end-effector travels between the pick-and-place locations in a straight line using joint reference trajectories that minimize the end-effector acceleration. On reaching the specified place location, the robot then returns to the starting location.

Position and velocity control loops have been implemented around each joint to provide the first layer of an overall control system, operated at 20 Hz. The overall system model is obtained by combining experimentally derived models of its constituent components as described by the

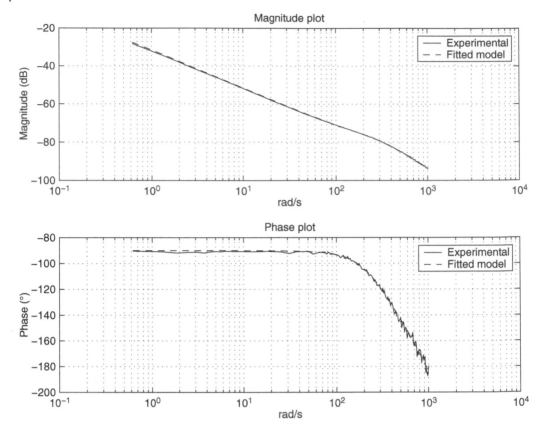

Figure 2.3 *Z*-axis Bode plot, experimental data, and fitted model.

transfer-function matrix description (2.2), where u_i and y_i, $i = 1, 2$, respectively, denote the inputs and outputs and the entries in this matrix are given in the Appendix.

$$\begin{bmatrix} y_1(s) \\ y_2(s) \end{bmatrix} = \begin{bmatrix} G_{11}(s) & G_{12}(s) \\ G_{21}(s) & G_{22}(s) \end{bmatrix} \begin{bmatrix} u_1(s) \\ u_2(s) \end{bmatrix} \tag{2.2}$$

2.2 Electro-Mechanical Systems

Many applications for ILC are electro-mechanical in operation, either entirely or at the subcomponent level. In this book, three such systems are used. The first of which replicates nonminimum phase behavior, the second is a multiple-input multiple-output (MIMO) system where the interaction between the two inputs and outputs can be varied, and the third is a rack feeder system that has time-varying dynamics.

2.2.1 Nonminimum Phase System

The experimental test-bed shown in Figure 2.7 consists of a rotary mechanical system of inertias, dampers, torsional springs, a timing belt, pulleys, and gears and the nonminimum phase

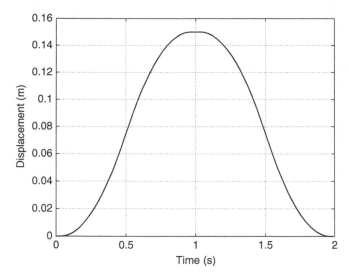

Figure 2.4 The reference trajectory for the Y-axis of the gantry robot.

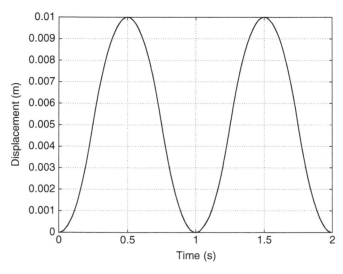

Figure 2.5 The reference trajectory for the Z-axis of the gantry robot.

characteristic is achieved using the arrangement shown in Figure 2.8, where θ_i and θ_o are the input and output positions, J_r and J_g are inertias, B_r is a damper, K_r is a spring, and G_r represents the gearing.

One other spring-mass-damper system is connected to the input to increase the relative degree and complexity of the system. A 1000 pulse/rev encoder records the output shaft position and a standard squirrel cage induction motor supplied by an inverter, operating in variable voltage variable frequency (VVVF) mode drives the load.

This system has been modeled using a least mean squares (LMS) algorithm to fit a linear model to frequency response test results. The resulting continuous-time transfer-function is

$$G(s) = \frac{1.202(4 - s)}{s^4 + 23s^3 + 164.25s^2 + 506.25s} \tag{2.3}$$

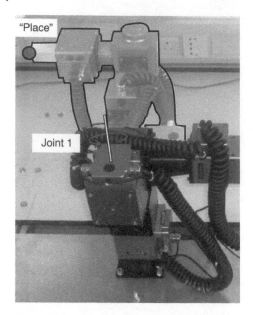

Figure 2.6 Six degree-of-freedom -and-place robot.

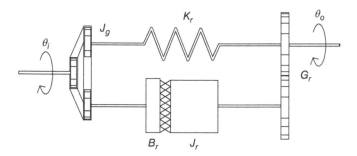

Figure 2.7 Nonminimum-phase electro-mechanical system.

Figure 2.8 Schematic of the nonminimum-phase element in the system of Figure 2.7.

and hence, a right-half plane zero at $s = 4$. A proportional, integral, and derivative (PID) feedback control loop around the plant is employed to act as a prestabilizer and provide greater stability, where the PID gains used (proportional, integral, and derivative) are, respectively, $K_p = 137$, $K_i = 5$, and $K_d = 3$.

2.2.2 Multivariable Testbed

This electro-mechanical system was developed to address (at the time it was designed and commissioned) the relative lack of MIMO ILC experimental evaluation in the literature. This facility uses differential gearboxes, a standard mechanism used within many industries, and enables variable levels of coupling between variables to be specified. The gearboxes are connected via spring-mass-damper systems to capture dynamics of broad technical relevance and are driven by motors of different types. In this system, the differential gearboxes and spring-mass-damper sections can be combined in a variety of configurations, as analyzed in detail in [65, 66], where interaction was quantified using the relative gain array approach.

The input comprises the demand signals fed to the induction motors $\{u_1, u_2\}$, and the output consists of the angular displacements (radians) of the spring-mass-damper sections $\{y_1, y_2\}$. This configuration has a level of interaction that can be manipulated using only the lumped damping parameter $B_{A1} + B_{A2}$. Injection of noise and disturbance is realized mechanically using a DC motor coupled to output y_2 and fed by a demand input u_3.

Figure 2.9 shows a photograph of the system with components defined in Table 2.1. Note that this is only one possible configuration of inputs and outputs since the position and number of motors can be easily altered without changing the central structure.

The system model in the form of the transfer-function matrix $G(z)$ has been identified experimentally. Sinewaves of different frequencies and amplitudes were injected into the inputs and transfer-functions fitted to the resulting Bode plots, using the least mean squares (LMS) approach. The transfer-functions governing the noise/disturbance injected by the DC motor have

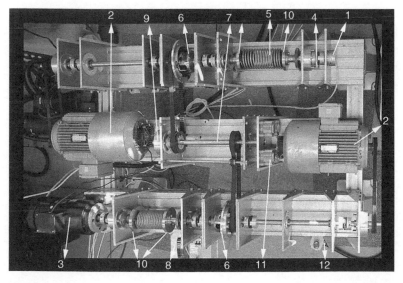

Figure 2.9 The complete MIMO system showing two induction motors and disturbance injection via a DC motor. Components are specified in Table 2.1. Source: University of Southampton.

Table 2.1 Components of the MIMO system.

System components			
1	Encoder	2	Induction motor
3	DC motor	4	Rotary damper
5	Torsional spring	6	Differential gearbox
7	Coupling shaft	8	DC motor controller
9	Clamp	10	Inertia
11	Adjustable interaction dampers	12	Emergency stop button

also been identified using the same technique. A variable parameter c is available to set the level of interaction, and the details of how this parameter can be varied are given in [65, 66].

The (sampled) transfer-function matrix description of the system is

$$\begin{bmatrix} y_1(z) \\ y_2(z) \end{bmatrix} = \begin{bmatrix} G_{11}(z) & G_{12}(z) & G_{13}(z) \\ G_{21}(z) & G_{23}(z) & G_{23}(z) \end{bmatrix} \begin{bmatrix} u_1(z) \\ u_2(z) \\ u_3(z) \end{bmatrix} \tag{2.4}$$

where the entries in this matrix are given in the Appendix for the coupling values $c = 0, 0.2, 0.4,$ 0.6, 0.8, 1. Also $G_{13}(z)$ and $G_{23}(z)$ do not significantly change depending on c and hence, a single representation is given for each.

2.2.3 Rack Feeder System

Figure 2.10 shows a photograph of a prototype high-speed rack feeder system. This system is used in Chapter 8 to validate a constrained ILC design for linear time-varying (LTV) systems experimentally.

This system consists of a carriage driven by an electric DC servo motor via a toothed belt, implementing the basic control input for the motion in the horizontal direction. An elastic double-beam structure is vertically mounted on this carriage. In addition, a cage (with variable load mass) can be moved in the vertical direction on the beam structure. The position of this cage, described by the coordinates $y_K(t)$ in horizontal and $x_K(t)$ in the vertical direction, represents the tool centre point of the rack feeder that should track desired trajectories as accurately as possible with small tracking errors in transient phases and without remaining oscillations at a standstill.

The movable cage is driven by a toothed belt with an electric DC servo motor in common with the carriage. Measured data are available from encoders for the actuator angles. Also, the horizontal carriage position is determined by a magnetostrictive transducer. Strain gauges are used to determine the beam deflection during system operation. The horizontal and vertical axes can be operated by fast underlying velocity controllers, running directly on the current converters. Consequently, the corresponding velocities can be taken as the control inputs.

In previous work [14, 15], it has been shown that a control-oriented elastic multibody model can be used to develop a model for control design of the rack feeder system shown in Figure 2.10.

(a)

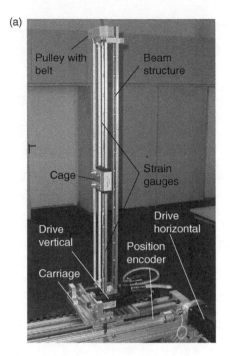

(b)

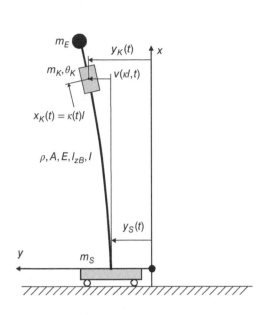

Figure 2.10 The high-speed rack feeder system (a) and its corresponding elastic multibody model (b).

To derive the control-oriented model, the rack feeder is represented by a multibody model (Figure 2.10b) with three rigid bodies – the carriage with mass m_S, the cage (mass m_K, mass moment of inertia θ_K, position $x_K(t)$) movable vertically on the beam structure, a point mass m_E at the tip of the beam, and an elastic Bernoulli beam (density ρ, cross section A, Young's modulus E, second moment of area I_{zB}, and length l). In the following, the vertical cage position is denoted by the dimensionless parameter $\kappa(t) := \frac{x_K(t)}{l}$.

The Ritz ansatz represents the elastic degrees of freedom of the beam with respect to its bending deflection

$$v\left(x_K(t), t\right) = v(\kappa(t)l, t) = \left(\frac{3}{2}\kappa^2(t) - \frac{1}{2}\kappa^3(t)\right) v_1(t) \tag{2.5}$$

in which only the first bending mode is considered. For use in ILC, the ansatz function for the bending deflection has been evaluated directly at the vertical cage position $x_K(t)$ to obtain a mathematical model of the desired system output.

The equations of motion for the rack feeder with the generalized coordinates $q := \left[y_S(t) \quad v_1(t)\right]^T$ (assuming that the vertical cage position $x_K(t)$ is determined a priori) and the input vector $h = \left[1 \quad 0\right]^T$ is

$$M\ddot{q}(t) + D\dot{q}(t) + Kq(t) = h\left(F_M(t) - F_F\left(\dot{y}_S(t)\right)\right) \tag{2.6}$$

where F_M and F_F denote, respectively, the motor and friction forces

$$M = \begin{bmatrix} m_S + \rho Al + m_K + m_E & m_{12} \\ m_{12} & m_{22} \end{bmatrix} \tag{2.7}$$

is the mass matrix, $m_{12} = \frac{3}{8}\rho Al + \frac{m_K \kappa^2}{2}(3 - \kappa) + m_E$ and $m_{22} = \frac{33}{140}\rho Al + \frac{6\rho I_{zB}}{5l} + \frac{m_K \kappa^4}{4}(3 - \kappa)^2 + \frac{90_K \kappa^2}{l^2}\left(1 - \kappa + \frac{\kappa^2}{4}\right) + m_E$. Also, the stiffness and damping matrices K and D, respectively, are given by

$$K = \begin{bmatrix} 0 & 0 \\ 0 & k_{22} \end{bmatrix}, \quad D = \begin{bmatrix} 0 & 0 \\ 0 & \frac{3k_d EI_{zB}}{l^3} \end{bmatrix} \tag{2.8}$$

with $k_{22} = \frac{3EI_{zB}}{l^3} - \frac{3}{8}\rho Ag - \frac{3m_K g\kappa^3}{l}\left(1 + \frac{3\kappa^2}{20} - \frac{3\kappa}{4}\right) - \frac{6m_E g}{5l}$.

For the experimental control implementation, it is assumed that the electric drive for the horizontal motion is operated in an underlying velocity control regime, applied using its current converter, with the resulting dynamics

$$T_{1y}\ddot{y}_S(t) + \dot{y}_S(t) = v_S(t) \tag{2.9}$$

Including the first-order lag dynamics in the overall system model gives the equations of motion as follows:

$$\dot{x}_y = \begin{bmatrix} 0 & I \\ -M_y^{-1}K_y & -M_y^{-1}D_y \end{bmatrix} x_y + \begin{bmatrix} 0 \\ M_y^{-1}h_y \end{bmatrix} v_S$$
$$=: A_y x_y + B_y u_y \tag{2.10}$$

with state vector

$$x_y = \begin{bmatrix} q \\ \dot{q} \end{bmatrix} \tag{2.11}$$

$u_y = v_S$ and modified mass and damping matrices

$$M_y = \begin{bmatrix} T_{1y} & 0 \\ \frac{3}{8}\rho Al + \frac{m_K \kappa^2}{2}(3 - \kappa) + m_E & m_{22} \end{bmatrix} \tag{2.12}$$

and

$$D_y = \begin{bmatrix} 1 & 0 \\ 0 & \frac{3k_d EI_{zB}}{l^3} \end{bmatrix} \tag{2.13}$$

The corresponding system output, i.e. the horizontal cage position, corresponds to $y(t) = y_S(t) + v(x_K(t), t)$ and the discrete-time state-space model is obtained numerically using a zero-order hold. To complete the model, the following data is used $Ts = 0.05$, $T = Ts$, $\Theta_K = 0$, $T_{1y} = 0.007$, $L = 1.07$, $E = 1.70 \times 10^9$, $I_z B = 1.2 \times 10^{-8}$, $g = 9.81$, $k_d = 1.1 \times 10^{-3}$, $\rho = 2.7 \times 10^3$, $A = 6 \times 10^{-4}$, $m_E = 0.90$, and $m_K = 0.95$.

2.3 Free Electron Laser Facility

Free electron lasers (FELs) use linear particle accelerators that increase the energy of the electrons by interaction with electromagnetic radio frequency (RF) fields. This section considers the control-oriented model of one such laser that has led to the successful implementation of ILC laws. For further background, see, e.g. [231] and the references given in this paper, including those that refer to uses for such lasers.

One form of operation is pulse mode, where every second, there is a pulse for approximately 1 ms. This pulsed system has the following properties: (i) any disturbances and uncertainties that arise only show small changes from pulse-to-pulse, (ii) the times between pulses is of the order of several hundred milliseconds, and this time can be used for computing optimal parameters and driving signals for the next pulse, (iii) the field program gate array (FPGA) structure of the digital intrapulse controller allows arbitrary input signals at a frequency of 1 MHz, and (iv) models can be identified by standard methods from measurement data.

Property (ii) strongly suggests that ILC can be applied, and the remainder of this section gives the relevant background and the model used for design in Chapter 5. In this application, the ILC design has been operating on the physical system for some years. This facility provides a service to groups of researchers who pay an access charge and, therefore, the service quality must be consistently high.

An FEL produces laser radiation with a tunable wavelength, and the particular case considered is an X-ray wavelength. The process uses a linear particle accelerator, which increases the energy of electrons by interaction with electromagnetic RF fields to a desired value. These fields are required to be very precise in amplitude and phase stability. Figure 2.11 shows the structure of the Vacuum Ultraviolet FEL. The linear accelerator consists of resonators for the RF fields housed in shape cryomodules, and the RF fields inside these superconducting resonator cavities are periodically supplied by an actuator system for a finite time interval and then turned off again periodically.

Figure 2.12 gives the amplitude of the desired envelope of the RF-field for one RF-pulse as a function of time. The field inside an accelerator cavity is required to be constant once the required amplitude for the appropriate energy gain of the electrons is reached at the end of the so-called "filling phase." During the flat-top phase, the electron beam is injected into the accelerator, and

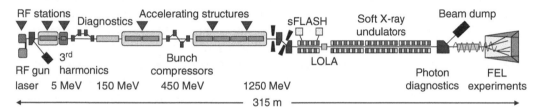

Figure 2.11 An FEL with frontend, accelerating structures, and undulators.

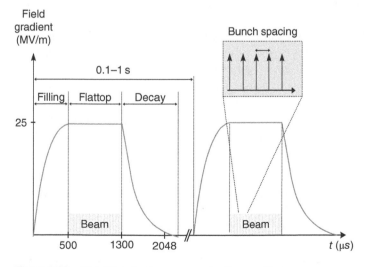

Figure 2.12 One RF-pulse in superconducting cavities.

when the electron beam has passed, the RF-field is turned off, and the field amplitude decays. The envelope of the RF-field oscillation must be kept constant in amplitude and phase during the flat top time interval to transfer a precise quantity of energy to the electrons.

Once the system is set to a desired operating point, the pulse trajectory remains unchanged for many pulses. Therefore, suppression of repetitive disturbances can be achieved by finding the optimal feedforward control signal to minimize the reference's deflection. The driving signal's adaptation is calculated using norm optimal iterative learning control (NOILC), covered in Chapter 5, for which a system model is required.

In the system, the acceleration takes place in the resonators where standing RF-waves (modes) provide the energy. The geometry determines the resonance frequency and for the case considered, the desired acceleration mode is 1.3 GHz. If the length of the cavity changes, the resonance frequency also changes. Due to the relatively thin walls, the resonators become susceptible to mechanical vibrations called microphonics, which detune the resonance frequency. The high-power RF-fields in the cavities lead to deformation of the cavity walls, and therefore, detuning also occurs. Induced currents cause Lorentz forces to act on the metal surroundings during the pulse sequence. Measurements have shown that detuning can occur up to $\Delta f \approx 500\,\text{Hz}$. Since the Lorentz force is induced every time the electric field is generated, the Lorentz force detuning is considered deterministic and repetitive, and amenable to ILC.

Another source of disturbance arises from the electron beam itself. As it passes through the accelerating structure, the charged particles gain energy from the current RF-field, leading to fluctuations in the present amount of energy stored in the system. The following bunches of charged particles will be influenced by these fluctuations, which have to be minimized by the control system. It can be assumed that the bunch arrival time is constant from pulse to pulse, thus having the properties of a repetitive disturbance. Next, the general architecture and related aspects of the digital control scheme used in this application are described.

The actuator system receives a precise RF signal of 1.3 GHz from the master oscillator (MO). The vector modulator can change this low-power sinusoidal in amplitude and phase. The output signal of the vector modulator is amplified by a klystron, a radio frequency amplifier. The amplified RF waves are transferred from the klystron to the cavities inside the cryomodules via a waveguide transmission and distribution system. For economic reasons, one high-power klystron supplies all 8–32 cavities of an RF station, thus RF fields cannot be influenced in each cavity individually, and the system is therefore underactuated.

The superconduction cavity simulator and controller (SIMCON) is based on FPGA structures that enable use of fast algorithms. A block diagram of the low-level radio frequency (LLRF) control system is shown in Figure 2.13, where the bottom part shows the digital FPGA controller. The LLRF control system's role is to keep the pulsed RF fields in the superconducting cavities of the RF station at the reference value during the flat-top phase of one RF pulse, as shown in Figure 2.12.

After measuring the actual RF-field by pickup antennas, the signals are downconverted to an intermediate frequency of 250 kHz. The real (I) and imaginary (Q) field components are digitalized in analog–digital-converters (ADC) with a sampling frequency of 1 MHz. An overview of the signals shown in Figure 2.13 expressed in terms of I and Q is given next.

- The input signals u_I, u_Q are produced by the actuator system and act directly on the vector modulator.
- The output signals y_I, y_Q are the real and imaginary parts, respectively, of the sum of the RF-field voltage vectors of eight cavities.
- The reference signals r_I, r_Q are the real and imaginary parts, respectively, of the vector sum of the RF-field's voltage vectors given by look-up tables for the specified field gradient.

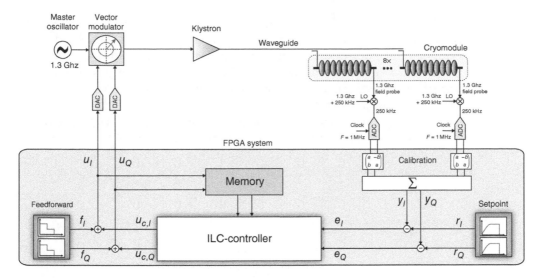

Figure 2.13 Structure of the RF system with master oscillator, vector modulator, klystron, cryomodule, measurement and calibration system, and the FPGA implemented control system.

- The feedforward signals f_I, f_Q are among the control signals determined by open-loop control.
- The control signals $u_{c,I}, u_{c,Q}$ are the ILC controller output signals, updated using the previous trial input signals.
- The control error signals e_I, e_Q are the deviations in real and imaginary parts, respectively, of the output signals from the corresponding reference.

Calibration of the measurement signals is undertaken for compensation of effects resulting, e.g. from different cable lengths. The control law usually uses the vector sums of all calibrated measurement signals of the individual cavities to be controlled because of the lack of individual action for each cavity.

To reach the desired set point (or reference) values requires, even in open-loop operation, that an adaptation of the feedforward signals is sufficient to track the reference trajectory. The disturbance is mainly the strong effect of the Lorentz forces that are deterministic from pulse-to-pulse. It is possible to compensate the drift away from the operating point by a smooth increase of the driving signal over the flat top.

Transients induced by the beam are predictable, and the arrival time is known. An increase in the driving power will keep this fluctuation low. Both compensation measures positively affect feedback control performance while keeping the control error signals small. It is essential to have an adequate model of the underlying system dynamics to predict the system behavior.

Additional external disturbances and several nonlinearities in the actuator system are relevant for a broad range of operation setpoints. However, standard identification procedures for linear time-invariant dynamics can be used to estimate models that can then be validated at specific setpoints in the manner outlined in, e.g. [148]. The subspace identification method *N4SID* from Matlab's System Identification Toolbox was used to estimate the matrices A, B, C, D of the state-space models

$$\dot{x}(t) = A\,x(t) + B\,u(t)$$
$$y(t) = C\,x(t) + D\,u(t) \tag{2.14}$$

where $u = \begin{bmatrix} u_I & u_Q \end{bmatrix}^T$ and $y = \begin{bmatrix} y_I & y_Q \end{bmatrix}^T$ denote, respectively, the system input and output vectors and the state vector x. The flat-top phase of the pulse is the primary interest for control and is also the system's operation point. Only measurements from this period are used for system identification. Persistent excitation signals can be injected into the accelerator system by superimposing random signals on standard feedforward signals with defined setpoints. Figure 2.14 gives a typical input sequence for the feedforward part.

The actuator system is operated at maximum power in the first 500 μs (filling phase). In the flat-top phase starting from 500 μs, the inputs are first reduced by a factor of 0.5 to reach the set point, and soon after, the excitation signal is added to both inputs, see Figure 2.14. A high amplitude leads to a good signal-to-noise ratio. Figure 2.15 shows measured versus simulated signals for an identified third-order model.

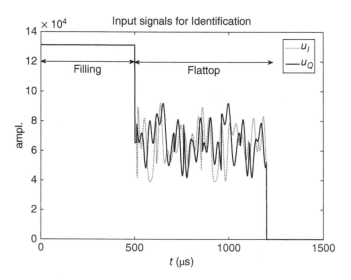

Figure 2.14 Input disturbances on both channels at flat top.

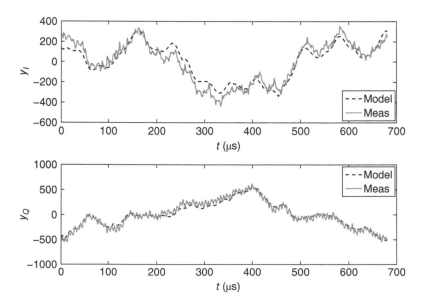

Figure 2.15 Measured versus simulated vector sum signals (third-order model).

2.4   ILC in Healthcare

In addition to engineering applications, recent years have seen innovative use of ILC in healthcare development, starting with robotic-assisted upper-limb stroke rehabilitation. This section gives a brief introduction to this application that has seen successful clinical trials, which is considered again in Chapters 4, 10, and 11. (Other applications of ILC in healthcare, following on from the stroke rehabilitation problem, are also briefly discussed in the last of these chapters.)

Annually, 110 000 people have a stroke in England, with associated costs estimated at 7 billion (UK pounds). Stroke is a leading cause of disability worldwide, with incidence set to rise, which, in turn, will place an increasing burden on healthcare and rehabilitation resources. If the capacity of health services is to meet future demand, novel approaches to rehabilitation are needed. Enabling rehabilitation outside the hospital, supported by mobile technology, may lead to (i) reduced cost, (ii) increased intensity of therapy, and (iii) shift the emphasis of responsibility for good health from healthcare professionals to the patients.

A stroke is usually caused when a blood clot blocks a blood vessel in the brain. It acts like a dam stopping the blood from reaching the brain downstream. As a result, some of the connecting nerve fibers die, and the person suffers partial paralysis on one side of the body, known as hemiplegia. These fibers cannot regrow. The brain, however, has spare capacity so new connections can be made. Moreover, the brain is continually and rapidly changing as a person learns new skills; new connections are formed, redundant ones disappear.

When people relearn skills after a stroke, they go through the same process as you do when you learn to play tennis, but they have a problem because they can hardly move at all, so they cannot practice, which means they do not get feedback. Muscles can be made to work by electrical stimulation (ES).

If ES is applied to a person's muscle, electrical impulses travel along the nerves in much the same way as the electrical impulses from the brain. If the stimulation is carefully controlled, beneficial movement can be made. This effect works better if the person is attempting the movement themselves and hence needs to combine a person's effort with just enough extra ES to achieve the movement.

Upper-limb impairment is very common poststroke and limits many daily living activities, especially those requiring reach to grasp actions such as picking up a drink. Functional electrical stimulation (FES) and robotic therapy have proven effective in reducing impairments in clinical trials. Factors affecting recovery include the intensity of treatment, repetition of functional tasks, and maximal voluntary actions combined with FES.

In recent years, collaboration between control engineers and health professions, see, e.g. [80, 113, 114] have successfully stimulated muscles in stroke patients' shoulders and elbows using ILC. This approach uses a biomechanical model of the combined arm support system and controls the applied FES to correct performance error in the next attempt and encourages and supports a voluntary effort by the participant, supplying just enough FES to achieve the movement. Central to this research is a rehabilitation robot which is a powered or mechanically supporting device that enables a person with limited physical ability to practice tasks.

Such practice and the resulting sensory feedback are associated with cortical changes that can bring about the recovery of beneficial movements. Measuring devices on the robots allow for objective measurement and evaluation of patients' performance. Studies have also shown increased motivation to participate in training exercises actively. After a stroke, relearning skills require a person to practice movements by repetition and use feedback from previous attempts to improve the next one.

In this research, ILC is used to regulate the FES by incorporating data from previous trials of the task. Then the FES is reduced in accordance with voluntary effort. The workstation in Figure 2.16a

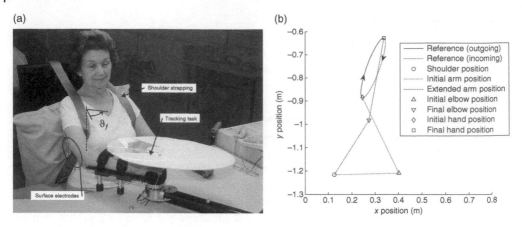

Figure 2.16 (a) Photograph of a patient using the robotic system and (b) ILC perspective.

uses a robotic arm to (i) constrain the arm to move in the horizontal plane, (ii) recreate the effect that the subject is moving a simple point mass with damping, and (iii) provide some assistance when FES is unable to. The task is reaching out from the body with the affected arm. The relationship between reaching and independence is reflected in functional independence measures, such as the Barthel index, where the ability to reach is required for over half of daily living tasks. Figure 2.16b shows the reference trajectory and the analogy with ILC as applied in the industrial domain is immediate. Clinical trials have confirmed the potential of this approach.

The system shown in Figure 2.16 is restrictive in the range of motions that can be replicated but has proved the basis for a realistic starting point for much further development. One area is reaching out and lifting actions, e.g. collect an item of clothing with the impaired limb, place it in a container, e.g. in an open drawer in a chest, and then close the drawer. Moreover, in Figure 2.16, the patient is required to follow a fixed path from start to end. Hence, a potential application area for point-to-point and terminal ILC, i.e. only the start and end locations, is specified in the latter case. In the former, some intermediate waypoints are specified.

This research on the upper limb has been followed through to successful small-scale clinical trials, as discussed in Section 4.2. Chapters 10 and 11 detail the generalization of this research to more realistic (3D) tasks where model-based linear and nonlinear ILC designs are required.

2.5 Concluding Remarks

This chapter has given the mathematical models of the engineering testbeds, developed by the authors or made available through coworkers, which are used to obtain the experimental results given in this book. Of course, this is not an exhaustive set, and others will be introduced as needed. The authors have also developed the set-up shown for robotic-assisted stroke rehabilitation, but commercial systems are also available. In the next chapter, the performance of ILC designs is considered.

3

An Overview of Analysis and Design for Performance

This chapter introduces the iterative learning control (ILC) design problem and gives an overview of the main results that form the basis for analysis and design, focusing on convergence, stability, and performance. As in other areas, these requirements can result in conflicts to be resolved, if possible, by the ILC designs treated in this book. A distinguishing feature is the finite trial length role, and it is not possible to cover all analysis settings. Therefore, the focus, in the main, is on those that have seen translation to applications. The following section deals with stability and convergence for discrete linear dynamics using the lifted model and draws on, in particular, [33, 176]. Subsequent sections then consider the repetitive process/2D systems setting, Chapter 7, and nonlinear dynamics, Chapters 10 and 11.

3.1 ILC Stability and Convergence for Discrete Linear Dynamics

For ease of presentation, the single-input single-output (SISO) case is considered with an obvious generalization to multiple-input multiple-output (MIMO) examples. The starting point is discrete linear systems with dynamics described in the ILC setting by (1.29), i.e.

$$y_k(p) = G(q)u_k(p) + d(p) \tag{3.1}$$

and control law (1.31), i.e.

$$u_{k+1}(p) = Q(q)(u_k(p) + L(q)e_k(p+1)) \tag{3.2}$$

and the following is the definition of stability given in [33] (for finite N).

Definition 3.1 An ILC system described by (3.1) and (3.2) is asymptotically stable if there exists a real scalar $\epsilon > 0$ such that

$$|u_k(p)| \leq \epsilon \tag{3.3}$$

and $\lim_{k \to \infty} u_k(p)$ exists $\forall\ 0 \leq p \leq N - 1,\ k \geq 0$.

The converged (or learned) control is defined as $u_\infty(p) = \lim_{k \to \infty} u_k(p)$.

The stability property for ILC can be defined for the finite and fixed trial length, which for discrete dynamics means that the number of samples along the trial, N, is finite. Alternatively, stability for all possible values of the finite trial length, i.e. for $N = \infty$, can be considered. The implications of these two properties for ILC design are recurring themes in the rest of this book.

Iterative Learning Control Algorithms and Experimental Benchmarking, First Edition.
Eric Rogers, Bing Chu, Christopher Freeman and Paul Lewin.
© 2023 John Wiley & Sons Ltd. Published 2023 by John Wiley & Sons Ltd.

Using the lifted representation (see (1.36)–(1.47)) of the controlled dynamics generated by (3.1) and (3.2) gives

$$U_{k+1} = Q(I - LG)U_k + QL(R - d) \tag{3.4}$$

This is a linear difference equation in the trial number k and conditions for asymptotic stability follow immediately. As one example, let $\rho(\cdot)$ denote the spectral radius of its matrix argument, i.e. if $h_i, 1 \leq i \leq h$, are eigenvalues of an $h \times h$ matrix, say H, then $\rho(H) = \max_{1 \leq i \leq h} |h_i|$.

Theorem 3.1 *An ILC system formed by combining (3.1) and (3.2) is asymptotically stable if and only if*

$$\rho(Q(I - LG)) < 1 \tag{3.5}$$

If Q and L are causal, the matrix $Q(I - LG)$ is a lower triangular and Toeplitz, all of its eigenvalues are $q_0(1 - l_0 g_1)$ and, hence, asymptotic stability provided

$$|q_0(1 - l_0 g_1)| < 1 \tag{3.6}$$

In the case of the z-transform representation, and hence, frequency domain analysis, consider again (1.49)–(1.51), which assumes that (see Section 1.5) $N = \infty$. Then

$$u_{k+1}(z) = Q(z)(1 - zL(z)G(z))u_k(z) + zQ(z)L(z)(r(z) - d(z)) \tag{3.7}$$

and a sufficient condition for stability is that $Q(z)(1 - zL(z)G(z))$ is a contraction mapping, see. e.g. [134] for the background functional analysis. Also for a given z-domain system, say $T(z)$, introduce (the H_∞ norm) $\|T\|_\infty = \sup_{\theta \in [-\pi,\pi]} |T(e^{j\theta})|$, and hence, the following result.

Theorem 3.2 *An ILC system described by (3.1) and (3.2) is asymptotically stable for $N = \infty$ if*

$$\|Q(z)(1 - zL(z)G(z))\|_\infty < 1 \tag{3.8}$$

In more familiar terms, this result requires that the Nyquist locus of $T(z)$ must lies inside the unit circle in the complex plane.

In the case when $Q(z)$ and $L(z)$ are causal functions, it can be shown [176] that (3.8) also implies asymptotic stability for the finite trial length ILC system. Moreover, (3.8) is a sufficient but not necessary condition and, in general, could be much more conservative than the necessary and sufficient condition of (3.5), see, e.g. [150].

One performance measure for an ILC design uses the error as $k \to \infty$. If the system is asymptotically stable, the resulting asymptotic error is

$$e_\infty(p) = \lim_{k \to \infty} e_k(p) = r(p) - G(q)u_\infty(p) - d(p) \tag{3.9}$$

and is often judged, for a given reference signal in terms of the difference between $e_\infty(p)$ and the initial error $e_0(p)$. Also, this comparison is made either qualitatively, see, e.g. [61], or quantitatively using a metric such as the mean square root error.

Suppose that the ILC system is asymptotically stable. Then in the lifted representation the asymptotic error is given by

$$E_\infty = (I - G(I - Q(I - LG))^{-1}QL)(R - d) \tag{3.10}$$

and in z-transform terms (with lifting not applied)

$$e_\infty(z) = \frac{1 - Q(z)}{1 - Q(z)(1 - zL(z)G(z))}(r(z) - d(z)) \tag{3.11}$$

These last two expressions can be obtained by replacing k by ∞ in the appropriate expressions given above and then solving for E_∞ and $e_\infty(z)$, respectively. The following result gives necessary and sufficient conditions for convergence to zero error [210].

Theorem 3.3 *Suppose that G and L are not identically zero. Then $e_\infty(p) = 0$ for all reference signals and disturbances in the ILC scheme defined by (3.1) and (3.2) if and only if the system is asymptotically stable and $Q(q) = 1$.*

The corresponding result for $N = \infty$ is given in [61].

In a significant number of applications, $Q(q) = 1$ is used, and hence no Q-filtering is employed, and by the above theorem, required for perfect tracking. However, choosing $Q(q) \neq 1$ can improve the transient learning and robustness properties of designs. Also in [60], the role of the Q filter in ILC performance was considered by assuming it to have an ideal low-pass characteristic with unity magnitude for, say, $\theta \in [0, \theta_c]$ and zero magnitude for $\theta \in (\theta_c, \pi]$. This ideal low-pass filter has no physical realization, and hence, its only use is for illustrative purposes.

Using (3.11) (and setting $z = e^{j\theta}$), $e_\infty(e^{j\theta})$ with the ideal low-pass filter chosen as the Q filter is zero for $\theta \in [0, \theta_c]$ and equal to $r(e^{j\theta}) - d(e^{j\theta})$ for $\theta \in (\theta_c, \pi]$. For frequencies at which the magnitude of the Q filter is unity, perfect tracking is achieved, and for those where the magnitude of this filter is zero, the ILC is effectively turned off. Consequently, the Q filter can be employed to determine the frequencies emphasized in the learning process. Emphasizing certain frequency bands helps control the trial domain transients in ILC. This topic is discussed again below, and in Chapter 7, the generalized Kalman–Yakubovich–Popov (KYP) lemma [117] will be applied to ILC design with experimental validation using the repetitive process setting for analysis.

3.1.1 Transient Learning

In [33], the following example in the ILC setting was considered

$$y_k(p) = \frac{q}{(q - 0.9)^2} u_k(p) \tag{3.12}$$

with the ILC law

$$u_{k+1}(p) = u_k(p) + 0.5e_k(p + 1) \tag{3.13}$$

For this case, $g_1 = 1$, $q_0 = 1$, and $l_0 = 0.5$. Also Q and L are causal and hence, the ILC system formed by applying (3.13) to (3.12) is asymptotically stable by (3.5). Moreover, $e_\infty = 0$ due to the choice for the Q filter. However, the error over the first 12 trials increases by nine orders of magnitude. Also, neither this transient growth nor the rate of increase is closely related to the stability conditions since the ILC design is stable.

Monotonic trial-to-trial error convergence avoids the problem highlighted by this last example and requires that for a given norm $\| \cdot \|$

$$\|e_\infty - e_{k+1}\| \leq \gamma \|e_\infty - e_k\|, \quad k \geq 0 \tag{3.14}$$

where $\gamma \in [0, 1)$ is the convergence rate. In the lifted representation,

$$E_\infty - E_{k+1} = GQ(I - LG)G^{-1}(E_\infty - E_k) \tag{3.15}$$

or, when G, Q, and L are causal,

$$E_\infty - E_{k+1} = Q(I - LG)(E_\infty - E_k) \tag{3.16}$$

Also, applying the z-transform to this last equation with $N = \infty$ gives

$$E_\infty(z) - E_{k+1}(z) = Q(z)(1 - zL(z)G(z))(E_\infty(z) - E_k(z)) \tag{3.17}$$

Let $\bar{\sigma}(\cdot)$ denote the maximum singular value of its matrix argument and also let $\|\cdot\|_2$ denote the 2-norm. Then (3.15) and (3.17) give the following results:

Theorem 3.4 *[176] If*

$$\gamma_1 := \bar{\sigma}(GQ(I - LG)G^{-1}) < 1 \tag{3.18}$$

then

$$\|E_\infty - E_{k+1}\|_2 \leq \gamma_1 \|E_\infty - E_k\|_2, \quad \forall k \geq 0 \tag{3.19}$$

holds for the ILC scheme defined by (3.1) and (3.2).

Theorem 3.5 *[61] If for the ILC scheme defined by (3.1) and (3.2) with $N = \infty$*

$$\gamma_2 := \|Q(z)(1 - zL(z)G(z))\|_\infty < 1 \tag{3.20}$$

then

$$\|E_\infty(z) - E_{k+1}(z)\|_\infty \leq \gamma_2 \|E_\infty(z) - E_k(z)\|_\infty, \quad k \geq 0 \tag{3.21}$$

If $Q(z)$ and $L(z)$ are causal functions, (3.21) also implies [176] that

$$\|E_\infty - E_{k+1}\|_2 \leq \gamma_2 \|E_\infty - E_k\|_2, \quad \forall k \geq 0 \tag{3.22}$$

for finite N. Moreover, the convergence condition (3.19) is equivalent to that of Theorem 3.2. Hence, the condition of Theorem 3.2 ensures stability and monotonic trial-to-trial error convergence independent of the trial length (N). In the lifted system description, the monotonic convergence condition (3.19) is more stringent than (3.5), and both these conditions depend on N (and hence, the trial length). Further analysis and discussion of these two cases, i.e. N finite and $N = \infty$, is discussed again in Section 3.2, which leads on to the design and experimental validation results of Chapter 7.

3.1.2 Robustness

Robustness (see also the brief discussion Section 1.8) is as much an issue in ILC design and implementation as in other areas. The critical question is: if an ILC design is asymptotically stable, does it retain this property under system model perturbations? This area is the subject of Chapter 6, and in the preliminary treatment below, stability robustness is considered together with monotonic trial-to-trial error convergence. This analysis provides insights into the limitations of safe operation requirements, e.g. stability and acceptable transients, to place on an ILC design. Stability robustness to time delay error in the system model is then briefly discussed and also the effects of nonrepeating disturbances on performance.

Consider the case when $Q(q) = 1$ in the ILC law (3.2), which, by Theorem 3.3, results in zero converged error, and $L(q)$ is causal. Then the stability condition (3.6) in this case is $|1 - l_0 g_1| < 1$, and if $l_0 \neq 0$ and $g_1 \neq 0$, asymptotic stability holds if and only if

$$\text{sgn}(g_1) = \text{sgn}(l_0) \quad \text{and} \quad l_0 g_1 \leq 2 \tag{3.23}$$

Hence, ILC can ensure zero-converged error for a system using only knowledge of the sign of its first Markov parameter g_1 and an upper bound on its absolute value. Also, the second term in (3.23) holds for a sufficiently large upper bound on $|g_1|$ by choosing $|l_0|$ sufficiently small. Hence, ILC systems can have robust stability to all perturbations that do not change the sign of g_1, but this does not guarantee acceptable learning transients.

Consider multiplicative uncertainty in the form

$$G(q) = G_n(q)(1 + W(q)\Delta(q)) \tag{3.24}$$

where $G_n(q)$ is the nominal system model, $W(q)$ is known and stable, and $\Delta(q)$ is unknown but stable with $\|\Delta(z)\|_\infty < 1$. Then the following result (Theorem 6 in [33]) characterizes robust monotonic trial-to-trial error convergence.

Theorem 3.6 *If*

$$|W(e^{j\theta})| \leq \frac{\gamma^* - |Q(e^{j\theta})||1 - e^{j\theta}L(e^{j\theta})G_n(e^{j\theta})|}{|Q(e^{j\theta})||e^{j\theta}L(e^{j\theta})G_n(e^{j\theta})|} \tag{3.25}$$

$\forall\, \theta \in [-\pi, \pi]$, *then for the ILC system formed by combining (3.1), (3.2), and (3.24) with $N = \infty$ monotonic trial-to-trial error convergence occurs with rate $\gamma^* < 1$.*

Unlike the case given by (3.23), (3.25) depends on the dynamics of $G(q)$, $Q(q)$, and $L(q)$. Also, from (3.25), decreasing the gain of the Q filter for a given θ increases robustness, but, from the Section 3.1.2, this action negatively impacts on convergence, i.e. a trade-off is required. See also [33] and the relevant cited references for discussion of the case when the system has an uncertain time delay that cannot be neglected.

The effects of noise, nonrepeating disturbances, and variations in the initial condition on the trials also affect performance. One way to combine the effects of these is to add a trial-dependent exogenous signal d_k to (3.2), which prevents error convergence to e_∞. If, however, d_k is bounded, error convergence to a ball centered on e_∞ occurs. Moreover, ideally, the ILC law should not attempt to learn from d_k, and hence, it is to be expected that slower learning would decrease the sensitivity to this combined disturbance signal.

Performance robustness to the features above will be covered in Chapters 6 and 7. In section 3.2, the repetitive processes/2D systems setting for convergence and performance analysis is introduced.

3.2 Repetitive Process/2D Linear Systems Analysis

3.2.1 Discrete Dynamics

Consider first the discrete 2D Roesser or Fornasini–Marchesini model approach. In particular, recall (1.67), i.e.

$$\begin{bmatrix} \eta_k(p+1) \\ e_{k+1}(p) \end{bmatrix} = \begin{bmatrix} A & BK \\ -CA & I - CBK \end{bmatrix} \begin{bmatrix} \eta_k(p) \\ e_k(p) \end{bmatrix} \tag{3.26}$$

This state-space model is of the Roesser form (see (1.62)), and hence, the systems theory for this 2D model can be applied to study trial-to-trial error convergence.

Convergence analysis for this representation of the ILC dynamics requires the general response formula for 2D discrete linear systems described by the Roesser model. This formula [125] for (3.26) is

$$
\begin{bmatrix} \eta_k(p) \\ e_k(p) \end{bmatrix} = \sum_{i=0}^{p-1} \Phi^{i,k} \begin{bmatrix} 0 \\ e_0(p-i) \end{bmatrix}, \quad p > 0
\tag{3.27}
$$

where the state transition matrix $\Phi^{i,k}$ is defined by

$$
\Phi^{i,k} = \Phi^{1,0}\Phi^{i-1,k} + \Phi^{0,1}\Phi^{i,k-1}
\tag{3.28}
$$

and $\Phi^{0,0} = I$. In the case considered

$$
\Phi^{1,0} = \begin{bmatrix} A & BK \\ 0 & 0 \end{bmatrix}, \quad \Phi^{0,1} = \begin{bmatrix} 0 & 0 \\ -CA & I - CBK \end{bmatrix}
\tag{3.29}
$$

with $\Phi^{i,j} = 0$, $i < 0$ or $j < 0$. Also, the following result, Lemma 1 in [136], is required in the proof that the ILC system converges.

Lemma 3.1 *The state transition matrix for the Roesser state-space model satisfies*

$$
\Phi^{i,k} \to 0, \quad k \to \infty
$$

and any given i if and only if $\rho(\Phi^{i,k}) < 1$.

Further development leads to the following result (Theorem 1 in [136]).

Theorem 3.7 *The IlC dynamics (3.26) converge as $k \to \infty$ if and only if $\rho(I - CBK) < 1$.*

In control law design terms, the existence of a stabilizing K is equivalent to controllability of the pair (I, CB). Hence, such a K exists if and only if the matrix $\begin{bmatrix} CB & CB & \dots & CB \end{bmatrix}$ has full row rank. Hence, CB must have full row rank.

Theorem 3.7 is, at first sight, somewhat surprising as convergence occurs independent of the system matrix A and, in particular, its eigenvalues. Hence, convergence of the error from trial to trial will occur even if the matrix A is unstable, and high-error values can result for small values of k and large values of p if the trial length is long.

As an example, consider the X-axis of the gantry robot of Section 2.1.1 with state-space model matrices given by (2.1), where A matrix has all eigenvalues inside the unit circle except for one of value unity on the real axis of the complex plane.

Consider the case when K is chosen such that $\rho(I - CBK) < 1$, where one choice is $K = 50$. Figure 3.1 shows the input, error, and output progression for this design, and Figure 3.2 the response on trial 4. Although ILC convergence occurs, the oscillatory behavior present in the dynamics of the early trials may not be acceptable, especially for physical implementation. These plots confirm the problems that can arise with the above design in the form of, in relative terms, large error and very oscillatory along the trial responses for, in this case, the early trials.

In cases such as above, one option is to apply feedback control to give the required performance along the trials and then proceed, via the lifted representation, to control law design for trial-to-trial error convergence. Another alternative in the 2D systems setting is to use a control law of the form (1.68), i.e.

$$
\Delta u_k(p) = -K_1 \eta_k(p+1) + K_2 e_k(p+1)
\tag{3.30}
$$

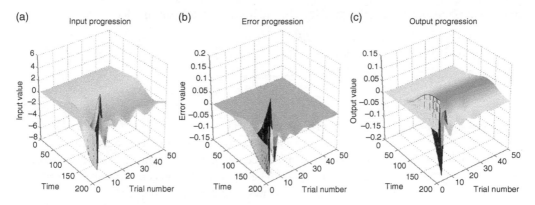

Figure 3.1 Simulation of the input (a), error (b), and output (c) progression for the Roesser model-based design with $K = 50$ for the X-axis of the gantry robot.

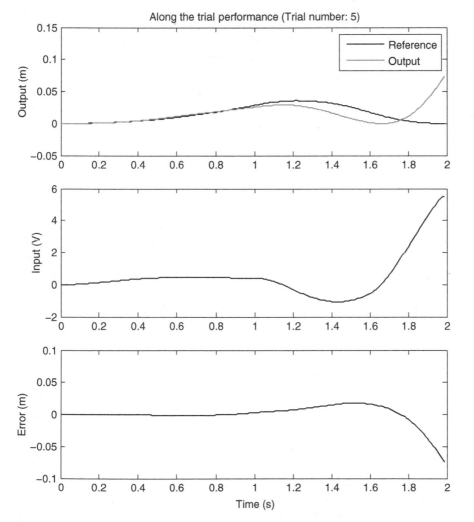

Figure 3.2 The output on trial 5 (light gray line) compared to the reference (dark gray line) together with the input (middle plot) and the error (bottom plot) (for the example of the previous figure).

with resulting controlled dynamics described by the Roesser state-space model

$$\begin{bmatrix} \eta_k(p+1) \\ e_{k+1}(p) \end{bmatrix} = \begin{bmatrix} A - BK_1 & BK_2 \\ -CA + CBK_1 & I - CBK_2 \end{bmatrix} \begin{bmatrix} \eta_k(p) \\ e_k(p) \end{bmatrix} \tag{3.31}$$

where it is assumed that C has full rank. Moreover, selecting

$$K_2 = (CB)^T(CB(CB)^T)^{-1}, \quad K_1 = K_2 CA \tag{3.32}$$

gives $e_1(p) = 0$, i.e. convergence to zero error after one trial.

It is also possible to define Roesser (and Fornasini–Marchesini) state-space models where a linear differential equation governs the dynamics in one direction of information, and there has also been research on using this setting in ILC see, e.g. [143]. The Section 3.2.2 considers the repetitive process setting for the analysis of the performance of ILC laws (and design in Chapter 7).

3.2.2 Repetitive Process Stability Theory

The repetitive process setting for ILC was introduced in Section 1.6.2 and to proceed to design; it is first necessary to introduce the stability theory for linear repetitive processes. This theory is based on an abstract model in a Banach space setting that includes many examples, including the state-space models used in linear model-based ILC, as special cases. The linear repetitive process state-space models of use in ILC, lead to stability tests that extend naturally to allow control law design. (In Chapters 8 and 12, the more recently developed stability theory for nonlinear repetitive processes will be used for ILC analysis and design, including, in Chapter 8, supporting experimental results.)

Suppose that the trial profile $y_k \in E_\alpha$, where E_α is a suitably chosen Banach space, see, e.g. [134] for the relevant background, with norm $\| \cdot \|$. Then the dynamics of a linear repetitive process of constant trial length $\alpha > 0$ are described by

$$y_{k+1} = L_\alpha y_k + b_{k+1}, \quad k \geq 0 \tag{3.33}$$

where $b_{k+1} \in W_\alpha$, W_α is a linear subspace of E_α, and L_α is a bounded linear operator mapping E_α into itself. In this model, the term $L_\alpha y_k$ represents the contribution of trial k to trial $k + 1$, and b_{k+1} represents terms that enter on trial $k + 1$, i.e. control inputs, trial state initial conditions, and disturbances.

Consider discrete linear repetitive processes described by (1.72) with associated boundary conditions, i.e.

$$x_{k+1}(p+1) = Ax_{k+1}(p) + Bu_{k+1}(p) + B_0 y_k(p)$$

$$y_{k+1}(p) = Cx_{k+1}(p) + Du_{k+1}(p) + D_0 y_k(p) \tag{3.34}$$

Then it can be shown that for this case L_α is the convolution operator for the standard, also termed 1D in the multidimensional systems literature, system described by the state-space quadruple $\{A, B_0, C, D_0\}$.

The unique control problem for repetitive processes is that the sequence of trial profiles generated $\{y_k\}_{k \geq 0}$ can contain oscillations that increase in amplitude in the trial-to-trial direction, i.e. in k. This problem arose in the original work on longwall coal cutting [70], where the trial profile is the height of the stone/coal interface above some datum line, and the cutting machine rests on the previous trial profile during the production of the current one. The result can be severe undulations/oscillations in the trial profiles that have to be removed after only a few trials, and the downtime involved is lost production. Consequently, the stability theory for linear repetitive processes is of the bounded-input, bounded-output (BIBO) form, i.e. a bounded initial trial profile is required to produce a bounded sequence of trial profiles, where boundedness is defined in terms of the norm on E_a.

Asymptotic stability in the trial-to-trial direction, or asymptotic stability for short, demands this BIBO stability property over the finite and fixed trial length, i.e. over $(k, p) \in [0, \infty] \times [0, \alpha]$. The mathematical formulation in terms of the abstract model is that this property requires that there exist real scalars $M_\alpha > 0$ and $\lambda_\alpha \in (0, 1)$ such that

$$\|L_\alpha^k\| \leq M_\alpha \lambda_\alpha^k, \quad k \geq 0 \tag{3.35}$$

where $\| \cdot \|$ also denotes the induced norm. Using standard results from Banach space theory [134], it can be shown that (3.35) holds if and only if $\rho(L_\alpha) < 1$.

Suppose that asymptotic stability holds for (3.33), then the strong limit

$$y_\infty := \lim_{k \to \infty} y_k \tag{3.36}$$

is termed the limit profile. Letting I denote the identity operator in E_a, application of standard results from the theory of Banach spaces [134] shows that the limit profile is the unique solution of the linear equation:

$$y_\infty = L_\alpha y_\infty + b_\infty \quad \text{or} \quad y_\infty = (I - L_\alpha)^{-1} b_\infty \tag{3.37}$$

where b_∞ is the strong limit of the sequence $\{b_k\}_{k \geq 1}$. In application, the limit profile can be used to characterize what happens after a "suitably large" number of trials has elapsed.

In the case of discrete linear repetitive processes described by (3.34), computing the spectral values of L_α, as detailed in [228], shows that asymptotic stability holds if and only if $\rho(D_0) < 1$. Also, the corresponding limit profile is described by a 1D discrete linear systems state-space model with state matrix $A_{\text{lp}} := A + B_0(I - D_0)^{-1}C$.

Asymptotic stability is independent of the state dynamics, and hence, the possibility that damping out the trial-to-trial oscillations, i.e. only in the k direction, may not be enough to ensure acceptable along-the-trial dynamics. As an example, consider the case when $A = -0.5$, $B = 1$, $B_0 = 0.5 + \beta$, $C = 1$, $D = D_0 = 0$, where β is a real scalar. This example is asymptotically stable with $A_{\text{lp}} = \beta$ and hence, the limit profile in this case is unstable if $|\beta| \geq 1$.

The most obvious way to exclude this last possibility is to demand the BIBO stability property uniformly with respect to the trial length, i.e. over $(k, p) \in [0, \infty] \times [0, \infty]$, and this property is termed stability along-the-trial. The abstract model-based characterization of this property can again be found in [228], and the following result gives the necessary and sufficient conditions for this property when applied to processes with dynamics described by (3.34).

Theorem 3.8 *[228] A discrete linear repetitive process described by the state-space model (3.34) with the pair $\{A, B_0\}$ controllable and the pair $\{C, A\}$ observable is stable along the trial if and only if*

(a) $\rho(D_0) < 1$
(b) $\rho(A) < 1$
(c) all eigenvalues of the transfer-function matrix

$$G(z) = C(zI - A)^{-1}B_0 + D_0 \tag{3.38}$$

have modulus strictly less than unity for all $|z| = 1$.

Before discussing these conditions and their implications, the use of differential linear repetitive processes in ILC analysis and design is considered.

Differential linear repetitive processes are described by (1.70), i.e.

$$\dot{x}_{k+1}(t) = Ax_{k+1}(t) + Bu_{k+1}(t) + B_0 y_k(t)$$
$$y_{k+1}(t) = Cx_{k+1}(t) + Du_{k+1}(t) + D_0 y_k(t) \tag{3.39}$$

with boundary conditions that are the counterpart of those for the discrete processes considered above. In this case, it can be shown that L_α is the convolution operator for the 1D differential linear system defined by the state-space quadruple $\{A, B_0, C, D_0\}$. The following result is the differential counterpart of Theorem 3.8.

Theorem 3.9 *[228] A differential linear repetitive process described by the state-space model (3.39) with the pair $\{A, B_0\}$ controllable and the pair $\{C, A\}$ observable is stable along the trial if and only if*

(a) $\rho(D_0) < 1$
(b) *all eigenvalues of the matrix A have strictly negative real parts*
(c) *all eigenvalues of the transfer-function matrix*

$$G(s) = C(sI - A)^{-1}B_0 + D_0 \tag{3.40}$$

have modulus strictly less than unity for all $s = j\omega$, $\omega \geq 0$.

Consider a discrete linear system described by the state-space triple $\{A, B, C\}$ and written in the ILC setting as follows:

$$x_{k+1}(p+1) = Ax_{k+1}(p) + Bu_{k+1}(p)$$
$$y_{k+1}(p) = Cx_{k+1}(p), \quad x_{k+1}(0) = 0, \quad 0 \leq p \leq N-1, \quad k \geq 0 \tag{3.41}$$

Write the state equation in (3.41) as follows:

$$x_{k+1}(p) = Ax_{k+1}(p-1) + Bu_{k+1}(p-1) \tag{3.42}$$

and introduce the following vector defined in terms of the difference between the current and previous trial state vectors in the system state-space model:

$$\eta_{k+1}(p+1) = x_{k+1}(p) - x_k(p) \tag{3.43}$$

Also select the control law as

$$u_{k+1}(p) = u_k(p) + \Delta u_{k+1}(p) \tag{3.44}$$

where $\Delta u_{k+1}(p)$ is the correction to the input used on the previous trial.

In the case when

$$\Delta u_{k+1}(p) = K_1 \eta_{k+1}(p+1) + K_2 e_k(p+1) \tag{3.45}$$

the resulting controlled system dynamics are described by

$$\eta_{k+1}(p+1) = (A + BK_1)\eta_{k+1}(p) + BK_2 e_k(p)$$
$$e_{k+1}(p) = -C(A + BK_1)\eta_{k+1}(p) + (I - CBK_2)e_k(p) \tag{3.46}$$

(where the second equation follows from considering $e_{k+1}(p) - e_k(p) = y_k(p) - y_{k+1}(p)$). This state-space model is a particular case of the discrete linear repetitive process state-space model (3.34) with zero input, current trial state vector $\eta_{k+1}(p)$ and previous trial profile vector $e_k(p)$.

Lifted model-based ILC design for differential dynamics would start from an infinite-dimensional systems model. In contrast, the repetitive process setting is directly applicable to, e.g. applications where design in the differential domain is required or design by emulation is the preferred route to implementation.

Consider a differential linear system described by the state-space triple $\{A, B, C\}$ and written in the ILC setting as follows:

$$\dot{x}_{k+1}(t) = Ax_{k+1}(t) + Bu_{k+1}(t)$$
$$y_{k+1}(t) = Cx_{k+1}(t), \quad x_{k+1}(0) = 0, \quad 0 \le t \le \alpha, \quad k \ge 0 \tag{3.47}$$

Also, introduce for analysis purposes the following vector defined in terms of the difference between the current and previous trial state vectors in the system state-space model:

$$\eta_{k+1}(t) = \int_0^t \left(x_{k+1}(\tau) - x_k(\tau) \right) d\tau \tag{3.48}$$

and let the ILC law compute the input on trial $k+1$ as follows:

$$u_{k+1}(t) = u_k(t) + \Delta u_{k+1}(t) \tag{3.49}$$

where this latter term is the correction to the input used on the previous trial and as one possible choice set

$$\Delta u_{k+1}(t) = K_1 \dot{\eta}_{k+1}(t) + K_2 \dot{e}_k(t) \tag{3.50}$$

Then the controlled system dynamics are described by

$$\dot{\eta}_{k+1}(t) = (A + BK_1)\eta_{k+1}(t) + BK_2 e_k(t)$$
$$e_{k+1}(t) = -C(A + BK_1)\eta_{k+1}(t) + (I - CBK_2)e_k(t) \tag{3.51}$$

This state-space model is a particular case of the differential linear repetitive process state-space model (3.39) with zero input, current trial state vector $\eta_{k+1}(t)$, and previous trial profile vector $e_k(t)$.

In the case of higher-order iterative learning control (HOILC), a detailed analysis for first- and second-order ILC, where order, in this case, refers to the number of previous trials from which the control law uses information, see, e.g. [174]. Moreover, the analysis is supported by experimental verification. An alternative setting for HOILC analysis and design is to use the models of differential and discrete nonunit memory repetitive processes [228]. In the case of differential dynamics, the state-space model is

$$\dot{x}_{k+1}(t) = Ax_{k+1}(t) + Bu_{k+1}(t) + \sum_{j=1}^M B_{j-1} y_{k+1-j}(t)$$
$$y_{k+1}(t) = Cx_{k+1}(t) + Du_{k+1}(t) + \sum_{j=1}^M D_{j-1} y_{k+1-j}(t) \tag{3.52}$$

and the rest of the notation and boundary conditions are the same as in the unit memory case, and the integer $M > 0$ is termed the memory length.

The dynamics of this model and its discrete counterpart can be expressed in an abstract model form using a Cartesian product space formulation [228], and the following result gives the necessary and sufficient conditions for stability along the trial.

Theorem 3.10 *A differential linear repetitive process described by (3.52) is stable along the trial if and only if*

(a) $r(\tilde{D}) < 1$, *where*

$$\tilde{D} = \begin{bmatrix} 0 & I & 0 & \cdots & 0 \\ 0 & 0 & I & \cdots & 0 \\ 0 & 0 & 0 & \cdots & 0 \\ \vdots & \vdots & \vdots & \ddots & I \\ D_{M-1} & D_{M-2} & D_{M-3} & \cdots & D_0 \end{bmatrix} \tag{3.53}$$

(b) *all eigenvalues of the matrix A have strictly negative real parts*

(c) *all eigenvalues of the block-companion transfer-function matrix*

$$\tilde{G}(s) = \begin{bmatrix} 0 & I & 0 & \cdots & 0 \\ 0 & 0 & I & \cdots & 0 \\ 0 & 0 & 0 & \cdots & 0 \\ \vdots & \vdots & \vdots & \ddots & I \\ G_M(s) & G_{M-1}(s) & G_{M-2}(s) & \cdots & G_1(s) \end{bmatrix} \tag{3.54}$$

where

$$G_j(s) = C(sI - A)^{-1}B_{j-1} + D_{j-1}, \quad 1 \le j \le M \tag{3.55}$$

have modulus strictly less than unity for all $s = j\omega$, $\omega \ge 0$.

Return to the result of Theorem 3.9 and consider the SISO case for simplicity. Then condition (c) of this result has a Nyquist-based interpretation requiring that each frequency component of the initial profile to be attenuated at a geometric rate from trial-to-trial. Frequency attenuation over the complete frequency spectrum is, in general, a very stringent condition that makes control law design harder/impossible. In many applications, the design specifications will require shaping the frequency response over finite frequency intervals. This requirement is the basis for some of the ILC designs of Chapter 7 with experimental support based on the generalized KYP lemma.

To introduce an H_∞ setting for ILC design, for differential dynamics with a natural extension to the discrete case, start with

$$\rho(G(j\omega)) \le \bar{\sigma}(G(j\omega)) \tag{3.56}$$

Then condition (c) of Theorem 3.9 can be replaced by the sufficient condition:

$$\|G(j\omega)\|_\infty \triangleq \sup_{\omega \in [0,\infty)} \bar{\sigma}(G(j\omega)) \tag{3.57}$$

where $\bar{\sigma}(\cdot)$ denotes the maximum singular values of its matrix argument. Hence, H_∞ analysis, see, e.g. [33] and Chapters 6 and 7 can be applied to ILC design. Successful design for a particular application will ensure trial-to-trial (in k) error convergence and minimizing $\|G(j\omega)\|_\infty$ will increase the convergence speed. Also, the transient response along the trials will be controlled. The result will be a class of designs that can be computed using linear matrix inequalities (LMIs) [31]. Moreover, the generalized KYP lemma can be used to impose different performance requirements over finite frequency ranges.

Next, trial-to-trial error convergence and along the trial performance in the repetitive process setting is considered. One outcome will be that the results on the equivalence of ILC and feedback control for ILC laws with no noncausal action were already known in the literature before the publication of [92].

3.2.3 Error Convergence Versus Along the Trial Performance

Fast trial-to-trial error convergence is a natural objective in ILC design. Even if zero error occurs as $k \to \infty$, only a finite number of trials will be completed in any application; therefore, the consequences of this need to be considered. The subsequent analysis shows that enforcing "fast" trial-to-trial error convergence can result in a design conflict. The analysis is in the differential setting, starting from the state-space model (3.47), where an alternative description of the dynamics in terms of the current trial error, obtained by solving for the state dynamics, is

$$e_{k+1}(t) = r(t) - (Gu_{k+1})(t), \quad 0 \le t \le \alpha \tag{3.58}$$

where

$$(Gu)(t) = C \int_0^t e^{A(t-\tau)} Bu(\tau) d\tau \tag{3.59}$$

Consider the class of HOILC laws described by the following static and dynamic combination of previous input vectors, the current trial error, and the errors on a finite number of previous trials:

$$u_{k+1}(t) = \sum_{i=1}^{M} \gamma_i u_{k+1-i}(t) + \sum_{i=1}^{M} K_i(e_{k+1-i})(t) + K_0(e_{k+1})(t) \tag{3.60}$$

In addition to the memory length M, the design parameters in this control law are the static scalars γ_i, $1 \le i \le M$, the linear operator K_0 that describes the current trial error contribution and the linear operator K_i, $1 \le i \le M$, which describes the contribution of the error on trial $k + 1 - i$.

The error dynamics on trial $k + 1$ generated with the HOILC law applied can be written over $0 \le t \le \alpha$ as

$$e_{k+1}(t) = (I + GK_0)^{-1} \left\{ \sum_{i=1}^{M} (\gamma_i I - GK_i)(e_{k+1-i})(t) + \left(1 - \sum_{i=1}^{M} \gamma_i \right) r(t) \right\} \tag{3.61}$$

or, equivalently, as a linear repetitive process in the abstract model setting with dynamics described by

$$\hat{e}_{k+1} = L_\alpha \hat{e}_k + b \tag{3.62}$$

where

$$\hat{e}_k(t) = \begin{bmatrix} e_{k+1-M}^T(t) & \cdots & e_k^T(t) \end{bmatrix}^T \tag{3.63}$$

and

$$L_\alpha = \begin{bmatrix} 0 & I & \cdots & 0 \\ \vdots & \vdots & \ddots & \vdots \\ 0 & \cdots & 0 & I \\ E_0 E_M & \cdots & E_0 E_2 & E_0 E_1 \end{bmatrix} \tag{3.64}$$

where

$$E_0(y)(t) = (I + GK_0)^{-1}(y)(t), \quad E_i(y)(t) = (\gamma_i I - GK_i)(y)(t), i \le i \le M$$

(and the structure of b plays no role in the following analysis and hence, set $b = 0$).

Direct application of the linear repetitive process stability theory now yields the following result.

Theorem 3.11 *An ILC scheme with dynamics of the form (3.62) is asymptotically stable and, hence, trial-to-trial error convergence occurs if and only if all roots of*

$$z^M - \gamma_1 z^{M-1} - \cdots - \gamma_{M-1} z - \gamma_M = 0 \qquad (3.65)$$

have modulus strictly less than unity.

If the condition of Theorem 3.11 holds, the ILC dynamics converge (in the norm topology of the underlying function space) to

$$e_\infty = (I + GK_{eff})^{-1} r \qquad (3.66)$$

where K_{eff}, termed the effective controller, is given by

$$K_{eff} = \frac{K}{1 - \gamma_s}, \quad \gamma_s = \sum_{i=1}^{M} \gamma_i, \quad K = \sum_{i=0}^{M} K_i \qquad (3.67)$$

(The simplest way to obtain (3.66) is to replace all variables in (3.62) by their strong limits and rearrange.)

The following result, whose proof again follows by direct application of repetitive process stability theory and can be found in [228], gives a bound on the error sequence under Theorem 3.11.

Theorem 3.12 *Suppose that the condition of Theorem 3.11 holds. Then an expression of the following form bounds resulting error sequence:*

$$\|\hat{e}_k - \hat{e}_\infty\| \leq M_1 (max(\|e_0\|, \ldots, \|e_{M-1}\|) + M_2) \lambda_e^k \qquad (3.68)$$

where M_1 and M_2 are positive real scalars, and $\lambda_e \in (max|\mu_i|, 1)$ where μ_i, $1 \leq i \leq M$, is a solution of (3.65).

The result of Theorem 3.11 is counterintuitive in the sense that stability is mainly independent of the system and the controllers used. This is because trial length α is finite, and over such an interval, a linear system can only produce a bounded output irrespective of its stability properties, and in this definition of stability, unstable outputs of this kind are acceptable. Hence, even if the error sequence generated is guaranteed to converge to a limit, this terminal error may be unstable and or possibly worse than the first trial error, i.e. the use of ILC has produced no improvement in performance. To guarantee an acceptable, i.e. stable, as the most basic requirement, limit error the stronger concept of stability along the trial has to be used.

In terms of performance and design Theorems 3.11 and 3.12 give the following conclusions:

1. Convergence is predicted to be rapid if λ_e is small and will be geometric in form, converging approximately with λ_e^k.
2. The limit error is nonzero and can be described by a standard linear systems unity negative feedback system with effective control K_{eff} defined by (3.67). If $max_i(|\mu_i|) \to 0+$, then the limit error is essentially the first learning iterate, i.e. use of ILC has little benefit and will lead to the normal large errors encountered in simple feedback loops. Choosing $max_i|\mu_i|$ close to unity results in a high-gain controller K_{eff} and will lead to small limit errors.
3. Zero-limit error requires that

$$\sum_{i=1}^{M} \gamma_i = 1,$$

which is not possible if $\rho(L_\alpha) < 1$, but is possible for the case of $\rho(L_\alpha) = 1$. This situation follows that in classical control where the inclusion of an integrator, on the stability boundary, in the controller results in zero steady-state error in response to constant reference signals.

The conflict in the above conclusions has implications on the systems and control structure from both the theoretical and practical points of view. Consider, without loss of generality, the case when $K_i = 0$, $1 \leq i \leq M$. Then small learning errors will require high-effective gain, yet GK_0 should be stable under such gains.

The effects of high-gain feedback systems can be described by the system root-locus. In the SISO case, the results can be summarized as the requirement that GK_0 is minimum phase and relative degree one or two and also the following:

1. Convergence is "rapid" for well-conditioned relative degree one, minimum phase systems, particularly if GK_0 has interlaced poles and zeros and has positive high-frequency gain, where this result is proved by a detailed pole-zero and residue analysis of the limit error dynamics.
2. Convergence is guaranteed for well-conditioned relative degree two, minimum-phase systems with the properties of positive high-frequency gain and negative intercept/pivot of the root locus of GK_0. Convergence in this case is, however, almost always slower than in the relative degree one case that is nonuniform in the derivative, and theoretically leads to limit errors of amplitude of the order of, considering $M = 2$ without loss of generality, $1 - \gamma_1 - \gamma_2$ and in the form of an exponentially decaying oscillation with a high frequency of the order of $(1 - \gamma_1 - \gamma_2)^{-\frac{1}{2}}$.

Two problems arise in the application of asymptotic stability of repetitive processes to ILC analysis. The first of these is that only statements about the situation after an infinite number of trials have occurred are possible, and little information is available concerning performance from trial-to-trial. The second problem is that although a limit error is guaranteed to exist, it could well have unacceptable dynamic characteristics resulting from the possible presence of exponentially growing signals due to the finite trial length.

Consider the unit memory case when the error dynamics of an example are defined by the integral operator $L_\alpha : e_k \to e_{k+1}$

$$e_{k+1}(t) = D_0 e_k(t) + C \int_0^t e^{A(t-\tau)} B_0 \, e_k(\tau) d\tau, \quad 0 \leq t \leq \alpha \tag{3.69}$$

(where the state-space model matrices must be interpreted in terms of the state-space model for the trial-to-trial error dynamics). Also choose $E_\alpha = C([0, \alpha]; \mathbb{R}^m)$ the space of bounded continuous real-valued functions in the interval $0 \leq t \leq \alpha$ on which the norm is defined as $\|e(t)\| = \sup_{0 \leq t \leq \alpha} \|e(t)\|_m$, where $\| \cdot \|_m$ is any convenient norm in $\mathbb{R}^m$, e.g. $\|y\|_m = \max_{1 \leq i \leq m} |y_i|$. In this case, asymptotic stability holds if and only if $\rho(D_0) < 1$ and to provide a physical explanation of this property suppose that $\rho(D_0) \leq \|D_0\| < 1$ (a sufficient condition) holds. Then $e_{k+1}(0) = D_0 e_k(0)$ is reduced from trial-to-trial, i.e. the sequence of initial errors is reduced from trial to trial, since this quantity only depends on D_0.

By continuity of e_{k+1}, this error reduction also occurs for t slightly greater than zero. In this way, the matrix D_0 squeezes the error to zero, starting from $t = 0$ and working to $t = \alpha$. Unfortunately, depending on the state space triple (A, B_0, C), it could be that for $t \gg 0$, the error increases over the first few trials, and it takes a large number of trials before the error is small everywhere, where the discussion that follows uses the fact [228] that asymptotic stability of a linear repetitive process described by (3.62) with $b = 0$ guarantees the existence of real scalars $\tilde{M}_\alpha > 0$ and $\tilde{\lambda}_\alpha \in (0, 1)$ such that

$$\|e_k - e_\infty\| \leq \tilde{M}_\alpha \tilde{\lambda}_\alpha^k \|e_0\|, \quad k \geq 0 \tag{3.70}$$

The term involving $\tilde{\lambda}_\alpha$ in this last equation relates to the error reduction due to D_0 and the term $\tilde{M}_\alpha$ relates to and depends on the system structure defined by the state space triple $\{A, B_0, C\}$. This whole process can be visualized as squeezing something out of a tube, e.g. toothpaste, where when the end is already flat, and a bulge develops in the middle, and after squeezing long enough, everything drops out at the end.

The analysis of this case can be extended by adding a disturbance term $b_{k+1}(t)$, $k \geq 0$, to the right-hand side of (3.69) arising, e.g. from nonzero state initial conditions on each trial of the form $x_{k+1}(0) = d_{k+1}$, $k \geq 0$, resulting in (3.69) with the term:

$$b_{k+1}(t) = Ce^{At}d_{k+1}, \quad k \geq 0 \tag{3.71}$$

added. If the limit disturbance $b_\infty(t)$ is nonzero, the limit error is given by the state-space model,

$$\dot{x}_\infty(t) = (A + B_0(I - D_0)^{-1}C)x_\infty(t) + B(I - D_0)^{-1}b_\infty(t), \quad x_\infty(0) = d_\infty$$
$$e_\infty(t) = (I - D_0)^{-1}Cx_\infty(t) + (I - D_0)^{-1}b_\infty(t) \tag{3.72}$$

where d_∞ is the strong limit of the sequence $\{d_k\}_{k\geq 1}$.

In (3.72), $(I - D_0)^{-1}$ exists since $\rho(D_0) < 1$, but the system is unstable in the standard sense unless all eigenvalues of $A + B_0(I - D_0)^{-1}C$ have strictly negative real parts. An exponentially growing limit error with an unacceptable magnitude for most practical applications could arise if this is not the case. Such cases require the stronger concept of stability along the trial and, in particular, all of the conditions of Theorem 3.9. Stability along the trial also guarantees [228] the existence of real scalars $M_\infty > 0$ and $\lambda_\infty \in (0, 1)$, which are independent of α, such that

$$\|e_k - e_\infty\| \leq M_\infty \lambda_\infty^k \left\{ \|e_0\| + \frac{\|b_\infty\|}{1 - \lambda_\infty} \right\}, \quad k \geq 0 \tag{3.73}$$

The physical interpretation associated with the conditions of Theorem 3.9 can also be applied, where (3.69) with the disturbance term $b_{k+1}(t)$ included has the transfer-function matrix description:

$$e_{k+1}(s) = G(s)e_k(s) + b_{k+1}(s) \tag{3.74}$$

Moreover, when Theorem 3.9 holds, $G(s)$, also termed the error transmission operator in some of the literature, is a stable transfer-function matrix with minimal realization that satisfies

$$\rho(G(j\omega)) < 1, \quad \forall \omega \geq 0 \tag{3.75}$$

i.e. the spectral radius condition must be satisfied at all frequencies. In contrast, asymptotic stability alone requires this at only one point, i.e. infinity, since $D_0 = \lim_{\omega \to \infty} G(j\omega)$.

In the SISO case, (3.75) reduces to

$$|G(j\omega)| < 1, \quad \forall \omega \geq 0 \tag{3.76}$$

Also if $b_k = 0$, the error $|e_k(j\omega)|$ is pointwise bounded by

$$|e_k(j\omega)| \leq |L(j\omega)|^k |e_0(j\omega)| \tag{3.77}$$

and the error in the frequency domain is monotonically reduced at geometric rate α_∞, where

$$\alpha_\infty = \sup_{\omega \geq 0} |L(j\omega)| \tag{3.78}$$

This connects stability along the trial in this case to (3.73) with $M_\infty = 1$ and $\lambda_\infty = \alpha_\infty$ easily computable.

The original research [195] on differential linear ILC gave a convergence analysis adapted to a specific algorithm where the definition of convergence is similar to that above, i.e. the system is convergent if the error with $b_{k+1}(t) = 0$, $k \geq 0$, goes to zero as $k \to \infty$ with respect to the supremum-norm in $[0, \alpha]$ and showed that the system is convergent if $\|D_0\| < 1$. This analysis and, in particular, the convergence proof uses the λ norm defined by (1.19) and proceeds by showing that if $\|D_0\| < 1$, then there exists a $\lambda > 0$ such that

$$\|e_{k+1}\|_\lambda \leq \beta \|e_k\|_\lambda \tag{3.79}$$

holds with $\beta < 1$. Hence, for $\lambda > 0$

$$e^{-\lambda t}\|e_k(t)\| \leq \|e_k(t)\|_\lambda \leq \|e_k(t)\| \tag{3.80}$$

and therefore, if (3.79) holds convergence also occurs in the supremum-norm since, using (3.79) and (3.80),

$$\lim_{k \to \infty} \|e_k(t)\| \leq \lim_{k \to \infty} e^{\lambda \alpha} \|e_k(t)\|_\lambda \leq e^{\lambda \alpha} \beta^k \lim_{k \to \infty} \|e_0\|_\lambda$$

$$= e^{\lambda \alpha} \|e_0\|_\lambda \lim_{k \to \infty} \beta^k = 0 \tag{3.81}$$

This last result provides a connection, missing from other ILC stability results developed in the 2D systems setting, to the error bound (3.70), where the constants $\tilde{M}_\alpha$ and $\tilde{\lambda}_\alpha$ are given by

$$\tilde{M}_\alpha \leq e^{\lambda \alpha}, \quad \tilde{\lambda}_\alpha = \beta \tag{3.82}$$

In application $\tilde{M}_\alpha$ could be very large, for large λ and α, and hence, it could take a very large number of trials before the error is reduced in terms of the usual supremum-norm.

The analysis in [195] developed a convergence condition for the continuous-time case by frequency domain methods, where the system is said to be L_2-convergent if $\|e(t)\|_{L_2[0,\infty)} \to 0$ as $k \to \infty$. This result is given next

Theorem 3.13 *[195] Suppose that the pair $\{C, A\}$ is observable and the pair $\{A, B\}$ is controllable. Then the ILC system (3.69) is L_2-convergent if the initial error is bounded, i.e. $e_0(t) \in L_2[0, \infty)$, the disturbance vector b_k is independent of k, i.e. $b_k(t) \equiv b_\infty(t) \in L_2[0, \infty)$, and*

$$\|C(sI - A)^{-1}B_0 + D_0\|_\infty \leq 1 \tag{3.83}$$

and the set of ω such that $\|C(j\omega I_n - A)^{-1}B_0 + D_0\|_\infty = 1$ is of measure zero [134].

Theorem 3.13 gives a sufficient condition for stability as opposed to the necessary and sufficient conditions of Theorem 3.9. The definitions of stability in these two theorems are slightly different, where the former includes a degree of robustness from a model perturbation in the abstract model-based theory. This property is not present in Theorem 3.13, and this result may allow $\|G(j\omega)\|_\infty = 1$.

3.3 Concluding Remarks

This chapter has covered the basic requirements in terms of design and performance for systems regulated by an ILC law. The focus has been on linear dynamics. The Chapter 4 builds on these results by addressing ILC design using tuning methods.

4

Tuning and Frequency Domain Design of Simple Structure ILC Laws

As in the standard systems case, there will be linear and nonlinear examples, where a simple structure iterative learning control (ILC) law can be tuned using "guidelines" to produce a good design. This chapter first gives a brief coverage of such designs, starting with early methods. Next, the application of phase-lead and adjoint ILC laws to the stroke rehabilitation area, based on the problem specification in Section 2.4, is described, focusing on the ILC design and following through to clinical trials. Also, a tuning-based ILC design for nonminimum phase systems is considered, where the experimental results in this chapter use the electromechanical system of Section 2.2.1. These results are for finite-dimensional systems, and Section 9.1 describes the tuning-based design of simple structure ILC laws for gust load management for wind turbines where a partial differential equation describes the dynamics.

4.1 Tuning Guidelines

The P, D, and PD-type ILC laws introduced in Chapter 1 for differential dynamics (see after (1.12)) have, of course, discrete counterparts and have been widely considered, including for nonlinear dynamics, see, e.g. [270]. Moreover, ILC has a natural integrator action from one trial to the next, and hence, in comparison to the standard control systems, integral action is less often used.

Consider, following [33], the discrete-time PD ILC law

$$u_{k+1}(p) = u_k(p) + K_p e_k(p+1) + K_d(e_k(p+1) - e_k(p)) \tag{4.1}$$

Then with $K_p = 0$ phase-lead ILC is obtained where the advance term in $p+1$ is causal because it is a previous trial term. It is also easy to show that an ILC PD design for linear time-invariant single-input single-output (SISO) dynamics is asymptotically stable if and only if $|1 - (K_p + K_d)g_1| < 1$, where g_1 is the first Markov parameter. Also, if g_1 is known, it is always possible to find K_p and K_d such that asymptotic stability holds.

Monotonic trial-to-trial error convergence is not always possible with PD type ILC. Moreover, the most commonly used approach to achieving monotonic convergence is to modify the law to include a low-pass Q filter. The idea is to use this filter to block learning at high frequencies, and this is of use in satisfying the robust monotonic error condition of Theorem 3.6. This form of the filter also has advantages of added robustness and the filtering of high-frequency noise.

In common with classical controllers, such as proportional plus derivative, the most commonly employed method for selecting the gains of simple structure ILC controllers is by tuning, where early examples include [105, 132]. If a Q filter is used, the filter type, e.g. Butterworth or Chebyshev,

Iterative Learning Control Algorithms and Experimental Benchmarking, First Edition.
Eric Rogers, Bing Chu, Christopher Freeman and Paul Lewin.
© 2023 John Wiley & Sons Ltd. Published 2023 by John Wiley & Sons Ltd.

and the filter order can be specified and the bandwidth used as a tuning variable. However, the important point is that tuning rules or guidelines comparable to the Ziegler Nichols rules for tuning three-term controllers in the standard case are unavailable. In [33] the following suggestions for tuning PD ILC laws are given:

- The goals of any tuning design include both good learning transients and low error.
- For each set of gains K_p and K_d, the learning should be reset and run sufficient times to determine the transient behavior and asymptotic error. Initially, small values for the ILC law coefficients and the filter bandwidth should be chosen.
- After stable transient behavior and error performance has been achieved, the control law coefficients and bandwidth can be increased.
- In applications, the control law coefficients influence the convergence rate, and the Q-filter influences the converged error performance.
- The signals should be monitored closely for several trials beyond the apparent convergence for signs of high-frequency growth.

A two-step approach is given in [150] based on the availability of the experimentally determined frequency response of the system. This approach is based on achieving the stability and monotonic convergence condition (3.20), which can be written as follows:

$$|1 - e^{j\theta}L(e^{j\theta})G(e^{j\theta})| < \frac{1}{|Q(e^{j\theta})|} \qquad (4.2)$$

$\forall\ \theta \in [-\pi, \pi]$. Using a Nyquist plot of $e^{j\theta}L(e^{j\theta})G(e^{j\theta})$ the control law gains are tuned to maximize the range of $\theta \in [0, \theta_c]$ over which $e^{j\theta}L(e^{j\theta})G(e^{j\theta})$ lies inside the unit circle centered at 1. The bandwidth of the Q-filer is selected last to satisfy the stability condition.

Next, the application of the phase-lead and adjoint ILC laws and popular simple structure ILC laws in engineering applications is considered. The coverage is for a relatively recent application area of robotic-assisted stroke rehabilitation, for which Section 2.4 gives the background. This work has extended to supporting clinical trials.

4.2 Phase-Lead and Adjoint ILC Laws for Robotic-Assisted Stroke Rehabilitation

Phase-lead and adjoint ILC laws have been successfully used in the rehabilitation application area described in Section 2.4, Figure 2.16b and followed through to clinical evaluation. This research was the first in this area, focusing on an in-plane (2D) task such as reaching out to a cup over a tabletop. In this section, ILC design, simulation evaluation, and the results from clinical trials are discussed. Subsequently, Chapters 10 and 11 will discuss extensions to more practical tasks (3D), where model-based linear and nonlinear ILC designs are required.

The background on the development of the experimental system and the modeling of the subsystems involved is given in [80, 114]. This section focuses on the ILC design and implementation, starting with the reference trajectory choice, which is a critical feature in any application of ILC. In this application area, the robotic arm's action must make the task feasible yet productive for the patients. Therefore, the following points concern the choice of trajectory and role of the robot during the completion of the task:

1. The trajectories will be elliptical reaching tasks for each subject's dominant arm and should be achievable given their identified arm model.

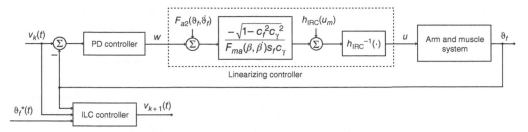

Figure 4.1 Block diagram representation of the ILC scheme applied as in Figure 2.16.

2. The triceps muscle will provide the only actuating torque about the elbow. Moreover, the robotic arm will use the control scheme to make the dynamics about the elbow feel "natural" to the subject.
3. The robotic arm will provide a torque acting about the subject's shoulder to track the reference in a manner that is entirely governed by the angle of the forearm. Consequently, the task is feasible without lessening the role played by the triceps.

These considerations lead to the reference trajectory shown in Figure 2.16b.

Figure 4.1 shows the block diagram of the ILC scheme used in this application area. The details of the experimental facility's construction are in [80, 114], together with the representations of the various blocks. As essential background, each of these is considered in turn, starting with the arm and muscle system.

In the case of Figure 2.16, the triceps muscle is stimulated. This stimulation, in turn, produces a torque about the elbow and, hence, the motion of the patient's hand. The objective is, therefore, to control the angle ϑ_f, which requires a model for the response of the triceps muscle to the stimulation. In the literature, see, e.g. [242], an accepted model of the torque, T_β, generated by electrically stimulated muscle acting about a single joint, is

$$T_\beta(\beta, \dot{\beta}, u, t) = g(u, t) \times F_{\mathrm{ma}}(\beta, \dot{\beta}) + F_{\mathrm{mp}}(\beta, \dot{\beta}) \tag{4.3}$$

where u denotes the stimulation pulsewidth applied, and β is the joint angle.

The first term in this model, $g(u, t)$, is formed from a Hammerstein structure incorporating a static nonlinearity, $h_{\mathrm{IRC}}(u)$, representing the isometric recruitment curve, cascaded with linear activation dynamics, $h_{\mathrm{LAD}}(t)$. Moreover, the activation dynamics can be adequately described by a critically damped second-order linear system, see, e.g. [21], for the details. Finally, the term $F_{\mathrm{ma}}(\beta, \dot{\beta})$ models the multiplicative effect of the joint angle and joint angular velocity on the active torque developed by the muscle, and the term $F_{\mathrm{mp}}(\beta, \dot{\beta})$ represents the passive properties of the joint. Details of the procedures used to establish the parameters appearing in the model and additional information relating to the subjects whose experimental results are given later in this section are in [79].

To develop the robotic control scheme to achieve the goals set out as 1–3 above, the human arm is treated as a two-link system, and this part of the model building process is detailed in [80, 114]. The control scheme, Figure 4.1, applied in this case, can implement two strategies, either a feedback control scheme using the linearizing controller or the feedback scheme augmented by an ILC feedforward controller. Consequently, it is also possible to examine the relative merits of ILC and standard feedback control in this application area.

The first component of the linearizing controller is $h_{\mathrm{IRC}}^{-1}(\cdot)$, which is the inverse of the isometric recruitment curve, a static map that has to be obtained for each subject. (See, e.g. [141] for a treatment of system identification-based modeling of the response of a human muscle to electrical stimulation.) Moreover, the remaining terms in the linearizing controller arise from the stimulated arm's approximate model, whose values can be shown to vary only slowly when the trajectories considered are followed perfectly (see [79] for the details). Overall, the control objective is to remove these effects to produce a system that approximates the linear activation dynamics.

This model is developed by linearization along a trajectory, a well-established approach. Moreover, properties of the nonlinear system can be inferred from the linearized representation when its state variables and input are close to those of the linearized system. In particular, if the resulting linear time-varying (LTV) system is stable, then the nonlinear system also has this property in the neighborhood of the trajectory [240]. This technique can be applied to examine the proposed control system's local stability and robustness by using trajectories comprising experimental test data or those resulting from simulations using the system model.

The stability of the linearized system can be assessed using well-known methods for LTV systems [127]. This task can then be repeated while varying the model parameters to obtain a broader picture of the system robustness and performance properties. Therefore, the system's stability can be examined through variation of these quantities, together with the variation of the remaining model parameters. This approach can then be extended to provide local convergence properties for the ILC laws considered later in this section.

Once the above analysis is complete, the following transfer-function description approximation of the dynamics linking ϑ and w in Figure 4.1 is obtained

$$\frac{\vartheta_f(s)}{w(s)} = \frac{\omega_n^2}{s^2 + 2\omega_n s + \omega_n^2} \times \frac{1}{s((b_{a3} + K_{M_2})s + K_{B_2})} = G(s) \qquad (4.4)$$

where ω_n is discussed below and the other parameters in the transfer-function are detailed in [79].

The next stage is to choose a feedback controller to supply the torque demand, w, necessary for the system to track the intended reference trajectories. In the current case, a PD controller was used, with transfer-function:

$$K(s) = \frac{K_d s + 1}{\epsilon K_d s + 1} \qquad (4.5)$$

where $\frac{1}{6} \geq \epsilon \geq \frac{1}{20}$. Also, the stimulation level that first produces a response from the triceps, u_m, is used to supply an offset so that the feedback system operates within the muscle's torque-generating capabilities. Next, the feedback controller is tuned for each subject, emphasizing robustness since stability is more significant than accurate tracking. Therefore, a conservative bandwidth and high gain and phase margins are desired.

Bode plots of the unity negative feedback system formed by $K(s)$ in series with $G(s)$ are shown in Figure 4.2 in the case when $\frac{b_{a3} + K_{M_2}}{K_{B_2}} = 0.03, 0.07, 0.4, 0.8$, and 1.5, respectively, illustrate the end-effector dynamics. These plots have been constructed using an experimentally identified value of $\omega_n = 0.85\pi$, and, for ease of comparison, they have all been tuned using the standard Zeigler–Nichols rules. The corresponding closed-loop systems are denoted by $P_a, P_b, \ldots, P_e$, and their respective bandwidths correspond to 0.53, 0.45, 0.42, 0.23, and 0.16 Hz. Changing $\frac{b_{a3} + K_{M_2}}{K_{B_2}}$ cannot produce bandwidths much in excess of these for the given controller and tuning method due to the limiting factor of the muscle dynamics.

It is desirable to select a ratio that makes the task feel natural to the patient, but the design does not lead to too narrow a system bandwidth, which would necessitate an excessive controller effort and correspondingly large levels of muscle torque to accomplish the task. Next, the effect of the end-effector dynamics on the ILC controller performance is considered.

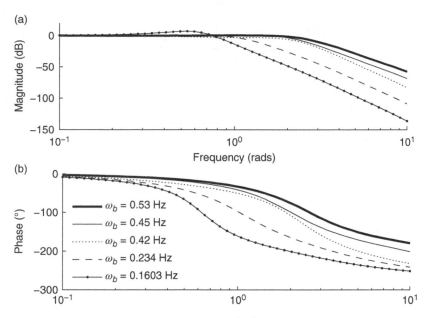

Figure 4.2 Bode plots generated by the linearized feedback system for different end-effector dynamics.

4.2.1 Phase-Lead ILC

The first ILC law considered is phase-lead to construct the signal v_{k+1} in Figure 4.1 as

$$v_{k+1}(z) = v_k(z) + Lz^\lambda e_k(z) \tag{4.6}$$

where L is a scalar gain, λ is the phase-lead in samples, and $e_k = \vartheta_f^*(t) - \vartheta_f(t)$. Hence,

$$e_{k+1}(z) = e_k(z) - Lz^\lambda P(z)e_k(z) \tag{4.7}$$

with

$$P(z) = \frac{K(z)G(z)}{1 - K(z)G(z)} \tag{4.8}$$

where $G(z)$ is the sampled version of (4.4) and $K(z)$ is the sampled version of (4.5). Moreover, the condition for monotonic trial-to-trial error convergence is

$$\left|1 - LP(e^{j\omega T_s})e^{j\omega\lambda T_s}\right| < 1 \tag{4.9}$$

for ω up to the Nyquist frequency. Satisfying this condition at a given frequency is a sufficient condition for monotonic convergence at that frequency. Also, the convergence speed is dictated by the magnitude of the left-hand side; if it is close to zero, convergence will occur in a single trial, while if it is greater than one, divergence is likely to occur at that frequency. Therefore, L and λ are chosen so that the left-hand side is minimized to provide the greatest convergence over those frequencies present in the reference signal. Frequencies above this, or those at which system uncertainty may cause the criterion to be violated, are removed by using a noncausal zero-phase filter applied to the error.

Further discussion relating to the choice of parameters is given in [79]. Moreover, the sampling period used is $T_s = \frac{1}{40}$ seconds, which corresponds to the frequency at which stimulation pulses are applied to the patient. This frequency is synchronized with the robotic control system, and each

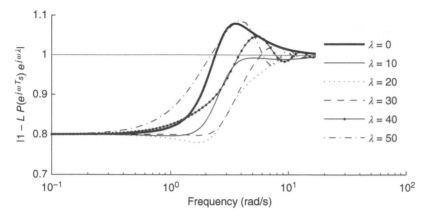

Figure 4.3 Monotonic convergence criterion for the system for the case corresponding to 0.45 Hz for $L = 0.2$ and various λ.

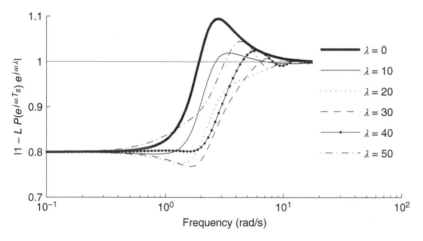

Figure 4.4 Monotonic convergence criterion for the system for the case corresponding to 0.53 Hz for $L = 0.2$ and various λ.

pulse is produced with a delay of less than 0.01 seconds. Next, the effects of varying the phase-lead term are considered.

Figure 4.3 shows the monotonic convergence criterion (4.9) for the case when the bandwidth corresponds to 0.45 Hz and the phase-lead λ takes a range of values, and Figure 4.4 shows the same data for the case corresponding to 0.53 Hz.

A value of $L = 0.2$ has been chosen in both cases to produce an extremely robust system at the expense of convergence speed. Reducing L can be shown to increase robustness to model uncertainty over all frequencies. It can also be shown that choosing the phase-lead to maximize the convergence at a given frequency also achieves maximum robustness to gain and phase uncertainty at that frequency [111], which means that there is no compromise between robustness and convergence speed with respect to choosing this parameter.

Based on these last considerations, the phase-lead ILC law was selected to maximize the convergence over a suitable frequency range. A zero-phase filter was implemented to ensure stability at all higher frequencies. For general application, it is natural for this frequency range to correspond to the system's bandwidth. The system corresponding to 0.45 Hz can be seen to have maximum convergence over its bandwidth using a phase-lead of 30 samples (although a phase-lead of 20 samples produces greater stability over higher frequencies, which is an essential factor if a filter is not used). Similarly, the system corresponding to 0.53 Hz has a maximum convergence over its bandwidth of 20 samples. When using these choices of the phase-lead term, the system with the higher bandwidth has the property of monotonic convergence at greater frequencies than the system with lower bandwidth; see, e.g. [82] for further discussion of this aspect.

4.2.2 Adjoint ILC

The adjoint ILC law is given in discrete form by

$$v_{k+1}(z) = v_k(z) + \beta P^*(z)e_k(z) \tag{4.10}$$

where $P^*(z)$ is the adjoint of the system model used. An attractive feature of the design is that with a sufficiently small scalar multiplier, β, the condition for monotonic convergence is satisfied at all frequencies and ensures a satisfactory transient response [47]. The monotonic convergence criterion in this case is

$$\left| 1 - \beta P(e^{j\omega T_s})P^*(e^{j\omega T_s}) \right| < 1 \tag{4.11}$$

which leads to

$$0 < \beta |P(e^{j\omega T_s})|^2 < 2 \tag{4.12}$$

for ω up to the Nyquist frequency. The law can provide a high level of robustness to system uncertainty (see, e.g. [82] in the general case).

To examine the effects of varying the gain β on performance, plots of (4.11) constructed for various choices of β can be used. The conclusion is that both cases result in monotonic trial-to-trial error convergence for any value of β such that $0 < \beta < 1$. For frequencies where the controlled system has a gain close to unity, approximate convergence will occur in a single trial for values of β close to one. In application, however, a significantly lower value of this parameter is chosen to increase robustness at the expense of convergence rate.

As with the phase-lead law, the convergence rate at a given frequency is connected to the system bandwidth. The convergence rate can be increased within a given range by convolving $P^*(z)$ with a zero-phase, band-pass filter and substituting in (4.10) (see [82] for details). The same effect could be achieved by replacing $P^*(z)$ with the plant inverse $P^{-1}(z)$, and then using a low-pass filter. However, the increase in convergence speed is obtained at the price of reduced robustness. Hence, a zero-phase, band-pass filter is not used unless the convergence rate is unacceptably low for the desired application.

Next, the results of tests with humans are given and discussed. In this area, prior ethical approval for studies with stroke patients must be obtained. Obtaining such approval involves making a case to the regulatory body, and one part of the evidence can be results with unimpaired subjects. Next, the results from a study with 18 unimpaired subjects are given and discussed.

4.2.3 Experimental Results

The arm model of each of the 18 unimpaired subjects tested was first identified using the tests and procedures described in [79]. These involved taking measurements of the arm, establishing its maximum range of movement, and then fitting a circle to the trajectory traced out by the elbow to provide γ and the shoulder joint position. The arm was then held stationary while functional electrical stimulation (FES) was applied to the triceps using a ramp signal to produce the functions, $h_{\text{IRC}}(u)$ and $h_{\text{LAD}}(t)$, in the muscle model. These tasks used deconvolution and a nonlinear optimization procedure, where the reference trajectory used is of the form of Figure 2.16b.

Stimulation sequences and kinematic trajectories imposed on the arm by the robot were then applied, and least mean squares optimization was subsequently used to determine the remaining model parameters. The reference was set at an angle of 20° from the y axis but was individually calculated for each subject so that it extended their arm from 55% to 95% of their total arm length throughout the movement.

Two trajectories have been created by moving along this reference at two different speeds. Each trajectory started with a waiting period when it was set equal to the reference's starting point. The "slow" trajectory lasted for 12.5 seconds in total (a 5 seconds waiting period and a 7.5 seconds movement along the reference). The "fast" trajectory lasted for 10 seconds (a 5 seconds waiting period followed by a 5 seconds movement). The waiting period was included to allow the ILC update to begin before the arm was required to move. Before each trial began, the subject's arm was moved to the robot's initial position and then released when the trajectory started. The subjects were not shown the trajectory before or during the test.

The values of K_{B_2} and K_{M_2} that dictate the end-effector characteristic were set at 5.78 N m/(rad/s) and 0.29 N m/(rad/s^2), respectively, since these created a natural feel to the system and allowed the chosen trajectories to be accomplished with moderate effort while not limiting the bandwidth excessively. For a typical identified value of b_{a3}, this choice of gains produces $\frac{b_{a3}+K_{M_2}}{K_{B_2}} \approx 0.1$ which leads to a closed-loop system bandwidth of approximately 0.45 Hz. The values of $\omega_n = 0.85$ πrad/s and $b_{a3} = 0.271$ were identified for the subject whose experimental results are given below. Finally, the feedback controller gains were tuned as described previously in this section, resulting in the use of $K_p = 10$ and $K_d = 2$, and the effect of the linearizing feedback controller in the absence of ILC follows by inspection of the first trial of the ILC results given below. Next, the results obtained for one of the subjects are given.

Figure 4.5 shows the results obtained using phase-lead ILC law for various values of λ. These results are for the slow trajectory and $L = 0.2$. This learning gain has been chosen

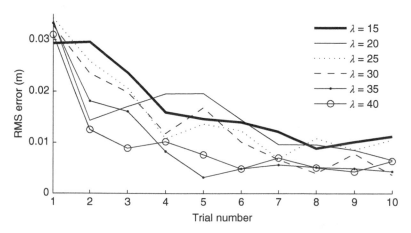

Figure 4.5 Single subject phase-lead ILC law results, slow trajectory, $L = 0.2$, and various λ.

conservatively so that the speed of convergence has been sacrificed for greater robustness. The zero-phase filter discussed above was not implemented since no signs of instability were observed throughout any experimental tests.

These results confirm that the best performing phase-lead corresponds to the use of 35 samples, which is similar to that found by inspection of the monotonic convergence criterion for P_b, a system with a similar bandwidth, shown in Figure 4.4. The root-mean-square (RMS) error corresponding to this phase-lead converges to approximately 5 mm and has a minimum value of 3.2 mm. Figure 4.6 shows the results obtained using $L = 0.2$ and $\lambda = 35$ in greater detail.

Overall, these results exhibit monotonic trial-to-trial error convergence during the first five trials, and the reference is tracked extremely well. In general terms, this reflects the improvement in tracking accuracy that simple ILC schemes can provide compared with feedback controllers alone. A critical factor in this application area is the level of stimulation required. The focus must be on the levels that can be applied to the subject's arm under ethical approval and not those imposed by the actuators. Figure 4.7 shows the associated stimulation input and ILC update.

The results is Figure 4.7a show not only that the stimulation applied on $k = 1$ saturates at 300 μs, but also that the effect of further trials removes this problem and produces lower levels of stimulation throughout the trial. Moreover, the ILC law also leads to stimulation during the initial five seconds waiting period before movement is required. Considering the results given under Figure 4.7b shows that the updated reference, v_k, is exhibiting convergence to a fixed trajectory over repeated trials.

Figure 4.8 shows single subject results obtained using the adjoint ILC law, again for the slow trajectory with $\beta = 0.2$. The system model is labeled P, and to examine the robustness of the algorithm, the models P_a, P_c, P_d, and P_e, given above (second paragraph before Section 4.2.1) have also been used in its place. The model P_b has not been used due to its close similarity with P.

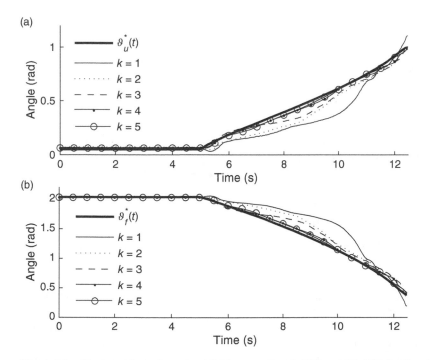

Figure 4.6 Single subject phase-lead ILC law results, (a) $\vartheta_u^*(t)$ and (b) $\vartheta_f^*(t)$ for the slow trajectory with $L = 0.2$ and $\lambda = 35$.

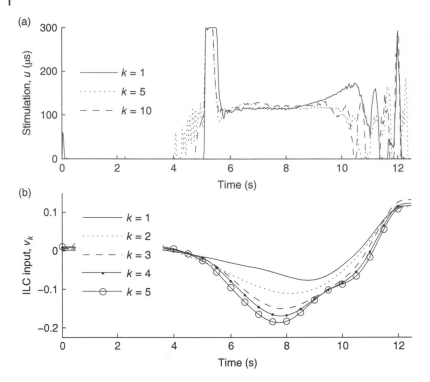

Figure 4.7 Single subject experimental results, (a) stimulation and (b) updated input results for the slow trajectory using phase-lead ILC law with $L = 0.2$ and $\lambda = 35$.

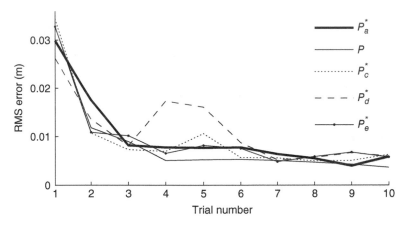

Figure 4.8 Single subject adjoint ILC law results for the slow trajectory with $\beta = 0.2$ and a range of system models.

These results indicate that the adjoint design can exhibit robustness to significant model uncertainty and produces minimum error and convergence rates comparable to the performance of the phase-lead ILC law. Using the model of the actual system results in convergence to ≈ 5 mm after four trials, a level of error maintained over the remaining trials. The tracking results using P are very similar to those for the phase-lead ILC law in terms of both monotonic trial-to-trial error convergence and, critically, the level of stimulation applied.

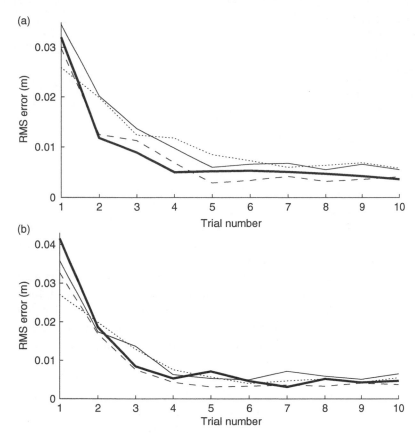

Figure 4.9 Single subject comparison of the phase-lead and adjoint ILC law results: (a) the slow trajectory, and (b) the fast trajectory.

Given the similarity in the phase-lead and adjoint ILC laws performance, group results are only given for the adjoint law implementation. Figure 4.9 shows results from the first four subjects tested. They relate to both (Figure 4.9a) the slow trajectory and (Figure 4.9b) the fast trajectory. The results indicate that convergence can be achieved within 5 trials for both trajectories. An RMS tracking error of less than 10 mm throughout the movement (and in some instances less than 5 mm) is possible in all cases. A comparison with an norm optimal iterative learning control (NOILC) design is given in chapter 5.

Table 4.1 shows the mean and standard deviation of the RMS error obtained during the last trial for all subjects tested. Trajectory 1 refers to the reference previously described, while trajectory 2 is a reference whose inclination from the y axis is increased to 40°.

These experimental results show that superior tracking performance is achieved compared with alternative control methods that have been applied to the upper limb. These include the previous application of ILC [67], and also to the use of both open-loop controllers [213] and to the limited number of closed-loop controllers that have been experimentally applied in this area [139]. However, the significant advantage of the new design is the simplicity of tuning and the absence of any training experiments. These simple structure ILC laws also permit the degree of assistance to easily be changed by using a forgetting factor (see [83]), which can also promote voluntary effort when applied to stroke patients.

Table 4.1 Mean of last trial RMS error for all 18 subjects (standard deviation in brackets).

Type	Mean of last trial RMS error /mm	
	Trajectory 1	Trajectory 2
Slow	9.41 (5.67)	10.22 (6.07)
Fast	12.18 (6.94)	12.93 (6.87)

A further advantage is that these laws are easy to tune, which is a significant concern when conducting tests with patients. This feature was essential in a study [113] that aimed to investigate whether ILC mediated by FES is a feasible intervention in upper limb stroke rehabilitation. In this first work, five hemiparetic participants with reduced upper-limb function who were at least six months poststroke were recruited from the community.

Participants undertook supported tracking tasks using 27 different trajectories augmented by responsive FES to their triceps muscle, with their hand movement constrained in a two-dimensional plane by a robot. Eighteen 1-hour treatment sessions were used. The FES level was adjusted in response to the user's performance at tracking trajectories and decreased throughout the intervention. The study results demonstrated significant improvements in unassisted tracking under accepted measures in the health sciences literature.

Comparison to other control actions for this application area, ILC is alone in encouraging maximal voluntary effort from the patient, which is a critical requirement in stroke rehabilitation. The results in this section are for planar motion only and require, see Figure 2.16a, constraints on the arm. Chapters 10 and 11 detail the generalization of this research to more realistic (3D) tasks where model-based linear and nonlinear ILC designs are required.

4.3 ILC for Nonminimum Phase Systems Using a Reference Shift Algorithm

Many ILC designs have been developed for (linear) nonminimum phase systems and [82, 83] report the results from a comprehensive comparative treatment of a range of these with experimental support using the nonminimum phase system of Section 2.2.1. This system also provides experimental support for the reference shift approach [39] considered next.

In [82] it was established, based on a significant number of experimental tests, that a simple phase-lead algorithm with a carefully selected filter can be used to control the nonminimum phase system considered successfully. In this case, an intuitive justification for using this form of ILC law arises from observing that a typical feature of nonminimum phase plants is a pronounced lag in the output response to an arbitrary input. To determine a single-time delay representative of this effect for a given input signal and system, the input can be shifted in time relative to the output. A minimum norm of the difference is sought for this shift.

Intuitively, the duration of this time shift could be used as a parameter, τ, to model the systems as a simple time delay, i.e. $G(z) = \gamma z^{-\tau/T_s}$, where γ is a scalar gain and T_s denotes the sampling period applied to the continuous-time dynamics. Moreover, the corresponding system inverse is given by

$G^{-1}(z) = \gamma^{-1} z^{\tau/T_s}$. Hence, the z-transformed inverse ILC law can be written as follows:

$$u_{k+1}(z) = u_k(z) + G^{-1}(z) e_k(z) = u_k(z) + Lz^{\tau/T_s} e_k(z) \tag{4.13}$$

which corresponds to the phase-lead ILC law

$$u_{k+1}(p) = u_k(p) + Le_k(p + \tau) \tag{4.14}$$

with $L = \gamma^{-1}$.

Since the reference signal is defined over $[0, \alpha]$ and $e_k(p + \tau)$ is defined over $[-\tau, \alpha - \tau]$, application of this ILC law involves discarding the error for $p \in [-\tau, 0)$, whilst continuing its last value until the end of the trial. The error signal used in (4.14) therefore is

$$e_k(p + \tau) = \begin{cases} e_k(p + \tau), & p \in [0, \ \alpha - \tau] \\ e_k(\alpha), & p \in (\alpha - \tau, \ \alpha] \end{cases} \tag{4.15}$$

The truncation of the error can then result in undesirable transients at the start of each trial. These are exacerbated if the reference is initially large or rapidly changing since these conditions typically generate a significant error early in the trial. Another unresolved difficulty is the choice of phase-lead, τ, to use in the law, see [82, 83] for further discussion of both issues.

One solution to the problem of truncation of the error is to shift the time interval, over which the input is applied to the system, forward in time. Following the reasoning given above, it is reasonable to apply the input to the system solely over $p \in [-\tau, \alpha - \tau]$. In this case, $\tau = 0$ on the first trial, and the input applied to the system is simply the reference signal. At the completion of this trial τ is calculated according to the criterion detailed below, and $e_k(p + \tau)$ must be added to the reference to produce the next input, as in (4.14).

The initial input was defined over $[0, \alpha]$ and therefore, in this procedure, it is shifted forward in time by τ seconds so that both signals are defined over $[-\tau, \alpha - \tau]$. This procedure is then repeated at each trial. If τ does not change significantly between trials, very little of the previous input is discarded in the construction of the current trial input. This, in turn, helps minimize the transients at the start of each trial, which may interfere with the learning process. In this procedure, the reference signal is unaltered and is still defined over $p \in [0, \alpha]$.

A final problem is the selection of τ, which is addressed by assuming that, as discussed above, the system dynamics are adequately modeled by $G(z) = \gamma z^{-\tau/T_s}$ and that at low frequencies γ is approximately unity. In this case the response of the system to an input, $u_k(p)$, is a delayed copy, $u_k(p - \tau)$. The method of calculating τ is then to minimize the norm of the difference between this and the experimentally measured output on the same trial, $y_k(p)$.

Both these methods of treating the problems that arise when implementing phase-lead ILC are combined to produce the following update structure

$$u_{k+1}(p) = r(p + \tau_k) + f_{k+1}(p) \tag{4.16}$$

$$f_{k+1}(p) = f_k(p) + Le_k(p + \tau_k) \tag{4.17}$$

$$e_k(p) = r(p) - y_k(p) \tag{4.18}$$

$$\tau_k = \arg \min_\tau \{ \| u_{k+1}(p - \tau) - y_k(p) \|_2^2 \} \tag{4.19}$$

where τ_k is the time shift calculated immediately following completion of trial k, f_k is the feed-forward signal applied to the system in addition to the shifted reference, and the system output is $y_k = r - e_k$.

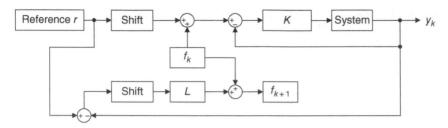

Figure 4.10 Controller structure for the reference shift algorithm.

Variations of the phase-lead law (4.17) for a non-minimum phase system are given in [119], although problems with initial tracking were experienced. In [248] a similar approach to the control of a minimum phase system was considered. Next, the problem considered is how to minimize the transient response at the start of the trial to ensure a high level of tracking in this region.

Figure 4.10 shows the controlled structure of the reference shift algorithm, where the shifted reference and the feedforward signal are fed into the feedback loop together, as given by (4.16). The cost function (4.19) is used to choose a shift value based on the representation of the system as a pure delay, which is applied to both the update given by (4.17) and the reference applied to the system input. These actions set the time interval over which the system input, $u_{k+1}(p)$, is defined.

The transient response component of the system's output used in the minimization is reduced due to the shifted reference signal but may still form a significant part of the whole. Since control of the transient response is an important consideration in ILC design, (4.19) is still a valid method of identifying a representative time delay to be applied to a system that is used over the same time interval and with similar input.

The action of recomputing the shift on each trial ensures that this input is as close as possible to that used in the identification. Furthermore, results showing the evolution of τ_k calculated using the same cost function for the system considered are given below and confirm that it does not vary significantly over the convergence period.

The algorithm considered is designed to produce convergence over the original, unshifted, reference and hence the error is calculated using (4.18). See [244] for a similar approach to repetitive control. The relationship between e_{k+1} and e_k is required to analyze the convergence of the algorithm. Starting from (4.17), let $e'_k(p)$ denote the error after shifting forward in time, i.e. $e'_k(p) = e_k(p + \tau_k)$, and let $r'_k(p)$ be the shifted reference for trial k, i.e. $r'_k(p) = r(p + \tau_k)$. Let G denote the open loop transfer function of the feedback loop. Then from Figure 4.10 (and dropping the dependence on p in some places for ease of presentation):

$$(r'_k + f_k)G = r - e_k$$
$$f_k = \frac{r - e_k - Gr'_k}{G}$$
$$f_{k+1} = \frac{r - e_{k+1} - Gr'_{k+1}}{G} \tag{4.20}$$

and using (4.17), f can be eliminated to give

$$\frac{r - e_{k+1} - Gr'_{k+1}}{G} = \frac{r - e_k - Gr'_k}{G} + Le'_k$$
$$e_{k+1} = e_k - GLe'_k + G(r'_k - r'_{k+1}) \tag{4.21}$$

It is expected that $G(r'_k - r'_{k+1})$ will be very small and potentially will tend to zero since there will be no difference between subsequent values of the calculated shift as the error reduces to zero.

However, it cannot be discounted entirely, and a revised algorithm is therefore necessary. This new algorithm follows, and then the convergence analysis is resumed.

In the algorithm given above an input of duration α seconds has been applied and during trial k the input is defined over $p \in [-\tau_{k-1}, \alpha - \tau_{k-1}]$ seconds and incorporated $r(p + \tau_{k-1})$ and $f_k(p)$. If the computed τ_k is not equal to τ_{k-1}, part of $f_k(p)$ must be discarded to form the new update $f_{k+1}(p)$, which is defined over $p \in [-\tau_k, \alpha - \tau_k]$. Discarding this data is likely to produce an unsatisfactory transient response as the applied input is no longer smooth and increases the possibility that a design that satisfies the theoretical convergence conditions does not result in an acceptable implementation. Also, no input is supplied during $t \in (\alpha - \tau_k, \alpha]$, which may lead to an unsatisfactory response in this interval. The new algorithm given next resolves these problems by expanding the interval in which the input is applied so that no data is discarded in computing each updated input, and Figure 4.11 gives a block diagram representation of its structure.

The system of Figure 4.11 is described by

$$u_{k+1}(p) = r'_{k+1}(p) + f_{k+1}(p) \tag{4.22}$$

$$f_{k+1}(p) = f_k(p - \tau''_{k+1}) + Le_k(p - \tau''_{k+1} + \tau_k) \tag{4.23}$$

$$r'_{k+1}(p) = r(p - \tau'_{k+1}) \tag{4.24}$$

$$\tau'_{k+1} = \max(\tau_k, \tau'_k) \tag{4.25}$$

$$\tau''_{k+1} = \max(0, \tau_k - \tau'_k) \tag{4.26}$$

$$\tau_k = \arg\min_{\tau} \{\| r'_k(p - \tau) + f_k(p - \tau) - y_k(p) \|_2^2\} \tag{4.27}$$

with $f_1(p) = 0$ and $\tau'_1 = 0$. As in the previous algorithm on trial k the feedforward signal, f_k, is summed with the shifted reference signal, r'_k, to produce the control input, u_k, which is applied to the system. The cost function (4.27) is identical to (4.19) since $r'_k(p - \tau_k) + f_k(p - \tau_k) = u_k(p - \tau)$ and so inherits the same motivation.

The most important difference between these algorithms is that, in order to express the time over which each trial extends in a meaningful way, each signal is now defined over $p \in [0, \alpha + \tau'_k]$. Moreover, the value of τ'_{k+1} is the largest value of the shift that has been calculated at the commencement of trial $k + 1$, i.e. $\tau'_{k+1} = \max\{\tau_k, \tau_{k-1}, \ldots, \tau_0\} \geq 0$. Another important difference is that the reference signal, $r'_{k+1}(p)$, which is tracked by the output of the system on trial $k + 1$ is now a copy of the original reference shifted backward in time, with a period of being set to zero concatenated to the start. Hence

$$r'_{k+1}(p) = \begin{cases} 0, & p \in [0, \ \tau'_{k+1}) \\ r(s), & s = p - \tau'_{k+1}, \quad t \in [\tau'_{k+1}, \ \alpha + \tau'_{k+1}] \end{cases} \tag{4.28}$$

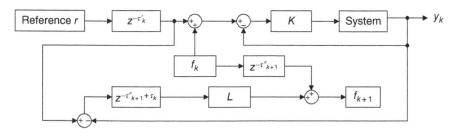

Figure 4.11 Controller structure of the revised reference shift algorithm.

The error is only defined over the original reference duration, and hence no truncation occurs when used in (4.23). Hence

$$e_{k+1}(p) = \begin{cases} 0, & p \in [0, \ \tau'_{k+1}) \\ r'_{k+1}(p) - y_{k+1}(p), & p \in [\tau'_{k+1}, \ \tau'_{k+1} + \alpha] \end{cases} \tag{4.29}$$

To describe the operation of the algorithm clearly, Figure 4.11 used the relationship $e_k(p) = r'(p) - y_{k+1}(p)$ and is therefore only strictly valid for the interval $p \in [\tau'_{k+1}, \tau'_k + \alpha]$. Further manipulation, see [39] for the details, produces the single control input update

$$u_{k+1}(p) = u_k(p - \tau''_{k+1}) + Le_k(p - \tau''_{k+1} + \tau_k) \tag{4.30}$$

providing the initial condition is changed from $f_1(p) = 0$ to $f_1(p) = r(p)$. Figure 4.12a shows an example of the input, $u_k(t)$, and reference, $r'_k(p)$, which are used during trial k.

The original reference, $r(p)$, has been shifted backward by τ'_k, and no data has been discarded in forming $u_k(p)$, which is applied over $p \in [0, \alpha + \tau'_k]$. Once this trial is complete, τ_k is calculated using (4.27). Figure 4.12b shows the components of the new input, $u_{k+1}(p)$, for the case when $\tau_k \leq \tau'_k$. In this case, the maximum shift, τ'_{k+1}, is set equal to the value used in the previous trial, τ'_k. The difference in these values is $\tau''_{k+1} = \tau'_{k+1} - \tau'_k = 0$. Consequently, neither u_k nor e_k need to be shifted to keep their relative position with respect to the reference and the latter is therefore merely shifted by τ_k and added to u_k to form the new input, $u_{k+1}(p)$.

Figure 4.12c shows the alternative case where $\tau_k > \tau'_k$. Once shifted by τ_k, the error will start before any of the other signals, and hence they all must be shifted backward in time such that it instead starts at $p = 0$. The value of the reference shift is accordingly changed from τ'_k to τ_k and both

(a)

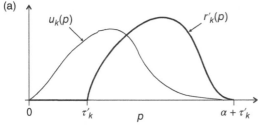

(b)

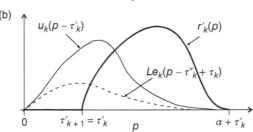

(c)

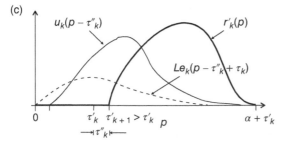

Figure 4.12 Update mechanism showing (a) signals during trial k, (b) the case where $\tau_k \leq \tau'_k$, and (c) the case where $\tau_k > \tau'_k$.

components used to construct the new input, u_{k+1}, are shifted by the difference, τ''_{k+1} to maintain their relative temporal position to the reference. The substitution $s = p - \tau_k$ can be used and then restricted to ensure that the original reference is tracked over the intended range, i.e. $s \in [0, \alpha]$.

The next stage is to determine the relationship between $e_{k+1}(p)$ and $e_k(p)$, required to examine the algorithm's trial-to-trial error convergence properties. Manipulation of (4.30) results in

$$e_{k+1}(p) = e_k\left(p - \tau''_{k+1}\right) - LGe_k\left(p - \tau''_{k+1} + \tau_k\right) \tag{4.31}$$

and applying the z-transform gives

$$e_{k+1}(z) = \left(1 - LG(z)z^{\tau_k/T_s}\right)z^{-\tau''_{k+1}/T_s}e_k(z) \tag{4.32}$$

Hence monotonic trial-to-trial error convergence occurs, see also [150, 244], provided

$$\left|\left(1 - LG(e^{j\omega T_s})e^{j\omega\tau_k}\right)e^{-j\omega\tau''_{k+1}}\right| = \left|1 - LG(e^{j\omega T_s})e^{j\omega\tau_k}\right| < 1 \tag{4.33}$$

and the convergence properties for any given frequency of interest can be examined.

Assuming that $r'_k(p) + f_k(p) - y_k(p)$ is periodic, a reference dependent assumption which is usually justified in practice, Parseval's relation can be used to write the cost function (4.27) as

$$\| r'_k(p - \tau_k) + f_k(p - \tau_k) - y_k(p) \|_2^2 = \sum_{n=1,2,\ldots,H} |\hat{\eta} - y_k(n)|^2$$

$$= H \sum_{k=1,2,\ldots,H} |a_k|^2$$

$$\hat{\eta} = r'_k(n - \tau_k/T_s) + f_k(n - \tau_k/T_s) \tag{4.34}$$

where the a_k are the Fourier Series coefficients (using capital letters to denote transformed variables) of

$$e^{-j\omega\tau_k}\left(R'_k(e^{j\omega T_s}) + F_k(e^{j\omega T_s})\right) - Y_k(e^{j\omega T_s}) \tag{4.35}$$

and $H = \alpha/T_s + 1$. Since

$$e^{-j\omega\tau_k}\left(R'_k(e^{j\omega T_s}) + F_k(e^{j\omega T_s})\right) - Y_k(e^{j\omega T_s}) = e^{-j\omega\tau_k}U_k(e^{j\omega T_s})\left(1 - e^{j\omega\tau_k}G(e^{j\omega T_s})\right) \tag{4.36}$$

the cost function (4.27) is equivalent to

$$\min_{\tau_k} \sum_{\omega = \frac{2\pi}{H}, \frac{4\pi}{H} \ldots 2\pi} \left|e^{-j\omega\tau_k}U_k(e^{j\omega T_s})\left(1 - e^{j\omega\tau_k}G(e^{j\omega T_s})\right)\right|^2 \tag{4.37}$$

Hence the problem can be written directly in terms of the monotonic convergence criterion with $L = 1$ as

$$\min_{\tau_k} \sum_{\omega = \frac{2\pi}{H}, \frac{4\pi}{H} \ldots 2\pi} \left|U_k(e^{j\omega T_s})\right|^2 \left|1 - LG(e^{j\omega T_s})e^{j\omega\tau_k}\right|^2 \tag{4.38}$$

and $\left|U_k(e^{j\omega T_s})\right|$ can be treated as a weighting function for the frequency-wise minimization of the monotonic convergence criterion. The appearance of u_k in the weighting function might not seem a natural choice in the design stage before implementing the algorithm. In practical terms, however, it is a most natural selection as there is no value to be gained in satisfying the monotonic convergence criterion for frequencies that are not present in the system output. The importance of satisfying the criterion is strongly related to the magnitude of each frequency present in the system output, again dictated by u_k.

The criterion restricts $LG(e^{j\omega T_s})e^{j\omega \tau_k}$ to lie in the unit circle centered on +1. Since the phase of a time-delay approaches zero at low frequencies, this effectively restricts the phase of $G(\cdot)$ to be less than 90° at low frequencies. This frequency-domain condition is well known in the ILC literature and forms the basis for the stability analysis of many control algorithms, see, e.g. [195]. Moreover, [150] have shown that in practice, it can produce good learning transients throughout each trial. Since the present algorithm has been specially formulated to reduce these transients, satisfying the criterion should also ensure the property of monotonic convergence exists over as large a section of each trial as possible. Experience has shown that in many applications of ILC, convergence and the transient performance along the trials can be conflicting requirements.

4.3.1 Filtering

Figure 4.13a shows a typical Nyquist plot of $1 - LG(z)z^{\tau_k/T_s}$ for the unfiltered algorithm in which $\tau_k = 1.1$. The values $T_s = 0.001$ seconds and $L = 0.3$ have been selected and G is nonminimum phase. In this figure, only a section of the plot is outside the unit circle centered at the origin and therefore does not satisfy the criterion for monotonic convergence. A filter is therefore required to ensure monotonic convergence by removing higher frequency components.

The filter is applicable to both the error signal and the input signal, the only difference being whether the reference signal itself is filtered. If applied to the error signal, the monotonic

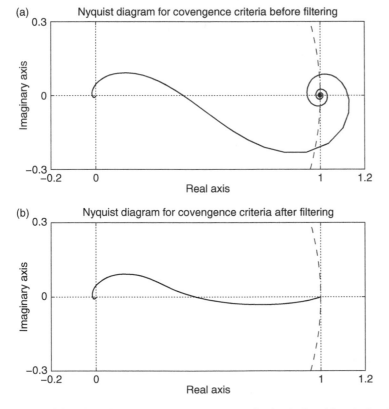

Figure 4.13 Nyquist plot for the convergence criterion before (a) and after (b) the filter is applied.

convergence criterion given by (4.33) becomes

$$\left|1 - L(z)G(z)z^{-\tau_k/T_s}\right| < 1 \tag{4.39}$$

where L is now a function of z. Initially, a low-pass finite impulse response (FIR) filter was designed to reduce the plot at those frequencies that do not satisfy the monotonic convergence criteria. When using FIR filters, one approach is to design a causal linear phase filter with the required magnitude properties and an odd number of coefficients that are symmetrical about a central term, $a_M z^{-M}$. Then this filter can then be multiplied by z^M to create a noncausal filter with the zero-phase characteristic. For further details relating to applying zero-phase filtering in ILC, see, e.g. [211].

Alternatively, an infinite impulse response (IIR) filter can be used in a procedure that requires fewer coefficients and consequently permits the computational speed to be increased. The error signal is filtered using an IIR filter with a gain, which is half the desired value at each frequency. The resulting output is then time-reversed producing another signal that is then fed into the same filter again. By reversing the output again, the required filtered signal is obtained.

After application of the noncasual filter, Figure 4.13b shows the resulting Nyquist plot of $1 - LG(z)z^{\tau_k/\alpha_s}$. The condition for monotonic convergence is satisfied using this fixed value of τ_k. Nyquist diagrams corresponding to different τ_k values using the noncausal filter are shown in Figure 4.14 and hence too much variation in τ_k does not result in monotonic trial-to-trial error convergence.

4.3.2 Numerical Simulations

A series of simulations have been used to test the algorithm's performance and compare with others using the same system model and reference trajectory. The two reference trajectories used in all the tests are shown in Figure 4.15a,b, where the first is a single period of a sinusoid with some bias and the second is a trapezoidal sequence. In these simulations $\alpha_s = 0.001$ seconds and $L = 0.3$ have been used.

Figure 4.16a gives the results for tracking performance for both the unfiltered and filtered systems, where the data is plotted as the normalized error recorded over 200 trials. After initial convergence, the unfiltered algorithm's start to increase after approximately 50 trials, and the system becomes unstable. Similar simulations using smaller values of gain have been performed, but this merely postpones this error increase. However, the use of a zero-phase IIR filter ensures good performance.

The reference shift algorithm aims to achieve faster convergence and smaller final error than the other comparable ILC laws. Simulations are used to examine relative performance, where a significant advantage is that the reference shift algorithm does not require any such system model. For the other cases, transfer-functions have been generated using a method in which a linear frequency response is fitted to experimental data. The results are shown in Figure 4.16b, and the performance advantage gained using the new design is evident.

4.3.3 Experimental Results

The reference shift algorithm has been applied to the nonminimum phase experimental test facility of Section 2.2.1 to assess the convergence properties of the algorithm in practice and compare its relative performance. As with the simulations, the values $T_s = 0.001$ seconds and $L = 0.3$ have

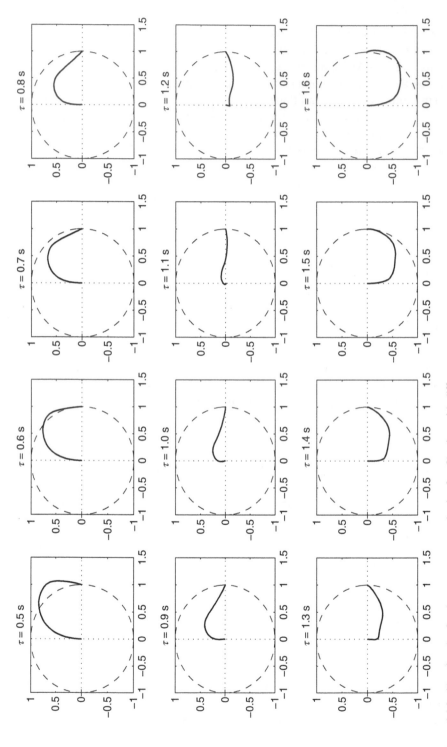

Figure 4.14 Nyquist plot for the convergence criteria after filtering with different τ_k.

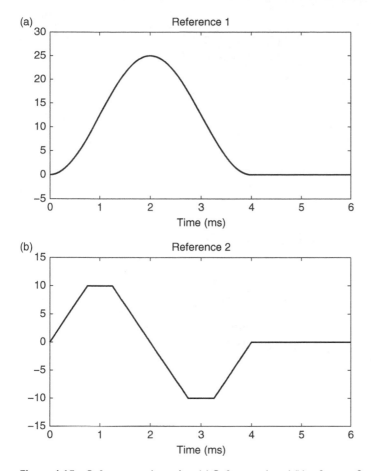

Figure 4.15 Reference trajectories. (a) Reference 1 and (b) reference 2.

been used in all the experiments conducted. Figure 4.17a shows the error results and the tracking performance of a 200 trial experiment using reference shift ILC for tracking the sinusoid shown in Figure 4.15a.

It can be seen that the error reduces rapidly and remains at a very low value for the remainder of the test. Further, there is no sign that the error will increase if the test were to be continued. The error results for tracking the second reference is shown in Figure 4.17b. Due to the abrupt transition in the second reference, instead of the smooth wave in the first reference, the mean squared error (for a SISO signal, the sum of the squares of the error values along a trial divided by the number of data points), and normalized error is greater than that shown in Figure 4.17a.

The system output on trial $k = 200$ of the test for both reference signals is shown in Figure 4.17c,d, in which the time variable "s" has been used so that the output follows the original unshifted reference, as discussed above. All error results refer to the error between these two signals. The values of the time shift, in this case, are, respectively, 0.807 and 0.977 seconds. The y axis of the error plots are in log scale to make differences visible.

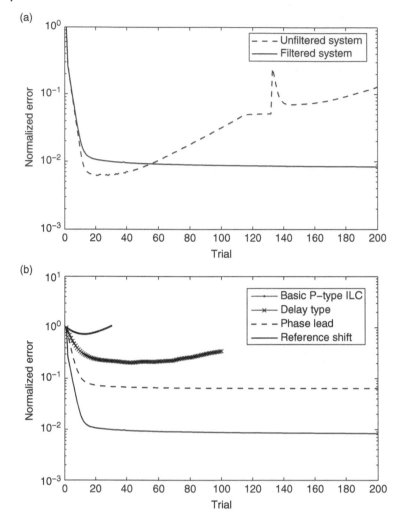

Figure 4.16 Simulation results for the reference shift algorithm. (a) Filtering comparison and (b) comparison of different algorithms.

It is of interest to test the reference shift algorithm's long-term behavior. Consequently, experiments over 400 trials for both references have were performed, and the error results and tracking progression are shown in Figure 4.17e,f. These plots indicate that the reference shift algorithm's long-term behavior is satisfactory as the error shows no signs of increasing. In Figure 4.17f, due to some unknown system disturbances, the error suddenly grows but quickly reduces again after only a few trials, which also shows the algorithm's relearning ability. Figure 4.17g,h give the evolution of τ_k over 400 trials for both reference signals, showing that it rapidly converges to a value which does not alter significantly over the course of the experiments.

Further development of ILC for nonminimum phase systems can be found in, e.g. [182, 187, 188], where [188] also gives experimental verification results on the nonminimum phase system used to obtain the experimental results given in this section.

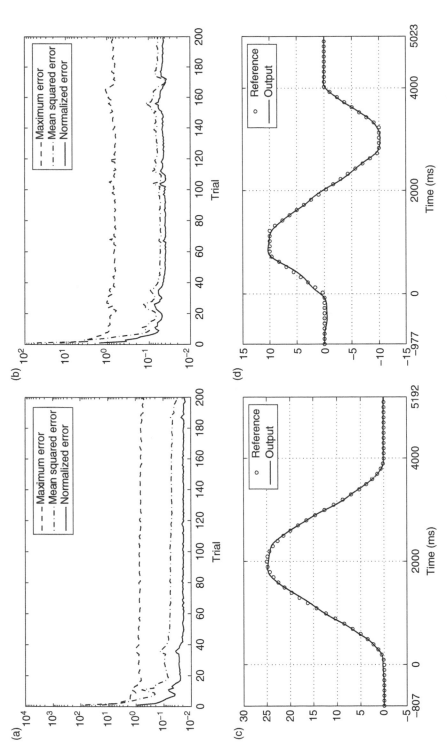

Figure 4.17 Experimental results for the reference shift algorithm. (a) Error results for the reference shift algorithm with reference 1. (b) Error results for the reference shift algorithm with reference 2. (c) The tracking on the 200th trial for reference 1 in one experiment. (d) The tracking on the 200th trial for reference 2 in one experiment. (e) Stability test of reference shift algorithm for reference 1. (f) Stability test of reference shift algorithm for reference 2. (g) Evolution of τ_k for reference 1. (h) Evolution of τ_k for reference 2.

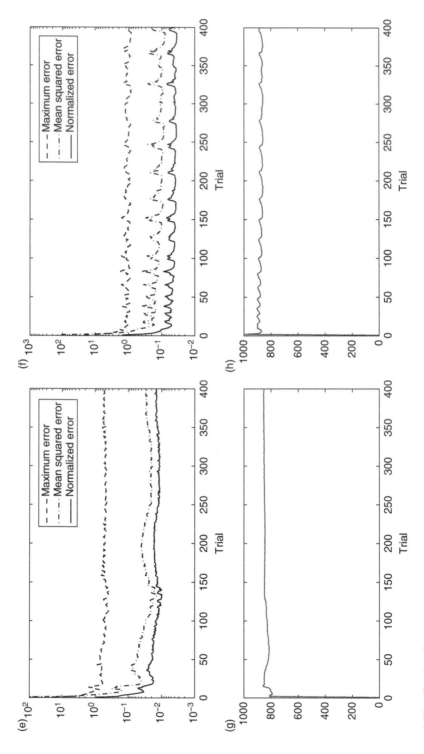

Figure 4.17 (Continued)

4.4   Concluding Remarks

This chapter has given an overview of tuning-based ILC design. Of course, the effectiveness of simple "three-term like" ILC laws is limited, but, as the rehabilitation example demonstrates, they can be very effective even in relatively new application areas. The next chapter begins treating model-based ILC, starting with design to minimize a suitably chosen cost function.

5

Optimal ILC

In common with many other applications, control law design based on minimizing a suitable cost function has continued to be an active area in iterative learning control (ILC) research and applications. This chapter first considers norm optimal iterative learning control (NOILC), where the cost function penalizes the difference between the control input vectors on successive trials. The application's inspired reasoning behind this choice is to decrease the error from trial-to-trial in an optimal manner without a possibly unacceptably large change in the control input from one trial to the next.

This chapter begins with the development of NOILC in an abstract Hilbert space setting, following [5], and subsequent conversion to a computational implementation algorithm as in [4]. Experimental verification on the gantry robot of Section 2.1.1, including a criterion for comparing the results obtained, is given, followed by physical application results from the free-electron laser facility of Section 2.3. Parameter optimal ILC [192] and terminal optimal ILC are also treated.

Prediction in ILC is possible in the trial-to-trial and along the trial directions, and both cases are considered in this chapter following, in the main, [6, 275], respectively, where results from the application of the latter design to the anthropomorphic robot arm of Section 2.1.2 are given. Finally, an overview of other experimental applications of NOILC is given, and constrained designs are motivated as a prelude to Chapter 8.

5.1 NOILC

5.1.1 Theory

Many optimization problems can be formulated in an abstract setting, and the solution derived, which is then specialized to the application area. This approach has seen much profitable research in optimal control and applications, and NOILC has this feature based on a Hilbert space setting. This section first details the formulation and solution in the general case, drawing mainly on results from [5] with similar, but later, analysis in [266].

This section uses the following definition of ILC in a Hilbert space setting, where for background on Hilbert spaces, see, e.g. [286].

Definition 5.1 Consider a dynamic system with input u and output y. Let $\mathcal{Y}$ and $\mathcal{U}$, respectively, be the output and input function spaces and let $r \in \mathcal{Y}$ be a desired reference trajectory for the system. An ILC law is successful if and only if it constructs a sequence of control inputs $\{u_k\}_{k \geq 0}$

Iterative Learning Control Algorithms and Experimental Benchmarking, First Edition.
Eric Rogers, Bing Chu, Christopher Freeman and Paul Lewin.
© 2023 John Wiley & Sons Ltd. Published 2023 by John Wiley & Sons Ltd.

which, when applied to the system (under identical experimental conditions), produces an output sequence $\{y_k\}_{k \geq 0}$ with the following properties of convergent learning

$$\lim_{k \to \infty} y_k = r, \quad \lim_{k \to \infty} u_k = u_\infty \tag{5.1}$$

with convergence interpreted in terms of the topologies assumed, respectively, in $\mathcal{Y}$ and $\mathcal{U}$.

This definition is sufficiently general to allow a simultaneous description of linear and nonlinear dynamics, continuous or discrete systems with either time-invariant or time-varying dynamics. Also, by way of notation, let the space of output signals $\mathcal{Y}$ be a real Hilbert space and $\mathcal{U}$ also be a real (and possibly distinct) Hilbert space of input signals, with corresponding inner products, denoted by $\langle \cdot, \cdot \rangle$ and norms $\| \cdot \|^2 = \langle \cdot, \cdot \rangle$. Also, e.g. $\|x\|_{\mathcal{Y}}$ denotes the norm of $x \in \mathcal{Y}$.

The systems considered are assumed to be linear and represented in operator form as follows:

$$y = G_{io}u + z_0 \tag{5.2}$$

where $G_{io} : \mathcal{U} \to \mathcal{Y}$ is the system input/output operator, which is assumed to be bounded, and z_0 represents the associated initial conditions.

The tracking error in this setting is

$$e = r - y = r - G_{io}u - z_0 = (r - z_0) - G_{io}u \tag{5.3}$$

No loss of generality arises in replacing r by $r - z_0$ and, hence, it can be assumed that $z_0 = 0$.

The optimization problem is to construct a minimizing input u_∞ from

$$\min_u \{ \|e\|^2 : \quad e = r - y, \quad y = G_{io}u \} \tag{5.4}$$

Many iterative solutions are possible for this problem, but a descent algorithm that maximizes the error reduction with each step is the most suitable.

The gradient-based class of ILC laws construct the control input for trial $k + 1$ as follows:

$$u_{k+1} = u_k + \epsilon_{k+1} G_{io}^* e_k \tag{5.5}$$

where $G_{io}^* : \mathcal{Y} \to \mathcal{U}$ is the adjoint operator of G_{io} and ϵ_{k+1} is a step length selected at each trial. This approach suffers from the need to choose a step length and the feedforward structure from trial-to-trial, which takes no account of current trial effects, including disturbances and systems modeling errors.

Relative to the general subject area, NOILC has the following two essential properties:

- Automatic choice of step size.
- The possibility of improved robustness by the use of causal feedback of current trial data and feedforward of data from previous trials.

These objectives are achieved by computing the control input on trial $k + 1$ as the solution of the minimum norm optimization problem:

$$u_{k+1} = \arg \min_{u_{k+1}} \{ J_{k+1}(u_{k+1}) : \quad e_{k+1} = r - y_{k+1}, \quad y_{k+1} = G_{io}u_{k+1} \} \tag{5.6}$$

where the cost function is

$$J_{k+1}(u_{k+1}) := \|e_{k+1}\|_{\mathcal{Y}}^2 + \|u_{k+1} - u_k\|_{\mathcal{U}}^2 \tag{5.7}$$

In theory, the initial control $u_0 \in \mathcal{U}$ can be arbitrary, but, in most applications, it will be a good first guess at the solution of the problem. Research on initial input selection for ILC has also been reported, see, e.g. [76, 77]. In this previous work, various strategies for choosing the initial input are developed and supported by experimental results from the gantry robot system of Section 2.1.1.

The NOILC design problem is the construction of the control input on trial $k + 1$ such that

- optimal reduction of the error at the end of the previous trial occurs and
- the control input for the next trial remains "close" to that applied on the previous trial.

The relative weighting of these two objectives has to be decided on an application to application basis. However, for the development of the underlying theory, this factor can be absorbed into the definitions of the norms on $\mathscr{Y}$ and $\mathscr{U}$.

By optimality and the fact that the (nonoptimal) choice of $u_{k+1} = u_k$ would lead to $J_{k+1}(u_k) = \|e_k\|^2$, the interlacing result

$$\|e_{k+1}\|^2 \leq J_{k+1}(u_{k+1}) \leq \|e_k\|^2, \quad \forall \ k \geq 0 \tag{5.8}$$

must hold. This condition, in turn, means that the resulting ILC law is of the descent form as the norm of the error is monotonically nonincreasing in k. Also, equality holds if and only if $u_{k+1} = u_k$, i.e. when convergence has occurred and no further input-updating takes place.

The control law on trial $k + 1$ is, see [5] for the derivation,

$$u_{k+1} = u_k + G_{io}^* e_{k+1}, \quad \forall \ k \geq 0 \tag{5.9}$$

or, on substituting for the error,

$$e_{k+1} = (I + G_{io} G_{io}^*)^{-1} e_k, \quad \forall k \geq 0 \tag{5.10}$$

and also

$$u_{k+1} = (I + G_{io}^* G_{io})^{-1}(u_k + G_{io}^* r), \quad \forall k \geq 0 \tag{5.11}$$

By the monotonicity property (5.8), the following limits exist:

$$\lim_{k \to \infty} \|e_k\|^2 = \lim_{k \to \infty} J_k(u_k) =: J_\infty \geq 0 \tag{5.12}$$

Also, induction and the inequality $\|y\| \leq \|G_{io}\| \|u\|$ establish that

$$\sum_{k \geq 0} \|u_{k+1} - u_k\|^2 < \|e_0\|^2 - J_\infty < \infty \tag{5.13}$$

$$\sum_{k \geq 0} \|e_{k+1} - e_k\|^2 < \|G_{io}\|^2 (\|e_0\|^2 - J_\infty) < \infty \tag{5.14}$$

and hence,

$$\lim_{k \to \infty} \|u_{k+1} - u_k\|^2 = 0, \quad \lim_{k \to \infty} \|e_{k+1} - e_k\|^2 = 0 \tag{5.15}$$

The properties of (5.15) show that NOILC has an implicit choice of step size, the first of the specific properties given above, as the incremental input converges to zero. This asymptotic slow variation is a prerequisite for convergence. Also, given (5.13), the sum of the energy costs from the first to the last trial is bounded and implicitly contains information on convergence rates.

The following result, Theorem 1 in [5], formally establishes the convergence properties of NOILC, where the notation $\mathscr{R}(\mathscr{A})$ denotes the range of an operator $\mathscr{A}$.

Theorem 5.1 *Suppose that either $r \in \mathscr{R}(G_{io})$ or $\mathscr{R}(G_{io})$ is dense in $\mathscr{Y}$. Then the ILC error sequence $\{e_k\}_k$ resulting from application of the control input sequence that solves the NOILC problem (5.6) converges in norm to zero in $\mathscr{Y}$.*

The guaranteed convergence and the monotonic trial-to-trial error convergence are critical properties of NOILC. The next stage is to develop a causal implementation for applications (see the next section), and the remaining task in this section is to establish that the input sequence converges, i.e. the second entry in (5.1).

Mathematically, the trial-to-trial error always converges to zero, but this does not imply convergence of the input sequence in $\mathcal{U}$ unless this space is chosen appropriately. This aspect is analyzed in detail in [5], and it is established that convergence of the input sequence requires additional assumptions on this sequence or the system considered. The following is Theorem 3 in [5].

Theorem 5.2   *The sequence $\{u_k\}_{k \geq 0}$ satisfies*

$$\lim_{k \to \infty} \|G_{io}^*(r - G_{io}u_{k+1})\|_{\mathcal{U}} = 0 \tag{5.16}$$

Also, if $G_{io}^* G_{io}$ has a bounded inverse in $\mathcal{U}$, the input sequence converges in norm to

$$u_\infty = (G_{io}^* G_{io})^{-1} G_{io}^* r \in \mathcal{U} \tag{5.17}$$

Moreover, if $\underline{\sigma} := \frac{1}{\|G_{io}^{-1}\|}$, (where $\underline{\sigma}(\cdot)$ denotes the smallest singular value) the convergence is bounded by a geometric relation of the form:

$$\|u_{k+1} - u_\infty\| \leq \frac{1}{1 + \underline{\sigma}^2} \|u_k - u_\infty\| \tag{5.18}$$

This last result only holds with the boundedness assumption imposed on the system inverse by the requirement that $\underline{\sigma}^2 > 0$. The following result, Theorem 4 in [5], relaxes this assumption.

Theorem 5.3   *If the sequence $\{u_k\}_{k \geq 0}$ is bounded in $\mathcal{U}$, the learned control input $u_\infty \in \mathcal{U}$ and $G_{io}^* G_{io}$ has range dense in $\mathcal{U}$, then this sequence converges to u_∞ in the weak topology in $\mathcal{U}$.*

For finite-dimensional spaces, weak convergence is equivalent to convergence in the norm. In such cases, convergence in the norm has also been established.

For applications, the results given above have to be converted into computational procedures or algorithms that are causal in the ILC sense, i.e. satisfy the conditions of Definition 1.1. Next, it is shown that this is possible for a form of the cost function used in most, if not all, reported experimental and actual implementations of NOILC.

5.1.2   NOILC Computation

In the abstract setting, the spaces $\mathcal{Y}$ and $\mathcal{U}$ are taken, respectively, as ℓ_2 spaces of $m \times 1$ and $l \times 1$ vectors on $[1, N]$ and $[0, N-1]$, respectively. Consider also the cost function:

$$J_{k+1}(u_{k+1}) = \frac{1}{2} \left(\sum_{p=1}^{N} (r(p) - y_{k+1}(p))^T Q(r(p) - y_{k+1}(p)) \right.$$
$$\left. + \sum_{p=0}^{N-1} (u_{k+1}(p) - u_k(p))^T R(u_{k+1}(p) - u_k(p)) \right) \tag{5.19}$$

where $Q > 0$ and $R > 0$, where > 0 denotes a symmetric positive definite matrix, are weighting matrices to be selected based on knowledge of the particular application considered. This cost function mirrors that for linear-quadratic optimal control theory and is a combination of the optimal tracking (of $r(p)$) and the disturbance accommodation problems (interpreting $u_k(p)$ as a known disturbance on trial $k+1$). The definitions of the inner products in $\mathcal{Y}$ and $\mathcal{U}$ are

$$\langle y_1, y_2 \rangle_{\mathcal{Y}} = \sum_{p=1}^{N} y_1^T(p) Q y_2(p) \tag{5.20}$$

$$\langle u_1, u_2 \rangle_{\mathcal{U}} = \sum_{p=0}^{N-1} u_1^T(p) R u_2(p) \tag{5.21}$$

The system dynamics are described by the state-space triple $\{A, B, C\}$ and hence, in this case G_{io} is the convolution operator defined by these matrices and it the remainder of this section is denoted by G. Moreover, the control input on trial $k+1$ that minimizes the cost function is

$$u_{k+1} = u_k + G^* e_{k+1}, \quad \forall\, k \geq 0 \tag{5.22}$$

where G^* is given by $R^{-1} G^T Q$ and R is invertible by construction. This form of the control law cannot, however, be implemented, but, as detailed below, it can be converted to an implementable form.

The first part of Theorem 5.2 specialized to this case shows that the input sequence minimizes the error in a least squares sense, even if G is singular or nonsquare. Moreover, if G is square and nonsingular, there exists a scalar $\underline{\sigma} > 0$ (the smallest singular value of G) such that $\|Gu\|_{\mathcal{U}} \geq \underline{\sigma}^2 \|u\|_{\mathcal{U}}$ holds.

If G has inverse with norm $\frac{1}{\underline{\sigma}}$, then

$$\|e_{k+1}\| \leq \frac{1}{1 + \underline{\sigma}^2} \|e_k\| \tag{5.23}$$

This is a corollary of the second part of Theorem 5.2 for this case and establishes that an exponential convergence rate is possible for "regular" systems. Also, $\underline{\sigma}(G)$ and hence, the rate of convergence of $\|u_{k+1} - u_\infty\|$ and $\|e_k\|$ can be arbitrarily changed by selection of Q and R, since

$$u^T G^T Q G u \geq \underline{\sigma}^2 u^T R u, \quad \forall\, u \in \mathcal{U} \tag{5.24}$$

and then if $R = h R_0$, where R_0 is fixed and the scalar h is a variable parameter, it follows that $\sigma = \frac{\sigma_0^2}{h}$ where $\underline{\sigma}_0$ denotes the smallest singular value of R_0. Hence, the parameter h provides complete control over the convergence rate, whereas h decreases the convergence rate of the input increases. As an example, to obtain a guaranteed reduction of the error of approximately $\frac{1}{2}$ on each trial h should be chosen to be of the order of magnitude of $\underline{\sigma}_0$.

The smallest singular value in the above analysis arises from a worst-case consideration, and it could be that for some practically relevant reference trajectories, trial-to-trial error convergence is much faster than that guaranteed by the bound of (5.23). Control over the trial-to-trial convergence rate is one of the significant advantages of NOILC over alternatives, where, in most cases, an exponential rate of convergence is not possible, nor is there any possibility of increasing the convergence speed.

To obtain a form of NOILC that can be implemented, the starting point is the adjoint (or transpose) for the class of systems considered, which involves time-reversal operations and a change of the state-space variables. In the case considered, the adjoint operator G^* is the following costate system (for background on this fundamental concept in optimal control, see, e.g. [10])

$$\psi_{k+1}(p) = A^T \psi_{k+1}(p+1) + C^T Q e_{k+1}(p+1), \quad \psi_{k+1}(N) = 0$$
$$u_{k+1}(p) = u_k(p) + R^{-1} B^T \psi_{k+1}(p), \quad N > p \geq 0 \tag{5.25}$$

This system has a terminal condition at $p = N$ instead of an initial condition, and it is a noncausal representation of the solution. Hence, NOILC cannot be implemented in this form.

Remark 5.1 This adjoint operator also arises in constrained ILC, see Chapter 8.

To obtain a causal implementation, suppose that all entries in the state vector are available. Also, write for the costate (ψ)

$$\psi_{k+l}(p) = -(K(p)(I + BR^{-1}B^T K(p)))^{-1} A(x_{k+1}(p) - x_k(p)) + \xi_{k+1}(p) \tag{5.26}$$

where $\xi_{k+1}(p)$ is known as the predictive or feedforward term. Then using (5.25) and (5.26) and standard techniques in optimal control theory, again see, e.g. [10], it follows that the matrix $K(p)$ is the solution of the following discrete matrix Riccati equation over the interval $[0, N-1]$

$$K(p) = A^T K(p+1)A + C^T QC$$
$$- \left(A^T K(p+1)B \left(B^T K(p+1)B + R \right)^{-1} B^T K(p+1)A \right) \tag{5.27}$$

with terminal condition $K(N) = 0$. Also, the predictive or feedforward term is generated by

$$\xi_{k+1}(p) = (I + K(p)BR^{-1}B^T)^{-1}(A^T \xi_{k+1}(p+1) + C^T Qe_k(p+1))$$
$$\xi_{k+1}(N) = 0 \tag{5.28}$$

and the input update equation is

$$u_{k+1}(p) = u_k(p) - ((B^T K(p)B + R)^{-1} B^T K(p)A(x_{k+1}(p) - x_k(p)))$$
$$+ R^{-1} B^T \xi_{k+1}(p) \tag{5.29}$$

In combination, (5.28) and (5.29) form a causal implementation of NOILC, since the computation of these two terms can be completed offline, between trials, by reverse time simulation using available previous trial information. The overall law consists of current trial full state feedback combined with feedforward from the previous trial error. Also, for time-invariant dynamics, the Riccati equation matrix $K(p)$, $0 \le p < N$, can be computed before the sequence of trials begins, i.e. no need to solve (5.27) for each new trial. Next, efficient implementation is considered.

For a given application, $K(p)$ does not contribute to the real-time processing load and the predictive term (5.28) must be computed between each trial and in descending sample order. The input (5.29) must be calculated at each sample instant, and it therefore significantly contributes to the real-time processing load.

To develop an efficient implementation, the variables in (5.28) are e_k and ξ_{k+1} and all of the other terms can be combined to produce constant matrices. In particular, introduce

$$\alpha(p) = \left(I + K(p)BR^{-1}B^T \right)^{-1}, \quad \beta(p) = \alpha(p)A^T, \quad \gamma(p) = \alpha(p)C^T Q \tag{5.30}$$

Hence, a computationally simpler computation of the predictive component is

$$\xi_{k+1}(p) = \beta(p)\xi_{k+1}(p+1) + \gamma(p)e_k(p+1) \tag{5.31}$$

and likewise for the input

$$u_{k+1}(p) = u_k(p) - \lambda(p)(x_{k+1}(p) - x_k(p)) + \omega(p)\xi_{k+1}(p) \tag{5.32}$$

where

$$\omega(p) = R^{-1}B^T, \quad \lambda(p) = (B^T K(p)B + R)^{-1} B^T K(p)A \tag{5.33}$$

Table 5.1 NOILC computation.

First level (before operation)

$$K(p) = A^T K(p+1)A + C^T QC$$
$$-[A^T K(p+1)B\big(B^T K(p+1)B + R\big)^{-1}$$
$$B^T K(t+1)A]$$
$$\alpha(p) = \big(I + K(p)BR^{-1}B^T\big)^{-1}$$
$$\beta(p) = \alpha(p)A^T$$
$$\gamma(p) = \alpha(p)C^T Q$$
$$\omega = R^{-1}B^T$$
$$\lambda(p) = (B^T K(p)B + R)^{-1}B^T K(p)A$$

Second level (between trials)

$$\xi_{k+1}(p) = \beta(t)\xi_{k+1}(p+1) + \gamma(t)e_k(p+1)$$

Third level (between sampling instants)

$$u_{k+1}(p) = u_k(p) - \lambda(t)\big(x_{k+1}(p) - x_k(p)\big) + \omega(p)\xi_{k+1}(p)$$

In this form, the resulting implementation requires seven matrices (A, B, C, β, γ, λ, and ω) to be supplied to the real-time controller. Further analysis, given in [218], results in Table 5.1, which gives the computations required.

5.2 Experimental NOILC Performance

In NOILC, the weighting matrices Q and R adjust the balance, respectively, between trial-to-trial error convergence speed and robustness. For the gantry robot of Section 2.1.1 applied to each axis, Q and R are positive scalars (the development below extends naturally for multiple-input multiple-output (MIMO) examples).

It is generally recognized that there are three variables that are of particular importance when describing the performance of an ILC law [283], i.e. trial-to-trial error convergence speed, minimum tracking error, and long-term performance. Although the instantaneous data recorded during each trial, such as input voltage and output error, are helpful in analyzing the learning process and its performance, it is necessary to develop a general measure of the tracking accuracy for each trial, and observe how this changes as the number of trials increases. This measure can specifically indicate, for a particular application, the minimum error, convergence speed, and any sign of performance degradation.

In an implementation, the error may decrease from trial-to-trial and then, after a large number of trials have elapsed, starts to grow again. This feature is sometimes termed long-term instability [47, 150] but is a performance issue that is still not completely understood, and for the remainder of this book, is referred to as long-term performance. It must be stressed that this is an implementation issue and not a failing of the underlying theory.

As for standard linear systems, popular tracking accuracy measures are in terms of a suitable error norm. Figure 5.1 shows the typical plot that arises for an ILC law with long-term performance

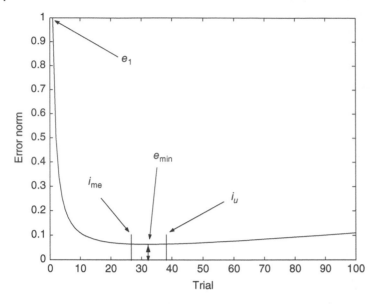

Figure 5.1 Typical error norm curve for an ILC system with long-term performance degradation.

degradation, where key parameters used to describe performance are also indicated. In particular, e_1 is the initial error norm value, i_{me} is the number of trials required to reach minimum error, e_{min} is the minimum error norm value, and i_u is the number of trials until the error norm begins to increase. The typical plot for a system without long-term performance degradation is similar, except that the i_u point is never reached, and the error norm does not increase as the number of trials increases. Of these parameters, i_{me} and e_{min} are most commonly used to describe ILC performance.

The performance index adopted involves finding the area under the error norm curve for the first H trials, where H is selected based on knowledge of the application area. This consideration results in the performance index, PI_H, given by

$$PI_H = \sum_{k=1}^{H} p(e)_k \tag{5.34}$$

where $p(e)_k$ denotes the norm of the error on trial k.

To allow a fair comparison of algorithm performance, test parameters such as the system model parameters and reference trajectory must be held constant. If the ILC law does not involve parameter-dependent current trial feedback, the value of e_1 will be theoretically parameter-invariant, and PI_H can be normalized by setting $e_1 = 1$. Since this does not hold for NOILC, PI_H will instead be normalized using the error norm produced in the absence of ILC current trial feedback. Approximate upper and lower bounds on the value of PI_H are, then given by $PI_H(\max) = He_1$ (assuming that no improvement occurs for $k > 1$) and $PI_N(\min) = e_1$, respectively, assuming that the error is zero for $k > 1$.

5.2.1 Test Parameters

A 2 seconds stoppage-time exists between each trial for the gantry robot of Section 2.1.1. During this time, the input for the next trial is calculated. The stoppage-time also allows vibrations induced on

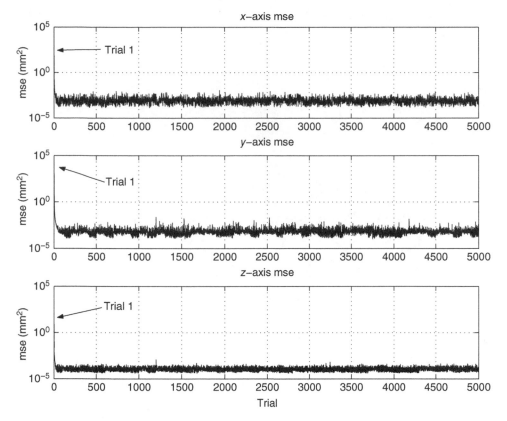

Figure 5.2 NOILC error norm over 5000 trials.

the previous trial to die away and prevents their propagation between trials. Before each trial, the axes are homed to within ±30 μm of a known starting location to minimize the effects of initial state error.

The system input voltage for the first trial is zero. Therefore, the ILC law must learn to track the reference trajectory in its entirety. A full-order observer estimates the entries in the state vector.

As representative results from the experiments performed, Figure 5.2 shows the error norm calculated for each axis during a 5000 trial test, where for the X and Y-axes $Q = 100$ and $R = 0.01$ and for the Z-axis $Q = 1000$ and $R = 0.1$. The mean square error (mse) has been used to enable comparison with other control methodologies that have been implemented on the gantry robot [218, 221] and the most important feature in this respect is that there is no sign of the error norm increasing. These features indicate that the ILC law can achieve long-term performance compared to other laws implemented on the same system, which resulted in an error divergence occurring after 100 or, in severe cases, just 3 trials.

To investigate the effect of varying Q and R, a batch of tests was performed using different combinations of these parameters, where from a starting value of $Q = 0.1$ increases by a factor of 10, in turn, were applied to a final value of 10^6. Starting from $R = 0.0001$ increases by a factor of 10, in turn, were applied to a final value of $R = 100$. This resulted in a total of 56 combinations.

Each combination was implemented for 100 trials and the PI_{100} performance index of the previous section were calculated. Given that there are 2 tuning parameters, it is particularly suitable

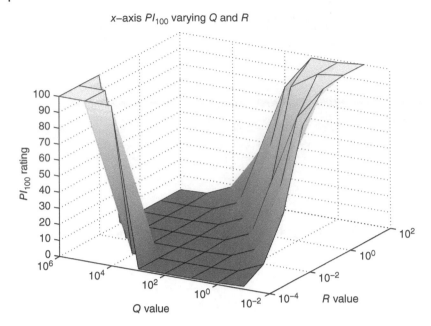

Figure 5.3 *X*-axis PI_{100} for various Q and R.

to plot performance on a 3D surface chart, as shown in Figure 5.3 for the X-axis. The performance plots corresponding to the other two axes are very similar, particularly the Y-axis, where the linear motor's low-frequency gain is practically identical to that of the X-axis.

Given that the cost function weighting Q affects the rate of error reduction and the input R limits the input change, the right-hand side of this last figure represents a region of poor tracking performance, where the $PI_{100} \approx 100$, indicating inferior learning performance. As expected, this corresponds to, respectively, small and large values for Q and R. These choices of Q and R result in an overly conservative control law.

As the ratio of Q to R increases, PI_{100} gradually reduces, indicating that the performance is improving. The slope represents this feature to the right-hand side of Figure 5.3. As the Q/R ratio continues to increase, PI_{100} is reduced to values very close to 1, indicating that the reference trajectory is learned to very high accuracy after one trial. Moreover, the balance of error reduction to input change is now approaching optimality.

Temporarily increasing Q/R has little effect on performance until the system degrades below an acceptable level and PI_{100} jumps back to 100. This feature is represented by the channel and then the steep slope to the left in Figure 5.3. Moreover, the ratio of Q to R determines the performance achieved rather than the absolute values of either Q or R.

The results given in this section are from a comprehensive set of experimental validations of the performance of NOILC reported in [221]. These confirm that NOILC can deliver excellent performance in convergence speed, minimum error, and long-term performance. The approach has proven well suited to the gantry robot due to the accuracy of each axis's linear model and the low level of interaction between axes. In Section 5.3, NOILC is applied to an actual physical system, and in Chapter 11, NOILC will be used as part of a solution of a nonlinear ILC design for stroke rehabilitation with supporting results from clinical trials.

5.3 NOILC Applied to Free Electron Lasers

In this section, the results from experimental application of NOILC to the free-electron laser system described in Section 2.3 are given, where $e_k = \begin{bmatrix} e_I & e_Q \end{bmatrix}^T$. To prevent damage to system components when the NOILC is implemented on the actual system, the input signals are limited, as shown in Figure 5.4. The limits are set to the maximum and minimum values of the input signals during the filling and decay phases.

This case is an example of constrained ILC where the control inputs' limits are applied based on the system's physical (or operator/engineer) knowledge. The development of such designs is the subject of Chapter 8. For the experimental results given below, the NOILC law was designed for the case when the 2×2 weighting matrices are selected as $Q = 100I$ and $R = I$.

The state variables required for the current state feedback component of the NOILC were obtained by simulation using the system model (a state observer was not used due to computational limitations when the experiments were run that now do not apply). Including input and output disturbances in the simulation, the results for the flat-top phase are shown in Figure 5.5, where, for simplicity and clarity, only the signals for the first input and output are given. The shape of the simulated disturbances can be computed as the deviation from the first trial's smooth trajectory.

As the number of trials increases, the output signals approach the desired setpoint (SP) trajectory. Rejection of the input and output disturbances can also be observed. Since the input signal reaches the given limits at the beginning of the phase, the output signal slowly approaches the setpoint in the first 100 μs.

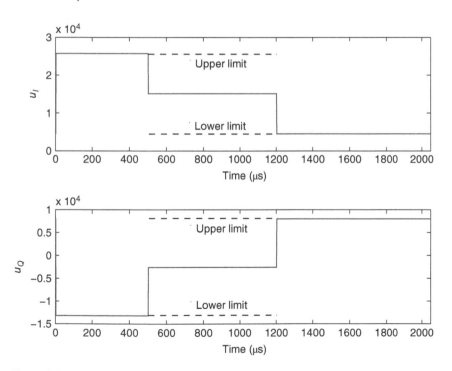

Figure 5.4 Input signals limits.

The NOILC design based on the identified model discussed in Section 2.3 and with the weighting matrices given above was successfully implemented and continues to run on the laser system at the DESY test facility. Sample results are shown in Figures 5.6–5.8. In these experiments, the number of trials was limited to 10 (based on discussions with the operating engineers).

Figures 5.6 and 5.7 show an increasing and decreasing trend in the first and the second output signals, respectively, during the flat-top phase, an example of the general behavior caused by the detuning effects known to arise in this application area. As the number of trials increases, both output signals approach the desired setpoint. After 10 trials, the output signals show only small deviations from the reference trajectory. Also, since only the signals during the flat-top phase are controlled, the filling and decay phase's input signals are kept constant, as illustrated in Figure 5.8.

Considering the output signals y_I and y_Q as the real and imaginary part of an output vector, respectively, the vector's amplitude and phase can be computed. To evaluate the performance of the NOILC in this application, the peak-to-peak error of the amplitude (A) is computed as follows:

$$e_{A,p2p} = \frac{\|\max(e_A(t)) - \min(e_A(t))\|}{\|\text{setpoint}_A\|} \tag{5.35}$$

where the subscript A on variables denotes that the amplitude value is used. Also, the root-mean-square (RMS)-error is computed as follows:

$$e_{A,\text{rms}} = \sqrt{\frac{1}{T_{ft}} \int_{t_{0,ft}}^{t_{0,ft}+T_{ft}} e_A^2 \, dt} \tag{5.36}$$

where $t_{0,ft}$ defines the beginning of the flat-top phase and T_{ft} the time duration of this phase. For the phase (ϕ), the errors are computed similarly, and the results are given in Table 5.2.

For comparison, the defined objectives and the values for the performance achieved using the already implemented PI feedback controller are also given in this table. The performance of

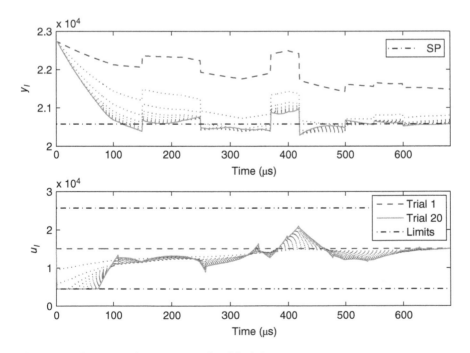

Figure 5.5 Output and input signals after 20 trials.

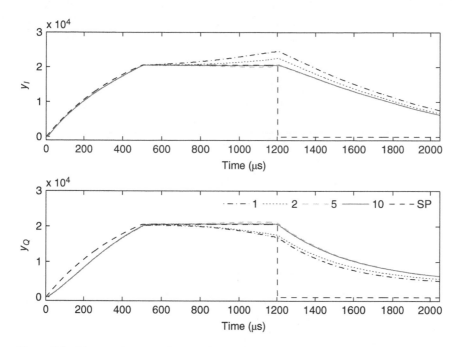

Figure 5.6   Measured output signals with the NOILC law applied.

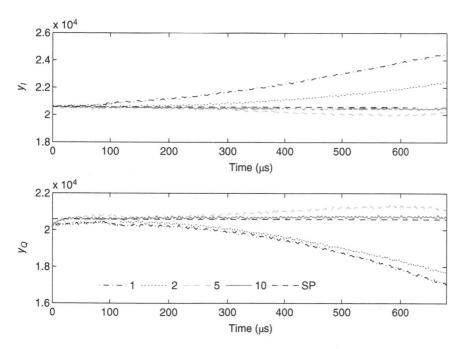

Figure 5.7   Measured output signals (flat top phase) with the NOILC law applied.

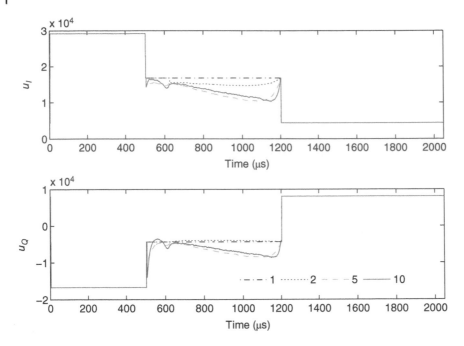

Figure 5.8 Input signals for the NOILC design.

Table 5.2 Performance objectives and performance achieved with the NOILC design.

	Amplitude A		Phase ϕ	
	$e_{A,p2p}$ (%)	$e_{A,\mathrm{rms}}$	$e_{\phi,p2p}$ (%)	$e_{\phi,\mathrm{rms}}$
NOILC	1.06	31.76	2.42	0.2879°
P-controller	0.5	0.1	1	0.1°
Objectives	≤0.1	≤0.01	≤0.1	≤0.01°

the NOILC design does not reach the specified objectives open-loop without the feedback action. However, the tracking errors of the NOILC after only 10 trials are close to those achieved by the proportional controller and the desired values defined by the performance specifications.

More recent work on ILC for free-electron lasers includes [224–226]. See also, [140] where an efficient implementation of NOILC applicable to examples where the trial length is very long using tensor products is developed. Free-electron lasers as one possible application area.

5.4 Parameter Optimal ILC

The convergence properties of the ILC laws in this chapter have, with others, been extensively analyzed. For some laws, there is limited analysis to guide the choice of the free parameters, i.e. the weighting matrices, in NOILC, to achieve fast or monotonic trial-to-trial convergence. In [192], a cost function was developed for the discrete-time systems with phase-lead ILC applied of the form

$$u_{k+1}(p) = u_k(p) + \gamma e_k(p+1) \tag{5.37}$$

where an optimal value of γ is computed by minimizing a cost function. Also, monotonic trial-to-trial error convergence in the ℓ_2 topology occurs if the example considered satisfies a positivity condition. If not, two methods have been developed that can be used to modify the example such that the positivity condition is satisfied.

This approach is known as parameter optimal ILC and treated further next for the phase-lead ILC law, where the scalar gain on the previous trial error term is trial varying, i.e.

$$u_{k+1}(p) = u_k(p) + \gamma_{k+1} e_k(p+1) \tag{5.38}$$

Consider the discrete case with the ILC dynamics written in lifted form (see (1.37)) $Y_k = GU_k$ (setting $d = 0$ causes no loss of generality). Then on completion of trial k, γ_{k+1} is chosen as the solution of the quadratic optimization problem:

$$\gamma_{k+1} = \arg\min_{U_{k+1}} \{J_{k+1}(\gamma_{k+1}) : E_{k+1} = R - Y_{k+1}, \ Y_{k+1} = GU_{k+1}\} \tag{5.39}$$

where (with weighting parameter w)

$$J(\gamma_{k+1}) = \|E_{k+1}\|^2 + w\gamma_{k+1}^2, \quad w \geq 0 \tag{5.40}$$

and

$$E_{k+1} = (I - \gamma_{k+1}G)E_k, \quad k \geq 0 \tag{5.41}$$

Routine analysis gives the optimal γ_{k+1} as follows:

$$\gamma_{k+1} = \frac{\langle E_k, GE_k \rangle}{w + \|GE_k\|^2} \tag{5.42}$$

It has been shown in [192] that the ILC law defined by (5.38) and (5.40) has the following properties.

(a) The performance index satisfies the interlacing monotonicity condition

$$\|E_{k+1}\|^2 \leq J(\gamma_{k+1}) \leq \|E_k\|^2 \tag{5.43}$$

with equality occurring if and only if $\gamma_{k+1} = 0$.

(b) The parameter sequence satisfies

$$\sum_{k=0}^{\infty} \gamma_{k+1}^2 < \infty \tag{5.44}$$

(c) As a consequence of (b)

$$\lim_{k \to \infty} \gamma_{k+1} = 0 \tag{5.45}$$

Property (a) states that this simple structure ILC law is a member of the descent class since the norm of the error is monotonically nonincreasing in k, and the "energy costs" from the first to the last trial are bounded. Properties (b) and (c) show that the learning rate becomes slower as the law progresses to convergence. The following result proved as Theorem 2 in [192] formally establishes this property.

Theorem 5.4 *Under the properties (5.43)–(5.45) and the additional assumption that the representation of the system dynamics G is positive, i.e.*

$$G + G^T > 0 \tag{5.46}$$

then

$$\lim_{k \to \infty} \|E_k\| = 0 \tag{5.47}$$

Hence (5.38) and (5.40), together with the positivity assumption on the system dynamics, ensure that the sequence $\{E_k\}_k$ converges in k to zero, i.e. guaranteed convergence of learning. Also, the error norm sequence $(\{\|E_k\|\}_k)$ is monotonically convergent.

The importance of the positivity requirement is re-enforced by Proposition 1 in [192], which establishes that if $G + G^T$ is not positive-definite, but G is invertible, then $\{\|E_k\|\}_k$ does not necessarily converge to zero. Also, the "strength" of the positive definite assumption can be weakened. Even if the system is not positive-definite, this property can be achieved by either replacing all signals by exponentially weighted signals or by "slow" sampling for differential dynamics, for the details again see [192].

5.4.1 An Extension to Adaptive ILC

Given (5.38) and (5.40), it follows that if $w > 0$, E_k decreases to a small value as $k \to \infty$ and γ_{k+1} must tend to zero, i.e. a possible reduction in the convergence rate. An obvious way of improving this situation is to make the weight w adaptive as detailed next. This formulation is one form of adaptive ILC, where this topic is also considered in Chapter 10.

Introduce the new performance index

$$J(\gamma_{k+1}) = \|E_{k+1}\|^2 + (w_1 + w_2 \|E_k\|^2)\gamma_{k+1}^2 \ : \ w_1 \geq 0, \ w_2 \geq 0 \tag{5.48}$$

i.e. $w_1 + w_2 \|E_k\|^2$ varies from trial-to-trial. Then the optimal choice for γ_{k+1} is

$$\gamma_{k+1} = \frac{\langle E_k, GE_k \rangle}{w_1 + w_2 \|E_k\|^2 + \|GE_k\|^2} \tag{5.49}$$

and it is shown in Proposition 4 of [192] that if $w_1 = 0$, $w_2 \geq 0$ and $G + G^T \succcurlyeq \sigma^2 I > 0$, then $\gamma_{k+1} \in [a, b]$, where $0 < a < b$. Also Proposition 5 in [192] shows that if $w_1 > 0$, $w_2 \geq 0$ and $G + G^T > 0$, then the singular values of $I - \gamma_{k+1} G$ lie inside the open unit circle in the complex plane. Moreover, $E_k \to 0$ geometrically, i.e.

$$\|E_{k+1}\|^2 \leq \lambda \|E_k\|^2 \tag{5.50}$$

where

$$0 \leq \lambda \leq 1 - a^2(2w_2 + \underline{\sigma}(G)^2) < 1 \tag{5.51}$$

In the case of $w_1 \neq 0$, the above inequality suggests that this ILC law converges geometrically when $w_2 \|E_k\|^2 \gg w_1$ but the buildup of small errors results in slow convergence. Qualitatively, geometric convergence can be obtained until $\|E_k\|^2 \approx \delta^2$ if $w_1 \approx w_2 \delta^2$. This optimization based analysis extends to the design of higher-order iterative learning control (HOILC) laws, considered in Chapter 7, as also detailed in [192]. Also, consult this reference for a supporting numerical example. Adaptive parameter optimal ILC is considered again in a robustness context in the next chapter.

5.5 Predictive NOILC

Predictive action in ILC design can be from trial-to-trial or along the trials (or, possibly, both). This section first briefly considers the former case, and then for the latter, control law development is considered with supporting experimental validation.

A natural extension of NOILC is termed predictive NOILC [6], where the cost function takes future predicted error signals into account. The cost function used to compute the input u_{k+1} on trial $k + 1$ is

$$J_{k+1}(u_{k+1,H}) := \sum_{i=1}^{H} \lambda^{i-1} \left(\|e_{k+i}\|^2 + \|u_{k+i} - u_{k+i-1}\|^2 \right) \qquad (5.52)$$

and includes the error not only of the next trial but of the next H, together with the corresponding changes in input. The weighting parameter $\lambda > 0$ determines the importance of more distant (future) errors and incremental inputs.

In this case, the ILC law is obtained from minimization of the cost function and the minimizing input, which is unique, is computed using dynamic programming, as detailed in [6]. Once the input has been constructed, a recursive formulation for the error (and input) evolution is computed.

The design developed below uses a similar cost function to that used in NOILC but embeds the reference signal/disturbance model in the controller and employs the receding horizon control principle. The idea of embedding the reference signal information in the controller has been successfully used in model predictive control, e.g. [271] and in other ILC-related research [168]. Such a design allows for the practically motivated case where the reference trajectory has dominant frequencies, and it is decided to only include these in control design as opposed to all frequencies. Also, it is assumed that the system dynamics can be adequately modeled, at least for initial control-related studies, as linear and time-invariant. See also Chapter 7 for another design in this general setting that uses the generalized Kalman–Yakubovich–Popov (KYP) lemma.

The design in this section is based on a frequency domain decomposition of the supplied reference signal or vector, respectively, in the SISO and MIMO cases. Once these are selected, they are embedded in the system state-space model in accordance with the internal model principle as described next.

Suppose that the frequency components of the reference signal to be included in the design have been selected. For details, see, e.g. [273, 274], where the notation in these two publications is used in the analysis that follows. This action results in the annihilator polynomial (for the SISO case with an immediate generalization to MIMO examples)

$$D(z) = (1 - z^{-1}) \prod_{i=1}^{l} (1 - 2\cos(i\omega)z^{-1} + z^{-2})$$
$$= 1 + d_1 z^{-1} + d_2 z^{-2} + d_3 z^{-3} + \cdots + d_\gamma z^{-\gamma} \qquad (5.53)$$

where 0 and $i\omega$, $i = 1, 2, \ldots, l$, for some chosen positive integer l, denote the frequencies to be included.

The control law is designed to track the reference signal, and hence, by the internal model principle [72], the corresponding $D(z)$, i.e. a particular case of (5.53) must be included in the denominator of the z transfer-function description of the controller dynamics. In the design method followed in this section, the route used is to add a vector term $(\mu(p)$ in the state-space model (5.54) below). The state dynamics in the system state-space model are described next (but alternative settings could be used).

For analysis and design, the system state-space model with the vector $\mu(p)$ added is

$$x_m(p+1) = A_m x_m(p) + B_m u(p) + \mu(p)$$
$$y(p) = C_m x_m(p) \tag{5.54}$$

where $x_m(p) \in \mathbb{R}^{n_1}$ is the state vector, $u(p) \in \mathbb{R}^{m_u}$ is the input vector and $y(p) \in \mathbb{R}^{m_y}$ is the output vector of the system. Also, each entry in the $n_1 \times 1$ vector $\mu(p)$ is the inverse z-transform of $\frac{1}{D(z)}$ and let q^{-1} denote the backward shift operator and $D(q^{-1})$ the shift operator interpretation of $D(z)$. Then applying $D(q^{-1})$ to $x_m(p)$ and $u(p)$ gives

$$x^s(p) = D(q^{-1})x_m(p), \quad u^s(p) = D(q^{-1})u(p) \tag{5.55}$$

Also, $D(q^{-1})\mu(p) = 0$ (since $D(z)$ contains all frequencies in $\mu(p)$) and from (5.54)

$$x^s(p+1) = A_m x^s(p) + B_m u^s(p)$$
$$D(q^{-1})y(p+1) = C_m A_m x^s(p) + C_m B_m u^s(p) \tag{5.56}$$

Introducing the state vector

$$x(p) = \left[\ (x^s)^T(p) \ \ y^T(p) \ \ \cdots \ \ y^T(p-\gamma+1) \ \right]^T$$

gives the following augmented state-space model for design:

$$x(p+1) = Ax(p) + Bu^s(p)$$
$$y(p) = Cx(p) \tag{5.57}$$

where

$$A = \begin{bmatrix} A_m & 0 \\ \hat{C} & A_d \end{bmatrix}, \quad \hat{C} = \begin{bmatrix} C_m A_m \\ 0 \end{bmatrix}$$

$$A_d = \begin{bmatrix} -d_1 I & -d_2 I & \cdots & -d_{\gamma-1} I & -d_\gamma I \\ I & 0 & 0 & \cdots & 0 \\ 0 & I & 0 & \cdots & 0 \\ \vdots & \ddots & \ddots & \ddots & \vdots \\ 0 & 0 & \cdots & I & 0 \end{bmatrix}$$

$$B = \begin{bmatrix} B_m^T & (C_m B_m)^T & 0 & \cdots & 0 & 0 \end{bmatrix}^T$$
$$C = \begin{bmatrix} 0 & I & 0 & \cdots & 0 & 0 \end{bmatrix} \tag{5.58}$$

The poles of (5.57) are the union of those of the system and those arising from the structure of $\mu(p)$.

Suppose that the frequency domain decomposition is applied to $r(p)$ and $D(z)$ of (5.53), where the latter polynomial is constructed from the frequencies to be included. Then the ILC problem can be formulated by following identical steps to those used to obtain (5.57), resulting in a state-space model for design of the form:

$$x_k(p+1) = Ax_k(p) + Bu_k^s(p)$$
$$y_k(p) = Cx_k(p) \tag{5.59}$$

where

$$x_k(p) = \left[\ (x^s)_k^T(p) \ \ e_k^T(p) \ \ e_k^T(p-1) \ \ \cdots \ \ e_k^T(p-\gamma) \ \right]^T \tag{5.60}$$

and $(x^s)_k^T(p)$ and $u_k^s(p)$ are formed using $D(z)$ on trial k as per their counterparts $x^s(p)$ and $u^s(p)$ in (5.56). The matrices A, B, and C have identical structures to their counterparts in (5.56), and by the structure of C, the output vector in this state-space model is the current trial error. If it is assumed that the system has reached the steady-state before a trial commences, the state initial vector on each trial can be assumed to be zero.

On trial $k + 1$ and sampling instant p, the future state vector at sample $p + m$, written $x_{k+1}(p + m \mid p)$ for the model (5.59), is predicted as follows:

$$x_{k+1}(p + m \mid p) = A^m x_{k+1}(p) + \sum_{i=0}^{m-1} A^{m-i-1} B u_{k+1}^s(i) \tag{5.61}$$

where $m > 0$ is a future sampling instant, and this prediction is performed along each trial. In designs that require modeling of the future control trajectory, one approach is to embed an integrator in the design and the incremental control trajectory is then directly computed within an optimization window. For the ILC design considered in this section, the signal to be optimized is the filtered control vector and the design can be undertaken by modeling this signal using pulse functions. The main drawback is the requirement to optimize a large number of parameters if fast sampling is required and or the system has a relatively complex dynamic response.

Fast sampling is typically required for mechanical and electro-mechanical systems because the time constants arising in the various subcomponents can vary in duration and a smaller sampling interval is needed to capture the effects of the smaller of these. One approach to reducing the number of parameters requiring optimization online is to parameterize the future trajectory of the filtered control signal using a set of Laguerre functions, where a scaling factor is used to reflect the time scale of the predictive system.

Laguerre functions are well known in system identification, see, e.g. [267], and their use in model predictive control is detailed in, e.g. [271]. The design developed in this section should only be used in applications where the number of optimized parameters must be reduced to construct the control law efficiently. This section considers direct digital control, but for cases where the design is completed in the continuous-time domain, they, or equivalents, must be used to parameterize the trajectories. In the SISO case, a summary is given next of the background on these functions relevant to the design in this section. The ILC notation, i.e. the trial number subscript, is dropped for ease of presentation.

The basis of design using these functions is the use of a set of discrete orthonormal functions to describe the filtered future control signal $u^s(m)$ within a moving horizon window, $0 \le m \le N_p$. Assume that N_l is the number of terms in the expansion and let $l_i(m)$, $1 \le i \le N_l$, be a set of Laguerre functions, which are orthonormal. Then

$$u^s(m) \approx \sum_{h=1}^{N_l} c_h l_h(m) \tag{5.62}$$

In this application, the z transfer-function of the hth Laguerre function is given by

$$\Gamma_h(z) = \frac{\sqrt{1 - a^2}}{1 - az^{-1}} \left(\frac{z^{-1} - a}{1 - az^{-1}} \right)^{h-1} \tag{5.63}$$

where $0 \le a < 1$ is the scaling factor. Also, the network structure of the z transfer-function for this system can be exploited to show that the set of discrete Laguerre functions satisfies the difference equation:

$$L(m + 1) = A_L L(m) \tag{5.64}$$

where $L(m) = \begin{bmatrix} l_1(m) & l_2(m) & \dots & l_{N_l}(m) \end{bmatrix}^T$ and

$$A_L = \begin{bmatrix} a & 0 & \dots & \dots & 0 \\ \beta & a & \ddots & \vdots & 0 \\ -a\beta & \beta & \ddots & 0 & 0 \\ \vdots & \vdots & \ddots & \ddots & 0 \\ -a^{N_l-2}\beta & a^{N_l-3}\beta & \dots & \beta & a \end{bmatrix} \qquad (5.65)$$

$\beta = (1 - a^2)$ and

$$L(0) = \sqrt{\beta}\begin{bmatrix} 1 & -a & a^2 & -a^3 & \dots & (-1)^{N_l-1}a^{N_l-1} \end{bmatrix}^T \qquad (5.66)$$

Setting $a_h = 0$ and $\delta_i(m) = \delta(i)$, where $\delta(i)$ is the Dirac delta function, recovers the standard formulation of model predictive control.

Laguerre's function-based design's basic idea represents each input by a set of Laguerre functions and their unknown coefficients. This design is illustrated by

$$u^s(i) = L^T(i)\eta \qquad (5.67)$$

where the Laguerre function vector $L(i) = \begin{bmatrix} l_1(i) & l_2(i) & \dots & l_{N_l}(i) \end{bmatrix}^T$ and the Laguerre coefficient vector $\eta = \begin{bmatrix} c_1 & c_2 & \dots & c_{N_l} \end{bmatrix}^T$. Moreover, N_l is the dimension of the Laguerre function vector and is also the number of terms used in the approximation. The Laguerre functions are predetermined in the design once the scaling factor $0 \le a < 1$ and the number of terms N_l are chosen. Moreover, the number of terms and the scaling factor used in this last construction can be chosen independently for each input.

With the control trajectory represented by a Laguerre polynomial, the predicted state vector (5.61) in the ILC case can be written as follows:

$$x_{k+1}(p + m \mid p) = A^m x_{k+1}(p) + \phi^T(m)\eta_{k+1} \qquad (5.68)$$

where $\phi^T(m) = \sum_{i=0}^{m-1} A^{m-i-1} B L^T(m)$, and this term is invariant from trial-to-trial. The ILC design problem is to find the sequence of current trial inputs that minimize the cost function:

$$J = \sum_{m=1}^{N_p} x_{k+1}(p + m \mid p)^T Q x_{k+1}(p + m \mid p)$$

$$+ \sum_{m=0}^{N_p} (u_{k+1}^s(m) - u_k^s(m))^T R(u_{k+1}^s(m) - u_k^s(m)) \qquad (5.69)$$

where $Q > 0$ and $R > 0$ are weighting matrices to be selected, and also, the difference between the control signals on the current and previous trials is penalized. The motivation for this last choice is, as in NOILC, to achieve trial-to-trial error reduction without unduly large changes in the amplitudes of the control signals required.

The previous trial filtered input vector is also parameterized in the form detailed above with a long prediction horizon N_p and, hence,

$$\sum_{m=0}^{N_p} u_k^s(m)^T R u_k^s(m) = \eta_k^T R_L \eta_k \qquad (5.70)$$

$$\sum_{m=0}^{N_p} u_{k+1}^s(m)^T R u_k^s(m) = \eta_{k+1}^T R_L \eta_k \tag{5.71}$$

$$\sum_{m=0}^{N_p} u_{k+1}^s(m)^T R u_{k+1}^s(m) = \eta_{k+1}^T R_L \eta_{k+1} \tag{5.72}$$

where the orthonormal property of the Laguerre functions has been used, i.e.

$$\sum_{m=0}^{N_p} L^T(m)L(m) = I$$

and $R_L = I$.

Substituting (5.68) and (5.70)–(5.72) into (5.69) gives

$$\begin{aligned} J = \eta_{k+1}^T \Omega \eta_{k+1} &+ 2\eta_{k+1}^T \Psi x_{k+1}(p) \\ &- 2\eta_{k+1}^T R_L \eta_k + \eta_k^T R_L \eta_k \end{aligned} \tag{5.73}$$

where

$$\Omega = \sum_{m=1}^{N_p} \phi(m)Q\phi^T(m) + R_L, \quad \Psi = \sum_{m=1}^{N_p} \phi(m)QA^m \tag{5.74}$$

The minimum value of this cost function occurs when

$$\eta_{k+1} = -\Omega^{-1}(\Psi x_{k+1}(p) - R_L \eta_k) \tag{5.75}$$

Under receding horizon control, only the first sample of the optimal control trajectory is implemented, which is constructed in the SISO case with an obvious generalization to MIMO examples, as the filtered control signal on trial $k+1$ at sample p

$$u_{k+1}^s(p) = L^T(0)\eta_{k+1} \tag{5.76}$$

By combining (5.75) and (5.76), the predictive ILC law can be written as follows:

$$u_{k+1}^s(p) = -L^T(0)\Omega^{-1}\Psi x_{k+1}(p) + L^T(0)\Omega^{-1}R_L \eta_k \tag{5.77}$$

The control law (5.77) is the linear sum of terms in the current trial state vector $x_{k+1}(p)$ and the previous trial term η_k. Here, the first term introduces state feedback on trial $k+1$, with gain matrix K_{mpc}, defined as follows:

$$K_{\mathrm{mpc}} = L^T(0)\Omega^{-1}\Psi \tag{5.78}$$

and the second term introduces feedforward information from the previous trial (compare with the NOILC structure). The prediction of the future state at time τ, i.e. $x(p + \tau|p)$ can be written as follows (for MIMO examples):

$$x(p + \tau \mid p) = A^\tau x(p) + \phi(\tau)\eta \tag{5.79}$$

where

$$\eta = \begin{bmatrix} \eta_1^T & \eta_2^T & \cdots & \eta_{m_u}^T \end{bmatrix}^T \tag{5.80}$$

and if B_i denotes the ith column of the state-space model input matrix, then

$$\phi(\tau) = \sum_{j=0}^{\tau-1} A^{\tau-j-1} \begin{bmatrix} B_1 L_1^T(j) & \cdots & B_{m_u} L_{m_u}^T(j) \end{bmatrix} \tag{5.81}$$

Also, the ith input is given by $L_i^T \eta_i$. where L_i is generated by applying (5.64) for this input. Moreover, the number of terms and the scaling factor used in this last construction can be chosen independently for each input.

In many applications, some entries in the state vector of (5.75) will not be measurable and in such cases an observer is required. The state vector $x_k(p)$ of (5.60) is formed from $x_k^s(p)$ and the feedback errors $e_k(p), e_k(p-1), \ldots, e_k(p-\gamma)$, where the latter are measurable, and hence, it is effective to design a reduced order observer to estimate the system state dynamics only.

Let $\hat{x}_k^s(p)$ denote the estimated state vector at trial k, then the observer dynamics are described by

$$\hat{x}_k^s(p+1) = A_m \hat{x}_k^s(p) + B_m u_k^s(p) + K_{ob}(y_k^s(p) - C_m \hat{x}_k^s(p)) \tag{5.82}$$

where $y_k^s(p)$ is the filtered output on trial k and sampling instant p, defined by $y_k^s(p) = D(q^{-1})y_k(p)$. The observer gain matrix K_{ob} is selected such that the closed-loop observer system state matrix $A_m - K_{ob}C_m$ has all eigenvalues strictly inside the unit circle of the complex plane. After obtaining $\hat{x}_k^s(p)$, the state vector for the ILC law, $x_k(p)$, is constructed using the estimated filtered state vector and the feedback errors as follows:

$$\hat{x}_k(p) = \left[\ (\hat{x}_k^s)^T(p) \ \ e_k^T(p) \ \ e_k^T(p-1) \ \ \ldots \ \ e_k^T(p-\gamma) \ \right]^T \tag{5.83}$$

Finally, a fundamental difference between this control law and NOILC is that the latter does not have the reference vector embedded (the polynomial $D(z)$) in the design model. Hence, there is no equivalence even as the Laguerre parameter $N \to \infty$.

5.5.1 Controlled System Analysis

In ILC, the trial length is finite, and hence, trial-to-trial error convergence (i.e. in k) can occur even if the system dynamics are unstable, i.e. all eigenvalues of the state matrix do not have modulus strictly less than unity. As the Laguerre parameter N_l and the prediction on p increases, it can be shown, see, e.g. [271], that the control input (5.77) can be computed from the solution to the infinite-time discrete quadratic regular problem for the state-space model triple $\{A, B, C\}$ of (5.59) with weighting matrices $Q > 0$ and $R > 0$. Hence,

$$K_{mpc} = (R + B^T P_\infty B)^{-1} B^T P_\infty A \tag{5.84}$$

where P_∞ is the solution of the algebraic Riccati equation:

$$A^T P_\infty A - A^T P_\infty B(R + B^T P_\infty B)^{-1} B^T P_\infty A + Q - P_\infty = 0 \tag{5.85}$$

Moreover, by discrete quadratic regulator theory [10], all eigenvalues of $A - BK_{mpc}$ lie inside the open unit circle in the complex plane.

This last fact, in turn, guarantees that for any matrix norm $\| \cdot \|$, there exist real numbers $M > 0$ and $0 < \lambda < 1$ such that $\|(A - BK_{mpc})^p\| \leq M\lambda^p$, $p > 0$. Also, for the first trial, with the assumption that η_{-1} is the zero vector, the filtered control signal is

$$u_1^s(p) = -K_{mpc}x_1(p) \tag{5.86}$$

and with the control law applied

$$x_1(p+1) = (A - BK_{mpc})x_1(p) = A_{cl}x_1(p) \tag{5.87}$$

where $A_{cl} = A - BK_{mpc}$. For a given $x_1(0)$, (5.87) gives

$$x_1(p) = (A - BK_{mpc})^p x_1(0) = A_{cl}^p x_1(0) \tag{5.88}$$

and also

$$\|x_1(p)\| \leq M\lambda^p \|x_1(0)\| \tag{5.89}$$

On the second trial, the filtered control signal is

$$u_2^s(p) = -K_{\text{mpc}}x_2(p) - K_1 x_1(p) \tag{5.90}$$

where $K_1 = L^T(0)\Omega^{-1}R_L\Omega^{-1}\Psi$, and

$$x_2(p+1) = (A - BK_{\text{mpc}})x_2(p) - BK_1 x_1(p) \tag{5.91}$$

or, using (5.88),

$$x_2(p+1) = (A - BK_{\text{mpc}})x_2(p) - BK_1(A - BK_{\text{mpc}})^p x_1(0) \tag{5.92}$$

Hence, for given $x_2(0)$,

$$x_2(p) = (A - BK_{\text{mpc}})^p x_2(0)$$
$$-\sum_{i=0}^{p-1}(A - BK_{\text{mpc}})^{p-i-1}BK_1(A - BK_{\text{mpc}})^i x_1(0) \tag{5.93}$$

and

$$\|x_2(p)\| \leq M\lambda^p \|x_2(0)\| + \sum_{i=0}^{p-1} M\lambda^{p-i-1}\|BK_1\|M\lambda^i\|x_1(0)\|$$
$$= M\lambda^p \|x_2(0)\| + M^2 p\lambda^{p-1}\|BK_1\|\|x_1(0)\| \tag{5.94}$$

By an inductive argument, for trial k,

$$\|x_k(p)\| \leq M\lambda^p \|x_k(0)\| + M^2 p\lambda^{p-1}\|BK_1\|\|x_{k-1}(0)\|$$
$$+ M^2 p\lambda^{p-1}\|BK_2\|\|x_{k-2}(0)\| + \cdots + M^2 p\lambda^{p-1}\|BK_{k-1}\|\|x_1(0)\| \tag{5.95}$$

where the matrices $K_1, K_2, \ldots, K_{k-1}$ are bounded from the predictive ILC design. Hence, the dynamics along each trial are influenced by the initial state vector on this and all previous trials.

In (5.95), the contributions from the previous trial initial state vectors are critical, and their influence decreases as λ decreases. One design approach is to select a value for λ and place the eigenvalues of $A - BK_{\text{mpc}}$ inside a circle with this radius. Such a design also regulates the transient dynamics along the trial by regulating the maximum magnitude that can arise in the response (i.e. in p). The choice of λ is application-specific, depending on the form of the required trial dynamics.

One way to complete this last design exercise for the case where the eigenvalues are to be inside a circle of radius λ is the following procedure [271]:

1. For the selected $0 < \lambda < 1$, solve the following modified version of the Riccati equation (5.85) for given weighting matrices $Q > 0$ and $R > 0$

$$\frac{A^T}{\lambda}P_\infty\frac{A}{\lambda} - \frac{A^T}{\lambda}P_\infty\frac{B}{\lambda}\left(R + \frac{B^T}{\lambda}P_\infty\frac{B}{\lambda}\right)^{-1}\frac{B^T}{\lambda}P_\infty\frac{A}{\lambda} + Q - P_\infty = 0 \tag{5.96}$$

2. Select $\mu > 1$ such that the matrix $\mu^{-1}A$ has all eigenvalues inside the unit circle in the complex plane and compute

$$\gamma = \frac{\lambda}{\mu}, \quad Q_\mu = \gamma^2 Q + (1 - \gamma^2)P_\infty, \quad R_\mu = \gamma^2 R \tag{5.97}$$

Table 5.3 Design parameters.

Laguerre pole a_1 $(= a_2)$	0.6
Laguerre term N_1 $(= N_2)$	8
Prediction horizon for each output N_p	100
Weighting matrix Q	$C^T C$
Weighting matrix R	I
μ	1.1
λ	0.7

3. For the selected γ from the last step, replace the state-space model matrices A and B, respectively, by $\mu^{-1} A$ and $\mu^{-1} B$. Then complete the design with the cost function:

$$J = \sum_{m=1}^{N_p} x_{k+1}(p + m \mid p)^T Q_\mu x_{k+1}(p + m \mid p)$$
$$+ \sum_{m=0}^{N_p} (u_{k+1}^s(m) - u_k^s(m))^T R_\mu (u_{k+1}^s(m) - u_k^s(m)) \qquad (5.98)$$

using the Q_μ and R_μ of the previous step and a sufficiently large N_p.

5.5.2 Experimental Validation

Consider the two-input, two-output anthromorphic robot arm of Section 2.1.2, where the robot's task is represented in joint space by the r_1 and r_2 reference signals shown in Figure 5.9. The parameters for the control design are given in Table 5.3, and the Riccati equation-based procedure of Section 5.5.1 was used with $\lambda = 0.7$ to place all poles of $A - BK_{\mathrm{mpc}}$ inside a circle of radius 0.7. The state observer has been designed as detailed in the previous section with a gain of 100.

Figure 5.9 shows the reference, output, and error signals for trials $k = 1, 2, 5$ and confirms that the controlled outputs closely track their reference signals for all three trials. The control signals for the three trials are shown in Figure 5.10, and the sum of the squared errors, i.e.

$$E_k = \sum_p (r(p) - y_k(p))^2$$

for the first five trials also confirms close tracking of the reference signals for each trial as shown in Figure 5.11. Given the existence of measurement noise in the system and the magnitudes of the errors in Figure 5.11, it is reasonable to conclude that the controlled performance is close to optimal for each trial.

Figure 5.12 shows the reference, output, and error signals for the case where Laguerre term N_1 $(= N_2)$ is reduced to 6. It can be seen that this increases output oscillation of both y_1 and y_2, leading to the increased error norms shown in Figure 5.13.

To examine the effect of increasing the weighting on the difference between successive control signals in (5.69), weighting matrix R is modified to $10 \times I$. The Laguerre term N_1 $(= N_2)$ is returned to its original value of 8. Figure 5.14 shows the reference, output, and error signals in this case

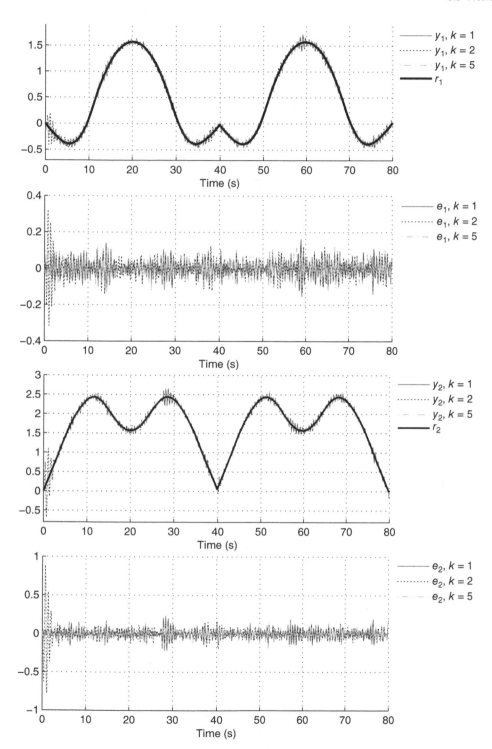

Figure 5.9 Reference, output, and error signals for $k = 1, 2, 5$.

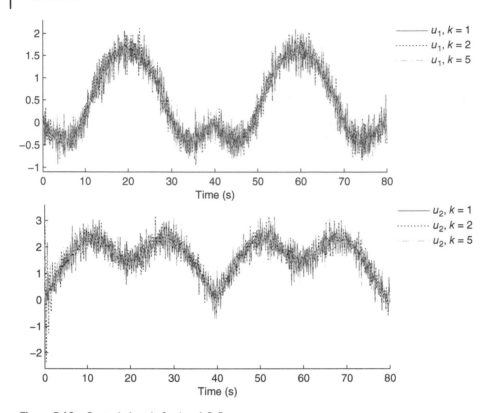

Figure 5.10 Control signals for $k = 1, 2, 5$.

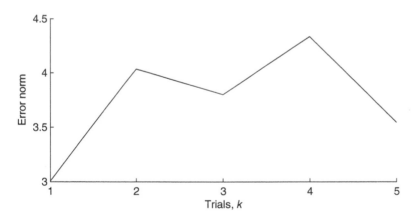

Figure 5.11 E_k for $k = 1, 2, 3, 4, 5$.

and confirms that oscillations on outputs y_1 and y_2 are reduced compared with those of Figure 5.9. This improvement in performance is reflected by the reduced error norms shown in Figure 5.15.

It is also of interest to compare the new design's performance with the NOILC design of Section 5.1. Figures 5.16–5.19 show the experimental results obtained when Q and R correspond to those used in the initial predictive ILC test (i.e. $R = I$, $Q = C^T C$). The NOILC trial-to-trial error convergence is much slower, and there are significant disturbances on the output and input signals.

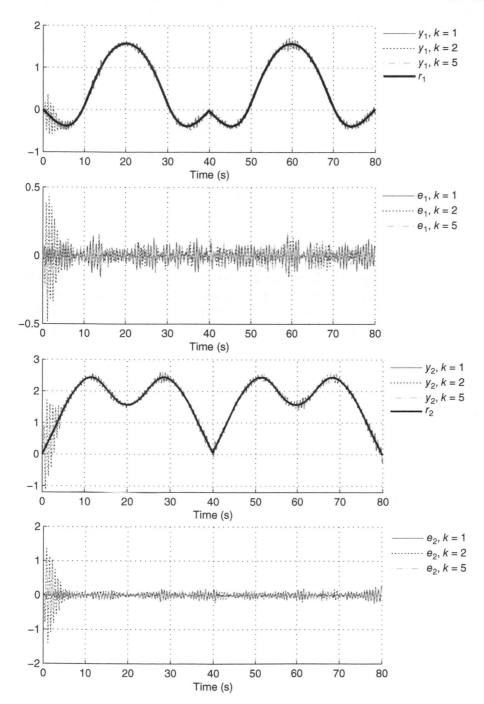

Figure 5.12 Reference, output, and error signals for $k = 1, 2, 5$ using $N_1 = N_2 = 6$.

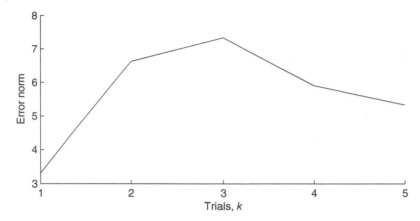

Figure 5.13 E_k for $k = 1, 2, 3, 4, 5$ using $N_1 = N_2 = 6$.

Finally, the NOILC weighting matrix Q is chosen as $Q = 100 \times C^T C$ to provide a comparison with a robust-inverse-based design, see (also the next chapter for further treatment of this latter design). Figures 5.20–5.23 show results for this case. It is evident that error norm convergence speed has increased but is still far lower than the predictive ILC design developed in the previous section. Tests with other choices of Q have also been performed. They show no increase in convergence speed, confirming that modeling error and disturbance degrade the nominal performance properties of inverse-based ILC designs. Hence, these results demonstrate that the design tested in this section can outperform a NOILC design in at least some cases.

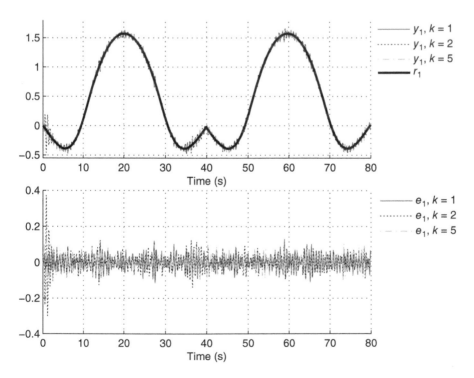

Figure 5.14   Reference, output, and error signals for $k = 1, 2, 5$ using $N_1 = N_2 = 8$ and $R = 10\,l$.

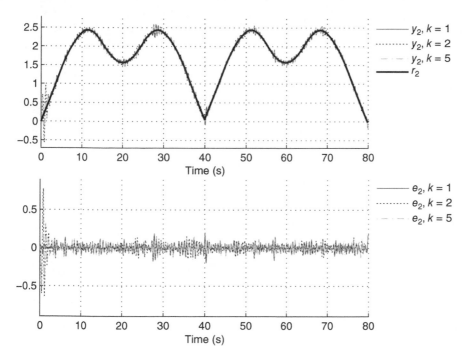

Figure 5.14 (*Continued*)

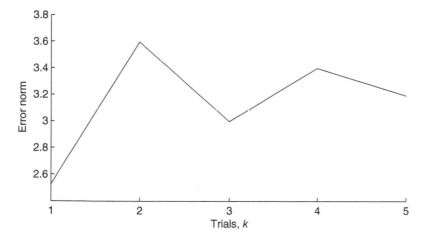

Figure 5.15 Sum of the squared error signals of $k = 1, 2, 3, 4, 5$ using $N_1 = N_2 = 8$ and $R = 10\,I$.

As a comparison with a tuning based design, a phase-lead ILC design has been experimentally tested, i.e. the input for each channel on trial $k + 1$ is constructed as follows:

$$u_{k+1}(p) = u_k(p) + \beta e_k(p + \lambda) \tag{5.99}$$

where β is the gain and $\lambda > 0$ is the phase-lead. A series of experiments with different phase-leads were completed, and the choice of $\lambda = 3$ gave the best results, see Figure 5.24. Overall, these results show that (i) the lowest possible error norm for any gain or phase-lead value is 1.72 and (ii) faster convergence is not possible as instability occurs earlier for the higher gain (i.e. β) required. These

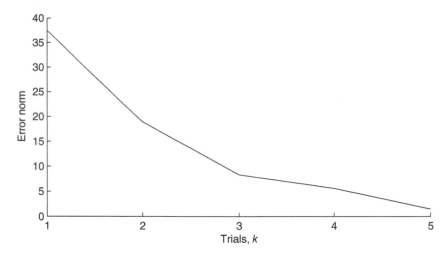

Figure 5.16 Error norm for the NOILC design with $R = I$ and $Q = 100\, C^T C$.

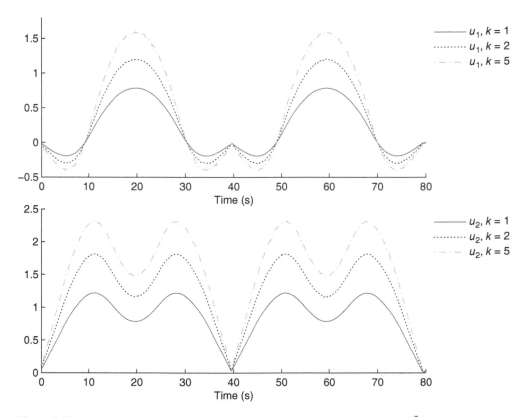

Figure 5.17 Control input signals for the NOILC design with $R = I$ and $Q = 100\, C^T C$.

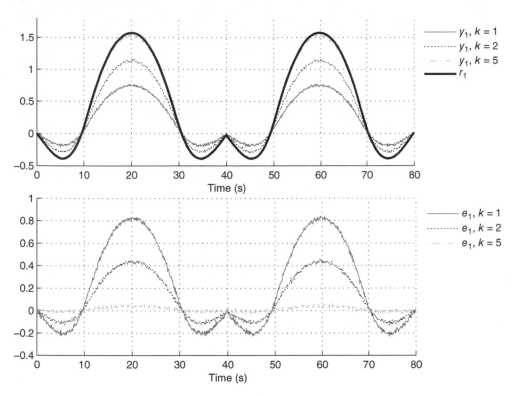

Figure 5.18 Reference, output, and error signals for output y_1 in the NOILC design with $R = I$ and $Q = C^T C$.

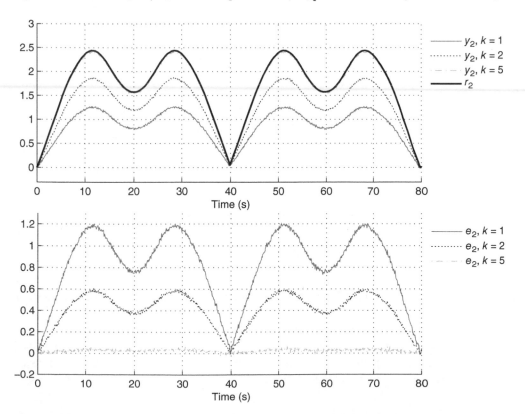

Figure 5.19 Reference, output, and error signals for output y_2 in the NOILC design with $R = I$ and $Q = C^T C$.

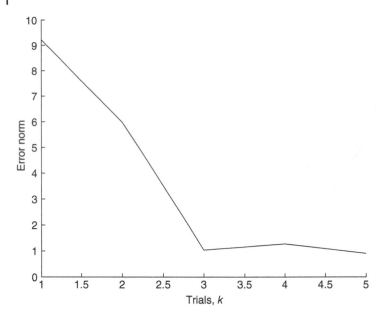

Figure 5.20 Error norm for the NOILC design with $R = I$ and $Q = 100\,C^T C$.

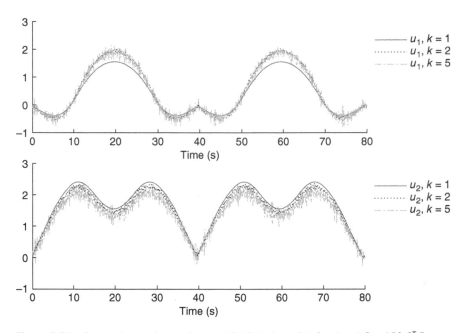

Figure 5.21 Control input signals for the NOILC design with $R = I$ and $Q = 100\,C^T C$.

results are expected given the relatively simple control law structure and the dynamics of the system considered.

The basic premise of this last design is that the reference signal will have a finite number of dominant frequencies in many cases, and it suffices to enforce tracking of these frequencies instead of the complete frequency spectrum. This design can be interpreted as selecting basis functions to approximate the reference signal, and there has been other work on such approaches for ILC, see,

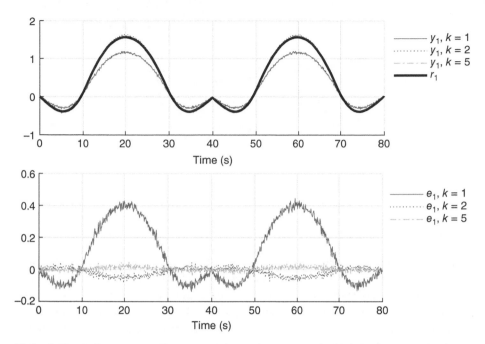

Figure 5.22 Reference, output, and error signals for output y_1 in the NOILC design with $R = I$ and $Q = 100\,C^T C$.

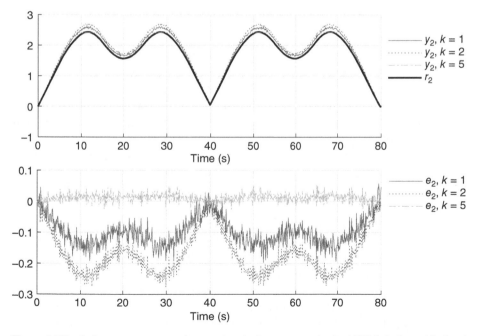

Figure 5.23 Reference, output, and error signals for output y_2 in the NOILC design with $R = I$ and $Q = 100\,C^T C$.

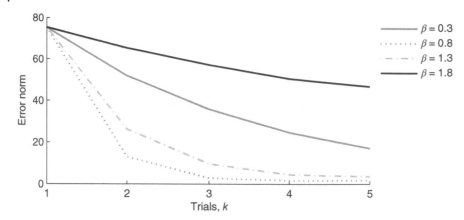

Figure 5.24 Error norm for the phase-lead ILC design (5.99) with $\lambda = 3$ and various β gains.

e.g. [102, 246, 261]. In [261] the problem considered is that the learned command signal is optimal for the specific fixed task only and, in general, extrapolation of the learned command signal to other tasks leads to significant performance deterioration. Basis functions are used to enhance the extrapolation to a class of reference signals. The approach in [102, 246] is to restrict the input/output space to an appropriate finite-dimensional space spanned by basis functions derived from the reference signal. These are valid alternatives, and detailed comparative studies of the merits of these two design settings are required.

5.6 Concluding Remarks

Optimization-based ILC design is a very well-developed area both in terms of theory and applications. Wide-ranging coverage of this area can be found in [186]. The role of NOILC for time-varying systems subject to input constraints is considered in Chapter 8, with experimental validation on the rack feeder system of Section 2.2.3. The particular case of terminal ILC, one form of point-to-point ILC, is considered in Chapter 13 with experimental results from an application to broiler chicken growth. Nonlinear optimization-based ILC is considered in Section 10.5.

Time-varying NOILC arises in the further development of the stroke rehabilitation results given in Chapter 4 to 3D tasks, see Section 11.4. In some applications, the trade-off between using a tuning-based design as opposed to a model-based design arises, i.e. a tuning-based design can achieve, or better, the performance obtained from a model-based design method. This area is of considerable interest in applications, and the rehabilitation application gives one supporting example.

In this last application area, the performance gained using NOILC cannot exceed that of the simpler ILC methods for overall convergence over the 10 trials, despite its more complex structure. This advantage is not present for the 3D tasks, see Section 11.4, and the choice of which approach, tuning, or model-based designs, is application dependent. Learning along the trials in ILC has been considered in Section 5.5 with experimental support. This approach also arises in nonlinear ILC based on an adaptive Lyapunov approach in Section 10.4. The next chapter considers robust ILC.

6

Robust ILC

Robust design to account for model uncertainty and other undesirable performance limitations, such as the effects of disturbances, is as relevant to iterative learning control (ILC) as other areas. In this chapter, robust control based on a linear approximate or nominal model is considered with both the frequency and time domain analysis, starting from the preliminary discussion in Chapter 3. Robustness of the simple structure inverse and adjoint ILC laws is considered first with supporting experimental results, and then the use of an H_∞ setting is considered.

6.1 Robust Inverse Model-Based ILC

The coverage in this section follows, in the main, [104] where for discrete linear time-invariant dynamics, it is assumed in the single-input single-output (SISO) case, with an obvious generalization to multiple-input multiple-output (MIMO), that the system transfer-function $G(z) = C(zI - A)^{-1}B$ has relative degree h^* (i.e. a pole-zero excess of h^*) and hence, the associated Markov parameters satisfy $CA^{j-1}B = 0$, $1 \leq j \leq h^* - 1$ and $CA^{h^*-1}B \neq 0$. Also, assume that the reference signal satisfies $r(j) = CA^j x_0$ for $0 \leq j < h^*$. Moreover, the case of $h^* > 1$ is a natural generalization of $h^* = 1$ and hence, the latter is considered in this section.

The dynamics are considered in the lifted form of (1.37), i.e.

$$Y_k = GU_k + d \tag{6.1}$$

Also, without loss of generality, the state initial vector on each trial is taken to be the zero vector and, hence, $d = 0$. Moreover, the assumption that $CB \neq 0$ ensures that G is invertible and the inverse ILC law in the lifted representation has the form:

$$U_{k+1} = U_k + \beta G^{-1}E_k \tag{6.2}$$

where β is a real scalar.

Remark 6.1 In this and Section 6.2, the dependence on p is omitted from the relevant equations for ease of notation (except where it is required for clarity).

Suppose that the arbitrary choice of initial control input U_0 produces an initial error E_0, and then in the case when $\beta = 1$

$$E_1 = R - GU_1 = R - G\left(U_0 + G^{-1}E_0\right) = 0 \tag{6.3}$$

If $\beta \neq 1$, then $E_k = (1 - \beta)^k E_0$, and hence, trial-to-trial error convergence to zero is guaranteed for $0 < \beta < 2$.

Iterative Learning Control Algorithms and Experimental Benchmarking, First Edition.
Eric Rogers, Bing Chu, Christopher Freeman and Paul Lewin.
© 2023 John Wiley & Sons Ltd. Published 2023 by John Wiley & Sons Ltd.

Consider the case where a nominal model of the system to be controlled, denoted by G_0, is available, including when this nominal model has been chosen to reduce the computational load for implementation. Suppose also that the relative degree of the system and its nominal model, i.e. G and G_0, respectively, is the same, and then for the ILC law,

$$U_{k+1} = U_k + \beta G_0^{-1} E_k \tag{6.4}$$

performance can be analyzed using the following description of the trial-to-trial error updating:

$$E_{k+1} = \left(I - \beta G G_0^{-1}\right) E_k \tag{6.5}$$

Trial-to-trial error convergence depends on the matrix $\beta G G_0^{-1}$, and for stability, the relationship between G and G_0 and also β is critical. It is also routine to show that trial-to-trial error convergence occurs if and only if

$$\rho\left(I - \beta G G_0^{-1}\right) < 1 \tag{6.6}$$

or, equivalently, all eigenvalues of $G G_0^{-1}$ lie in the open right-half of the complex plane, and $\beta > 0$ is sufficiently small.

The robustness of this ILC law is considered under the multiplicative uncertainty description, i.e. $G = U G_0$, where U is a square matrix. The basic questions to be answered are the following: (i) What is the maximum level of uncertainty possible for a given value of β? (ii) What is the range of β that can be used for a given uncertainty? When U represents a proper causal linear time-invariant (LTI) system, this matrix must have a similar structure to G. This, in turn, requires that the entries satisfy $U_{i,j} = U_{i+1,j+1}$, $1 \leq i,j \leq N-1$. This form of uncertainty is well established in the theory and application of LTI systems, see, e.g. [243].

In the transfer-function domain, suppose that $G(z)$ has relative degree greater than or equal to that of $G_0(z)$ and that $G(z) = U(z)G_0(z)$. Then if G, G_0, and U are the corresponding matrix representations of these systems, $G = U G_0$, where the relative degree assumption ensures that U is causal. The basic requirements are trial-to-trial error convergence to zero independent of the initial control input and monotonicity of the error's Euclidean norm (or mean squared value).

Using (6.5)

$$\|E_{k+1}\|_2^2 - \|E_k\|_2^2 = -2\beta E_k^T U E_k + \beta^2 E_k^T U^T U E_k \tag{6.7}$$

Suppose that $U + U^T > 0$ and $\beta > 0$ and hence, for an arbitrary nonzero E_k, the two terms on the right-hand side of this last equation are, respectively, negative and positive. Hence, $\|E_{k+1}\|^2 < \|E_k\|^2$ if and only if

$$2\beta E_k^T U E_k > \beta^2 E_k^T U^T U E_k \tag{6.8}$$

for all nonzero E_k. Also, since the left- and right-hand sides in (6.8) are of the order β and β^2, respectively, it follows that this inequality holds for all sufficiently small $\beta > 0$. Consequently, there exists a value of β, say β^*, independent of E_k, such that monotonic trial-to-trial error convergence occurs for $0 < \beta < \beta^*$. Moreover, routine algebraic manipulations applied to (6.7) establish that any β such that $0 < \beta < \beta^*$ satisfies

$$\left(\frac{1}{\beta}I - U\right)^T \left(\frac{1}{\beta}I - U\right) < \frac{1}{\beta^2}I \tag{6.9}$$

This last condition (Proposition 3 in [104]) combined with $U + U^T > 0$ shows that trial-to-trial error convergence is monotonic independent of the detailed structure of the model uncertainty U. The next two results (Propositions 4 and 5, respectively, in [104]) establish that, under the conditions associated with (6.9), the error sequence $\{E_k\}_k$ converges geometrically to zero for all E_0.

Lemma 6.1 *Under the assumptions made in establishing (6.9) and $0 < \beta < \beta^*$, the tracking error sequence $\{E_k\}_{k \geq 0}$ converges monotonically to zero.*

Lemma 6.2 *If $0 < \beta < \beta^*$ and $U + U^T > 0$, then there exists a real scalar $\gamma \in (0,1)$, which depends on β only, such that $\|E_{k+1}\| \leq \gamma \|E_k\|, \forall k \geq 0$.*

It is also possible to give a transfer-function interpretation of the convergence result for the inverse ILC law. This interpretation is based on the underlying relationship between the matrix U and the transfer-function, $U(z)$, of the LTI system it represents. The following result is Proposition 6 in [104], where Re denotes the real part of a complex number.

Lemma 6.3 *Suppose that $U(z)$ is a stable system. Then*

(a) $U + U^T > 0$ if $\mathrm{Re}(U(z)) > 0, \forall |z| = 1$
(b) monotonic trial-to-trial error convergence occurs if

$$\left| \frac{1}{\beta} - U(z) \right| < \frac{1}{\beta}, \quad \forall |z| = 1 \tag{6.10}$$

As one interpretation of this last result, suppose that the Nyquist plot of $U(z)$ is constructed. Then it is sufficient for a monotonic trial-to-trial error convergence that this plot lies within a circle in the complex plane of radius $\frac{1}{\beta}$ and center $(\frac{1}{\beta}, 0)$. As $\beta \to 0+$, this circle expands to include the whole of the complex plane. This case demonstrates that robust monotonic convergence is possible for any strictly positive uncertainty $U(z)$ by using a sufficiently small $\beta > 0$.

The condition (6.10) requires that the modeling error $U(z)$ must be positive real, and hence, the associated phase shift must lie within $\pm 90°$ for all $|z| = 1$. Consequently, this ILC law can tolerate a system uncertainty of $\pm 90°$, but the gain tolerance is phase-dependent. An example to illustrate this analysis is given in [104], which also develops other extensions.

The analysis extends to adaptive parameter optimal ILC, and refers to Section 5.4, with

$$U_{k+1} = U_k + \beta_{k+1} F_k \tag{6.11}$$

where the variable β_{k+1} is chosen to minimize the cost function:

$$\min_{\beta_{k+1}} J(\beta_{k+1}) = \|E_{k+1}\|^2 + w_{k+1} \beta_{k+1}^2 \tag{6.12}$$

and

$$w_{k+1} = w_1 + w_2 \|E_k\|^2, \quad w_1 \geq 0, \quad w_2 \geq 0, \quad w_1 + w_2 > 0 \tag{6.13}$$

Also, monotonic trial-to-trial error convergence to zero occurs for a positive real system G and the convergence is geometric if $w_1 = 0$.

Consider an ILC law of the form:

$$U_{k+1} = U_k + \beta_{k+1} G_0^{-1} E_k \tag{6.14}$$

which is an inverse law in the parameter optimal ILC setting. The error evolution is given by

$$E_{k+1} = \left(I - \beta_{k+1} G G_0^{-1} \right) E_k \tag{6.15}$$

with β_{k+1} chosen to minimize (6.12). In the case of no modeling error, i.e. $G = G_0$, the solution of this optimization problem is

$$\beta_{k+1} = \frac{\|E_k\|^2}{w_{k+1} + \|E_k\|^2} \tag{6.16}$$

and if $w_1 = 0$, $0 < \beta_{k+1} = \frac{1}{1+w_2} < 1$ is a constant, and the previous case is recovered. Hence, $w_1 > 0$ is assumed from this point onward.

The following result, Proposition 9 in [104], establishes when monotonic trial-to-trial error convergence occurs for the ILC law considered.

Lemma 6.4 *Suppose that the ILC law (6.14) is applied in the case when $G = G_0$ and $w_1 > 0$ and then $\|E_{k+1}\| < \|E_k\|$ if $E_k \neq 0$ and*

$$\lim_{k \to \infty} \|E_k\| = 0 \quad \text{and} \quad \lim_{k \to \infty} \beta_{k+1} = 0 \tag{6.17}$$

In the case when $G \neq G_0$, i.e. modeling errors are present, the following result is established as Proposition 10 in [104].

Lemma 6.5 *Suppose that $U + U^T > 0$., then $\|E_{k+1}\|^2 < \|E_k\|^2$ if $E_k \neq 0$ and*

$$\left(\frac{1}{\beta_{k+1}} I - U \right)^T \left(\frac{1}{\beta_{k+1}} I - U \right) < \frac{1}{\beta_{k+1}^2} I \tag{6.18}$$

Also, if this inequality holds for $k = 0$, it holds $\forall k > 0$ and the error sequence $\{E_k\}_k$ converges to zero.

This result shows that a smaller $\|E_0\|$ (and hence β_1) increases the monotonic convergence robustness of this ILC law. Thus, the first trial is critical, and therefore, a design strategy based on the application of a high-quality feedback loop with good tracking performance reduces E_0.

The frequency-domain equivalent of this last result is stated next.

Lemma 6.6 *Suppose that $U(z)$ is stable and $w_1 > 0$, and then the parameter optimal ILC law considered has the robust monotonic trial-to-trial convergence property if*

$$\sup_{|z|=1} \left| \frac{1}{\beta_1} - U(z) \right| < \frac{1}{\beta_1} \tag{6.19}$$

where

$$\beta_1 = \frac{\|E_0\|^2}{w_1 + w_2 \|E_0\|^2 + \|E_0\|^2} \tag{6.20}$$

This result shows that an increase in the accuracy of U_0 increases robustness. Also, any trial can be regarded as the initial for subsequent ones. Hence, the result can be expressed in terms of any β_{k+1}, for some $k \geq 0$. In particular, the design becomes more robust as the trials evolve and β_{k+1} reduces to zero.

The model inverse optimal ILC law has been implemented [219] on all three axes of the gantry robot test facility of Section 2.1.1, using the cost function (6.12) with the same weighting on each trial, i.e. $w_{k+1} = w$. For each test, the input signal to the first trial was set to zero, and hence, the ILC law must learn the required input from zero. Figures 6.1–6.3 show the mean square error (MSE) (on a log scale) against trial number data for each axis when $w = 0, 0.0001, 0.001, 0.01, 0.1$, and 1 (where this quantity on any trial is the sum of the squared error values over the trial length).

These experimental results confirm that this ILC law can improve tracking accuracy for all values of w. Figure 6.4 shows the end effector's three-dimensional motion as it follows the reference trajectory (dots). For trial 1, there is no motion as the input is set to zero, and once two trials are complete, the error is minimal. On completion of the next trial, the tracking of the reference trajectory is excellent.

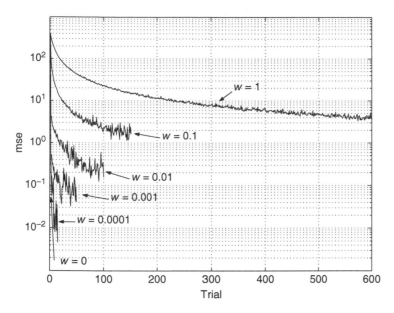

Figure 6.1 X-axis MSE for various w.

If w is varied, the results can be significantly different, where for a larger value of w, the convergence rate is significantly slower. This effect is expected as increasing w forces the learning gain to be smaller and produces more cautious adjustment to the system input. For certain values of w, particularly for the X-axis (Figure 6.1) the MSE plot ends after a few trials. In these tests, the ILC law begins to diverge very quickly, and following an initial reduction, the MSE then begins to increase. This feature is most pronounced for the Z-axis when $w = 0.0001$ (Figure 6.3).

As the ILC law diverges, the system's vibrations steadily grow until the system must be switched off to prevent damaging the robot. It is generally accepted that high-frequency noise has a destabilizing effect on any ILC law. In cases where explicit action is required zero-phase filtering of the previous trial error signal may limit this problem's effects, but if not then a stochastic ILC design is required, where this topic is the subject of Chapter 12.

The inverse ILC law considered in this section with $0 < \beta < 1$ gives excellent results for the case of no model uncertainty. Again, considering the SISO case, write $G(z) = U(z)G_0(z)$, where $U(z)$ is the model uncertainty and related to multiplicative modeling error $\Delta(z)$ via $U(z) = 1 + \Delta(z)$. The nominal value is $U(z) \equiv 1$, corresponding to $\Delta(z) \equiv 0$. Consider also the condition for robust monotonic trial-to-trial error convergence in the case of multiplicative uncertainty, i.e. (6.10) holds, and the following general points arise:

(i) If significant high-frequency errors are present, only small values of β are allowed, and hence, slow trial-to-trial error convergence will result.

(ii) The phase of the uncertainty must lie in the range $(-\frac{\pi}{2}, \frac{\pi}{2})$, which constrains the form of uncertainty that can be allowed. This requirement arises from the monotonic trial-to-trial error convergence requirement and is equivalent to strict positive-realness of $U(z)$.

(iii) $U(z)$ is positive-real if $\|\Delta(z)\|_\infty < 1$. For a more general case, let γ be a real scalar and write $U(z) = \gamma + \Delta_\gamma(z)$. Then $U(z)$ has the strictly positive real property if $\|\Delta_\gamma(z)\|_\infty < \gamma$. Also, γ should ideally be chosen close to the center of the uncertainty set for $U(z)$.

(iv) In general, this last condition must be checked against the known uncertainty set to determine if it holds or not. The simplest form of this set is when $U(z)$ is not known, but it is known to

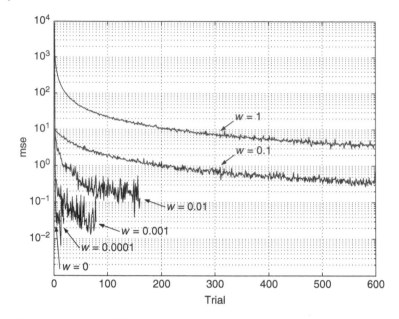

Figure 6.2 *Y*-axis MSE for various *w*.

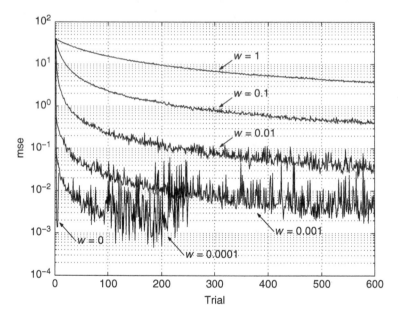

Figure 6.3 *Z*-axis MSE for various *w*.

belong to a set characterized by an inequality of the form

$$\left| \frac{1}{\beta^*} - U(z) \right| < \frac{1}{\beta^*}, \quad \forall \, |z| = 1 \tag{6.21}$$

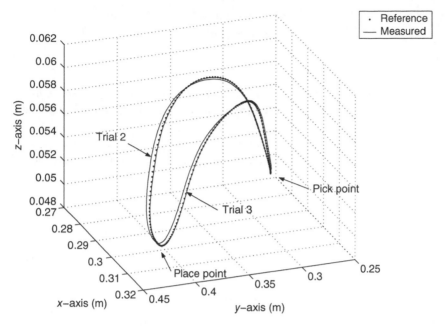

Figure 6.4 3D tracking. $w = 0$.

and robust monotonic trial-to-trial error convergence is guaranteed for all β such that $0 < \beta < \beta^*$. This set is the maximal for the uncertainty that can be tolerated for robust monotonic trial-to-trial error convergence with a specified β^*.

6.2   Robust Gradient-Based ILC

Gradient-based ILC was considered for continuous time systems in [87] and for discrete time systems in [189] and [220]. This section primarily follows [189] and [220], where the latter includes experimental validation.

Consider the gradient-based ILC law

$$U_{k+1} = U_k + \beta G^T E_k \tag{6.22}$$

where again G is the matrix in the lifted form of the system dynamics. Then consideration of $\|E_k\|^2 = E_k^T E_k$ gives the steepest descent direction as $G^T E_k$ and, hence, the possibility that (6.22) could ensure robust monotonic trial-to-trial error convergence. Moreover, $G^T E_k$ can be computed from a state-space model of G, see [189] for the background. Hence, the lifted form is not needed for implementation.

Under the ILC law (6.22),

$$E_{k+1} = (I - \beta GG^T)E_k, \quad k \geq 0 \tag{6.23}$$

Also, since $\beta > 0$ by assumption, and

$$\|E_{k+1}\|^2 = \|E_k\|^2 - 2\beta E_k^T GG^T E_k + \beta^2 E_k^T GG^T GG^T E_k \tag{6.24}$$

the following result, proved as Theorem 1 in [189], follows since G is nonsingular by construction (if the relative degree is greater than unity then a modification is needed before constructing the lifted model, see [189, 220]).

Theorem 6.1 *For $\beta > 0$, the ILC law (6.22) has the monotonic and trial-to-trial error convergence properties, i.e. $\|E_{k+1}\| < \|E_k\|$ and $\lim_{k\to\infty} E_k = 0$ in some range, $0 < \beta < \beta'$, and for all E_0, provided*

$$2I > \beta G^T G > 0 \qquad (6.25)$$

An immediate consequence of this condition is that robust monotonic trial-to-trial error convergence to zero holds if and only if $0 < \beta \bar{\sigma}^2(G) < 2$, where $\bar{\sigma}(\cdot)$ denotes the maximum singular value of its matrix argument.

Under the multiplicative uncertainty structure considered in Section 6.1, the following result, proved as Theorem 2 in [189], characterizes robust monotonic trial-to-trial error convergence of the gradient-based ILC law (6.22).

Theorem 6.2 *The ILC law (6.22) gives robust monotonic trial-to-trial error convergence in the presence of multiplicative modeling error $U(z)$ with lifted representation U if and only if*

$$U + U^T > \beta G_0^T U^T U G_0 > 0 \qquad (6.26)$$

A necessary condition for this last property is that $U + U^T > 0$.

The following result, proved as the Theorem on page 645 of [189], on robust monotonic trial-to-trial error convergence can now be stated.

Theorem 6.3 *The gradient-based ILC law for dynamics with nominal model $G_0(z)$ and multiplicative modeling error with transfer-function $U(z)$ results in robust monotonic trial-to-trial error convergence if*

$$\left| \frac{1}{\beta} - |G_0(z)|^2 U(z) \right| < \frac{1}{\beta}, \quad \forall |z| = 1 \qquad (6.27)$$

The difference, with the inverse ILC law robustness result of Section 6.1, is the replacement of $U(z)$ by $|G_0(z)|^2 U(z)$ and the following points on the relative merits of these two laws are relevant:

(i) Both laws require a strictly positive real $U(z)$ for robust monotonic trial-to-trial error convergence. This condition is closely connected to the monotonicity property of the MSE. If this property does not hold, then stability and convergence for both laws may not hold.

(ii) For both laws, the positive-real requirement on $U(z)$ most often requires that G and G_0 have the same relative degree.

(iii) It is required to check $U(z)$ for positive realness, where, e.g. if $U(z) = \gamma + \Delta_\gamma$, where Δ_γ is the multiplicative modeling error, a sufficient condition for this property is that $\|\Delta_\gamma\|_\infty < \gamma$.

(iv) Checking this last condition for all systems within the uncertainty set will depend on its structure. For the uncertainty set considered in the last point, i.e. the set of $U(z)$ satisfying $\|\Delta_\gamma\|_\infty < \gamma$, it remains to check that

$$\left| \frac{1}{\beta} - |G_0(z)|^2 \gamma_0 \right| < \frac{1}{\beta}, \quad \forall |z| = 1 \qquad (6.28)$$

and for all γ_0 in the ball in the complex plane with center γ and radius $\|\Delta_\gamma\|_\infty$. This requirement has an obvious graphical interpretation.

(v) The gradient-based ILC law will reduce the performance limitations resulting from high-frequency errors in G not modeled in G_0, e.g. high-frequency resonances. In such a case, $U(z)$ will often have large gain values at frequencies close to any resonant frequency. These will then mandate the use of small values of β in the inverse law. Such a behavior does not arise for the gradient law as often in applications G will have low-pass filter characteristics. This fact, in turn, means that $G(z)$ and G_0 will have small magnitudes at high frequencies. Hence, the magnitude of $|G_0|^2 U$ will be substantially reduced compared to U and permit increased learning gains and hence, improved convergence rates.

(vi) The previous point's beneficial effects could be lost (at least partially) if G and hence, G_0 has a substantial resonant peak within its bandwidth. Therefore, the allowable learning gains will be reduced relative to the inverse law. A feedback control loop should be applied to the system and ILC designed for the resulting dynamics to limit this effect. The feedback control law design should aim to remove or at least reduce the resonant peak. If successful, this should give superior performance and robustness for the gradient instead of the inverse law.

An extension of the above analysis and discussion to multiple uncertainties is possible [189]. This characterization in terms of inequalities between matrix representations of the system and associated uncertainty model into simple-frequency domain conditions for robustness transfers to this more general case. In the particular case when $U = 1$, i.e. no uncertainty, robust monotonic trial-to-trial error convergence to zero occurs if $0 < \beta \|G\|_\infty^2 < 2$.

It is also possible to use a gradient-based parameter optimal ILC law, and the analysis extends to the use of weighted norms. Consider the ILC law:

$$U_{k+1} = U_k + \epsilon_{k+1} G^T E_k \tag{6.29}$$

where ϵ_{k+1} is the solution of the optimization problem

$$\min_{\epsilon_{k+1}} J(\epsilon_{k+1}), \quad J(\epsilon_{k+1}) := \|E_{k+1}\|^2 + w\epsilon_{k+1}^2 \tag{6.30}$$

where $w \in \mathbb{R}$ and $w > 0$. In this cost function, the first term aims to make the tracking error small during each trial and the second to make the magnitude of ϵ_{k+1} small. The solution of the optimization problem (6.30) is

$$\epsilon_{k+1} = \frac{\|G^T E_k\|^2}{w + \|GG^T E_k\|^2} \tag{6.31}$$

and the following result, Proposition 2 in [220], characterizes the trial-to-trial error convergence properties of this law.

Lemma 6.7 *If $w > 0$ in (6.30), then $\|E_{k+1}\| \leq \|E_k\|$, where equality holds if and only if $\epsilon_{k+1} = 0$. Moreover,*

$$\lim_{k \to \infty} \|E_k\| = 0 \quad \text{and} \quad \lim_{k \to \infty} \epsilon_k = 0 \tag{6.32}$$

and hence, monotonic trial-to-trial error convergence to zero holds.

Using the optimal ϵ_{k+1} and $G = G_0 U$ gives, after routine manipulations,

$$\|E_{k+1}\|^2 - \|E_k\|^2 = -2\frac{\|G_0^T E_k\|^2}{w + \|G_0 G_0^T E_k\|^2} E_k^T G_0 \left(\frac{U + U^T}{2}\right) G_0^T E_k$$
$$+ \frac{\|G_0^T E_k\|^4}{(w + \|G_0 G_0^T E_k\|^2)^2} \|GG_0^T E_k\|^2 \tag{6.33}$$

and monotonic trial-to-trial error convergence requires $\|E_{k+1}\|^2 - \|E_k\|^2 < 0$ for nonzero E_k. The following result, Proposition 3 in [220], shows how this property can be achieved by choosing $w > 0$ as a sufficiently large number.

Lemma 6.8 *Assume that $U + U^T > 0$ and w is chosen such that*

$$w > \frac{1}{2}\frac{\|G_0^T\|^2\|GG_0^T\|^2\|E_0\|^2}{\underline{\sigma}(G_0\frac{(U+U^T)}{2}G_0^T)} \tag{6.34}$$

where $\underline{\sigma}(G_0(\frac{U+U^T}{2})G_0^T)$ is the smallest eigenvalue of $G_0(\frac{U+U^T}{2})G_0^T > 0$. In this case, $\|E_{k+1}\| < \|E_k\|$ when $E_k \neq 0$.

The estimate for w can be very conservative since the term $E_k^T G_0(\frac{U+U^T}{2})G_0^T E_k$ is estimated using the smallest eigenvalue of $G_0(\frac{U+U^T}{2})G_0^T$. Moreover, if w is selected to be excessively large, this choice can have an undesirable effect on the convergence speed because a large w will result in a small ϵ_{k+1}, implying that $u_{k+1} \approx u_k$. Consequently, the last result should be interpreted as an existence result and in practice w can be selected by a trial and improvement approach.

Using (6.34), the sufficient value of w decreases with $\|E_k\|^2$ and hence, the possibility to reduce w with each successive trial k. Moreover, a reduced w could result in an increase in the convergence speed and a natural choice to exploit this fact would be $w = w_1\|E_k\|^2$, but as $\|E_k\|^2$ approaches zero dangerously high inputs could be applied to the system. A simple alternative is to use $w = w_0 + w_1\|E_k\|^2$ – see [189] for the detailed analysis for this choice.

The following result, (Proposition 4 in [220]) shows that in addition to monotonic trial-to-trial convergence, $U^T + U > 0$ also implies that $\lim_{k\to\infty} E_k = 0$ if w is selected to be sufficiently large.

Lemma 6.9 *Under the assumptions of Lemma 6.8, the ILC design results in monotonic trial-to-trial error convergence to zero.*

In summary, the modified steepest-descent ILC law will converge monotonically to zero tracking error if the multiplicative uncertainty U satisfies $U^T + U > 0$ and w is chosen to be sufficiently large. In the standard law $w = 0$ and therefore, the introduction of w in the modified law has resulted in a straightforward mechanism to find a balance between convergence speed and robustness.

The remainder of this section gives the results of experimental application of this section's design to the gantry robot of Section 2.1.1. In the robust control case, it is also necessary to construct U ($G = G_0 U$) for the particular application under consideration, there is an extensive range of admissible G_0 and U, and here, the objective is to demonstrate the capabilities of this uncertainty structure by selecting these parameters to resemble choices informed by input from a practicing engineer. The following cases are considered in turn: (i) there is an error only in the zero frequency gain of the transfer-function constructed from measured frequency response data, (ii) a model is produced based only on the low-frequency characteristics, and (iii) the axis dynamics are represented by a pure integrator with a tunable gain.

In the case of (i), the consequences of this choice can be investigated by constructing G_0 starting from the measured Bode plots in the continuous-time domain and then exploring what happens when the positive real scalar U is varied. For (ii), a standard form of continuous-time model for a linear motor drive is $\frac{\hat{k}}{s(s+a)}$, where $\hat{k}$ and a are positive scalars. In physical terms, this approximation is equivalent to assuming that there is no interaction between the robot's axes, and all higher-frequency dynamics are ignored. The tuned models used for the three axes are

$$G_x(s) = \frac{11}{s(s + 200)} \tag{6.35}$$

$$G_y(s) = \frac{17}{s(s + 300)} \tag{6.36}$$

$$G_z(s) = \frac{10}{s(s + 400)} \tag{6.37}$$

Finally, in (iii) no account of the individual axis frequency responses is used in constructing G_0.

6.2.1 Model Uncertainty – Case (i)

Results for Case (i) with $w = 0$ are shown in Figure 6.5 where U is selected in the range between 0.75 and 1.25. Also, it follows immediately from (6.31) that if $U < 1$, then this will result in an increase in the learning gain, and consequently, performance will be severely degraded because the ILC law will overcompensate for the error on each trial. For example, Figure 6.6 shows the poor tracking performance that results from a zero frequency gain error of 0.5.

As described above, one option to restore a high trial-to-trial error convergence rate is to allow the adaptation of the magnitude of w as a function of the current level of tracking error, i.e. use

$$w_{k+1} = w_0 + w_1 \|E_k\|^2 \tag{6.38}$$

where w_0 and w_1 are two new tuning parameters, which must be appropriately selected to match the operation of the controlled system. Parameter w_0 can be used to guarantee long-term performance by setting a baseline magnitude for w_{k+1}, while $w_1 \|E_k\|^2$ adapts w_{k+1} to match the change in tracking error.

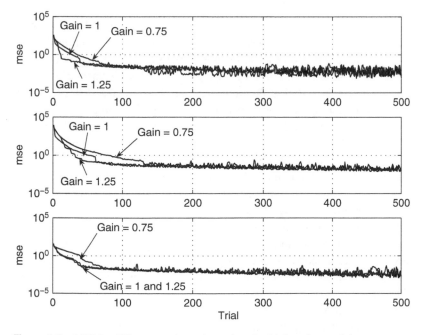

Figure 6.5 MSE for different scalar gains using the high-order model.

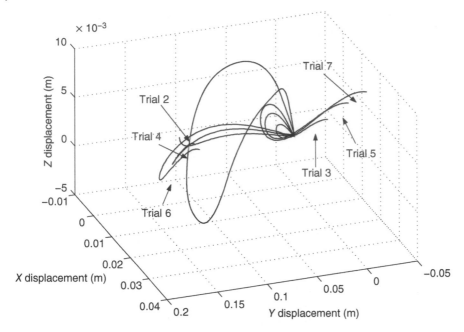

Figure 6.6 3D tracking progression $U = 0.5$ – high-gain model.

Figure 6.7 gives the X-axis displacement profiles for trials 490–500 with $U = 0.5$ and $w_{k+1} = 2 \times 10^{-7} \|E_k\|^2$. The adaptive variant of the gain tuning parameter has improved the convergence rate and the minimum tracking error.

6.2.2 Model Uncertainty – Cases (ii) and (iii)

For Case (ii), the representative log MSE results in Figure 6.8 demonstrate that the robustness structure assumed in this work can be used to design a controller based on a reduced-order model whose comparative performance is very close to that achieved with a full-order model. As expected, the difference is most noticeable for the X-axis, where the performance of the low-order model is noticeably worse than that of the high-order model. The X-axis has the most significant high-frequency dynamics and so is most affected by the simplification process.

The theory allows G_0 to be constructed from a continuous-time model of the form β/s, where β is the tunable gain, as in Case (iii). (This choice represents the effects of choosing any other continuous-time transfer-function whose phase lies between 0 and $-180°$.) Consider the transfer-function constructed in this way for a range of values of β from 0.01 to 1.0. Then Figure 6.9 compares, respectively, the convergence rate for β equal to 1.00, 0.05, and 0.03. These results confirm that a large value for β, reduces the rate of learning.

As β is reduced, the learning steadily becomes faster until an optimum β is reached. Increasing β any further then rapidly reduces the learning rate, and very quickly, the system performance degrades. The best values of β for each axis were found to be 0.05 for the X and Y-axes and 0.03 for the Z-axis. In all three cases, the zero frequency gain of the resulting simplified model is a close match to that of the higher-order model.

If β is too large, the ILC law will assume that a small change in controller effort will have a larger effect on the output and effectively reduces the convergence rate. If β is selected too small, large changes in the control signal will be needed to affect the output. Consequently, there will be

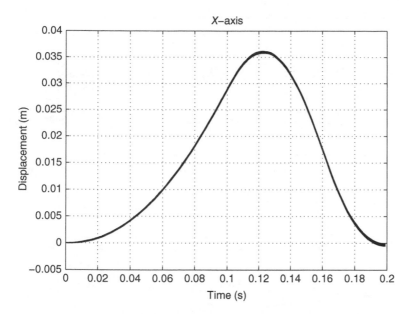

Figure 6.7 *X*-axis tracking performance for trials 490–500 ($w_{k+1} = 2 \times 10^{-7}\|E_k\|^2$, high-order model, $U = 0.5$).

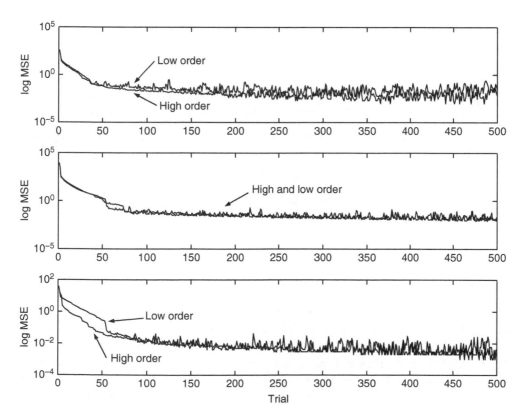

Figure 6.8 Comparison of high- and low-order system models, $w = 0$.

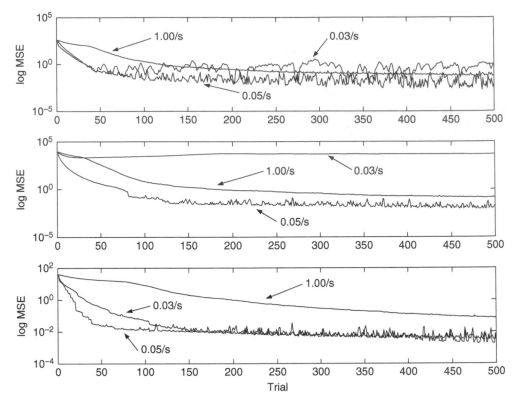

Figure 6.9 Comparison of the convergence rates for β/s.

an overcompensation for a given tracking error, and severely degraded performance will result. These conclusions suggest a straightforward tuning rule for the control law in this case.

1. Set β to a large value (if any form of system model is available, ensure that β generates a greater gain than that of the model at low frequency).
2. Operate the system and ensure that stability is achieved.
3. Reduce β.
4. Continue steps 2 and 3 until optimal convergence is achieved or until the system begins to degrade severely.
5. If the situation of the previous step occurs, increase β slightly.
6. The value of β is determined.

   To investigate the $\pm 90°$ stability boundary, the system model $1/s^2$ has been used to design and implement the control law on each axis. This model generates a phase shift of $-180°$ and at low frequency, each axis of the system has a phase shift of $-90°$. The system is therefore just on the theoretical stability boundary, and the gain of $1/s^2$ generates a conservative ILC law for each axis. Figure 6.10 shows the trajectory described by the X-axis from trial 1, where the trajectory is zero, to trial 201, in intervals of 20 trials and the thicker line denotes the desired reference trajectory. By trial 121, the axis initially travels in the opposite direction to what is required. This effect becomes steadily worse, and by trial 201, the rapid change of direction at the start of the trial is severe.

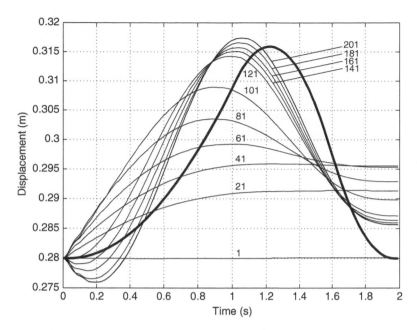

Figure 6.10 *X*-axis performance using model $1/s^2$.

The test was terminated at 203 trials to prevent damaging the gantry robot. Also, a growing oscillatory component of the displacement waveform occurs at time 0.1 seconds. This feature initially appears at trial 41 and is visible at trial 201. Small oscillations of this form can usually be seen in other ILC laws as their performance begins to degrade. These observations suggest that the use of a $1/s^2$ model will produce a poor control law as it introduces an additional phase shift of $-90°$ at low frequency.

The results above have only considered the case when $w = 0$, but the theory predicts that a nonzero w could increase the controller's robustness. The approximate model β/s with β selected too small is an example and is investigated further next.

Focusing on the Y-axis, consider first the case when $\beta = 0.2$. This is well below the best choice of 0.5 and causes the ILC law to degrade within the first few trials severely. For this model, a range of values of w was applied. When w is too small, performance does not improve. However, as w is increased, a point is reached at which improvement occurs but at the cost of a slow convergence rate. Increasing w further causes the convergence rate to slow even further.

Figure 6.11 shows the displacement recorded every 50 trials from 1 to 451 for the Y-axis with $w = 2.5 \times 10^{-7}$. If w is reduced to 2.0×10^{-7}, the gantry robot cannot operate for more than three trials. Therefore, the first choice here gives the better convergence rate without degrading performance but at the cost of a poor final tracking error bound. This outcome suggests that w becomes increasingly conservative as the error reduces and gradually prevents any learning from occurring. Consequently, w needs to be reduced in proportion to the error to prevent this unwanted feature from arising.

An alternative selection method has already been given in (6.38). If w_1 is nonzero, then w is linked to the magnitude of the error, which is exactly what is required. Figure 6.12 shows the same test as before, but with $w_0 = 0$ and $w_1 = 2 \times 10^{-7}$. Clearly, appropriate adaptation of w can greatly improve the convergence rate for the unstable system model.

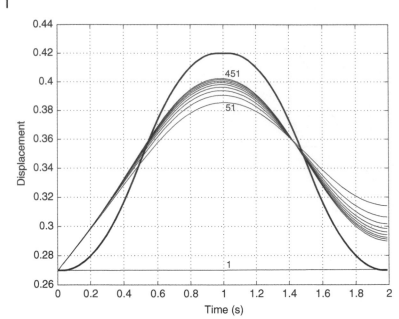

Figure 6.11 *Y*-axis, model $= 0.2/s$, $w = 2.5 \times 10^{-7}$.

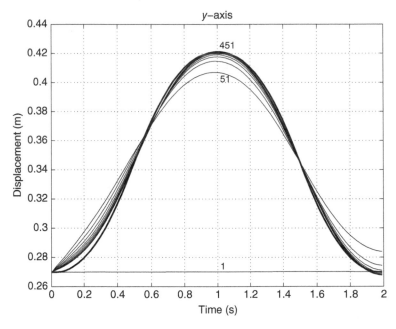

Figure 6.12 *Y*-axis, model $= 0.2/s$, $w = 0 + 2 \times 10^{-7} \|E_k\|^2$.

6.3 H_∞ Robust ILC

6.3.1 Background and Early Results

Early research contributions include [144, 195], but the control laws used were selected heuristically. The synthesis of ILC laws in an H_∞ setting was considered in, e.g. [7], using control action that combined current trial error feedback and feedforward from the previous trial. This research gave guidelines for choosing the weighting functions required for H_∞ design. Also, [61] developed a systematic design procedure and applied it to a semiconductor manufacturing experimental facility. This design is considered again below after the required background material is given.

Consider linear dynamics described in the ILC setting by the state-space model of (1.8) with an input contribution added to the output equation, i.e.

$$\dot{x}_k(t) = Ax_k(t) + Bu_k(t)$$
$$y_k(t) = Cx_k(t) + Du_k(t), \quad 0 \le t \le \alpha \tag{6.39}$$

with assumed zero state initial conditions on each trial. Applying the ILC law

$$u_{k+1}(t) = u_k(t) + \Gamma e_k(t) \tag{6.40}$$

results in the convolution description of the trial-to-trial error dynamics updating

$$e_{k+1}(t) = (I - D\Gamma)e_k(t) - \int_0^t Ce^{A(t-\tau)}B\Gamma e_k(\tau)d\tau \tag{6.41}$$

and hence, the error transmission is described by a Volterra integral operator of the second kind (acting on a vector, say, w)

$$L[w](t) = (I - D\Gamma)w(t) - \int_0^t Ce^{A(t-\tau)}B\Gamma w(\tau)d\tau \tag{6.42}$$

In [195], a sufficient condition for L_2 trial-to-trial error convergence of the ILC dynamics considered based on an infinite trial length, which implies convergence for any finite trial length, was developed. The ILC dynamics in this setting was termed L_2 convergent if the error converges to zero as $k \to \infty$ with respect to the L_2 norm on $[0, \infty)$, denoted by $|| \cdot ||_{L_2}$ and defined (for the SISO case) as

$$||e_k(t)||_{L_2} = \left(\int_0^\infty |e_k(t)|^2 dt \right)^{\frac{1}{2}} \tag{6.43}$$

The following result is from [195].

Theorem 6.4 *The ILC dynamics described by the linear operator L of (6.42) has the L_2-convergence property $\forall\, k$ if*

(a) *L is a stable system and therefore, $L \in H_\infty$*
(b) *$e_0 \in H_2$ and has finite norm*
(c)

$$||L(s)||_\infty := \sup_{0 \le \omega < \infty} \overline{\sigma}(L(j\omega)) \le 1 \tag{6.44}$$

and $\overline{\sigma}(L(j\omega)) = 1$ only at a countable number of frequencies.

If this result holds, monotonic trial-to-trial error convergence occurs in the H_2 or the L_2 norms. Based on this result, [7] gives the following "rules of thumb," or heuristics, for causal ILC laws, stated for the SISO case.

(i) Any ILC system satisfying $\sup_\omega |L(j\omega)| < 1$ is convergent with respect to the L_2-norm on any, finite or infinite, time interval (trial length) with monotonic trial-to-trial error reduction.

(ii) Any ILC design that satisfies

$$\lim_{\omega \to \infty} |L(j\omega)| < 1 \tag{6.45}$$

converges on a finite trial length, but the trial-to-trial error convergence is not guaranteed to be monotonic.

(iii) If the trial length is "long" and $\|L\|_\infty > 1$, then the rate of average or pointwise reduction of the error could be low but performance can still be assessed in the frequency domain.

(iv) If there are frequencies ω, where $|L(j\omega)| > 1$, then there will be, at least for the initial trials, an increase in the error depending on the content of $L(j\omega)$ at these frequencies.

(v) If $\lim_{\omega \to \infty} |L(j\omega)| = 1$ and $|L(j\omega)| < 1$, $\forall\, \omega < \infty$ the trial-to-trial error is reduced for all frequencies (but not "at infinity"), then convergence occurs for finite or infinite trial lengths.

A possible design method arising from these guidelines aims to achieve $|L(j\omega)| < 1$, for all ω and reduce the error for trials with finite or infinite length at all frequencies. More rapid reduction of the trial-to-trial error occurs when $|L(j\omega)|$ is smaller. This frequency-domain condition's implications are discussed again in the Chapter 7 on repetitive process-based design.

Consider the use of an ILC law of the following form (and notation) as in [7]

$$u_{k+1} = u_k + K_0[e_k] + K_1[e_{k+1}] \tag{6.46}$$

where $K_0[\cdot]$ and $K_1[\cdot]$ are controllers to be designed. Then assuming the same initial conditions on each trial

$$e_{k+1} = (I + GK_1)^{-1}(I - GK_0)e_k \tag{6.47}$$

At this stage, a connection with the repetitive process setting for ILC analysis arises. Consider again the abstract model for linear repetitive processes given by (3.33). Then in this case, $L_\alpha = L = (I + GK_1)^{-1}(I - GK_0)$ and setting $K_1 = 0$ recovers the ILC law first proposed in [12]. Moreover, if $K_0 = 0$, then L_α is equal to the sensitivity transfer-function matrix $S = (I + GK_1)^{-1}$.

The role of K_1 is to stabilize the overall system by shaping the maximum singular value of S, and the term defined by K_0 addresses error correction. In [12], it is chosen as the inverse of the direct transmission term, D in (6.39) (the current trial direct "feedthrough" term from input to output). Ideally, it should be equal to the exact inverse of G, or an approximation to this inverse.

If the system to be controlled is strictly proper and the controllers are proper, then

$$\lim_{\omega \to \infty} (I + G(j\omega)K_1(j\omega))^{-1}(I - G(j\omega)K_0(j\omega)) = I \tag{6.48}$$

and hence, any minimization can, at best, result in $\|L\|_\infty = 1$. Hence, this control law can, at best, achieve nonrobust learning as defined by Theorem 6.4. Robust convergence for infinite or finite trial lengths cannot be achieved as (6.48) contradicts (6.45). Suppose that $K_0 = 0$ and GK_1 is positive real for all ω, then $\|L_\infty\| = 1$ and for a K_1 with increasing gain, the rate of convergence (strictly pointwise) in the frequency domain increases. However, H_∞ minimization is required to achieve $\|L\| = 1$ and to give further reduction.

The control law entries K_0 and K_1 can be computed by a standard H_∞ optimization method as given below, for background, see, e.g. [243]. Weighting terms are included in the design, depending on the error's frequency content and chosen to improve overall performance. Suppose that for (6.46), the error on trial $k = 0$ is the reference signal (for which it is assumed the Laplace transform exists). Then the error on this trial will be minimized if, in the SISO case for simplicity and suppressing the transform variables (s and $j\omega$) for ease of presentation, $|(I + GK_1)^{-1}(I - GK_0)r|$ is minimized for all frequencies. Consequently, the reference trajectory is a natural candidate for the weighting function, and since there are two control terms to be designed, a two-step procedure is used, as detailed next.

Step 1. Compute an approximate inverse of G by solving

$$\min_{K_0} \|I - GW_1 K_0\|_\infty \tag{6.49}$$

where the weight W_1 is chosen to be nonproper and therefore, GW_1 is proper, and the overall system satisfies the rank conditions [243] for solving the H_∞ minimization problem. A common choice for the weighting is a diagonal matrix with unity amplification for low frequencies and derivative action at high frequencies. The meanings of "low" and "high" are a function of the reference trajectory's frequency content. This step aims to construct a control law that approximates the system's inverse for low-to-intermediate frequencies.

Step 2. Set $W_2 = (I - GK_0)r$, where K_0 is the result of the previous step and solves the sensitivity minimization problem:

$$\min_{K_1} \|(I + GK_1)^{-1} W_2\|_\infty \tag{6.50}$$

To satisfy the rank conditions and improve robustness, the following mixed H_∞ problem is preferred:

$$\min_{K_1} \left\| \begin{array}{c} (I + GK_1)^{-1} W_2 \\ K_1(I + GK_1)^{-1} W_3 \end{array} \right\|_\infty \tag{6.51}$$

where W_3 is a bound on the system uncertainty under the additive model of the uncertainty and is often a low-pass filter. Moreover, by varying the weights W_2 and W_3, a trade-off between performance and robustness can be sought.

The convergence condition (6.44) can be checked after the design of K_0 and K_1 by plotting $\bar{\sigma}(L(j\omega))$ as a function of ω.

Remark 6.2 The generalized KYP lemma can be used to develop ILC law designs where finite frequency ranges with varying weights are allowed. See the Chapter 7.

The results from applying this procedure to examples are given in [7]. In one of these, the system is unstable with transfer-function:

$$G(s) = \frac{s+1}{(s-2)(s+3)} \tag{6.52}$$

and a control law, where $K_0 = 0$ and H_∞ optimization is applied to design K_1. In this case, good convergence is achieved by pure feedback with "moderately high gain." The feedback controller is less sensitive to model errors than an inverse-based feedforward ILC law and is given by

$$K_1(s) = 3931 \frac{(s+3)(s+3.348)}{(s+0.7857)(s+2)(s+446.4)} \tag{6.53}$$

Simulation results are given in [7] where the design is compared for $G(s)$ and a perturbed system with transfer-function:

$$\tilde{G}(s) = 6\frac{(s+0.5)}{(s-3)(s+6)} \tag{6.54}$$

The results confirm that over 20 trials, the error is reduced monotonically and rapidly despite the difference (or mismatch) between the transfer-functions used for the design ($G(s)$) and the simulations ($\tilde{G}(s)$). The use of feedback control action ensures controlled system stability. Cases will exist, however, where (6.44) cannot hold, and the limitations to the design method are as follows:

(i) The first step in the above procedure results in an approximate inverse only over a limited frequency range. If the system is unstable, standard H_∞ analysis cannot be used to design K_0.

(ii) Minimizing the sensitivity function for nonminimum phase systems has inherent limitations and for systems with a high-relative degree. For such systems, the sensitivity function $S = (I + GK_1)^{-1}$ (and hence L) always has a frequency domain peak greater than unity, and therefore, $\|L\|_\infty < 1$ cannot hold.

(iii) Forming an $\|S, K_1 S\|_\infty$ problem limits the achievable value of $\|L\|_\infty$ since there is a compromise between stability and robustness. In particular, (6.44) can only be satisfied in the presence of no uncertainty.

Despite these limitations, an H_∞ setting for ILC design has a role for a broad class of systems. In particular, consider a frequency (or a range thereof) where the error is "large." Then error reduction in such cases is very easily achieved by H_∞ optimized controller, and the concept of "practical convergence" may be of use (for some examples at least). Using this approach, substantial error reduction over a useful number of initial trials can be achieved, and only after this apparent convergence does the expected divergence appear. In such cases, implementation of the ILC law can be terminated just before divergence appears, and hence a small tracking error results.

Justification for this procedure is based on the frequency-dependent maximum singular value of the error transmission operator and the norms of the signals involved. From (6.47)

$$e_{k+1}(j\omega) = L(j\omega)e_k(j\omega), \quad e_k(j\omega) = L^k(j\omega)e_0(j\omega) \tag{6.55}$$

with $e_0(j\omega) = r(j\omega)$ if $u_0(j\omega) = 0$. Also, the reference trajectory often has large and low amplitudes at low and high frequencies, respectively, since this trajectory must correspond to the output of a strictly proper system and hence, must have a large enough roll-off for high frequencies. Only if the reference times the inverse of the system is strictly proper can zero limit error be achieved. In particular, $|r(j\omega)|$ must decrease for $\omega \to \infty$ at a rate of at least 20 dB/decade times the relative degree of the system.

The error transmission operator (again considering the SISO case for simplicity) can normally be forced to have an amplification much less than unity for frequencies below its bandwidth by an appropriate choice of the weighting functions in the design criterion. If the bandwidth can be made sufficiently large such that the reference trajectory is already small at frequencies, where $|L(j\omega)| \gg 1$, then application of the ILC law reduces the first trial error $e_1(j\omega) = L(j\omega)r(j\omega)$ in the H_2 sense. Hence, by Parseval's theorem, the corresponding time-domain signal in $L_2[0, \infty)$ is substantially reduced. This situation will remain valid for all trials until $L^k(j\omega)r(j\omega)$ becomes large in the interval, where $|L(j\omega)| > 1$.

As an outcome of the above discussion, ILC can be applied until the error norm is below the desired tolerance or until it increases. In particular, achievable tolerance depends on the bandwidths of L and the reference trajectory. Suppose the bandwidth of L is denoted by ω_L, i.e. $|L(j\omega)| < 1$ for all $\omega \in [0, \omega_L]$. Also, let the bandwidth of the reference trajectory, ω_r, be the frequency, where $|r(j\omega)|$ is "large," i.e. before it starts to decrease monotonically. Then the following conclusions are valid:

(a) If $\omega_r \ll \omega_L$, then $L(j\omega)e_1(j\omega)$ is small in the H_2 sense and hence, good performance can be expected on applying the ILC law.

(b) If $\omega_r \gg \omega_L$, poor transient effects can be expected since the significant high-frequency content of the reference trajectory is rapidly amplified.

(c) If the frequency ranges $[0, \omega_r]$ and $[\omega_L, \infty)$ overlap, then the error can be expected to exhibit increasing amplitude for the overlapping frequencies.

As "rules of thumb" from this analysis, it is reasonable to target obtaining an error transition operator with a large bandwidth, a small gain, especially at low frequencies, and either no peak or only a small peak, where $\bar{\sigma}(L(j\omega) > 1$. These "rules" are in agreement with those for classical controller design to shape $\bar{\sigma}(S)$. The question of how well the objectives can be met in an H_∞ design depends on the structure of the system, e.g. its relative degree and the location of any nonminimum phase zeros. Moreover, performance can be examined in the frequency domain where the function $\bar{\sigma}(L(j\omega)^k)|r(j\omega)|$ is an upper bound for the error on trial k. The maximum number of trials over which the error is reduced can be determined from a plot of this bound. For a finite trial length, this is a good but conservative consideration.

To highlight the points given above, consider the following example from [7]:

$$G(s) = 25\frac{(s-100)}{(s+50)^2} \tag{6.56}$$

and in the control law $K_0 = 0$. The minimization algorithm (6.51) gives

$$K_1(s) = -556\frac{(s+50)^2(s+200)}{(s+2)(s+148.9)(s^2-677.7s+133\ 100)} \tag{6.57}$$

and the resulting H_∞ norm of L is 1.229, and hence, the infinite trial length stability condition does not hold. Consequently, undesirable transient effects can be expected.

For this design, simulations confirm that up to $k = 12$, the ILC law reduces the tracking error, which begins to increase again for $k > 13$. If the ILC law is switched out after $k = 12$, then its application results in a decrease in the maximum value of $|e_k(t)|$ over these 12 trials of approximately $\frac{1}{350}$, and the L_2-norm of the error is reduced by at least $\frac{1}{2500}$. If the ILC law is applied for $k > 12$, the tracking error deteriorates and oscillatory components with a frequency similar in magnitude to the frequencies where the magnitude of $L(j\omega)$ crosses over from less than to greater than unity will appear.

6.3.2 H_∞ Based Robust ILC Synthesis

In this section, the analysis first reported in [61], with preliminary results in [60], is considered, resulting in a synthesis procedure as an alternative to the heuristic guidelines given in Section 6.3.1. Consider, therefore, Figure 6.13, where the system G and the controller C are discrete linear time-invariant systems (with the transform variables z and $e^{j\omega}$ suppressed for ease of presentation in some places).

The map from $\begin{bmatrix} r \\ f \end{bmatrix}$ to $\begin{bmatrix} y \\ e \end{bmatrix}$ is given, after routine block diagram manipulations, by

$$(I+GC)^{-1}\begin{bmatrix} GC & G \\ I & -G \end{bmatrix} =: \begin{bmatrix} T & R \\ S & -R \end{bmatrix} \tag{6.58}$$

Figure 6.13 Feedback control system where f, r, u, y, and e denote, respectively, a forcing function, the reference trajectory, the system input, output, and the error.

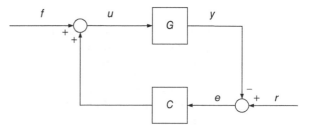

where

$$T = (I + GC)^{-1}GC$$
$$R = (I + GC)^{-1}G$$
$$S = (I + GC)^{-1} \tag{6.59}$$

Consider, as in [60, 61], the ILC law

$$f_{k+1}(p) = Qf_k(p) + Le_k(p) \tag{6.60}$$

where Q and L are linear filters designed such that

$$\lim_{k\to\infty} f_k(p) = f_\infty(p), \quad \lim_{k\to\infty} e_k(p) = e_\infty \tag{6.61}$$

where f_∞ and e_∞ are fixed points and e_∞ is minimal over the trial length in some signal norm. Use of a contraction mapping setting, for the functional analysis background again see [134], gives the following result [195].

Theorem 6.5 *For the control configuration of Figure 6.13, suppose that both f and e lie in $L_2(0, \infty)$. Then under the ILC law (6.60), this system converges to a fixed point $f_\infty(p)$ over the trial length if*

$$\|Q(z) - L(z)R(z)\|_\infty < 1 \tag{6.62}$$

where R is defined in (6.59).

Also, if (6.62) holds, it follows that

$$f_\infty(p) = (I - Q + LR)^{-1}LSr(p)$$
$$e_\infty(p) = \|(I - R(I - Q + LR)^{-1}L)Sr(p)\|_2 \tag{6.63}$$

The following result [61] is critical to the choice of $Q(z)$ and $L(z)$.

Theorem 6.6 *Suppose that $L, R \neq 0$ in the control system of Figure 6.13. Then under the ILC law (6.60) the fixed point $e_\infty(p)$ is zero if and only if (6.62) holds and $Q(z) = I$.*

This result is the main reason why many ILC laws are designed with $Q(z) = 1$, and in this case, the condition of Theorem 6.6 becomes

$$\|I - L(z)R(z)\|_\infty < 1 \tag{6.64}$$

Also, for this condition to hold for all z, (6.64) implies that $L(z) = R^{-1}(z)$ is required. Constructing this inverse is not possible when the system is nonminimum phase or too complex to describe. However, selecting $Q(z) \neq I$ in the control law increases robustness against uncertainty in the dynamics of $R(z)$, even though such a choice prevents perfect tracking.

Based on the considerations above, the following, see, e.g. [128], is a heuristic design procedure for the design of $Q(z)$ and $L(z)$.

(i) Set $L(e^{j\omega}) \approx R^{-1}(e^{j\omega})$, $\omega \in [0, \omega_c]$, i.e. select L as the inverse of R up to some frequency ω_c.
(ii) Select $Q(e^{j\omega})$ as a low-pass filter with cut-off frequency close to ω_c. In particular, $\|Q(e^{j\omega})\|_\infty = 1$, $\forall \omega \in [0, \omega_c]$, and $\|Q(e^{j\omega})\|_\infty \approx 0 \, \forall \omega > \omega_c$.

This procedure is motivated by the observation that many physical systems are well described at low frequencies, and any uncertainties present are at high frequencies. Hence, for frequencies

$\omega \in [0, \omega_c]$ and the choice of $Q(e^{j\omega})$ as above $e_\infty(p)$ will have zero frequency content by Theorem 6.6, and for $\omega > \omega_c$, the frequency content of this signal equals that of e_0, i.e. the error with no learning applied.

In application, the difficulty in applying the above procedure is to decide on the value of ω_c, i.e. the trade-off between performance, which requires $e_\infty = 0$ for $Q(z) = I$ and robustness, which requires $Q(z) \approx 0$ for all z such that $L(z) \neq R^{-1}(z)$. This heuristic approach can be replaced as described next where the ILC law (6.60) is modified to the following without loss of generality:

$$f_{k+1}(p) = Q(f_k(p) + Le_k(p)) \tag{6.65}$$

One of the two reasons for this modification is that most physical systems are strictly proper (more poles than zeros). In applications, L will behave as a differentiator at high frequencies. Hence, to avoid this action on high-frequency signals, L is also cut-off at some high frequency. The second reason is that, given (6.65), the convergence criterion of Theorem 6.5 is

$$\|Q(z)(I - L(z)R(z))\|_\infty < 1 \tag{6.66}$$

and in H_∞ design, the filter $Q(z)$ can be interpreted as a weighting function for the learning performance, i.e.

$$\|I - L(z)R(z)\|_\infty < \|Q^{-1}(z)\|_\infty \tag{6.67}$$

Given Theorem 6.6, $Q(z)$ can be interpreted as a measure for learning performance, i.e. choose the cut-off frequency ω_c as large as possible to guarantee zero tracking error up to ω_c and the following is an alternative design procedure to the one given above.

(i) Choose $Q(e^{j\omega})$ to a be a low-pass weighting filter with prespecified, cut-off frequency ω_c such that $\|Q(e^{j\omega})\|_\infty = 1 \, \forall \, \omega \in [0, \omega_c]$ and $\|Q(e^{j\omega})\|_\infty \approx 0 \, \forall \, \omega > \omega_c$.

(ii) For given Q and R, obtain L as the solution of the following suboptimal H_∞ synthesis problem:

$$L(z) = \arg \min_{L \in H_\infty} \|Q(z)(I - L(z)R(z))\|_\infty \tag{6.68}$$

In H_∞ analysis, see, e.g. [243], (6.68) is known as a "model matching" problem, i.e. for given Q, L is matched to the inverse of R. Let $L_{inv}(z)$ denote the solution of this problem. Then for ILC convergence, the design procedure should, by (6.66), result in

$$\|Q(z)(I - L_{inv}(z)R(z))\|_\infty = \gamma_\infty < 1 \tag{6.69}$$

Also,

$$\|f_{k+1} - f_k\|_2 \leq \gamma_\infty \|f_k - f_{k-1}\|_2 \leq \cdots \leq \gamma_\infty^k \|f_1 - f_0\|_2 \tag{6.70}$$

and hence, by minimizing $\|Q(I - LR)\|_\infty$, the highest convergence rate in L_2 is obtained.

A full treatment of the solution of (6.68) for standard linear systems is given in, e.g. [19, 68]. In the ILC case, the approach used is to reformulate the problem to the structure shown in Figure 6.14. This setting provides methods for computing a stabilizing K that minimizes $\|T_{zw}\|_\infty$, where T_{zw} is the transfer-function from w to z using γ iteration. Suppose that G is partitioned according to the inputs (w, u) and outputs (y, z) as follows:

$$G = \begin{bmatrix} G_1 & G_2 \\ G_3 & G_4 \end{bmatrix} \tag{6.71}$$

then

$$T_{zw} = \mathcal{F}_l(G, K) := G_1 + G_2 K(I - G_4 K)^{-1} G_3 \tag{6.72}$$

where $\mathscr{F}_l(G, K)$ is known as a lower-linear fractional transformation (LFT) of G and K. In this setting, (6.68) is a special case with

$$K = L, \quad G := \begin{bmatrix} Q & Q \\ -R & 0 \end{bmatrix} \tag{6.73}$$

This choice for G and K gives a stabilizing L such that $\|Q(I - LR)\|_\infty$ is minimized. Moreover, for the ILC case z and w represent the signals f_{k+1} and f_k, respectively, and the time delay, equal to the trial length, between these two signals has unity magnitude. Hence, by the small gain theorem $\|T_{zw}\|_\infty = \|Q(I - LR)\|_\infty$ must be less than unity to guarantee stability of the controlled system, i.e. the ILC convergence condition (6.66).

Figure 6.15 shows the ILC design problem in the format of Figure 6.14. Using the two coupled Riccati equation-based solution to the H_∞ synthesis problem in [68] will, in general, result (for the details see [60, 61]) in a singular H_∞ synthesis problem, but this difficulty can be overcome by a slight perturbation of the original problem [245].

An advantage of the above design procedure over the alternative given in Section 6.3.1 is that for a given Q, the best L_2-convergence is achieved because (6.66) is minimized. Hence, there is no other L that can obtain a higher convergence rate with respect to Q. A second advantage is that for MIMO design, interaction is explicitly accounted for in the design of Q and L. The most important

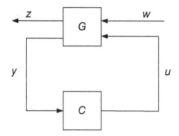

Figure 6.14 Standard configuration of the ILC design problem.

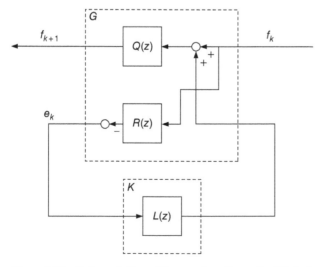

Figure 6.15 Reformulation of the previous figure for ILC design.

advantage, of course, is that this setting allows for robust control with respect to uncertainty in $R(z)$.

Suppose that the actual controlled system is described by a nominal transfer-function $R_0(z)$ and output multiplicative uncertainty as specified by a stable weighting function $W_0(z)$ that also has a stable inverse, i.e.

$$R(z) := \{(I + W_0(z)\Delta(z))R_0(z) \ : \ \|\Delta(z)\|_\infty \le 1\} \tag{6.74}$$

Then the last ILC design procedure extends naturally to this case on setting

$$K = L, \quad G = \left[\begin{array}{cc|c} Q & 0 & Q \\ R_0 & 0 & 0 \\ \hline -R_0 & W_0 & 0 \end{array}\right] \tag{6.75}$$

and again solving for K by minimizing $T_{zw} = \mathscr{F}_l(G, K)$, where in this case

$$\begin{aligned}
\mathscr{F}_l(G, K) &= \begin{bmatrix} Q & 0 \\ R_0 & 0 \end{bmatrix} + \begin{bmatrix} Q \\ 0 \end{bmatrix} L \begin{bmatrix} -R_0 & W_0 \end{bmatrix} \\
&= \begin{bmatrix} Q(I - LR_0) & QLW_0 \\ R_0 & 0 \end{bmatrix}
\end{aligned} \tag{6.76}$$

Hence, $\|T_{zw}\|_\infty \ge \|R_0\|_\infty$.

A difficulty that can arise is that for practical applications $\|R_0\|_\infty > 1$ may occur, and this problem cannot, in general, be solved using the standard H_∞ synthesis procedure. This situation is to be expected since the aim is robustness against specified uncertainty and the resulting robust performance design problem has, in general, no solution. An alternative is to use structured singular value, or μ, analysis, see Section 6.3.4. By exploiting the diagonal structure in the mapping from z to w, μ can be computed for a finite number of frequencies. Then a successive iteration can be performed by scaling the off-diagonal terms of the matrix on the right-hand side of (6.76) followed by solving the suboptimal H_∞ problem for the scaled system.

The performance specifications of Theorem 6.6 for a given example can be addressed by iterating over the bandwidth ω_c of the filter Q using the following design procedure:

(a) Model R as per (6.74).
(b) Chose $Q(e^{j\omega})$ as a low-pass filter with a cut-off frequency such that $\|Q(e^{j\omega})\| = 1 \ \forall \ \omega \in [0, \omega_c]$ and $\|Q(e^{j\omega})\|_\infty \approx 0$ for all $\omega > \omega_c$.
(c) For given Q and R, find an L that minimizes (6.76) using μ-synthesis.
(d) If an L can be found that results in $\|T_{z,w}\|_\infty < 1$, increase the bandwidth ω_c of Q and repeat Step 3. Otherwise, decrease ω_c and apply Steps (b) and (c) above iteratively until the maximum possible ω_c has been obtained.

In [61], the above design procedure has been applied in simulation and experimentally to a wafer-scale motion system with acceptable results. Many other publications, e.g. [90, 282] have considered robust convergence of ILC laws also using the frequency domain setting, i.e. H_∞ control and μ analysis. The general outcome of these and other publications is that the approach based on the results covered in this section does enable the inclusion of uncertainty models in ILC design. Still, only causal ILC laws can be designed. Moreover, the frequency domain representation of the ILC dynamics is only approximate when applied over the finite trial length. Hence, errors may result in the initial stages of the transient behavior along the trials. These limitations form the basis for the results given in Section 6.3.4.

6.3.3 A Design Example

As a detailed design example, consider the digital control scheme shown in Figure 6.16 from [155], which combines feedback control with an ILC law in the presence of an additive disturbance on the system output, which is assumed to be the same on each trial.

The SISO case is considered, where with q denoting the forward shift operator, the nominal plant dynamics is denoted by $G(q)$ and multiplicative uncertainty is assumed. Hence, the dynamics to be controlled are represented by

$$\tilde{G}(q) = \left(1 + \Delta(q)W_G(q)\right) G(q) \tag{6.77}$$

where the nominal dynamics are assumed to be stable, $W_G(q)$ is a known stable weighting function for the uncertainty and $\Delta(q)$ represents an unknown but stable transfer-function satisfying $\|\Delta(q)\|_\infty \leq 1$. The output disturbance is described by

$$\tilde{G}_d(q) = \left(1 + \Delta(q)W_{G_d}(q)\right) G_d(q) \tag{6.78}$$

where the terms in this equation have the corresponding meanings to those in (6.77).

In the case of Figure 6.16, a proportional plus integral plus derivative (PID) controller is used in the feedback loop and described by

$$C(q) = k_p + k_i \frac{T_s}{1 - q^{-1}} + k_d \frac{1 - q^{-1}}{T_s} \tag{6.79}$$

where T_s is the sampling period and k_p, k_i and k_d are the controller gains. Also, the dynamics of the internal feedback control loop can be written as follows:

$$y_{k+1}(p) = \tilde{T}(q)\left(r(p) + v_{k+1}(p)\right) + \tilde{S}(q)\tilde{G}_d(q)d(p) \tag{6.80}$$

where $\tilde{S}(q)$ and $\tilde{T}(q)$ are, respectively, the sensitivity and complementary sensitivity functions given by

$$\tilde{S}(q) = \left(1 + \tilde{G}(q)C(q)\right)^{-1}, \quad \tilde{T}(q) = \tilde{S}(q)\tilde{G}(q)C(q) \tag{6.81}$$

The ILC law is

$$v_{k+1}(p) = Q(q)\left(v_k(p) + L(q)e_k(p)\right) \tag{6.82}$$

As one possible choice, the learning or Q-filter is selected as a series connection of two identical first-order, low-pass infinite impulse response (IIR) filters, where one of them is applied in the forward time direction, and the other in the backward, resulting in a zero-phase filter (recall that

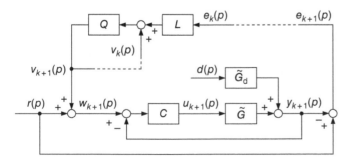

Figure 6.16 Combined feedback and ILC law with an output disturbance term.

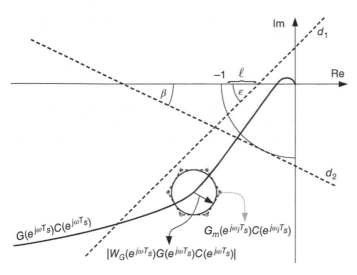

Figure 6.17 Linear constraint specifications in the frequency domain for the robust feedback control loop design.

zero-phase filtering can be used in the ILC setting due to the finite trial length).

$$Q(q) = \frac{1}{a_0 + a_1 q^{-1}} \times \frac{1}{a_1 q + a_0} = \frac{1}{q_1 q + q_0 + q_1 q^{-1}} \tag{6.83}$$

$$q_0 = a_0^2 + a_1^2, \quad q_1 = a_0 a_1 \tag{6.84}$$

Consider the case when $L(q)$ is an finite impulse response (FIR) filter, i.e.

$$L(q) = q^{s_L}(l_0 + l_1 q^{-1} + l_2 q^{-2} + \cdots + l_{n_L} q^{-n_L}) \tag{6.85}$$

Both of these filters ($Q(q)$ and $L(q)$) are noncausal, and for the latter, the forward shift s_L and the order n_L are design parameters. Also, the trial-to-trial error updating is described by

$$e_{k+1}(p) = Q(q)\left(1 - \tilde{T}(q)L(q)\right)e_k(p)$$
$$+ (1 - Q(q))\tilde{S}(q)\left(r(p) - \tilde{G}_d(q)d(p)\right) \tag{6.86}$$

Many methods are available for designing a robust PID controller. The main novelty in this section is the simultaneous design of the Q and L filters, respectively, of (6.83) and (6.85). Both design tasks are solved in the frequency domain by convex optimization. The Nyquist plot for the set of models $\tilde{G}(e^{j\omega T_s})C(e^{j\omega T_s})$, $\omega \in [0, \omega_n]$, where $\omega_N = \frac{\pi}{T_s}$ is the Nyquist frequency, is represented at a particular frequency by the circle centered at $\tilde{G}(e^{j\omega T_s})C(e^{j\omega T_s})$ with radius $|W_G(e^{j\omega T_s})G(e^{j\omega T_s})C(e^{j\omega T_s})|$ in Figure 6.17.

In this last figure, the straight line d_1 crossing the real axis between -1 and 0 at a distance $\ell \in {]}0,1[$ from $(-1,0)$ and slope $\epsilon \in (0°, 90°]$ acts to constrain the open-loop frequency response in the high-frequency range $\omega \in [\omega_x, \omega_N]$, where ω_x is application-dependent. Moreover, if the Nyquist plot lies to the right of d_1, then the robustness margins are ensured.

The constraint in the last figure is applicable over the complete frequency range if the system model contains at most one integrator. For the case of two or three integrators, i.e. a single or double integrator in the system model, a part of the frequency response $\tilde{G}(e^{j\omega T_s})C(e^{j\omega T_s})$ in the low-frequency range $\omega \in {]}0, \omega_x]$ lies above the straight line d_1. In this case, an additional constraint, given by the straight line d_2 in Figure 6.17, is required.

As seen, the straight-line d_2 crosses the negative real axis at an angle β and tangential to the unit circle boundary centered at the origin. It lies between the straight line d_1 and the imaginary axis. Adding this constraint keeps the open-loop Nyquist plot at a certain distance from the critical $(-1,0)$ point. Robust performance for the low-frequency range is achieved if the discs $W_G(e^{j\omega T_s})G(e^{j\omega T_s})C(e^{j\omega T_s})$ lie below the straight line d_2.

Consider the low- and high-frequency regions, whose specification is application-dependent. The linear constraints defined for these regions must be satisfied for an infinite number of frequencies. From a practical point of view, this may be relaxed to a finite but large enough set of logarithmically spaced angular frequencies ω_i. Moreover, this frequency grid is natural because, e.g. it arises from the experimental identification of system models in the frequency domain.

It is well known in robust control theory for standard linear systems, e.g. [243], that the unstructured uncertainty disc of the system model at a specified frequency ω_i can be approximated by a polygon with vertices:

$$G_m(e^{j\omega_i T_s}) = \left(1 + \frac{|W_G(e^{j\omega_i T_s})|}{\cos(\pi/n_v)} e^{j2\pi m/n_v}\right) G(e^{j\omega_i T_s}) \tag{6.87}$$

where $n_v > 2$ denotes the number of vertices and $m = 1, \ldots, n_v$ the vertex number. This results in an optimization problem where the linear constraints must be satisfied for all points specified by the vertices $G_m(e^{j\omega_i T_s})C(e^{j\omega_i T_s})$ at given frequencies ω_i in the specified frequency regions, as illustrated in Figure 6.17.

The robust feedback control design's goal is the fast rejection of step-like changes in the disturbance $d(p)$. In general, this can be achieved by maximizing the integral gain k_i of the PID controller (6.79) subject to the constraints (where the dependence of variables on $e^{j\omega_i T_s}$ is omitted from this point onward in some expressions for ease of notation, and Re and Im, respectively, denote the real and imaginary parts of a complex number)

$$\underset{k_p,\, k_i,\, k_d}{\text{minimize}} k_i$$

subject to

$$\cos(\beta)\,\text{Im}(G_mC) + \sin(\beta)\,\text{Re}(G_mC) \leq -1$$
$$\text{for } m \in \{1, \ldots, n_v\} \text{ and } \omega_i \in \{\omega_1 < \omega_2 < \cdots < \omega_x\} \tag{6.88}$$
$$\cot(\epsilon)\,\text{Im}(G_mC) - \text{Re}(G_mC) + \ell \leq 1$$
$$\text{for } m \in \{1, \ldots, n_v\} \text{ and } \omega_i \in \{\omega_x < \cdots < \omega_N\}$$
$$|C| \leq C_{\max}, \quad \omega_i \in \{\omega_c < \cdots < \omega_N\}$$

Solving this optimization problem gives the coefficients of the robust PID controller, where the design parameters to be specified are ϵ, ℓ, β defining the straight lines d_1 and d_2, the frequency ω_x separating the low- and high-frequency ranges, the maximum allowable gain $C_{\max}$ of the PID controller and the frequency ω_c, which applies the third constraint. This last set of constraints has been added to limit the PID controller's maximum gain, starting from the frequency ω_c, to prevent excessive amplification of measurement noise and or to limit the peak values of the control input.

The ILC law design aim is to reduce further the tracking error of the underlying closed-loop system with input $w_{k+1}(p)$ and output $y_{k+1}(p)$. Moreover, the dynamics of this uncertain system is described by the frequency response for $\omega_i \in [\omega_1 < \omega_2 < \cdots < \omega_N]$, i.e.

$$\tilde{T}(e^{j\omega_i T_s}) = \left(1 + \Delta(e^{j\omega_i T_s})W_T(e^{j\omega_i T_s})\right) T(e^{j\omega_i T_s}) \tag{6.89}$$

where

$$T(e^{j\omega_i T_s}) = \frac{G(e^{j\omega_i T_s})C(e^{j\omega_i T_s})}{1 + G(e^{j\omega_i T_s})C(e^{j\omega_i T_s})} \tag{6.90}$$

is the frequency response of the closed-loop feedback control system for the nominal plant model parameters and $W_T(e^{j\omega_i T_s})$ is a known uncertainty weighting function. Moreover, $\|\Delta(e^{j\omega_i T_s})\|_\infty \leq 1$.

In ILC design, the aim is to select the Q and L-filters, with the structure, respectively, of (6.83) and (6.85) to guarantee fast and robust monotonic convergence of the tracking error over a specified range of frequencies. The starting point in the frequency domain is the following condition for robust monotonic convergence of the tracking error for an ILC scheme of the form Figure 6.16, see, e.g. [33],

$$\left| Q(e^{j\omega T_s}) \left(1 - T(e^{j\omega T_s})L(e^{j\omega T_s}) \right) \right| + \left| Q(e^{j\omega T_s})T(e^{j\omega T_s})W_T(e^{j\omega T_s})L(e^{j\omega T_s}) \right| \leq \gamma < 1 \tag{6.91}$$

for all $\omega \in [0, \omega_N]$. In this setting, γ denotes the convergence rate and $Q(e^{j\omega T_s})$ and $L(e^{j\omega T_s})$ are the frequency responses, respectively, of the Q and L-filters.

The requirements for fast convergence and a large bandwidth conflict and hence, constant design parameters are introduced that address the robust monotonic convergence condition (6.91) in specified frequency regions, see Figure 6.18. These parameters M_1 and M_2 are to be chosen together with the frequencies ω_y, in the design. As the Q-filter is zero-phase, it does not introduce any phase shift, and its frequency response is given for $\omega_i \in [\omega_1 < \omega_2 < \cdots < \omega_N]$ by positive real numbers

$$Q(e^{j\omega_i T_s}) = \frac{1}{q_0 + 2q_1 \cos(\omega_i T_s)} = \frac{1}{Q_{\text{den}}(e^{j\omega_i T_s})} \tag{6.92}$$

Hence, (6.90) in this case (with the argument $e^{j\omega_i T_s}$ dropped from the notation for ease of presentation) is

$$Q|1 - TL| + Q|TW_T L| < 1 \tag{6.93}$$

$\forall \omega_i \in [\omega_1 < \omega_2 < \cdots < \omega_N]$ or, equivalently,

$$|1 - TL| + |TW_T L| < Q_{\text{den}} \tag{6.94}$$

Moreover, the frequency characteristics of T and W_T are known, and hence, this last condition is convex.

The design aims to achieve a high bandwidth for the ILC law with guaranteed convergence rates in the two specified frequency ranges shown in Figure 6.18. This objective is achieved by solving a constrained convex optimization problem formed by maximizing the cut-off frequency of the

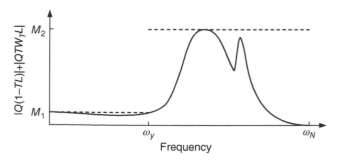

Figure 6.18 Constant value constraints in the robust ILC design to achieve a desired trial-to-trial error convergence rate.

Q-filter (6.83). A first-order, low-pass filter with a unity DC gain has the following form in the digital domain:

$$H(q) = \frac{1}{a_0 + a_1 q^{-1}} \tag{6.95}$$

with

$$a_0 = 1 + \frac{1}{\omega_c T_s}, \quad a_1 = -\frac{1}{\omega_c T_s} \tag{6.96}$$

By substitution of (6.96) in (6.84), the coefficients of the Q-filter (6.83) can be written as follows:

$$q_0 = 1 + \frac{2}{\omega_c T_s} + \left(\frac{\sqrt{2}}{\omega_c T_s}\right)^2, \quad q_1 = -\frac{1}{\omega_c T_s}\left(1 + \frac{1}{\omega_c T_s}\right) \tag{6.97}$$

Maximization of the cut-off frequency of the Q-filter results from minimization of q_0 or maximization of q_1, where both coefficients depend on the cut-off frequency ω_c. Combining the design criteria leads to the following convex optimization problem as a solution algorithm:

$$
\begin{aligned}
&\text{maximize } q_1 \\
&\text{subject to} \\
&\quad |1 - TL| + |TW_T L| \leq M_1 Q_{\text{den}} \\
&\qquad \text{for } \omega_i \in \{\omega_1 < \omega_2 < \cdots < \omega_y\} \\
&\quad |1 - TL| + |TW_T L| \leq M_2 Q_{\text{den}} \\
&\qquad \text{for } \omega_i \in \{\omega_{y2} < \omega_{y3} < \cdots < \omega_N\} \\
&\quad |L| \leq L_{\max} \text{ for } \omega_i \in \{\omega_1 < \omega_2 < \cdots < \omega_N\} \\
&\quad q_0 + 2q_1 = 1, \quad q_0 \geq 1
\end{aligned} \tag{6.98}
$$

In the optimization problem considered, the variables are the coefficients q_0 and q_1 of the Q-filter (6.83) and those of the L-filter (6.85). The design parameters to be chosen are the forward shift s_L and order n_L of the learning filter and the bounds $0 < M_1, M_2 < 1$ limiting the monotonic convergence condition in two frequency ranges separated by the parameter ω_y. Also, ω_{y2} and ω_{y3} represent the lowest discrete frequencies that are greater than ω_y.

A faster trial-to-trial error convergence rate in the low-frequency range can be achieved by enforcing $M_1 < M_2$. The optimization problem (6.98) is formulated for two frequency regions, but the number can be easily increased if necessary. Lastly, the design parameter is the maximum allowable gain $L_{\max}$ of the L-filter that affects the third group of constraints. This parameter is added to avoid an excessive amplification of high-frequency measurement noise by the learning filter. Finally, in the fourth group of constraints, the first condition enforces unity DC gain of the Q-filter, and the last is the requirement that this filter is stable.

To implement the Q-filter, its coefficients a_0 and a_1 are calculated from the values q_0 and q_1. The simplest method is

$$a_0 = \pm\sqrt{\frac{q_0 \pm \sqrt{q_0^2 - 4q_1^2}}{2}}, \quad a_1 = \frac{q_1}{a_0} \tag{6.99}$$

where the solution to be used out of the four possible from this equation corresponds to a stable pole and a positive DC gain. Only this solution is considered below.

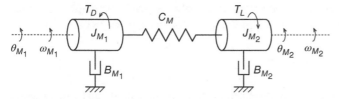

Figure 6.19 Block diagram of the drive system with a flexible shaft.

As a numerical case study, the model of a drive system consisting of two disks with a flexible shaft in between [215], see Figure 6.19, is used. Moreover, many elastic drive trains can be described by this or a similar model. In the case considered, the control input is the electromagnetic torque T_D generated by an electric motor acting on the first disc with a resulting mass moment of inertia J_{M1}. The first disk is connected to the second using a long shaft with torsional stiffness C_M. As one of several possible control objectives, the case considered is where it is required to accurately track the angular reference position of the second disk, which has a resulting mass moment of inertia J_{M_2}, in the presence of an external disturbance torque T_L.

The system of Figure 6.19 has the continuous-time state-space model

$$\dot{x}(t) = A_c x(t) + B_c T_D(t) + E_c T_L(t)$$

$$y(t) = C_c x(t) \tag{6.100}$$

where

$$x(t) = \begin{bmatrix} \theta_{M_1}(t) & \omega_{M_1}(t) & \theta_{M_2}(t) & \omega_{M_2}(t) \end{bmatrix}^T$$

In this model, θ_{M_1} and θ_{M_2} denote the angular positions, respectively, of the first and the second disk. Also, $\omega_{M_1} = \dot{\theta}_{M_1}$ and $\omega_{M_2} = \dot{\theta}_{M_2}$ denote the angular velocities, respectively, of the first and the second disk and

$$A_c = \begin{bmatrix} 0 & 1 & 0 & 0 \\ -\dfrac{C_M}{J_{M_1}} & -\dfrac{B_{M_1}}{J_{M_1}} & \dfrac{C_M}{J_{M_1}} & 0 \\ 0 & 0 & 0 & 1 \\ \dfrac{C_M}{J_{M_2}} & 0 & -\dfrac{C_M}{J_{M_2}} & -\dfrac{B_{M_2}}{J_{M_2}} \end{bmatrix}, \quad B_c = \begin{bmatrix} 0 \\ \dfrac{1}{J_{M_1}} \\ 0 \\ 0 \end{bmatrix}$$

$$C_c = \begin{bmatrix} 0 & 0 & 1 & 0 \end{bmatrix}, \quad E_c = \begin{bmatrix} 0 & 0 & 0 & -\dfrac{1}{J_{M_2}} \end{bmatrix}^T \tag{6.101}$$

where B_{M_1} and B_{M_2} are the coefficients of the velocity-proportional damping, respectively, for first and the second disks.

The numerical values of the system model parameters have been identified by least-squares methods, resulting in $J_{M_1} = 0.031\ 25$ kgm^2, $B_{M_1} = 0.07$ kgm^2/s, $C_M \in [8.1, 9.9]$ Nm/rad, $J_{M_2} \in [0.0313, 0.2373]$ kgm^2, $B_{M_2} = 0.2317$ kgm^2/s. In this case, the uncertain parameters are characterized by intervals. Also, the frequency response terms for the uncertain drive system model in the form (6.87) are calculated by first computing the nominal model frequency response $G(e^{j\omega_i T_s})$ with averaged parameter values using an exact discrete-time representation of the state-space model (6.100) with sampling period $T_s = 0.01$ seconds.

This computation was completed for 1000 logarithmically spaced frequency samples $\omega_i \in \{10^{-2}, \omega_N = \pi/T_s\}$. The uncertainty weighting function $W_G(\omega_i)$ for a given frequency is

$$W_G(\omega_i) = \max \left| \frac{G_g(e^{j\omega_i T_s})}{G(e^{j\omega_i T_s})} - 1 \right| \quad \text{for } g = 1, \dots, n_g \tag{6.102}$$

where $G_g(e^{j\omega_i T_s})$ is the frequency response at ω_i of the discrete-time models calculated for all possible combinations of 10 linearly spaced values of each uncertain parameters, i.e. C_M and J_{M_2}.

In the case considered, the number of $G_g(e^{j\omega_i T_s})$ for all ω_i is $n_g = 10^{n_p} = 100$, where n_p denotes the number of uncertain parameters. Figure 6.20 gives the Bode plots of the nominal discrete-time model of the drive system (left-hand side, dark gray plot) and the uncertainty weighting function W_G (right-hand side, light gray). Finally, the uncertainty circles with radii $W_G(e^{j\omega_i T_s})G(e^{j\omega_i T_s})$ have been approximated for all considered frequencies by the polygons (6.87) each with $n_v = 8$ vertices.

The set of the frequency characteristics for each vertex forms the nonparametric model of the drive system. This model is used for robust PID control design based on the optimization algorithm (6.88), where the following numerical values of the design parameters were used: $\beta = \pi/6$ rad, $\omega_x = 0.8$ rad/s, $\epsilon = \pi/4$ rad, $\ell = 0.5$, $C_{\max} = 10$, and $\omega_c = \omega_x$. Figure 6.21 shows the

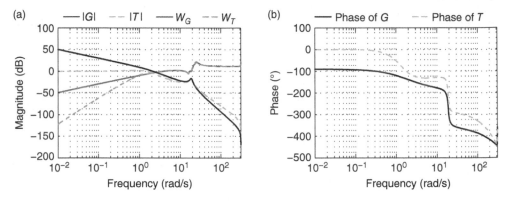

Figure 6.20 Bode plots of $G(e^{j\omega_i T_s})$ and $T(e^{j\omega_i T_s})$ together with the uncertainty weighting functions for both (a) the open- and (b) closed-loops.

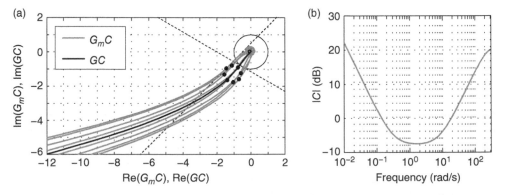

Figure 6.21 (a) Nyquist plots of the open-loop frequency response, and (b) frequency response of the PID controller.

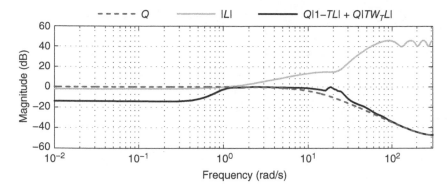

Figure 6.22 Frequency responses of the Q-filter, the learning filter, and the ILC system convergence speeds.

Nyquist plots of both the open-loop system for nominal plant model parameters (light gray plot) and $G_m(e^{j\omega_i T_s})C(e^{j\omega_i T_s})$ (dark gray plots).

A magnitude plot of the frequency response for the designed PID controller is also shown in Figure 6.21. In this last figure, the black dashed lines represent the chosen linear constraints, and the values of the blue Nyquist plots for frequency ω_x are marked by black points. These plots confirm that all the design requirements have been met.

The coefficients of the Q and L-filters are calculated based on the convex optimization algorithm (6.98). This algorithm is based on the nonparametric model of the underlying controlled system given in the form (6.89) and the frequency response of the underlying closed control loop for averaged model parameters has been calculated using (6.90), see Figure 6.20 (gray-dashed plot). The magnitude plot of the frequency response of the uncertainty weighting function $W_T(\omega_i)$ is determined in an analogous manner to $W_G(\omega_i)$, i.e. using (6.102).

In this part of the design $G(e^{j\omega_i T_s})$ is replaced by $T(e^{j\omega_i T_s})$ and $G_g(e^{j\omega_i T_s})$ by the frequency responses of the controlled system, determined for all possible combinations of 10 linearly spaced values for each of the two uncertain model parameters, covering the whole parameter intervals. The uncertainty weighting function W_T is represented by the dashed line plot in Figure 6.20. Also, the forward shift and the order of the learning filter were chosen, respectively, as $s_L = 25$ and $n_L = 10$.

The remaining design parameters of the optimization algorithm (6.98) have been selected as: $M_1 = 0.2, M_2 = 0.99, \omega_y = 0.1$ rad/s, $L_{\max} = 200$. Figure 6.22 shows the frequency responses of the Q-filter (dark gray dashed plot), the learning filter (light gray plot), and ILC system convergence speed (black plot). These results confirm that all design requirements have been met and that a fast convergence rate has been obtained up to a frequency of ≈ 0.8 rad/s.

The performance of the design was evaluated by simulating the controlled system over 300 trials. Figure 6.23a (where y^{ref} denotes the reference signal) shows the reference trajectory (light gray plot) used together with the disturbance signal (dark grey dashed plot), and Figure 6.23b the single-sided amplitude spectrum of the reference trajectory. During the simulations, the uncertain model parameters and the external disturbance torque were changed according to the data in Table 6.1. Moreover, to assess the robustness of measurement noise, normally distributed pseudorandom noise was added to the system output, with a standard deviation of 0.0016 rad.

The root mean square tracking error, denoted by $RMS(e_k)$, and given by

$$RMS(e_k) = \sqrt{\frac{1}{N}\sum_{p=0}^{N-1}\|e_k(p)\|^2} \qquad (6.103)$$

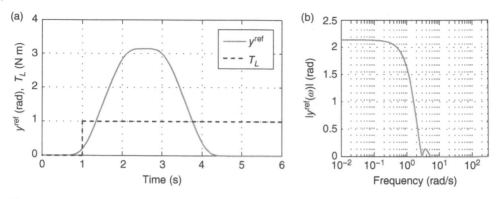

Figure 6.23 Reference and disturbance signals (a). Frequency spectrum of the reference trajectory (b).

Table 6.1 Variation of J_{M_2}, C_M, and T_L during the robust ILC application.

Trials	J_{M_2} (kgm^2)	C_M (Nm/rad)	$T_L(p)$ (Nm)
1–40	0.1343	9	0
41–80	0.0313	8.1	0
81–120	0.2373	8.1	0
121–160	0.0313	9.9	0
161–200	0.2373	9.9	0
201–240	0.1343	9	0
241–300	0.1343	9	As in Figure 6.23a

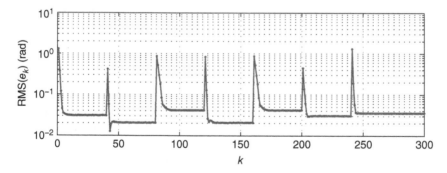

Figure 6.24 $RMS(e_k)$ values of the tracking errors.

resulting from the simulation is shown in Figure 6.24. This result confirms that the serial ILC significantly improves the tracking accuracy. The solution is robust to the considered changes in the model parameters and the impact of external disturbance torques.

The convex optimization algorithm requires only the specification of the design parameters for desired ILC performance. This robust ILC design allows simultaneous zero-phase IIR Q-filter and FIR learning filter design. (Another method for the design of a zero-phase filter in ILC is given in Chapter 8.) Moreover, the associated simulations confirm the expected performance and provide a realistic basis for experimental validation.

6.3.4 Robust ILC Analysis Revisited

This section removes some of the difficulties and limitations of the analysis so far in this chapter. In particular, the following issues are addressed, where this section follows, in the main, [262]:

1. The implications of the finite trial length on results derived under the mathematical assumption of an infinite trial length.
2. The ability to include uncertainty models in design and thereby aim for reduced conservativeness, i.e. the "gap" between a necessary and sufficient design and sufficient only.
3. Deliver the possibility of analyzing the robustness properties of a wide range of noncausal ILC laws.

The infinite trial length representation of the ILC dynamics for discrete dynamics is assumed to be defined by a real rational discrete-time transfer-function matrix $G(z)$ that is causal in z, i.e. $y(p + 1) = zy(p)$. On truncation to $p \in [0, N - 1]$, the finite trial length mapping $G : u \mapsto y$ has the form

$$
\begin{bmatrix} y(0) \\ \vdots \\ y(N-1) \end{bmatrix} = \begin{bmatrix} g(0) & \cdots & g(N-1) \\ \vdots & \ddots & \vdots \\ g(N-1) & \cdots & g(0) \end{bmatrix} \begin{bmatrix} u(0) \\ \vdots \\ u(N-1) \end{bmatrix}
\tag{6.104}
$$

where $g(0), \dots, g(N - 1)$ are the Markov parameters of the system.

As earlier in this chapter, the finite time interval model J_Δ of the uncertain dynamics is modeled as per the standard linear systems case, see, e.g. [243]. Moreover, for ease of presentation, the dimensions of the matrices and vectors arising are not explicitly stated. Suppose that $J_a(z)$ denotes a nominal model for the dynamics, which is assumed to be causal. Then the set of uncertain systems, denoted by $\prod_z$, is given by

$$
\prod_z = \{J_\Delta(z) \ : \ J_\Delta(z) = J(z) + W_i(z)\Delta(z)W_o(z), \quad \|\Delta(z)\|_2 < 1, \quad \Delta \quad \text{structured}\}
\tag{6.105}
$$

where the filters $W_i(z)$ and $W_o(z)$ are user-defined and represent the specific uncertainty in the example under consideration. These filters are designed to be stable and causal with inverses that also have these properties. Finally, $\Delta(z)$ is a causal system with a known structure.

For a given infinite time system, the finite interval mapping J_Δ maps the input to the output. For causal linear time-invariant dynamics, the matrix representing the map is a lower block triangular Toeplitz matrix. Moreover, the finite interval representations of the filters $W_i(z)$ and $W_o(z)$ are also represented by matrices, W_i and W_o, respectively, with the same structure. If $\Delta(z)$ is causal and linear time-invariant, it can be represented by a lower-block triangular Toeplitz matrix Δ.

A standard result for a lower (block) triangular Toeplitz matrix Δ is that $\|\Delta\|_2 \leq \|\Delta(z)\|_2$, i.e. the 2-norm of a finite interval lower-block triangular Toeplitz matrix can never be greater than the induced 2-norm of the underlying transfer-function (or transfer-function matrix). Consequently, the set of uncertain systems over a finite time interval (in ILC, the trial length) can be described by

$$
\prod = \{J_\Delta : J_\Delta = J + W_i \Delta W_o, \ \Delta \in \Delta\}
\tag{6.106}
$$

where $\Delta = \{\Delta : \|\Delta\|_2 \leq 1, \ \Delta \quad \text{structured}\}$. The assumption of time-invariant dynamics can be removed under certain assumptions [263]. Also, multiplicative uncertainty can be considered in the same setting, again see [263] for the details.

Consider the robust ILC system of Figure 6.25, where $w^{-1} : u_k = w^{-1}u_{k+1}$ is the one trial shift operator. In this case

$$
\begin{aligned}
u_{k+1}(p) &= Qu_k(p) + L_0 e_k(p) \\
f_k(p) &= L_c u_k(p), \quad u_0 = 0
\end{aligned}
\tag{6.107}
$$

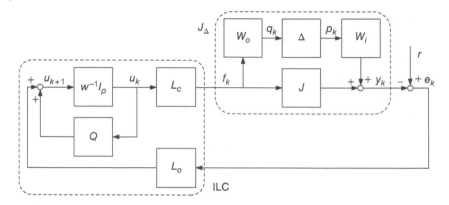

Figure 6.25 Robust ILC block diagram.

and substitution of $e_k(p) = r(p) - J_\Delta L_c u_k(p)$ gives

$$u_{k+1}(p) = (\mathcal{Q} - L_0 J_\Delta L_c)u_k(p) + L_0 r(p) \tag{6.108}$$

Two trial-to-trial error convergence conditions are of interest, i.e. for the nominal system when $W_i = 0$ and $W_o = 0$ and for the robust system. In the latter case, trial-to-trial error convergence occurs if and only if $\rho(\mathcal{Q} - L_0 J_\Delta L_c) < 1$ for all $J_\Delta \in \prod$. As discussed previously in this book, trial-to-trial error convergence may still result in undesirable transient behavior along the trials. To prevent such a case from arising in the analysis that follows, the following definition, which can be interpreted as robust monotonic convergence, is used.

Definition 6.1 Consider the ILC dynamics described by (6.107) and (6.108). Then the monotonic trial-to-trial error convergence property in f_k requires that there exists $0 \le \mathcal{K}_f < 1$ such that

$$\|f_{k+1} - f_\infty\|_2 < \mathcal{K}_f \|f_k - f_\infty\|_2 \tag{6.109}$$

$\forall k \ge 0$, $r(p)$ and $J_\Delta \in \prod$, with $f_\infty = \lim_{k\to\infty} f_k$. Also, robust monotonic trial-to-trial error convergence holds for e_k if there exists $0 \le \mathcal{K}_e < 1$ such that

$$\|e_{k+1} - e_\infty\|_2 < \mathcal{K}_e \|e_k - e_\infty\|_2 \tag{6.110}$$

$\forall k \ge 0$, $r(p)$ and $J_\Delta \in \pi$.

The reason for the inclusion of the converged error $e_\infty = \lim_{k\to\infty} e_k$ in this last definition is that, in general, for uncertain systems and arbitrary r, $e_\infty \ne 0$. If $e_\infty = 0$, this definition reduces to

$$\|e_{k+1}\|_2 < \mathcal{K}_e \|e_k\|_2 \Leftrightarrow \|y_{k+1} - r\|_2 < \mathcal{K}_e \|y_k - r\|_2 \tag{6.111}$$

In this section, $L_c \ne I$ is considered instead of $L_c = I$ as in some other published analysis. Moreover, a modification, see [260], is required in what follows if $\mathcal{Q} = I$, $L_c = I$ and the rank of J is less than the dimension of the input vector.

Nominal performance of the ILC dynamics considered is defined as follows.

Definition 6.2 The ILC dynamics described by (6.107) and (6.108) with J_Δ replaced by J, i.e. $\Delta = 0$, is said to have nominal performance $\|e_\infty\|_2$.

In the nominal case, $\|e_\infty\|_2 = 0$ for any r if and only if $\mathcal{Q} = I$ and the rank of J is equal to the dimension of the input vector.

To establish the condition for robust monotonic trial-to-trial error convergence in f_k, start from

$$f_{k+1} = L_c(Q - L_0 J_\Delta L_c)u_k + L_0 r$$
$$f_\infty = L_c(Q - L_0 J_\Delta L_c)u_\infty + L_0 r \tag{6.112}$$

Then setting $\tilde{u}_k = u_k - u_\infty$, $\|f_{k+1} - f_\infty\|_2 < \|f_k - f_\infty\|_2$ gives

$$\|L_c(Q - L_0 J_\Delta L_c)\tilde{u}_k\|_2 < \|L_c \tilde{u}_k\|_2 \tag{6.113}$$

Hence,

$$\tilde{u}_k^T(Q - L_0 J_\Delta L_c)^T L_c^T L_c(Q - L_0 J_\Delta L_c)\tilde{u}_k < \tilde{u}_k^T L_c^T L_c \tilde{u}_k \tag{6.114}$$

The result now follows and is stated next together with the condition for robust monotonic trial-to-trial convergence in e_k. For detailed proof, see Lemma 2 of [260].

Lemma 6.10 *The ILC dynamics described by (6.107) and (6.108) has the robust monotonic trial-to-trial error convergence property in f_k if all eigenvalues of*

$$m_{\text{rmc}} = (Q - L_0 J_\Delta L_c)^T L_c^T L_c(Q - L_0 J_\Delta L_c) - L_c^T L_c \tag{6.115}$$

(this matrix is symmetric and, hence, has only real eigenvalues) are strictly negative $\forall J_\Delta \in \prod$. Also, robust monotonic trial-to-trial error convergence in e_k holds if all eigenvalues of

$$m_{\text{rme}} = (Q - L_0 J_\Delta L_c)^T L_c^T J_\Delta^T J_\Delta L_c(Q - L_0 J_\Delta L_c) - L_c^T J_\Delta^T J_\Delta L_c \tag{6.116}$$

are strictly negative $\forall J_\Delta \in \prod$.

This last result holds independent of the dimension and rank of J_Δ. Also, setting $L_c = I$ and $\Delta = 0$ recovers the well-known nominal model monotonic convergence condition $\|Q - L_0 J\|_2 < 1$. Moreover, an immediate corollary of this last result is that if the matrix T is such that $T^T L_c^T L_c T = I$, then the ILC dynamics considered is has the robust monotonic trial-to-trial error convergence property in f_k if

$$\|T^{-1}(Q - L_0 J_\Delta L_c)T\|_2 < 1 \tag{6.117}$$

$\forall J_\Delta \in \prod$. This result states that for the current trial state vector transformation $z_k = T^{-1}u_k$, robust monotonic trial-to-trial error convergence in z_k implies the same property in f_k and also that T can be obtained, by, e.g. a Cholesky decomposition of $L_c^T L_c$.

6.4 Concluding Remarks

It is also possible to extend the analysis setting of this section to allow the use of μ analysis, where for a background on this approach to robust control, see, e.g. [19, 194]. The details for the ILC setting can again be found in [262, 263]. Robust ILC design is considered again in Chapter 7, where the 2D systems setting for ILC analysis and design is considered and also in Chapter 10 with a particular focus on the stroke rehabilitation application area, and the dynamics are nonlinear.

7

Repetitive Process-Based ILC Design

The analysis in this chapter begins with design in a repetitive process setting and experimental verification on the gantry robot of Section 1.6, where the analysis assumes an exact model of the dynamics. The extension to robust design is also covered again with experimental verification. Also, iterative learning control (ILC) design in the differential setting is treated, together with designs that use the generalized KYP lemma, extending to robust design and experimental validation for the latter case. The use of the repetitive process setting for higher-order iterative learning control (HOILC) design is also considered. Finally, a relatively recent approach termed inferential design is considered, which has also been experimentally verified.

7.1 Design with Experimental Validation

7.1.1 Discrete Nominal Model Design

Consider again the ILC problem for discrete linear dynamics, i.e.

$$x_{k+1}(p+1) = Ax_{k+1}(p) + Bu_{k+1}(p)$$
$$y_{k+1}(p) = Cx_{k+1}(p), \quad x_{k+1}(0) = 0, \quad 0 \le p \le N-1, \quad k \ge 0 \tag{7.1}$$

and write the state equation of this model in the form

$$x_k(p) = Ax_{k+1}(p-1) + Bu_k(p-1) \tag{7.2}$$

Also, introduce

$$\eta_{k+1}(p+1) = x_{k+1}(p) - x_k(p) \tag{7.3}$$

and

$$u_{k+1}(p) = u_k(p) + \Delta u_{k+1}(p) \tag{7.4}$$

where

$$\Delta u_{k+1}(p) = K_1 \eta_{k+1}(p+1) + K_2 e_k(p+1) \tag{7.5}$$

The resulting controlled system dynamics are described by

$$\eta_{k+1}(p+1) = (A + BK_1)\eta_{k+1}(p) + BK_2 e_k(p)$$
$$e_{k+1}(p) = -C(A + BK_1)\eta_{k+1}(p) + (I - CBK_2)e_k(p) \tag{7.6}$$

Iterative Learning Control Algorithms and Experimental Benchmarking, First Edition.
Eric Rogers, Bing Chu, Christopher Freeman and Paul Lewin.
© 2023 John Wiley & Sons Ltd. Published 2023 by John Wiley & Sons Ltd.

This last description of the ILC dynamics is a particular case of the discrete linear repetitive process state-space model (3.34) with zero input, current trial state vector $\eta_{k+1}(p)$, and previous trial profile vector $e_k(p)$. Hence, Theorem 3.8 can be applied to give necessary and sufficient conditions for stability along the trial. This result is stated again as follows, where the following notation is adopted.

$$\hat{A} = A + BK_1, \quad \hat{B}_0 = BK_2$$
$$\hat{C} = -C(A + BK_1), \quad \hat{D}_0 = I - CBK_2 \tag{7.7}$$

Theorem 7.1 *[228] Suppose that the pair $\{\hat{A}, \hat{B}_0\}$ is controllable and the pair $\{\hat{C}, \hat{A}\}$ observable. Then the ILC dynamics described by the discrete linear repetitive process state-space model (7.6) has the stability along the trial property if and only if*

(a) $\rho(\hat{D}_0) < 1$
(b) $\rho(\hat{A}) < 1$
(c) *all eigenvalues of the transfer-function matrix*

$$G(z) = \hat{C}(zI - \hat{A})^{-1}\hat{B}_0 + \hat{D}_0 \tag{7.8}$$

have modulus strictly less than unity for all $|z| = 1$.

The use of this last result in control design is returned to in Section 7.3 after the Lyapunov approach to the stability analysis and control law design for linear repetitive processes is considered. In a repetitive process/2D linear systems theory, there are two forms of Lyapunov equations for stability analysis, respectively, 1D and 2D [228]. In the former, the defining matrices have entries that are functions of a complex variable, and in the latter, the defining matrices have real constant entries, which are used below for ILC design.

Consider the following candidate Lyapunov function for (7.6)

$$V(k, p) = V_1(k, p) + V_2(k, p)$$
$$= \eta_{k+1}^T(p)P_1\eta_{k+1}(p) + e_k^T(p)P_2e_k(p) \tag{7.9}$$

where $P_1 > 0$ and $P_2 > 0$. Also, define the increment of this Lyapunov function as follows:

$$\Delta V(k, p) = V_1(k, p + 1) - V_1(k, p) + V_2(k + 1, p) - V_2(k, p) \tag{7.10}$$

This Lyapunov function is the sum of a term that measures the energy along the trial ($V_1(p, k)$) and a term that measures the energy from trial-to-trial ($V_2(k, p)$) and the corresponding increments measure the change in these quantities. Moreover, it can be shown [228] that a sufficient condition for stability along the trial is

$$\Phi^T P\Phi - P < 0 \tag{7.11}$$

where

$$\Phi = \begin{bmatrix} \hat{A} & \hat{B}_0 \\ \hat{C} & \hat{D}_0 \end{bmatrix}, \quad P = \text{diag}[P_1, P_2] \tag{7.12}$$

The 2D Lyapunov equation has the same structure as that for a standard discrete linear time-invariant system with state matrix Φ. However, it is, in general, a sufficient, as opposed to a necessary and sufficient, condition for stability along the trial. However, in some special cases, it is necessary and sufficient where the most relevant are single-input single-output (SISO) examples. For a detailed treatment, see, e.g. [228], which cites the original work.

Returning to the ILC setting, the following result, proved as Theorem 1 in [107], is an linear matrix inequality (LMI)-based characterization of stability along the trial with formulas for computing the stabilizing control law matrices. For background on LMIs, see, e.g. [31].

Theorem 7.2 *The ILC dynamics described by the discrete linear repetitive process state-space model (7.6) has the stability along the trial property if there exist compatibly dimensioned matrices $X_1 > 0, X_2 > 0, R_1,$ and R_2 such that the following LMI is feasible:*

$$M = \begin{bmatrix} -X_1 & 0 & X_1 A^T + R_1^T B^T & -X_1 A^T C^T - R_1^T B^T C^T \\ 0 & -X_2 & R_2^T B^T & X_2 - R_2^T B^T C^T \\ AX_1 + BR_1 & BR_2 & -X_1 & 0 \\ -CAX_1 - CBR_1 & X_2 - CBR_2 & 0 & -X_2 \end{bmatrix} < 0 \quad (7.13)$$

If this LMI is feasible, stabilizing control law matrices K_1 and K_2 are given by

$$K_1 = R_1 X_1^{-1}, \quad K_2 = R_2 X_2^{-1} \quad (7.14)$$

The control law (7.4) can be written as follows:

$$u_{k+1}(p) = u_k(p) + K_1(x_{k+1}(p) - x_k(p)) + K_2(r(p+1) - y_k(p+1)) \quad (7.15)$$

where the last term on the right-hand side is phase-lead ILC action. Moreover, the second term on the right-hand side involves the state vectors' difference on the current and previous trials. Consequently, if all state vector entries cannot be measured, one option is to use an observer to estimate them.

An alternative to state feedback is to use the trial output. Consider the case when in (7.5) is replaced by

$$\Delta u_{k+1}(p) = K_1 \mu_{k+1}(p+1) + K_2 \mu_{k+1}(p) + K_3 e_k(p+1) \quad (7.16)$$

where

$$\mu_k(p) = y_k(p-1) - y_{k-1}(p-1) = C \eta_k(p) \quad (7.17)$$

The additional term in this control law relative to (7.4) has been added as a means, if necessary, of compensating for the effects of not assuming that the state vector is available for use. By routine analysis, (7.16) can be written as follows:

$$\Delta u_{k+1}(p-1) = K_1 C \eta_{k+1}(p) + K_2 C \eta_{k+1}(p-1) + K_3 e_k(p) \quad (7.18)$$

and on introducing

$$\tilde{\eta}_{k+1}(p+1) = \begin{bmatrix} \eta_{k+1}(p+1) \\ \eta_{k+1}(p) \end{bmatrix} \quad (7.19)$$

the controlled dynamics can be written as follows:

$$\begin{aligned} \tilde{\eta}_{k+1}(p+1) &= \hat{A} \tilde{\eta}_{k+1}(p) + \hat{B}_0 e_k(p) \\ e_{k+1}(p) &= \hat{C} \tilde{\eta}_{k+1}(p) + \hat{D}_0 e_k(p) \end{aligned} \quad (7.20)$$

where

$$\hat{A} = \begin{bmatrix} A + BK_1 C & BK_2 C \\ I & 0 \end{bmatrix}, \quad \hat{B}_0 = \begin{bmatrix} BK_3 \\ 0 \end{bmatrix}$$

$$\hat{C} = \begin{bmatrix} -CA - CBK_1 C & -CBK_2 C \end{bmatrix}, \quad \hat{D}_0 = I - CBK_3 \quad (7.21)$$

and again, repetitive process stability theory can be applied. The following result is proved as Theorem 2 in [108] and, from this point onward, $(\star)$ denotes block entries in a symmetric matrix.

Theorem 7.3 *The ILC dynamics described by the discrete linear repetitive process state-space model (7.20) has the stability along the trial property if there exist compatibly dimensioned matrices $Y > 0$, $Z > 0$, N_1, N_2, and N_3 such that the following LMI with linear constraints is feasible:*

$$\begin{bmatrix} Z - Y & (\star) & (\star) \\ 0 & -Z & (\star) \\ \Omega_1 & \Omega_2 & -Y \end{bmatrix} < 0$$
$$CY_1 = PC, \quad CY_2 = QC \tag{7.22}$$

where

$$Y = \begin{bmatrix} Y_1 & 0 & 0 \\ 0 & Y_2 & 0 \\ 0 & 0 & Y_3 \end{bmatrix} \tag{7.23}$$

and

$$\Omega_1 = \begin{bmatrix} AY_1 + BN_1C & BN_2C & BN_3 \\ Y_1 & 0 & 0 \\ 0 & 0 & 0 \end{bmatrix}$$
$$\Omega_2 = \begin{bmatrix} 0 & 0 & 0 \\ 0 & 0 & 0 \\ -CAY_1 - CBN_1C & -CBN_2C & Y_3 - CBN_3 \end{bmatrix} \tag{7.24}$$

The matrices P and Q are additional decision variables. If the LMI with equality constraints of (7.21) is feasible, the control law matrices are given by

$$\hat{K}_1 = N_1P^{-1}, \quad \hat{K}_2 = N_2Q^{-1}, \quad \hat{K}_3 = N_3Y_3^{-1} \tag{7.25}$$

For application, the control law (7.16) can be written as follows:

$$u_k(p) = u_{k-1}(p) + K_1(y_k(p) - y_{k-1}(p)) + K_2(y_k(p-1) - y_{k-1}(p-1)) + K_3(r(p+1) - y_{k-1}(p+1)) \tag{7.26}$$

The last term in (7.26) is phase lead on the previous trial error and the second and third terms are proportional acting, on the error between the current and previous trials at, respectively, p and $p - 1$, which are novel additions to the class of simple structure ILC laws. Implementation of this control law does not require a state observer but does assume that the level of noise and other disturbances on the measurements is negligible (stochastic ILC is considered in Chapter 12).

Both of the control laws in this section have been experimentally verified on the gantry robot of Section 2.1.1. In this section, the results for (7.15) are given with those for (7.26) in [108]. Consider the X-axis of the gantry robot where the minimal state-space model matrices are given by (2.1). Suppose also that the following constraints are imposed (where for this particular case X_2 is a scalar) on the LMI of Theorem 7.2:

$$X_1^TX_1 < 1 \times 10^{-4}I, \quad X_2 < 10^{-4}, \quad R_1^TR_1 < 1 \times 10^{-4}I \tag{7.27}$$

resulting in the control law matrices:

$$K_1 = \begin{bmatrix} 7.3451 & -2.7245 & 0.1499 & 7.6707 & 2.7540 & -3.6088 & -20.4519 \end{bmatrix}$$
$$K_2 = 82.4119 \tag{7.28}$$

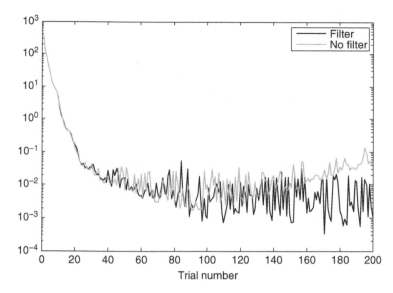

Figure 7.1 The effect of filtering – light gray line without, dark gray line – filter added.

In, e.g. [150], it is reported that ILC laws can exhibit higher-frequency noise buildup as the number of trials increases. Tracking of the reference signal then begins to diverge due to numerical problems in both computation and measurement. In this design, the higher-frequency component buildup was observed in some cases, resulting in vibrations that increase the trial error. One relatively simple option in such cases is to employ a zero-phase Chebyshev low-pass filter, and for this application consider the case when

$$H(z) = \frac{0.0002 + 0.0007z^{-1} + 0.0011z^{-2} + 0.0007z^{-3} + 0.0002z^{-4}}{1 - 3.5328z^{-1} + 4.7819z^{-2} - 2.9328z^{-3} + 0.6868z^{-4}} \tag{7.29}$$

with cut-off frequency 10 Hz and is applied to the current trial error. Figure 7.1 shows a case where the trial error without filtering starts to diverge after (approximately) 100 trials, but the addition of the filter (7.29) can maintain (this aspect of overall) performance. (An alternative would be to include the zero-phase filter design in the problem formulation, as in Section 6.3.3. Another design using this latter approach is given in Chapter 8.)

As a representative of the performance possible and using the tuning opportunities offered by the LMI solutions, Figure 7.2 shows the experimentally measured, with the zero-phase filter in place, input error, and output progression over 20 trials. These results show that the tracking error has been reduced to a minimal value within 20 trials. Figure 7.3 shows the input, error, and output signals, respectively, on trial 200, which are highly acceptable.

The design was repeated for the Y axis (perpendicular to the X-axis in the same plane) and Z (perpendicular to the X–Y plane). Figure 7.4 shows the mean square error (MSE) for all axes in comparison to those from a simulation study for each corresponding axis with the ILC control law of (7.15) applied.

As a comparison, consider again Figure 3.1, which gives the simulation results for a Roesser model-based design [136] for this example when $K_1 = 0$. The superiority of the repetitive process-based design is evident.

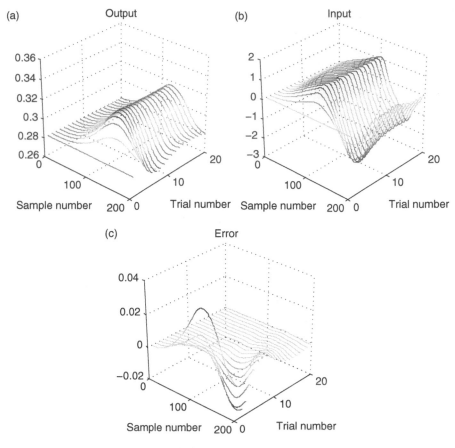

Figure 7.2 The input, error, and output progression for the design of (7.28).

7.1.2 Robust Design – Norm-Bounded Uncertainty

In this section, the focus is on discrete systems with additive uncertainty in the system state-space model matrices of the norm-bounded form, i.e.

$$A = \tilde{A} + \Delta A, \quad B = \tilde{B} + \Delta B \tag{7.30}$$

where $\tilde{A}$ and $\tilde{B}$ represent the nominal versions of A and B, respectively, and by assumption, ΔA and ΔB satisfy

$$\Delta A = H_1 F_1 E_1, \quad \Delta B = H_2 F_2 E_2 \tag{7.31}$$

where $F_1 = F_1^T$, $F_1^T F_1 \preccurlyeq I$, $F_2 = F_2^T$, $F_2^T F_2 \preccurlyeq I$, and $F_1, F_2, H_1, E_1, H_2, E_2$ are matrices of compatible dimensions, and $\preccurlyeq 0$ denotes a positive semi-definite matrix. Here H_1, H_2, E_1, and E_2 are assumed constant, but F_1 and F_2 can vary both from trial-to-trial and point-to-point along-the-trial, provided they are norm-bounded.

Suppose that the control law given by (7.4) and (7.16) is applied and write the state-space model of the controlled dynamics as

$$\begin{aligned} \tilde{\eta}_{k+1}(p+1) &= \widehat{A}\tilde{\eta}_{k+1}(p) + \widehat{B}_0 e_k(p) \\ e_{k+1}(p) &= \widehat{C}\tilde{\eta}_{k+1}(p) + \widehat{D}_0 e_k(p) \end{aligned} \tag{7.32}$$

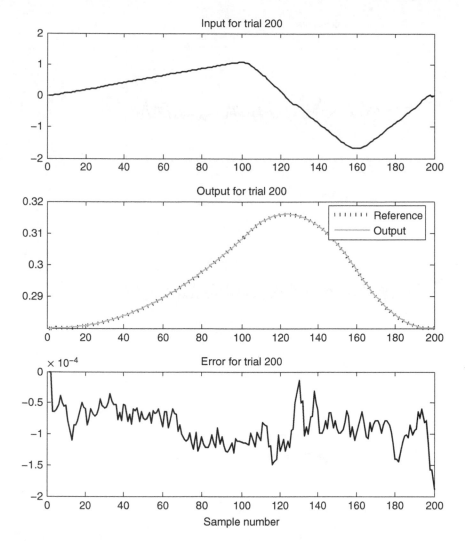

Figure 7.3 The output on trial 200 compared to the reference together with the input (middle plot) and the error (bottom plot) for the design of (7.28).

where

$$\widehat{A} = \begin{bmatrix} \tilde{A} + \Delta A + (\tilde{B} + \Delta B)K_1 C & (\tilde{B} + \Delta B)K_2 C \\ I & 0 \end{bmatrix}$$

$$\widehat{B}_0 = \begin{bmatrix} (\tilde{B} + \Delta B)K_3 \\ 0 \end{bmatrix} \tag{7.33}$$

$$\widehat{C} = \begin{bmatrix} -CA - CBK_1 C & -CBK_2 C \end{bmatrix}, \quad \widehat{D}_0 = I - CBK_3$$

which is of the form (7.20) and, hence, the repetitive process stability theory can also be applied to this case, leading to the following result proved as Theorem 3 in [108].

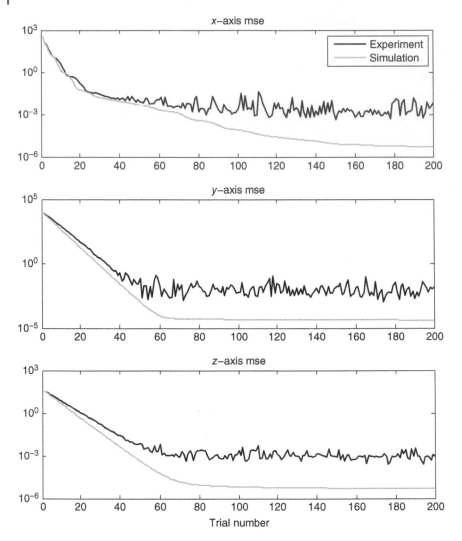

Figure 7.4 MSE for all axes generated by the design of (7.28).

Theorem 7.4 *The ILC dynamics described by the discrete linear repetitive process state-space model (7.32) has the stability along the trial property if there exist compatibly dimensioned matrices $Y > 0$, $Z > 0$, N_1, N_2, N_3, P, Q, and real scalars $\tilde{e}_1 > 0$ and $\tilde{e}_2 > 0$ such that the following LMI with linear constraints is feasible:*

$$
\begin{bmatrix}
Z - Y & (*) & (*) & (*) & (*) & (*) \\
0 & -Z & (*) & (*) & (*) & (*) \\
\Omega_{31} & \Omega_{32} & -Y & (*) & (*) & (*) \\
0 & 0 & \Omega_{43} & \Omega_{44} & (*) & (*) \\
\Omega_{51} & 0 & \Omega_{53} & 0 & \Omega_{55} & (*) \\
\Omega_{61} & \Omega_{62} & 0 & 0 & 0 & \Omega_{66}
\end{bmatrix} < 0
$$

$$
CY_1 = PC, \quad CY_2 = QC
$$

(7.34)

where

$$Y = \begin{bmatrix} Y_1 & 0 & 0 \\ 0 & Y_2 & 0 \\ 0 & 0 & Y_3 \end{bmatrix}, \quad \Omega_{31} = \begin{bmatrix} \tilde{A}Y_1 + \breve{B}N_1C & \breve{B}N_2C & \breve{B}N_3 \\ Y_1 & 0 & 0 \\ 0 & 0 & 0 \end{bmatrix}$$

$$\Omega_{32} = \begin{bmatrix} 0 & 0 & 0 \\ 0 & 0 & 0 \\ -C\tilde{A}Y_1 - C\breve{B}N_1C & -C\breve{B}N_2C & Y_3 - C\breve{B}N_3 \end{bmatrix}$$

$$\Omega_{43} = \begin{bmatrix} \tilde{e}_2 H_2^T & 0 & 0 \\ 0 & \tilde{e}_2 H_2^T & 0 \\ 0 & 0 & -\tilde{e}_2 H_2^T C^T \end{bmatrix}, \quad \Omega_{44} = \begin{bmatrix} -\tilde{e}_2 I & 0 & 0 \\ 0 & -\tilde{e}_2 I & 0 \\ 0 & 0 & -\tilde{e}_2 I \end{bmatrix}$$

$$\Omega_{51} = \begin{bmatrix} 0 & 0 & 0 \\ 0 & 0 & 0 \\ E_2 N_1 C & E_2 N_2 C & E_2 N_3 \end{bmatrix}, \quad \Omega_{53} = \begin{bmatrix} \tilde{e}_1 H_1^T & 0 & 0 \\ 0 & 0 & -\tilde{e}_1 H_1^T C^T \\ 0 & 0 & 0 \end{bmatrix}$$

$$\Omega_{55} = \begin{bmatrix} -\tilde{e}_1 I & 0 & 0 \\ 0 & -\tilde{e}_1 I & 0 \\ 0 & 0 & -\tilde{e}_2 I \end{bmatrix}, \quad \Omega_{61} = \begin{bmatrix} 0 & 0 & 0 \\ E_1 Y_1 & 0 & 0 \\ 0 & 0 & 0 \end{bmatrix}$$

$$\Omega_{62} = \begin{bmatrix} E_2 N_1 C & E_2 N_2 C & E_2 N_3 \\ 0 & 0 & 0 \\ E_1 Y_1 & 0 & 0 \end{bmatrix}, \quad \Omega_{66} = \begin{bmatrix} -\tilde{e}_2 I & 0 & 0 \\ 0 & -\tilde{e}_1 I & 0 \\ 0 & 0 & -\tilde{e}_1 I \end{bmatrix}$$

If this LMI with equality constraints is feasible, the control law matrices are given by

$$K_1 = N_1 P^{-1}, \quad K_2 = N_2 Q^{-1}, \quad K_3 = N_3 Y_3^{-1} \tag{7.35}$$

For application, the control law is given by (7.26) (interpreted for this case).

This ILC design has also been tested on the gantry robot, where, for the X-axis, the matrices $E_1, E_2, H_1,$ and H_2 are given in [108] and $F_1 = F_2 = 0.99I$. Two possible sets of control law matrices are

$$K_1 = -121.5, \quad K_2 = -13.9, \quad K_3 = 57.5 \tag{7.36}$$

$$K_1 = -199.0, \quad K_2 = -29.9, \quad K_3 = 10.8 \tag{7.37}$$

and Figure 7.5 shows the progression of the input, output, and error as the trials are completed under the control law defined by (7.36). Figure 7.6 shows the MSE plotted against trial number for (7.36) compared against (7.37). These show that acceptable performance levels are possible with this design and that varying the control law matrices can influence performance.

As discussed in Chapter 6, many alternative approaches to robust ILC control design for discrete linear systems have been reported. One of these is to use the lifted model of the dynamics, but this will encounter difficulties due to the presence of matrix products in the resulting model. For example, the phase-lead ILC law, i.e.

$$u_{k+1}(p) = u_k(p) + K e_k(p+1) \tag{7.38}$$

results in error dynamics described by the lifted model:

$$E_{k+1} = \mathcal{Q} E_k \tag{7.39}$$

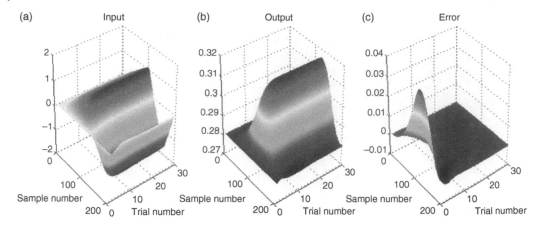

Figure 7.5 Experimentally measured input (a), output (b), and error (c) progression with trial number with the control law (7.36) applied.

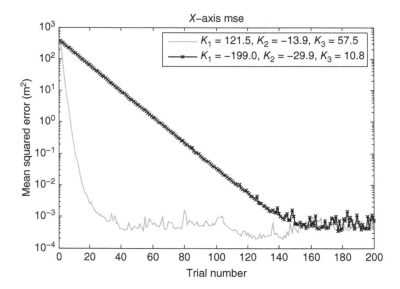

Figure 7.6 A comparison of experimentally measured MSE for the *X*-axis against trial number for the control law matrices of (7.36) and (7.37).

where

$$
\mathcal{Q} = \begin{bmatrix} I - KCB & 0 & \cdots & 0 \\ -KCAB & I - KCB & \cdots & 0 \\ \vdots & \vdots & \ddots & \vdots \\ -KCA^{N-1}B & -KCA^{N-2}B & \cdots & I - KCB \end{bmatrix} \tag{7.40}
$$

Hence, when the state-space model matrices A, B, C have uncertainty associated with them that belong to a convex set, $\mathcal{Q}$ does not belong to a convex set, and only a bound is possible that increases the conservativeness of the results.

7.1.3 Robust Design – Polytopic Uncertainty and Simplified Implementation

Under polytopic uncertainty for discrete dynamics, the state-space model of the uncontrolled dynamics in the ILC setting is of the form:

$$x_{k+1}(p+1) = A(\lambda)x_{k+1}(p) + B(\lambda)u_{k+1}(p)$$
$$y_{k+1}(p) = Cx_{k+1}(p) \tag{7.41}$$

where the notation and dimensions are as in (7.1) and the matrices $A(\lambda)$ and $B(\lambda)$ are assumed to belong to a convex polytopic given by

$$\begin{bmatrix} A(\lambda) & B(\lambda) \end{bmatrix} = \sum_{j=1}^{M} \lambda_j \begin{bmatrix} A_j & B_j \end{bmatrix}$$
$$\lambda = \begin{bmatrix} \lambda_1 \\ \vdots \\ \lambda_M \end{bmatrix}, \quad \lambda_j \geq 0, \quad \sum_{j=1}^{M} \lambda_j = 1 \tag{7.42}$$

Applying the ILC law given by (7.4) and (7.5) results in the controlled dynamics state-space model:

$$\eta_{k+1}(p+1) = \mathscr{A}(\lambda)\eta_{k+1}(p) + \mathscr{B}_0(\lambda)e_k(p)$$
$$e_{k+1}(p) = \mathscr{C}(\lambda)\eta_{k+1}(p) + \mathscr{D}_0(\lambda)e_k(p) \tag{7.43}$$

where

$$\mathscr{A}(\lambda) = A(\lambda) + B(\lambda)K_1, \quad \mathscr{B}_0(\lambda) = B(\lambda)K_2$$
$$\mathscr{C}(\lambda) = -C(A(\lambda) + B(\lambda)K_1), \quad \mathscr{D}_0(\lambda) = I - CB(\lambda)K_2 \tag{7.44}$$

The design can now proceed in the same way as for norm-bounded uncertainty case of Section 7.1.2.

Implementation of the ILC law (7.4) with the control update (7.5) requires storage of $u_k(p)$, $e_k(p+1)$, and the previous trial state vector $x_k(p)$ over the complete trial length, see Figure 7.7a.

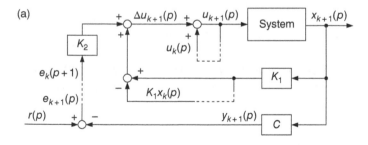

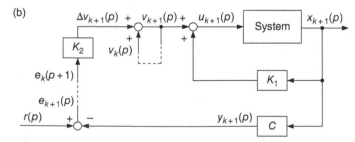

Figure 7.7 Block diagrams of ILC with the commonly used ILC law (a) and with the simplified equivalent ILC law (b).

As shown next, storage of $K_1 x_k(p)$ is not required for implementation, which is of significant practical importance since it allows the use of a control law that requires less memory allocation. (Especially relevant for examples with long trial lengths, where such examples are considered in [140] from the standpoint of the computations needed to compute a norm optimal iterative learning control (NOILC) law in such cases).

The ILC law considered requires storage, in real-time, of data from the current $(k+1)$ trial that realizes the state feedback stabilizing controller $K_1 x_{k+1}(p)$ and also feedforward data based on two signals from the previous trial $-K_1 x_k(p)$ and $K_2 e_k(p+1)$. Suppose, however, that the (current trial) state feedback control law

$$u_{k+1}(p) = K_1 x_{k+1}(p) + v_{k+1}(p) \tag{7.45}$$

with the reference input $v_{k+1}(p)$ is applied to (7.41). Then the resulting dynamics are described by

$$x_{k+1}(p+1) = \left(A(\lambda) + B(\lambda)K_1\right) x_{k+1}(p) + B(\lambda)v_{k+1}(p)$$
$$y_{k+1}(p) = Cx_{k+1}(p) \tag{7.46}$$

Next, apply the ILC law

$$v_{k+1}(p) = v_k(p) + \Delta v_{k+1}(p) \tag{7.47}$$

with

$$\Delta v_{k+1}(p) = K_2 e_k(p+1) \tag{7.48}$$

to (7.46). Then, after routine manipulations, and using (7.3)

$$\eta_{k+1}(p+1) = \left(A(\lambda) + B(\lambda)K_1\right) \eta_{k+1}(p) + B(\lambda)\Delta v_{k+1}(p-1)$$
$$y_{k+1}(p) = C\eta_{k+1}(p) + y_k(p) \tag{7.49}$$

The second equation of (7.49) can be written as follows:

$$e_{k+1}(p) = -C\eta_{k+1}(p) + e_k(p) \tag{7.50}$$

and it follows from routine manipulations that the controlled dynamics are again of the form of the uncontrolled dynamics. Also the following result is proved as Theorem 1 in [157].

Theorem 7.5 *The ILC dynamics described by the discrete linear repetitive process (7.49) is stable along the trial if for given $0 < \mu < 1$, there exist compatibly dimensioned matrices $Y_i > 0$, H, N, K_2 such that for $j = 1, \ldots, M$, the LMIs*

$$\begin{bmatrix} Y_j & A_jH + B_jN & B_jK_2 & 0 \\ (\star) & H + H^T - Y_j & 0 & (-CA_jH - CB_jN)^T \\ (\star) & (\star) & I & (I - CB_jK_2)^T \\ (\star) & (\star) & (\star) & \mu I \end{bmatrix} > 0 \tag{7.51}$$

are feasible. If these LMIs are feasible, the control law matrix K_1 can be computed as follows:

$$K_1 = NH^{-1} \tag{7.52}$$

A fast trial-to-trial error convergence rate requires that the term μ in (7.51) is minimized. However, this can lead to large valued entries in the matrix K_1 and consequently to a significantly noisy control signal $u_{k+1}(p)$. Hence, the values of the entries in the matrix K_1 should be limited, which leads to slower dynamics from the state feedback control system and consequently a slower convergence rate of the tracking error from trial-to-trial, and a compromise is necessary, and the following result [157] holds.

Theorem 7.6 *If for given ϕ_i, $i = 1, \ldots, M$, there exist matrices N, H, Z $\geqslant$ 0, and $Y_j \geqslant$ 0 such that $\forall j = 1, \ldots, M$ the LMIs (7.51) are feasible and*

$$\begin{bmatrix} Z & N \\ N^T & H + H^T - Y_j \end{bmatrix} \geqslant 0 \tag{7.53}$$

and each diagonal element $(Z)_{i,i}$ of Z satisfies

$$(Z)_{i,i} \leq \phi_i \tag{7.54}$$

then (7.49) is stable along the trial and the entries in the control law matrix K_1 calculated from (7.52) are limited by appropriately chosen values ϕ_i.

A balance between the dynamics along the trial and fast monotonic trial-to-trial terror convergence can be attempted using the following minimization problem:

$$\text{minimize}(\mu), \quad \text{subject to}: (7.51), \quad \mu < 1, (7.53) \text{ and } (7.54) \tag{7.55}$$

If a solution of (7.55) exists, K_1 can be calculated using (7.52). Also, K_2 is an optimization variable and is therefore obtained directly.

As a simulation-based case study, consider the following state-space model [154] of two drives with permanent magnet synchronous motors connected by a stiff shaft with an additional rotary mass attached:

$$\dot{x}(t) = A_c x(t) + B_c u(t) + E_c d(t)$$
$$y(t) = C_c x(t) \tag{7.56}$$

with

$$A_c = \begin{bmatrix} 0 & 1 \\ 0 & -\dfrac{b}{J} \end{bmatrix}, \quad B_c = \begin{bmatrix} 0 \\ \dfrac{k_t}{J} \end{bmatrix}, \quad E_c = \begin{bmatrix} 0 \\ -\dfrac{1}{J} \end{bmatrix}, \quad C_c = \begin{bmatrix} 1 & 0 \end{bmatrix}$$

$$x(t) = \begin{bmatrix} \theta(t) \\ \omega(t) \end{bmatrix}, \quad u(t) = i_q^{\text{ref}}(t), \quad d(t) = T_l(t)$$

where b denotes the resulting friction coefficient, J the overall mass moment of inertia, k_t the torque constant of the first drive, $\theta(t)$ the motor shaft angle, $\omega(t) = \dot{\theta}(t)$ the angular velocity of the motor shaft, $i_q^{\text{ref}}(t)$ is the reference motor current of the first drive, and $T_l(t)$ the load torque generated by the second drive. The model parameter values are known only approximately in the finite intervals $b \in [1.3 \times 10^{-3}, \ 1.5 \times 10^{-3}] \ \text{kg}\,\text{m}^2/\text{s}, J \in [5.9 \times 10^{-4}, \ 10.6 \times 10^{-4}] \ \text{kg}\,\text{m}^2, k_t \in [0.35, \ 0.39] \ \text{N}\,\text{m/A}$.

The dynamics of this system have been sampled using the impulse invariant discretization method with a sampling period 2.5×10^{-3} seconds. Also, standard software has been used to construct the six vertices of a minimum convex set that satisfies (7.42). Table 7.1 gives the numerical solutions (K_1, K_2, γ) of the robust ILC design, where $\gamma = \sqrt{\mu}$ for three cases, and, for comparison, (c) is the case when the constraints (7.53) and (7.54) are not enforced.

These results confirm that the constraints (7.53) and (7.54) on the entries in K_1 are enforced. By reducing ϕ_1, lower values of the entries in K_1 and K_2 are obtained. If these constraints are not applied (the third design), the entries in K_1 and K_2 are larger. As expected, the larger limitations on the entries in K_1 result in a larger γ, i.e. slower trial-to-trial error convergence.

Table 7.1 Numerical solutions of the three considered designs.

Variant (design)	K_1	K_2	γ
(a) $\phi_1 = 50\,000$	$[-150.061 - 0.740]$	62.274	0.957
(b) $\phi_1 = 20\,000$	$[-140.409 - 0.694]$	31.311	0.980
(c) Without (7.53) and (7.54)	$[-190.967 - 0.816]$	145.181	0.894

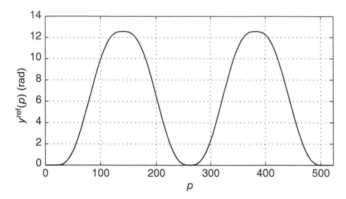

Figure 7.8 The reference trajectory for the system in [154].

The implementation developed above is compared in simulation against the law given by (7.4) and (7.5) with $v_0(p) = 0$, for the dynamics generated by the average of the system model parameters, i.e. $b = 1.4 \times 10^{-3}$ kg m^2/s, $J = 8.25 \times 10^{-4}$ kg m^2, $k_t = 0.37$ N m/A. The required reference position of the motor shaft, i.e. the reference trajectory, is shown in Figure 7.8. Moreover, to avoid comparing the numerical conditions for both ILC laws, normally distributed pseudorandom signals, with standard deviation 10^{-10}, are added to all elements of the state vector in the simulations.

The simulations were run over 250 trials, and the root mean square of the tracking error, $RMS(e_k)$, is used to evaluate the two designs, where this quantity is defined in (6.103), i.e.

$$RMS(e_k) = \sqrt{\frac{1}{N} \sum_{p=0}^{N-1} \|e_k(p)\|^2} \tag{7.57}$$

The required reference position of the motor shaft, i.e. the reference trajectory, is shown in Figure 7.8 and the simulation results are given in Figure 7.9. These confirm that (i) all designs are stable, (ii) the ILC schemes shown in Figure 7.7 are equivalent, and (iii) that the constraints imposed on the elements of the gain vector K_1 reduce the trial-to-trial error convergence speed (compare the slopes of $RMS(e_k)$ plots with the values of γ shown in the last column of Table 7.1).

As a check, the ILC system's robustness was examined by simulations with the simplified control law and occasional changes in the system model parameters and the load torque generated by the second drive. These simulations were run according to the scenario shown in Table 7.2 for the first design with K_1 and K_2 given in Table 7.1. The load torque $T_l(p)$ is generated from trial 1251, and the reference trajectory and initial conditions are the same as in the first set of simulations.

The $RMS(e_k)$ of the tracking error for the simulation study considered is shown in Figure 7.10. These results confirm that the ILC system is stable for all considered changes in the plant model parameters and the load torque. After each of these changes, the tracking error increases, but on

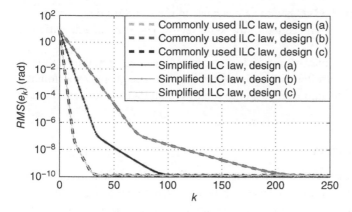

Figure 7.9 The $RMS(e_k)$ values of the tracking error for the first set of simulations.

Table 7.2 Variations of the system model parameters for robust ILC example (where $\mathbb{1}(p-200)$ denotes the unit step sequence applied at $p = 200$).

Trials	b (kg m^2/s)	J (kg m^2)	k_t (N m/A)	$T_l(p)$ (N)
1–250	1.3×10^{-3}	5.9×10^{-4}	0.35	0
251–500	1.5×10^{-3}	5.9×10^{-4}	0.35	0
501–750	1.3×10^{-3}	10.6×10^{-4}	0.35	0
751–1000	1.5×10^{-3}	10.6×10^{-4}	0.35	0
1001–1250	1.3×10^{-3}	5.9×10^{-4}	0.39	0
1251–1500	1.5×10^{-3}	5.9×10^{-4}	0.39	$0.5\,\mathbb{1}(p-200)$
1501–1750	1.3×10^{-3}	10.6×10^{-4}	0.39	$0.5\,\mathbb{1}(p-200)$
1751–2000	1.5×10^{-3}	10.6×10^{-4}	0.39	$0.5\,\mathbb{1}(p-200)$

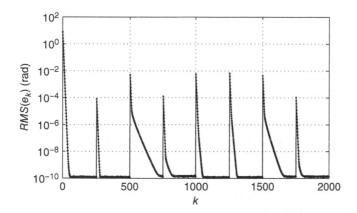

Figure 7.10 The $RMS(e_k)$ values of the tracking error for the second set of simulations.

subsequent trials is compensated by the ILC law to values close to the standard deviation of the normally distributed pseudorandom signals added to the state vector. The standard deviation of the added random signal is very small.

In some cases, the static control laws may be insufficient to control the dynamics. One option in such cases is to consider the use of a controller with its own dynamics. The design and experimental validation of one such controller with supporting experimental validation on the gantry robot of Section 2.1.1 are given in [109].

7.1.4 Design for Differential Dynamics

The lifting model designs for discrete dynamics do not extend to differential dynamics, or at least an infinite-dimensional model would be required. Consequently, the lifted model designs cannot be applied to applications where design based on differential dynamics is the only or preferred design setting, i.e. design by emulation. In contrast, the repetitive process setting extends naturally.

Consider a differential linear time-invariant system described in the ILC setting by

$$\dot{x}_{k+1}(t) = Ax_{k+1}(t) + Bu_{k+1}(t)$$
$$y_{k+1}(t) = Cx_{k+1}(t), \quad 0 \le t \le \alpha < \infty, \quad k \ge 0 \tag{7.58}$$

and introduce, for analysis purposes only, the following vector defined in terms of the difference between the current and previous trial state vectors:

$$\eta_{k+1}(t) = \int_0^t \left(x_{k+1}(\tau) - x_k(\tau) \right) d\tau \tag{7.59}$$

Also, let the ILC law compute the input on trial $k + 1$ as follows:

$$u_{k+1}(t) = u_k(t) + \Delta u_{k+1}(t)$$

and as one possible choice set

$$\Delta u_{k+1}(t) = K_1 \dot{\eta}_{k+1}(t) + K_2 \dot{e}_k(t)$$

Then the controlled system dynamics are described by

$$\dot{\eta}_{k+1}(t) = (\tilde{A} + \tilde{B}K_1)\eta_{k+1}(t) + (\tilde{B}K_2)e_k(t)$$
$$e_{k+1}(t) = -\tilde{C}(\tilde{A} + \tilde{B}K_1)\eta_{k+1}(t) + (I - \tilde{C}\tilde{B}K_2)e_k(t) \tag{7.60}$$

(where the matrices $\tilde{A}$, $\tilde{B}$, and $\tilde{C}$ have the same structure as their discrete counterparts) which is a special case of the differential linear repetitive state-space model (3.39). It is now possible to develop the equivalent results to those given so far in this chapter for discrete processes.

7.2 Repetitive Process-Based ILC Design Using Relaxed Stability Theory

Consider the necessary and sufficient conditions for stability along the trial given by Theorem 7.1. The third condition of this result requires attenuation of the previous trial error over the complete frequency spectrum. By analogy with the standard linear systems case, this condition could be very stringent in some applications. This section and the next address this issue, where below the use of a relaxed form of stability along the trial is used to produce an ILC design with experimental validation. In the following section, designs based on the generalized KYP lemma are developed, again with experimental validation.

Stability along the trial imposes the bounded input, bounded output property over the complete upper-right quadrant of the 2D plane, i.e. over $(k, p) \in [0, \infty] \times [0, \infty]$ for discrete dynamics. However, the trial length is always finite, and in any application, only a finite number of trials will be completed. In repetitive process systems theory, strong practical stability [59] aims to address this issue by relaxing the stability along the trial property. In particular, the uniform boundedness requirement as both $k \to \infty$ and $\alpha \to \infty$ is removed. However, this property is still required when (i) both k and α are finite, (ii) the trial number $k \to \infty$ and the trial length α is finite, and (iii) the trial number k is finite and the trial length $\alpha \to \infty$.

Cases (i) and (ii) have obvious practical motivation, i.e. the process of vertical and horizontal dynamics, when decoupled, are required to be stable. Case (iii) is the mathematical formulation of an application where the process completes a finite number of trials. Still, the trial length is very long, and there is a requirement to control the along the trial dynamics. The ILC design and experimental validation that follows is based, in the main, on [58].

Suppose that the discrete linear repetitive process described by (3.34) is asymptotically stable, i.e. $\rho(D_0) < 1$. Then as $k \to \infty$, it can be shown [228] that the dynamics converge in k for any p to a standard discrete linear system, termed the horizontal limit profile, with state matrix $A + B_0(I - D_0)^{-1}C$. If $\rho(A) < 1$, then it can be shown that the dynamics converge in p, for any k, to a standard discrete linear system with state matrix $C(I - A)^{-1}B_0 + D_0$, and the following result holds.

Theorem 7.7 *A discrete linear repetitive process described by (3.34) is strongly practically stable if and only if*

(a) asymptotic stability holds, i.e.

$$\rho(D_0) < 1 \tag{7.61}$$

(b) the state dynamics along any trial are stable, i.e.

$$\rho(A) < 1 \tag{7.62}$$

(c) the horizontal limit profile is stable, i.e.

$$\rho(A + B_0(I - D_0)^{-1}C) < 1 \tag{7.63}$$

and

(d) the vertical limit profile is stable, i.e.

$$\rho(C(I - A)^{-1}B_0 + D_0) < 1 \tag{7.64}$$

In the case of 2D discrete linear systems practical stability, the original references are cited in [228], was introduced in response to observations that the stability for these systems was too strong for some applications. This property required that the response in each direction of information propagation is stable, assuming no interaction between them. Practical stability can be extended to discrete linear repetitive processes and requires conditions (7.61) and (7.62). The example with $A = -05$, $B_0 = 0.5 + \beta$, $C = 1$, $D_0 = 0$ violates (7.63) for $|\beta| \geq 1$ and demonstrates that this 2D systems property is too weak for some discrete linear repetitive processes.

The following result from [57] characterizes strong practical stability in terms of LMIs and is the basis for ILC design.

Theorem 7.8 *A discrete linear repetitive process described by (3.34) is strongly practically stable if and only if there exist compatibly dimensioned matrices $W_1 > 0$, $W_2 > 0$, $Q_1 > 0$, $Q_2 > 0$ and nonsingular matrices S_1 and S_2 such that the following set of LMIs is feasible:*

$$\begin{bmatrix} -W_1 & W_1 D_0^T \\ D_0 W_1 & -W_1 \end{bmatrix} < 0 \tag{7.65}$$

$$\begin{bmatrix} -W_2 & W_2 A^T \\ A W_2 & -W_2 \end{bmatrix} < 0 \tag{7.66}$$

$$\begin{bmatrix} -Q_1 & S_1^T A_1^T \\ A_1 S_1 & Q_1 - E_1 S_1 - S_1^T E_1^T \end{bmatrix} < 0 \tag{7.67}$$

$$\begin{bmatrix} -Q_2 & S_2^T A_2^T \\ A_2 S_2 & Q_2 - E_2 S_2 - S_2^T E_2^T \end{bmatrix} < 0 \tag{7.68}$$

where

$$A_1 = \begin{bmatrix} A & 0 \\ C & 0 \end{bmatrix}, \quad A_2 = \begin{bmatrix} 0 & B_0 \\ 0 & D_0 \end{bmatrix}$$

$$E_1 = \begin{bmatrix} I & -B_0 \\ 0 & I - D_0 \end{bmatrix}, \quad E_2 = \begin{bmatrix} I - A & 0 \\ -C & I \end{bmatrix}$$

Consider now a phase-lead ILC law, i.e.

$$u_{k+1}(p) = u_k(p) + K e_k(p+1) \tag{7.69}$$

applied to (7.1). Then (using (7.3)) the controlled ILC dynamics are described by the discrete linear repetitive process state-space model:

$$\eta_{k+1}(p+1) = \hat{A}\eta_{k+1}(p) + \hat{B}_0 e_k(p)$$
$$e_{k+1}(p) = \hat{C}\eta_{k+1}(p) + \hat{D}_0 e_k(p) \tag{7.70}$$

where

$$\hat{A} = A, \quad \hat{B}_0 = BK, \quad \hat{C} = -CA, \quad \hat{D}_0 = I - CBK \tag{7.71}$$

Introduce the notation (redefining A_2 and E_2 from above):

$$A_2 = \begin{bmatrix} 0 & BK \\ 0 & I - CBK \end{bmatrix} = \tilde{A}_2 + \tilde{\Pi}\tilde{K}, \quad E_2 = \begin{bmatrix} I - A & 0 \\ CA & I \end{bmatrix} \tag{7.72}$$

where

$$\tilde{A}_2 = \begin{bmatrix} 0 & 0 \\ 0 & I \end{bmatrix}, \quad \tilde{\Pi} = \begin{bmatrix} B \\ -CB \end{bmatrix}, \quad \tilde{K} = \begin{bmatrix} 0 & K \end{bmatrix} \tag{7.73}$$

The following result proved as Theorem 3 in [58] allows strong practical stability based ILC design.

Remark 7.1 From this point onward, $\text{diag}[X_1, \ldots, X_h]$ denotes the block diagonal matrix formed by compatibly dimensioned matrices $X_1, \ldots, X_h$.

Theorem 7.9 *The ILC dynamics described by the discrete linear repetitive process state-space model (7.70) with $\rho(A) < 1$ and CBK nonsingular has the strong practical stability property if*

$$\begin{bmatrix} -Q_2 & S^T \tilde{A}_2^T + \tilde{N}^T \tilde{\Pi}^T \\ \tilde{A}_2 S + \tilde{\Pi}\tilde{N} & Q_2 - ES - (ES)^T \end{bmatrix} < 0 \tag{7.74}$$

is feasible for compatibly dimensioned matrices $Q_2 > 0$, $S = \text{diag}[S_1, S_2]$ and $\tilde{N} = \begin{bmatrix} 0 & \overline{N} \end{bmatrix}$. Also, if (7.74) is feasible, a stabilizing K is given by

$$K = \overline{N} S_2^{-1} \tag{7.75}$$

Theorem 7.8 is necessary and sufficient, but Theorem 7.9 is sufficient but not necessary since there is only one matrix K available for selection to satisfy two conditions simultaneously. Moreover, Theorem 7.9 requires that the system matrix A is stable. If this is not the case, one option is to apply the preliminary control law:

$$u_k(p) = \hat{K}_1 y_k(p) = \hat{K}_1 C x_k(p) \tag{7.76}$$

which is output-based to avoid the need for state vector measurements or the use of a state observer. The following lemma is an immediate consequence of the main result in [254].

Lemma 7.1 *Suppose that the control law (7.76) is applied to the uncontrolled ILC dynamics. Then $\rho(A + B\hat{K}_1 C) < 1$ in the resulting state-space model if there exist compatibly dimensioned matrices $V > 0$, F, M, and S such that the following LMI is feasible:*

$$\begin{bmatrix} -V & (AS + BFC)^T \\ AS + BFC & V - S - S^T \end{bmatrix} < 0 \tag{7.77}$$

$$MC = CS \tag{7.78}$$

If this LMI is feasible, a stabilizing $\hat{K}_1$ is

$$\hat{K}_1 = FM^{-1} \tag{7.79}$$

where the matrices $\hat{K}_1$ and K are obtained using Lemma 7.1 and Theorem 7.9 as appropriate. This control law was previously considered, see, e.g. [252], in the ILC literature, and the analysis here gives an alternative design method.

Next, results from experimental validation of this design using the gantry robot of Section 2.1.1 are given, for the X-axis only with the results for the other two axes in [58], and the control law for implementation is

$$u_{k+1}(p) = u_k(p) + \hat{K}_1(y_{k+1}(p) - y_k(p)) + K(r(p+1) - y_k(p+1)) \tag{7.80}$$

No experimental validation of the design in [252] has been reported. Moreover, in comparison to (7.26), the second term on the right-hand side of the latter is not present, where this term can be regarded as introducing phase lag.

Application of Lemma 7.1 gives one choice as $\hat{K}_1 = -45.5$, and the experimental results investigate the performance resulting from particular strategies for selecting $\hat{K}$ and K. The first case is when $\hat{K}_1$ is constant, i.e. the eigenvalues of the state matrix A are fixed once $\hat{K}_1$ is selected, and the design task is then to choose a corresponding K, see (7.75), where the LMIs to be solved provide a family of solutions.

Consider the case when $\hat{K}_1 = -45.5$ for which admissible values of K include 13.5, 112.59, 242.41, 325.69, and 415.69. Figure 7.11 shows the experimentally measured $MSE(e_k)$ of (7.57) for these choices and confirms that increasing the value of K leads to faster trial-to-trial error convergence. Moreover, performance begins to degrade again if K is too large. This outcome demonstrates that enforcing fast trial-to-trial error convergence can result in weaker along the trial performance.

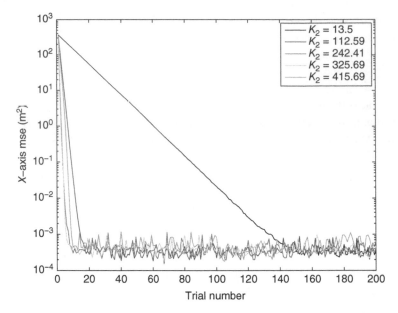

Figure 7.11 Experimentally measured $MSE(e_k)$ against trial number k for the control law (7.80) applied to the X-axis of the gantry robot of Section 2.1.1.

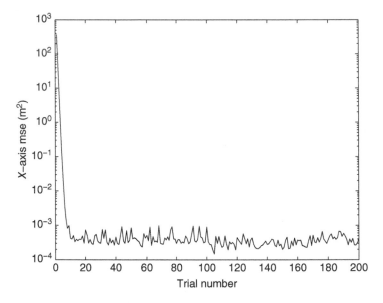

Figure 7.12 The $MSE(e_k)$ against trial number k for $\hat{K}_1 = -257.60$.

To examine the effects of focusing on the selection of $\hat{K}_1$, consider the case when $K = 190$ and, in turn, $\hat{K}_1 = -257.60, -157.47$, and -357. Then from inspecting the results of Figures 7.12–7.14, which show the MSE against the trial number, it is concluded that the speed of the trial-to-trial error convergence is unchanged (as expected with fixed K), but the levels of oscillation in the error vary and, hence, the possibility of potentially large-scale oscillations in the along the trial responses.

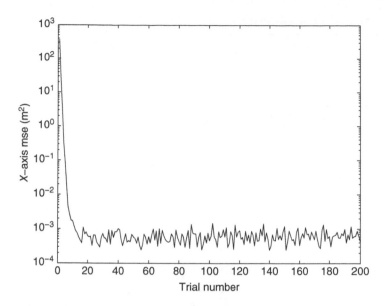

Figure 7.13 The $MSE(e_k)$ against trial number k for $\hat{K}_1 = -157.47$.

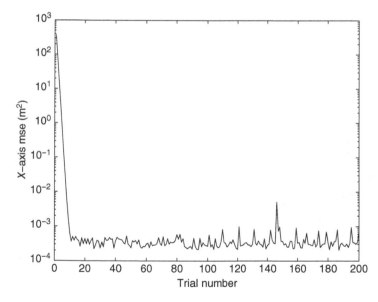

Figure 7.14 The $MSE(e_k)$ against trial number k for $\hat{K}_1 = -357$.

In applications, only a finite number of trials will ever be completed. Consequently, there is interest in selecting $\hat{K}_1$ and K to give the desired level of performance in terms of, e.g. the minimum $MSE(e_k)$ over a finite number of trials, say 200. Based on extensive numerical evaluations, the case is considered when $\hat{K}_1 = -300.29$ and $K = 325.69$. Figure 7.15 shows the output, input, and error on trial $k = 200$ and Figures 7.16–7.18, respectively, the progression with trial number k of the input and error, and Figure 7.19 shows the corresponding $MSE(e_k)$ against trial number k.

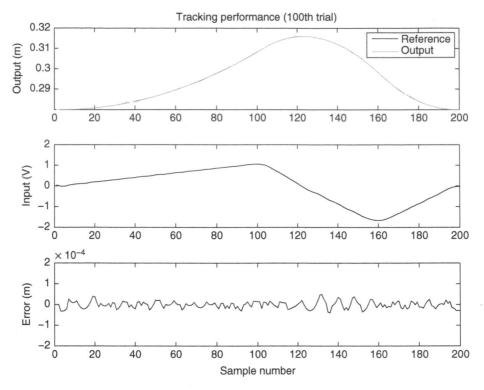

Figure 7.15 Output (top), input (middle), and error (bottom) the for ILC law on trial $k = 100$.

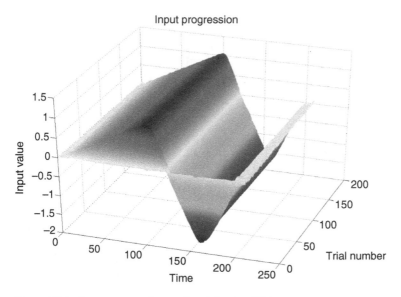

Figure 7.16 Input progression against k for the ILC law.

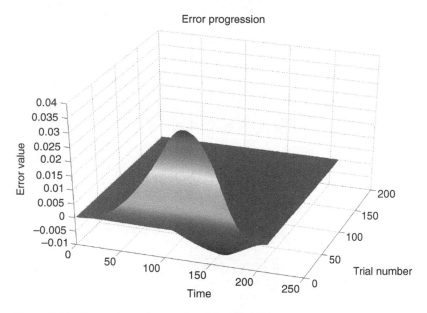

Figure 7.17 Error progression against *k* for the ILC law.

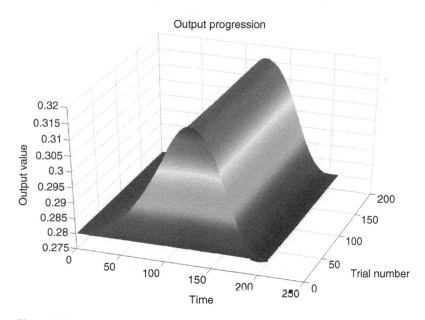

Figure 7.18 Output progression against *k* for the ILC law.

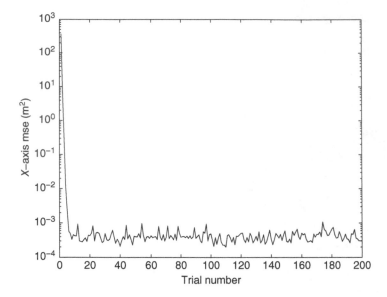

Figure 7.19 MSE against trial number k for the ILC law.

7.3 Finite Frequency Range Design and Experimental Validation

The motivation for the design given in this section is that, considering (temporally) the SISO case for simplicity, condition (c) of Theorem 7.1 for stability along the trial requires that the frequency response of the transfer-function $G(z)$ lies inside the unit circle in the complex plane. In control law design terms, this requires frequency attenuation over the complete frequency range and is a stringent condition, especially since many reference signals will have dominant frequency content over a finite range of values. Below, the generalized KYP lemma enables the development of designs where frequency attenuation is imposed only over finite frequency ranges. This lemma, stated next, converts frequency domain inequalities to LMIs.

This section follows, in the main, [203, 204] and begins with the property of stability along the trial over finite frequency ranges, including the case when uncertainty is present.

7.3.1 Stability Analysis

Lemma 7.2 *[117] For a discrete linear time-invariant system with transfer-function matrix $G(z)$ and frequency response matrix*

$$G(e^{j\theta}) = C(e^{j\theta}I - A)^{-1}B + D$$

the following inequalities are equivalent:

(a) the frequency domain inequality

$$\begin{bmatrix} G(e^{j\theta}) \\ I \end{bmatrix}^* \Pi \begin{bmatrix} G(e^{j\theta}) \\ I \end{bmatrix} \prec 0, \quad \forall \theta \in \Theta \tag{7.81}$$

where Π is a given real symmetric matrix, $*$ denotes the complex conjugate transpose operation and Θ the following frequency ranges

Θ	Low frequency range $\|\theta\| \leq \theta_l$	Middle frequency range $\theta_1 \leq \|\theta\| \leq \theta_2$	High frequency range $\|\theta\| \geq \theta_h$

(b) the LMI

$$\begin{bmatrix} A & B \\ I & 0 \end{bmatrix}^T \Xi \begin{bmatrix} A & B \\ I & 0 \end{bmatrix} + \begin{bmatrix} C & D \\ 0 & I \end{bmatrix}^T \Pi \begin{bmatrix} C & D \\ 0 & I \end{bmatrix} < 0 \tag{7.82}$$

where $Q > 0$, P is a symmetric matrix and the matrix Ξ is compatibly partitioned as follows:

$$\Xi = \begin{bmatrix} \Xi_{11} & \Xi_{12} \\ \Xi_{12}^* & \Xi_{22} \end{bmatrix} \tag{7.83}$$

and specified as follows:

- *for the low-frequency range*

$$\Xi = \begin{bmatrix} \Xi_{11} & \Xi_{12} \\ \Xi_{12}^* & \Xi_{22} \end{bmatrix} = \begin{bmatrix} -P & Q \\ Q & P - 2\cos(\theta_l)Q \end{bmatrix} \tag{7.84}$$

- *for the middle-frequency range*

$$\Xi = \begin{bmatrix} \Xi_{11} & \Xi_{12} \\ \Xi_{12}^* & \Xi_{22} \end{bmatrix} = \begin{bmatrix} -P & e^{j(\theta_1+\theta_2)/2}Q \\ e^{-j(\theta_1+\theta_2)/2}Q & P - (2\cos((\theta_2 - \theta_1)/2))Q \end{bmatrix} \tag{7.85}$$

- *and for the high-frequency range*

$$\Xi = \begin{bmatrix} \Xi_{11} & \Xi_{12} \\ \Xi_{12}^* & \Xi_{22} \end{bmatrix} = \begin{bmatrix} -P & -Q \\ -Q & P + 2\cos(\theta_h)Q \end{bmatrix} \tag{7.86}$$

The analysis also makes use of the following well-known results, where for a square matrix, say H, sym(H) denotes the matrix $H + H^T$ and the superscript $\perp$ represents the orthogonal complement.

Lemma 7.3 *[88] Given a $q \times q$ symmetric matrix Ψ and two matrices Λ, Σ of column dimension q, there exists a matrix W such that the following LMI holds*

$$\Psi + \text{sym}\{\Lambda^T W \Sigma\} < 0 \tag{7.87}$$

if and only if the following two inequalities with respect to Ψ are satisfied:

$$\Lambda^{\perp T}\Psi\Lambda^{\perp} < 0, \quad \Sigma^{\perp T}\Psi\Sigma^{\perp} < 0 \tag{7.88}$$

Lemma 7.4 *[130] Let Σ_1, Σ_2 be real constant matrices of compatible dimensions. Then for any real matrix function $F(p)$ satisfying $F(p)F^T(p) \leq I$ and scalar $\epsilon > 0$ the following inequality holds*

$$\text{sym}\{\Sigma_1 F(p)\Sigma_2\} \leq \epsilon^{-1}\Sigma_1\Sigma_1^T + \epsilon\Sigma_2^T\Sigma_2 \tag{7.89}$$

The primary use of the generalized KYP lemma in this area is to compute a control law that allows a designer to specify the desired shape of the controlled frequency response in specified frequency ranges without the use of frequency gridding and weighting filters. In the ILC application, it allows particular frequency ranges to be emphasized in the design. Since the bandwidth of the reference signal has the most effect on convergence speed, the main focus in applications will be on the low-frequency range.

If some frequencies above the bandwidth are considered, then better performance is a possibility. In the high-frequency range, well above the reference signal bandwidth, noise and nonrepeatable disturbances could be present. Moreover, they cannot be effectively attenuated by an ILC law designed in the deterministic setting (see Chapter 12 for ILC design in the stochastic setting).

Choose the matrix Π as

$$\Pi = \begin{bmatrix} I & 0 \\ 0 & -I \end{bmatrix} \tag{7.90}$$

Then (7.81) gives

$$G^*(e^{j\theta})G(e^{j\theta}) < I, \quad \forall \theta \in \Theta \tag{7.91}$$

or

$$\overline{\sigma}\left(G(e^{j\theta})\right) < 1, \quad \forall \theta \in \Theta \tag{7.92}$$

Also for all $\theta \in [0, 2\pi]$

$$\rho\left(G(e^{j\theta})\right) \le \overline{\sigma}\left(G(e^{j\theta})\right) \tag{7.93}$$

Suppose that conditions (a) and (b) of Theorem 7.1 hold for the controlled dynamics described by (7.6) and (7.7), where these are standard discrete linear systems stability conditions (for systems, respectively, with state matrices $\hat{D}_0$ and $\hat{A}$). Consider $G(e^{j\theta})$ of (c) of this theorem for a specified finite frequency range $\theta \in \Theta$ over which stability is required, where the selection of Θ is application-dependent and is discussed again in the experimental results given below. Then a sufficient condition for the ILC dynamics to be stable along the trial over the finite frequency range $\theta \in \Theta$ is that the inequality (7.92) holds for this $G(e^{j\theta})$. In the SISO case, equality holds in (7.93) and (7.92) is then a necessary and sufficient condition.

Applying Lemma 7.2 gives that condition (c) of Theorem 7.1 holds if

$$\begin{bmatrix} \hat{A} & \hat{B}_0 \\ I & 0 \end{bmatrix}^T \Xi \begin{bmatrix} \hat{A} & \hat{B}_0 \\ I & 0 \end{bmatrix} + \begin{bmatrix} \hat{C} & \hat{D}_0 \\ 0 & I \end{bmatrix}^T \Pi \begin{bmatrix} \hat{C} & \hat{D}_0 \\ 0 & I \end{bmatrix} < 0 \tag{7.94}$$

where Ξ (given by (7.83)) is the only matrix whose block entries depend on the chosen frequency range, i.e. low, middle, or high, as specified, respectively, by (7.84), (7.85), and (7.86). The condition of (7.94) cannot, however, be directly applied to control law design since it involves product terms in P, Q, and the controlled process state-space model matrices for any frequency range. The following result proved as Theorem 1 in [204], overcomes this difficulty.

Theorem 7.10 *Consider the ILC dynamic described by the discrete linear repetitive process state-space model (7.6) and (7.7). Then stability along the trial holds over the finite frequency range $\theta \in \Theta$ if there exist compatibly dimensioned matrices $S > 0$, $P > 0$, $Q > 0$, and W such that the following two LMIs are feasible*

$$\hat{A}^T S \hat{A} - S < 0 \tag{7.95}$$

$$\begin{bmatrix} \Xi_{11} & \Xi_{12} - W & 0 & 0 \\ \Xi_{12}^* - W^T & \Xi_{22} + \hat{A}^T W + W^T \hat{A} & W^T \hat{B} & \hat{C}^T \\ 0 & \hat{B}^T W & -I & \hat{D}_0^T \\ 0 & \hat{C} & \hat{D}_0 & -I \end{bmatrix} < 0 \tag{7.96}$$

where Ξ_{11}, Ξ_{12}, and Ξ_{22} are the block entries in the matrix Ξ with the structure (7.83), chosen according to the specific frequency range Θ of interest, i.e. the low-, middle-, or high-frequency ranges given, respectively, by (7.84), (7.85), and (7.86).

This last result also gives a direct method for improving the trial-to-trial error convergence by selecting the matrix Π in (7.90) as

$$\Pi = \begin{bmatrix} I & 0 \\ 0 & -\gamma^2 I \end{bmatrix}$$

and minimizing γ under the constraint $0 < \gamma \leq 1$. Using such a minimization procedure, the eigenvalues of $G(z)$ in Theorem 7.1 are assigned to locations inside the circle of radius γ in the complex plane.

Consider first the nominal model case. Then application of Theorem 7.10 yields matrix inequalities that are not convex. The following result, Theorem 6 in [204], removes this difficulty expressed in terms of LMIs that provides an algorithm for designing the control law, i.e. stabilizing matrices K_1 and K_2 for the chosen frequency range obtained, respectively, by selecting the matrix Ξ to have the structure of (7.84), (7.85), or (7.86).

Theorem 7.11 *The ILC dynamics described by (7.6) and (7.7) are stable along the trial over the finite frequency range $\theta \in \Theta$ if there exist compatibly dimensioned matrices X_1, X_2, $\hat{W}$, $\hat{S} > 0$, $\hat{Q} > 0$, $\hat{P} > 0$ together with real scalars ρ_1 and ρ_2 such that the following LMIs are feasible:*

$$\begin{bmatrix} \hat{S} + \rho_2 \hat{W} + \rho_2 \hat{W}^T & -\rho_2 A\hat{W} - \rho_2 BX_1 - \rho_1 \hat{W}^T \\ (\star) & -\hat{S} + \mathrm{sym}\left\{\rho_1 A\hat{W} + \rho_1 BX_1\right\} \end{bmatrix} < 0 \tag{7.97}$$

$$\begin{bmatrix} \hat{\Xi}_{11} & \hat{\Xi}_{12} - \hat{W}^T & 0 & 0 \\ (\star) & \hat{\Xi}_{22} + \mathrm{sym}\left\{A\hat{W} + BX_1\right\} & BX_2 & -\hat{W}^T A^T C^T - X_1^T B^T C^T \\ (\star) & (\star) & -I & I - X_2^T B^T C^T \\ (\star) & (\star) & (\star) & -I \end{bmatrix} < 0 \tag{7.98}$$

where for compatibly dimensioned matrices Ξ_{11}, Ξ_{12}, Ξ_{22} of (7.83) in this case, $\hat{\Xi}_{11} = \hat{W}^T \Xi_{11} \hat{W}$, $\hat{\Xi}_{12} = \hat{W}^T \Xi_{12} \hat{W}$, $\hat{\Xi}_{22} = \hat{W}^T \Xi_{22} \hat{W}$, and ρ_1 and ρ_2 satisfy

$$\rho_1^2 - \rho_2^2 < 0 \tag{7.99}$$

If the LMIs (7.97) and (7.98) are feasible, stabilizing control law matrices K_1 and K_2 are given by

$$K_1 = X_1 \hat{W}^{-1}, \quad K_2 = X_2 \tag{7.100}$$

In Theorem 7.11, the scalars ρ_1 and ρ_2 must be selected prior to solving the LMIs given by (7.97)–(7.98). Selecting ρ_1 and ρ_2 to increase trial-to-trial error convergence is one option, for which the only available method is trial and error as part of the design and simulation of the controlled system before physical implementation.

Consider the case when there is uncertainty associated with the available model of the dynamics. In particular, as one possible case, suppose that uncertainty is norm-bounded and, hence, the LMIs in the last theorem will include terms formed by multiplication of two matrices, e.g. $C + \Delta C$ and $A + \Delta A$, representing the uncertainty. Then the result of Lemma 7.4 cannot be directly applied and the following analysis removes this obstacle to robust ILC law design.

Introduce the notation

$$\Phi = \begin{bmatrix} \hat{\Xi}_{11} & \hat{\Xi}_{12} - \hat{W}^T & 0 & 0 \\ (\star) & \hat{\Xi}_{22} + \mathrm{sym}\left\{A\hat{W} + BX_1\right\} & BX_2 & 0 \\ (\star) & (\star) & -I & I \\ (\star) & (\star) & (\star) & -I \end{bmatrix}, \quad J = \begin{bmatrix} 0 \\ 0 \\ 0 \\ -C \end{bmatrix}, \quad N = \begin{bmatrix} 0 \\ \hat{W}^T A^T + X_1^T B^T \\ X_2^T B^T \\ 0 \end{bmatrix}$$

Then the LMI (7.98) for the nominal model can be written as follows:

$$
\begin{bmatrix} I & J \end{bmatrix} \begin{bmatrix} \Phi & N \\ N^T & 0 \end{bmatrix} \begin{bmatrix} I \\ J^T \end{bmatrix} \prec 0 \tag{7.101}
$$

Also, application of the result of Lemma 7.3 with

$$
\Psi = \begin{bmatrix} \Phi & N \\ N^T & 0 \end{bmatrix}, \quad W = \begin{bmatrix} W_1 \\ W_2 \end{bmatrix}, \quad \Lambda = I, \quad \Sigma = \begin{bmatrix} J^T & -I \end{bmatrix}
$$

gives that (7.101) holds for some W_1 and W_2 if and only if

$$
\begin{bmatrix} \Phi & N \\ N^T & 0 \end{bmatrix} + \mathrm{sym}\left\{ \begin{bmatrix} W_1 \\ W_2 \end{bmatrix} \begin{bmatrix} J^T & -I \end{bmatrix} \right\} \prec 0 \tag{7.102}
$$

Partitioning W_1 as

$$
W_1 = \begin{bmatrix} W_{11}^T & W_{21}^T & W_{31}^T & W_{41}^T \end{bmatrix}^T
$$

enables (7.102) to be rewritten as follows, where all terms involving the product of the nominal state-space model matrices are decoupled and $W_{11}, W_{21},$ and $W_{31}, W_{41},$ and W_2 are new slack matrix variables of compatible dimensions:

$$
\begin{bmatrix}
\hat{\Xi}_{11} & \hat{\Xi}_{12} - \hat{W}^T & 0 & -W_{11}C^T & -W_{11} \\
(\star) & \hat{\Xi}_{22} + \mathrm{sym}\left\{ A\hat{W} + BX_1 \right\} & BX_2 & -W_{21}C^T & \hat{W}^T A^T + X_1^T B^T - W_{21} \\
(\star) & (\star) & -I & I - W_{31}C^T & X_2^T B^T - W_{31} \\
(\star) & (\star) & (\star) & -I - \mathrm{sym}\left\{ W_{41}C^T \right\} & -W_{41} - CW_2^T \\
(\star) & (\star) & (\star) & (\star) & -W_2 - W_2^T
\end{bmatrix} \prec 0 \tag{7.103}
$$

If (norm-bounded) uncertainty is present in the state-space model, (7.103) can be written as follows:

$$
\begin{bmatrix}
\hat{\Xi}_{11} & \hat{\Xi}_{12} - \hat{W}^T & 0 & -W_{11}C^T & -W_{11} \\
(\star) & \hat{\Xi}_{22} + \mathrm{sym}\left\{ A\hat{W} + BX_1 \right\} & BX_2 & -W_{21}C^T & \hat{W}^T A^T + X_1^T B^T - W_{21} \\
(\star) & (\star) & -I & I - W_{31}C^T & X_2^T B^T - W_{31} \\
(\star) & (\star) & (\star) & -I - \mathrm{sym}\left\{ W_{41}C^T \right\} & -W_{41} - CW_2^T \\
(\star) & (\star) & (\star) & (\star) & -W_2 - W_2^T
\end{bmatrix}
$$
$$
+ \mathrm{sym}\left\{ \begin{bmatrix} 0 & 0 \\ 0 & H_1 \\ 0 & 0 \\ -H_2 & 0 \\ 0 & H_1 \end{bmatrix} \begin{bmatrix} \mathcal{F}(p) & 0 \\ 0 & \mathcal{F}(p) \end{bmatrix} \begin{bmatrix} E_1 W_{11}^T & E_1 W_{21}^T & E_1 W_{31}^T & E_1 W_{41}^T & E_1 W_2^T \\ 0 & E_1\hat{W} + E_2 X_1 & E_2 X_2 & 0 & 0 \end{bmatrix} \right\}
$$
$$
\prec 0 \tag{7.104}
$$

Moreover, there are no multiplications between the state-space model matrices describing the nominal system and those describing the uncertainty in (7.97) and this LMI can be written as follows:

$$
\begin{bmatrix}
\hat{S} + \mathrm{sym}\left\{ \rho_2 \hat{W} \right\} & -\rho_2 A\hat{W} - \rho_2 BX_1 - \rho_1 \hat{W}^T \\
(\star) & -\hat{S} + \mathrm{sym}\left\{ \rho_1 A\hat{W} + \rho_1 BX_1 \right\}
\end{bmatrix}
$$
$$
+ \mathrm{sym}\left\{ \begin{bmatrix} -\rho_2 H_1 \\ \rho_1 H_1 \end{bmatrix} \mathcal{F}(p) \begin{bmatrix} 0 & E_1\hat{W} + E_2 X_1 \end{bmatrix} \right\} \prec 0 \tag{7.105}
$$

where in the uncertainty description Eq. (7.30) and (7.31) F_1 and F_2 are both taken to be $\mathcal{F}(p)$.

The following result, Theorem 7 in [204], leads to experimental validation on one axis of the gantry robot of Section 2.1.1.

Theorem 7.12 *The ILC dynamics of the previous theorem but with norm-bounded uncertainty is stable along the trial over finite frequency range Θ if there exist compatibly dimensioned matrices $\hat{S} > 0$, $\hat{Q} > 0$, $\hat{P} > 0$, X_1, X_2, $\hat{W}$, W_2, W_{11}, W_{21}, W_{31}, W_{41} together with real scalars ρ_1, ρ_2, $\epsilon_1 > 0$, and $\epsilon_2 > 0$ such that the following LMIs are feasible*

$$
\begin{bmatrix}
\hat{S} + \text{sym}\{\rho_2\hat{W}\} + \epsilon_1\rho_2^2 H_1 H_1^T & F_1 & 0 \\
(\star) & F_2 & X_1^T E_2^T + \hat{W}^T E_1^T \\
(\star) & (\star) & -\epsilon_1 I
\end{bmatrix} \prec 0
\tag{7.106}
$$

$$
F_1 = -\rho_2 A\hat{W} - \rho_2 BX_1 - \rho_1\hat{W}^T - \epsilon_1\rho_1\rho_2 H_1 H_1^T
$$
$$
F_2 = -\hat{S} + \text{sym}\{\rho_1 A\hat{W} + \rho_1 BX_1\} + \epsilon_1\rho_1^2 H_1 H_1^T
$$

$$
\begin{bmatrix}
\hat{\Xi}_{11} & \hat{\Xi}_{12} - \hat{W}^T & 0 & -W_{11}C^T \\
(\star) & \hat{\Xi}_{22} + \text{sym}\{A\hat{W} + BX_1\} + \epsilon_2 H_1 H_1^T & BX_2 & -W_{21}C^T \\
(\star) & (\star) & -I & I - W_{31}C^T \\
(\star) & (\star) & (\star) & -I - \text{sym}\{W_{41}C^T\} + \epsilon_2 H_2 H_2^T \\
(\star) & (\star) & (\star) & (\star) \\
(\star) & (\star) & (\star) & (\star) \\
(\star) & (\star) & (\star) & (\star)
\end{bmatrix}
$$

$$
\begin{bmatrix}
-W_{11} & W_{11}E_1^T & 0 \\
\hat{W}^T A^T + X_1^T B^T - W_{21} + \epsilon_2 H_1 H_1^T & W_{21}E_1^T & -\hat{W}^T E_1^T + X_1^T E_2^T \\
X_2^T B^T - W_{31} & W_{31}E_1^T & X_2^T E_2^T \\
-W_{41} - CW_2^T & W_{41}E_1^T & 0 \\
-W_2 - W_2^T + \epsilon_2 H_1 H_1^T & W_2 E_1^T & 0 \\
(\star) & -\epsilon_2 I & 0 \\
(\star) & (\star) & -\epsilon_2 I
\end{bmatrix} \prec 0
$$

$$
\tag{7.107}
$$

and ρ_1 and ρ_2 satisfy (7.99). If the LMIs (7.106) and (7.107) are feasible, stabilizing control law matrices K_1 and K_2 can be computed as in the previous result.

As in all experimental implementations of robust control, the uncertainty modeling is a critical aspect and is application-specific. The uncertainty model in this work is chosen to demonstrate that a robust control design can be tested experimentally. Further development is required in this area, but it is impossible to produce explicit formulas for the terms in the uncertainty model used in this work or others.

One area that could be addressed is to follow the tuning-based ideas in [220] for an adjoint optimal ILC law, which started from an approximate axis model of an integrator and a gain. The purpose of the results given next, from [203, 204], is to establish that robust control with an ILC law designed with a finite frequency range error attenuation specification can be undertaken and experimentally validated, where for the norm-bounded uncertainty considered the defining matrices of (7.31) are

$$
E_1 = \begin{bmatrix} 0.3 & -0.1 & 0.1 & -0.1 & 0.05 & -0.05 & 0.1 \end{bmatrix}, \quad E_2 = 0.01
$$

$$
H_1 = \begin{bmatrix} 0.3 & 0 & 0 & 0 & 0 & 0 & 0 \end{bmatrix}^T, \quad H_2 = 0.01
$$

Consider again the X-axis of the gantry robot of Section 2.1.1, for which the frequency spectrum constructed by application of the fast Fourier transform (FFT) is given in Figure 7.20.

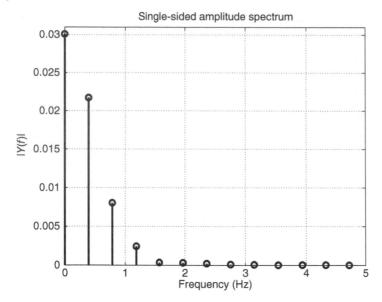

Figure 7.20 The frequency spectrum of the X-axis reference trajectory for the gantry robot of Section 2.1.1.

Inspecting the amplitudes in this frequency spectrum gives significant harmonics in the range from 0 to 5 Hz, and this is taken as the low-frequency range. Hence, $\theta_l = 0.3142$ and the matrix Ξ has the structure (7.84). Applying Theorem 7.12 with $\rho_1 = 1$ and $\rho_2 = -2$ gives the stabilizing robust control law matrices

$$K_1 = \begin{bmatrix} -0.3256 & -2.6079 & -0.7878 & -0.2023 & -1.6026 & 0.3601 & -10.5012 \end{bmatrix}$$

$$K_2 = 38.6513$$

As a preliminary to experimental verification, a simulation of the controlled system was performed over 200 trials, resulting in the $MSE(e_k)$ error data shown in Figure 7.21 against the trial number. This simulation confirms that the designed ILC control law is stabilizing and results in trial-to-trial error convergence.

This design has also been completed for the other two axes of the gantry robot. Again, 5 Hz is used to specify the low-frequency range. Moreover, the matrices defining the uncertainty model for the Y and Z-axes, respectively, are

$$E_1 = \begin{bmatrix} 0.5 & -0.1 & 0.1 \end{bmatrix}, \quad E_2 = 0.02, \quad H_1 = \begin{bmatrix} 0.2 & 0 & 0 \end{bmatrix}^T, \quad H_2 = 0.01$$

and

$$E_1 = \begin{bmatrix} 0.2 & -0.1 & 0.1 \end{bmatrix}, \quad E_2 = 0.01, \quad H_1 = \begin{bmatrix} 0.3 & 0 & 0 \end{bmatrix}^T, \quad H_2 = 0.01$$

Applying the design procedure of Theorem 7.12 for the low-frequency range [0,5] Hz gives the corresponding control law matrices:

$$K_1 = \begin{bmatrix} -4.5044 & -0.0198 & -27.8846 \end{bmatrix}, \quad K_2 = 23.3258$$

for Y-axis, when ρ_1, ρ_2, are chosen as $\rho_1 = 1, \rho_2 = -2$ and

$$K_1 = \begin{bmatrix} -2.0851 & -0.3364 & -51.3834 \end{bmatrix}, \quad K_2 = 102.8790$$

for the Z-axis, when $\rho_1 = 1, \rho_2 = -3$.

This design has been applied experimentally to the gantry robot of Section 2.1.1. The $MSE(e_k)$ for each axis plotted against the trial number is shown in Figure 7.22 and Figure 7.23, shows

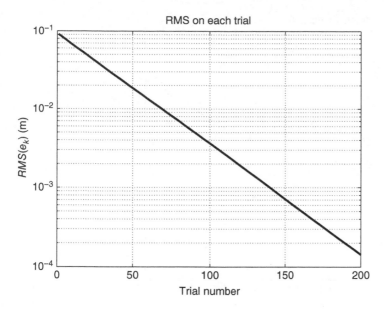

Figure 7.21 $MSE(e_k)$ values of the error for the X-axis over 200 trials.

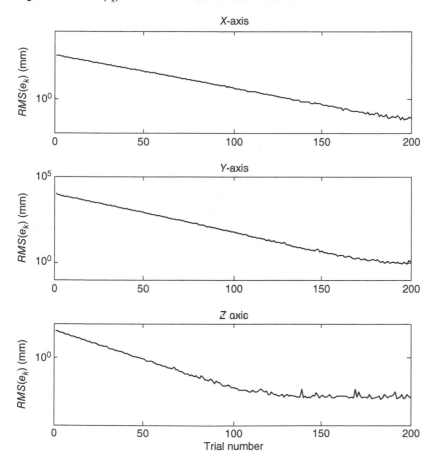

Figure 7.22 Experimentally measured $MSE(e_k)$ against trial number for each axis.

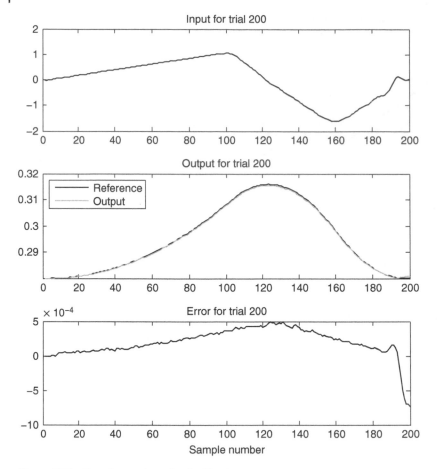

Figure 7.23 Experimental results for *X*-axis.

the input, output, and error, respectively, for the *X* axis, against the trial number. These experimentally confirm the simulation predictions, and this is also the case for the other two axes.

Figure 7.24 shows the control input on selected trials for the *X*-axis. These plots show that the control input is acceptable and converges as the trials proceed. This feature is also the case for the other two axes.

The remainder of this section follows [203] and considers strictly proper discrete linear time-invariant systems described by the state-space triple (state, input, and output) $\{A, B, C\}$ of relative degree greater than unity. Also, the ILC scheme has the form shown in Figure 7.25.

Given Figure 7.25, the ILC law is

$$F_{k+1}(z) = Q(z)\left(F_k(z) + L(z)E_k(z)\right)$$

and

$$E_k(z) = (I + G(z)C(z))^{-1}R(z) - (I + G(z)C(z))^{-1}G(z)F_k(z)$$

Hence, the previous trial error feedforward contribution to the current trial error is described by

$$E_k(z) = -\left[(I + G(z)C(z))^{-1}G(z)\right]F(z) = -S_P(z)F_k(z)$$

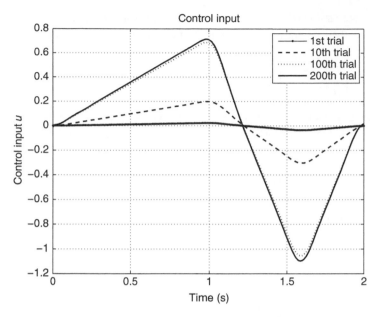

Figure 7.24 Control inputs for the *X*-axis on four representative trials.

Figure 7.25 ILC block diagram representation.

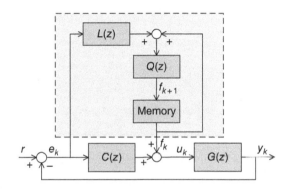

where $S_P(z) = (I + G(z)C(z))^{-1}G(z)$ denotes the sensitivity function and the propagation of the error from trial-to-trial is

$$E_{k+1}(z) = Q(z)\left(I - S_P(z)L(z)\right)E_k(z)$$

or, on introducing,

$$M(z) = Q(z)\left(I - S_P(z)L(z)\right) \tag{7.108}$$

$$E_{k+1}(z) - E_k(z) = M(z)\left(E_k(z) - E_{k-1}(z)\right)$$

To formulate the ILC design problem in the repetitive process setting, assume $Q = I$ and rewrite (7.108) as follows:

$$M(z) = (I + G(z)C(z))^{-1}(G(z)C(z)G(z)L(z) + (I + G(z)C(z))(I - G(z)L(z)))$$

or

$$M(z) = I - G(z)\begin{bmatrix} C(z) & L(z) \end{bmatrix}\left(\begin{bmatrix} (I + G(z)C(z)) & G(z)L(z) \\ 0 & I \end{bmatrix}\right)^{-1}\begin{bmatrix} 0 \\ I \end{bmatrix}$$

Moreover, $M(z)$ can be represented as a particular case of a general control configuration [243], where the generalized plant P, i.e. the interconnection system of the controlled dynamics, is

$$P(z) = \left[\begin{array}{c|c} P_{11}(z) & P_{12}(z) \\ \hline P_{21}(z) & P_{22}(z) \end{array}\right] = \left[\begin{array}{c|cc} I & -G(z) \\ \hline 0 & -G(z) \\ I & 0 \end{array}\right] \qquad (7.109)$$

which is the transfer-function matrix from $[e_k^T \ u_{k+1}^T]^T$ to $[e_{k+1}^T \ y_{k+1}^T]^T$, where $y_{k+1} = [(\bar{y}_{k+1}^T) \ (\hat{y}_{k+1}^T)]^T$, where the vectors on the right-hand side of this last formula are given below. The generalized ILC controller to be found is given by $K(z) = \begin{bmatrix} C(z) & L(z) \end{bmatrix}$ and, hence, $M(z)$ can be rewritten as follows:

$$M(z) = P_{11}(z) + P_{12}(z)K(z)\big(I - P_{22}(z)K(z)\big)^{-1}P_{21}(z)$$

This last equation shows that the transfer-function matrix from e_k to e_{k+1} is given by a linear fractional transformation (LFT) of $P(z)$ with $K(z)$ as the parameter. Furthermore, the state-space representation of (7.109) is

$$\begin{aligned}
x_{k+1}(p+1) &= Ax_{k+1}(p) + Bu_{k+1}(p) \\
\bar{y}_{k+1}(p) &= -Cx_{k+1}(p) \\
\hat{y}_{k+1}(p) &= e_k(p) \\
e_{k+1}(p) &= -Cx_{k+1}(p) + e_k(p)
\end{aligned} \qquad (7.110)$$

Consider also the controller

$$\begin{aligned}
\tilde{x}_{k+1}(p+1) &= A_K\tilde{x}_{k+1}(p) + \begin{bmatrix} B_{K1} & B_{K2} \end{bmatrix}\begin{bmatrix} \bar{y}_{k+1}(p) \\ \hat{y}_{k+1}(p) \end{bmatrix} \\
u_{k+1}(p) &= C_K\tilde{x}_{k+1}(p) + \begin{bmatrix} D_{K1} & D_{K2} \end{bmatrix}\begin{bmatrix} \bar{y}_{k+1}(p) \\ \hat{y}_{k+1}(p) \end{bmatrix}
\end{aligned} \qquad (7.111)$$

with transfer-function matrix $K(z) = C_K(zI - A_K)^{-1}B_K + D_K$ where

$$B_K = \begin{bmatrix} B_{K1} & B_{K2} \end{bmatrix}, \quad D_K = \begin{bmatrix} D_{K1} & D_{K2} \end{bmatrix}$$

and the ILC dynamics can now be written as the discrete linear repetitive process

$$\begin{bmatrix} x_{k+1}(p+1) \\ \tilde{x}_{k+1}(p+1) \\ \hline e_{k+1}(p) \end{bmatrix} = \left[\begin{array}{cc|c} A - BD_{K1}C & BC_K & BD_{K2} \\ -B_{K1}C & A_K & B_{K2} \\ \hline -C & 0 & I \end{array}\right]\begin{bmatrix} x_{k+1}(p) \\ \tilde{x}_{k+1}(p) \\ \hline e_k(p) \end{bmatrix} \qquad (7.112)$$

where the current trial state vector is $\begin{bmatrix} x_{k+1}^T(p) & \tilde{x}_{k+1}^T(p) \end{bmatrix}$, zero input vector and previous pass profile $e_k(p)$.

If an example considered has relative degree $h > 0$, an h-step delay between its input and output results. Consequently, the previous trial error is shifted by h samples to form the anticipative feedforward control law. In implementation, this structure means that the input signal modification $f_k(p)$ in Figure 7.25 at time instant p is paired with the error signal $e_k(p+h)$ at time instant $p+h$, this is possible since the error signal from trial k is available once this trial is complete. Hence, in this case, the transfer-function matrix of the learning controller is taken as $z^h L$ instead of L.

The generalized plant P resulting from the above modification is

$$P(z) = \left[\begin{array}{c|c} P_{11}(z) & P_{12}(z) \\ \hline P_{21}(z) & P_{22}(z) \end{array}\right] = \left[\begin{array}{c|cc} I & -G(z) \\ \hline 0 & -z^h G(z) \\ I & 0 \end{array}\right] \qquad (7.113)$$

where the block P_{22} includes the anticipatory operator, and the forward time shift is applied to the error signal transmitted through $\hat{y}_{k+1}$ – see the third entry in (7.110). Then, on introducing the following notation for (7.112)

$$\overline{A} = \begin{bmatrix} A - BD_{K1}C & BC_K \\ -B_{K1}C & A_K \end{bmatrix}, \quad \overline{B}_0 = \begin{bmatrix} BD_{K2} \\ B_{K2} \end{bmatrix}, \quad \overline{C} = \begin{bmatrix} -C & 0 \end{bmatrix}, \quad \overline{D}_0 = I \tag{7.114}$$

the transfer-function matrix $M(z)$ from e_k to e_{k+1} is

$$M(z) = \overline{C}(zI - \overline{A})^{-1}\overline{B}_0 z^h + \overline{D}_0 = \overline{C}z^h(zI - \overline{A})^{-1}\overline{B}_0 + \overline{D}_0 \tag{7.115}$$

Also, routine manipulations allow the last equation to be written as follows:

$$M(z) = \mathbb{C}(zI - \mathbb{A})^{-1}\mathbb{B}_0 + \mathbb{D}_0 \tag{7.116}$$

where the matrices in the state-space quadruple $\{\mathbb{A}, \mathbb{B}_0, \mathbb{C}, \mathbb{D}_0\}$ are

$$\mathbb{A} = (\mathscr{A} + \mathscr{B}\mathscr{K}) = \begin{bmatrix} A - BD_{K1}C & BC_K \\ -B_{K1}C & A_K \end{bmatrix}, \quad \mathbb{B}_0 = \begin{bmatrix} BD_{K2} \\ B_{K2} \end{bmatrix}$$

$$\mathbb{C} = \overline{C}\mathscr{A}^{h-1}(\mathscr{A} + \mathscr{B}\mathscr{K}) = \begin{bmatrix} -CA^h + CA^{h-1}BD_{K1}C & -CA^{h-1}BC_K \end{bmatrix}$$

$$\mathbb{D}_0 = \overline{D}_0 + \overline{C}\mathscr{A}^{h-1}\overline{B}_0 = I - CA^{h-1}BD_{K2} \tag{7.117}$$

The design below also uses auxiliary slack matrix variables (see, e.g. [181] for a detailed treatment). Suppose that the auxiliary slack matrix variable $\mathscr{W}$ and its inverse are partitioned into blocks as follows:

$$\mathscr{W} = \begin{bmatrix} X & ? \\ U & ? \end{bmatrix}, \quad \mathscr{W}^{-1} = \begin{bmatrix} N & ? \\ R & ? \end{bmatrix}$$

where "?" is used to denote block entries in these matrices that are not involved in what follows. Also, suppose, without loss of generality, that U and R are invertible and define the transformation matrices F_1 and F_2 as follows:

$$F_1 = \begin{bmatrix} X & I \\ U & 0 \end{bmatrix}, \quad F_2 = \begin{bmatrix} I & N \\ 0 & R \end{bmatrix} \tag{7.118}$$

where $\mathscr{W}F_2 = F_1$, and introduce the notation

$$\widehat{W} = F_2^T\mathscr{W}F_2 = \begin{bmatrix} X & I \\ Z^T & N^T \end{bmatrix}, \quad \widehat{A} = F_2^T\mathscr{W}^T\mathbb{A}F_2 = \begin{bmatrix} X^TA - \tilde{B}_1C & \tilde{A} \\ A - BD_{K1}C & AN + B\tilde{C} \end{bmatrix}$$

$$\widehat{B}_0 = F_2^T\mathscr{W}^T\mathbb{B}_0 = \begin{bmatrix} \tilde{B}_2 \\ BD_{K2} \end{bmatrix}, \quad \widehat{C} = \mathbb{C}F_2 = \begin{bmatrix} -CA^{h-1}(A - BD_{K1}C) & -CA^{h-1}(AN + B\tilde{C}) \end{bmatrix} \tag{7.119}$$

where

$$\tilde{A} = X^TAN - X^TBD_{K1}CN - U^TB_{K1}CN + X^TBC_KR + U^TA_KR, \quad Z = X^TN + U^TR$$

$$\tilde{B}_1 = X^TBD_{K1} + U^TB_{K1}, \quad \tilde{B}_2 = X^TBD_{K2} + U^TB_{K2}$$

$$\tilde{C} = C_KR - D_{K1}CN \tag{7.120}$$

The following result, Theorem 2 in [203], gives an LMI-based characterization for the existence of an ILC controller K represented by (7.111).

Theorem 7.13 *Consider ILC dynamics described as a discrete linear repetitive process with no input terms and state-space model matrices given by (7.117). Then stability along the trial property holds* *and, hence, monotonic trial-to-trial error convergence occurs if there exist matrices $\tilde{A}$, $\tilde{B}_1$, $\tilde{B}_2$, $\tilde{C}$, D_{K1}, D_{K2}, N, X, Z with dimensions defined by (7.119), a matrix $\hat{S} > 0$ and a symmetric matrix $\widehat{\mathscr{P}}$ such that the following LMIs are feasible:*

$$
\begin{bmatrix}
\hat{S} - \widehat{W} - \widehat{W}^T & \hat{A} \\
\hat{A}^T & -\hat{S}
\end{bmatrix} \prec 0 \tag{7.121}
$$

$$
\begin{bmatrix}
-\widehat{\mathscr{P}} & -\widehat{W} & 0 & 0 \\
-\widehat{W}^T & \widehat{\mathscr{P}} + \hat{A}^T + \hat{A} & \hat{B}_0 & \hat{C}^T \\
0 & \hat{B}_0^T & -I & \mathbb{D}_0^T \\
0 & \hat{C} & \mathbb{D}_0 & -I
\end{bmatrix} \prec 0 \tag{7.122}
$$

Suppose that the LMIs (7.121) and (7.122) are feasible, then the following is a systematic procedure for obtaining the corresponding controller matrices of (7.111):

1. Compute the singular value decomposition (SVD) of $Z - X^T N$ to obtain square and invertible matrices U_1, V_1 such that $Z - X^T N = U_1 \Sigma_1 V_1^T$.
2. Choose the matrices U and R as $U^T = U_1 \Sigma_1^{\frac{1}{2}}$, $R = \Sigma_1^{\frac{1}{2}} V_1^T$.
3. Compute the matrices of the ILC controller state space model (7.111) using

$$
\begin{aligned}
C_K &= (\tilde{C} + D_{K1} CN) R^{-1} \\
B_{K1} &= U^{-T}(\tilde{B}_1 - X^T B D_{K1}) \\
B_{K2} &= U^{-T}(\tilde{B}_2 - X^T B D_{K2}) \\
A_K &= U^{-T}(\tilde{A} - X^T AN + X^T B D_{K1} CN + U^T B_{K1} CN - X^T B C_K R) R^{-1}
\end{aligned} \tag{7.123}
$$

where $U^{-T} = \left(U^{-1}\right)^T = \left(U^T\right)^{-1}$.

In terms of applications, the critical problem is to achieve the desired shape of the maximum singular value of $M(z)$ given by (7.116) over the complete range to account for the spectra of exogenous signals, penalize regulated variables, and specify the level of uncertainty. See, e.g. [115, 251] for further discussion, respectively, on this requirement in the standard and ILC settings.

One common approach to ensuring that the resulting frequency specifications are satisfied is to include weighting filters in the general system description [203]. The choice and tuning of these weighting filters is a nontrivial task. Their inclusion increases the order of the general system and, hence, the ILC controller, consequently increasing the design's complexity. The objective considered next is to develop a method for shaping the maximum singular value of $M(e^{j\omega})$ to meet the control specifications in different frequency ranges without introducing weighting filters.

Instead of considering the entire frequency range $[-\pi, \pi]$, attention can be limited to $\omega \in [0, \pi]$. Moreover, divide this frequency range into H arbitrarily chosen intervals such that

$$
[0, \pi] = \bigcup_{h=1}^{H} [\omega_{h-1}, \omega_h] \tag{7.124}
$$

where $\omega_0 = 0$ and $\omega_H = \pi$, and then Lemma 7.2 can be applied at any frequency in any of these intervals. The possibility to specify different performance specifications has considerable practical significance since common performance issues occur over different frequency ranges. For example, the trial-to-trial error convergence rate is in the "low" frequency range, whereas the low sensitivity to disturbances and sensor noise is in the "high" frequency range.

These objectives are included in the requirement that

$$\overline{\sigma}(M(e^{j\omega})) < \mu_h, \quad \forall \omega \in [\omega_{h-1}, \omega_h], \quad h = 1, \dots, H \tag{7.125}$$

where the entire frequency range is partitioned according to (7.124) and

$$0 < \mu_h \leq 1, \quad h = 1, \dots, H \tag{7.126}$$

The choice of the frequency intervals and the μ_h is determined by knowledge of the application under consideration.

Making use of Lemma 7.2 gives the following result, Theorem 3 in [203], for design to meet the specifications of (7.125).

Theorem 7.14 *Consider ILC dynamics described as a discrete linear repetitive process with no input terms and state-space model matrices (7.117). Furthermore, suppose that the entire frequency range is arbitrarily divided into H possible different frequency intervals as in (7.124). Then (i) the resulting dynamics are stable along the trial, (ii) monotonic trial-to-trial error convergence occurs, and (iii) the finite frequency performance specifications (7.125) are satisfied if there exist compatibly dimensioned matrices $\tilde{A}$, $\tilde{B}_1$, $\tilde{B}_2$, $\tilde{C}$, D_{K1}, D_{K2}, N, X, Z, $\widehat{\mathbb{Q}}_h > 0$, $\widehat{S} > 0$, symmetric $\widehat{\mathcal{P}}_h$ and arbitrary chosen scalars μ_h satisfying (7.126) such that the following LMIs are feasible:*

$$\begin{bmatrix} \widehat{S} - \widehat{W} - \widehat{W}^T & \widehat{A} \\ \widehat{A}^T & -\widehat{S} \end{bmatrix} < 0$$

$$\begin{bmatrix} -\widehat{\mathcal{P}}_h - \widehat{W} - \widehat{W}^T & e^{j\omega_{ch}}\widehat{\mathbb{Q}}_h - \widehat{W} & 0 & 0 \\ e^{-j\omega_{ch}}\widehat{\mathbb{Q}}_h - \widehat{W}^T & \widehat{\mathcal{P}}_h - 2\cos(\omega_{dh})\widehat{\mathbb{Q}}_h + \widehat{A}^T + \widehat{A} & \widehat{B}_0 & \widehat{C}^T \\ 0 & \widehat{B}_0^T & -\mu_h^2 I & \mathbb{D}_0^T \\ 0 & \widehat{C} & \mathbb{D}_0 & -I \end{bmatrix} < 0 \tag{7.127}$$

$\mathbb{D}_0 = I - CBK_2, \forall h = 1, \dots, H, where$

$$\omega_{ch} = \frac{\omega_{h-1} + \omega_h}{2}, \quad \omega_{dh} = \frac{\omega_h - \omega_{h-1}}{2}$$

If these LMIs is feasible, the ILC controller (7.111) are obtained from (7.123).

This last design has been applied to a laboratory servomechanism system. The system consists of a DC motor and the inertial mass (brass cylinder), weight 2.03 kg, diameter 0.066 m, and length 0.068 m, which are connected through the rigid shaft. The mass's rotational motion is excited by the DC motor, an incremental encoder measures load position, and the whole system operates with a PC-based digital controller.

A brake-out box contains an interface module that amplifies the control signals transmitted from the PC to the DC motor. The hardware setup includes a data acquisition board, a command transmission unit, and a computer that implements the control laws. The system is fully integrated with MATLAB/SIMULINK and operates in real-time. Figure 7.26 shows a schematic diagram of the system and Table 7.3 gives its basic parameters.

Figure 7.26 The modular servo system.

Table 7.3 The DC motor parameters.

Rated voltage	24 V
Rated output power	47 W
Rated torque	15 N cm
No load speed	3900 rpm/min
Rated current	3.1 A

Commonly, the angular position of the mass is taken as the output and the armature voltage as the input. Then the system transfer-function description is

$$G(s) = \frac{\Theta(s)}{V(s)} = \frac{K}{s((Js + b)(Ls + R) + K^2)} \tag{7.128}$$

where K represents both the motor torque constant (K_t) and the back emf constant (K_e), J is the total moment of inertia of the rotor (J_r) and the mass (J_m), b is the viscous friction constant of the motor, L is the electric inductance, and R is the electric resistance. Moreover, the armature voltage of the DC motor is controlled by a pulse-width modulation (PWM) signal. As a result, the dimensionless control signal $u(t)$ is the scaled input voltage and the admissible controls satisfy $|u(t)| \leq 1$. Consequently, the gain of $G(s)$ in (7.128) has to be increased by a factor equal to the maximum armature voltage $v_{max} = 24$ V. Table 7.4 shows the physical parameters of the DC motor.

The transfer-function (7.128) has been discretized with a sampling period of $T_s = 0.01$ seconds to give a minimal discrete linear systems state-space model with the state, input, and output matrices:

$$A = \begin{bmatrix} 1.0 & 0 & 0 \\ 0 & 0.9860 & 0.0002 \\ 0 & -0.0002 & -2.481 \times 10^{-8} \end{bmatrix}, \quad B = \begin{bmatrix} 50.6240 \\ 2.0613 \\ 0.0119 \end{bmatrix}$$

$$C = \begin{bmatrix} 0.0845 & -2.0613 & 0.0119 \end{bmatrix}$$

The reference trajectory for the mass position is of duration 6.5 seconds as shown in Figure 7.27a and its frequency spectrum is given in Figure 7.27b, which consists of harmonics from 0 to 2 Hz. The reference signal consists of six revolutions in the positive direction, a return path, six revolutions in the negative direction, and a return to the starting position. Hence, the cut-off frequency of the Q-filter must be at least 2 Hz.

In the results given below, the Q-filter is chosen as a sixth order Butterworth filter, with a cut-off frequency of 2 Hz. Also, 2 frequency ranges are chosen, i.e. from 0 to 1.2 Hz and from 1.2 to 2 Hz, respectively. The first of these covers the reference signal bandwidth and is of primary interest

Table 7.4 Physical parameters of the DC motor.

J_r	Moment of inertia of the rotor	18×10^{-6} kg m^2
J_m	Moment of inertia of the mass	0.0011 kg m^2
b	Motor viscous friction constant	3.5077×10^{-6} N m s
K_e	Electromotive force constant	0.056 V/(rad s)
K_t	Motor torque constant	0.056 N m/A
R	Electric resistance	2 Ω
L	Electric inductance	0.001 H

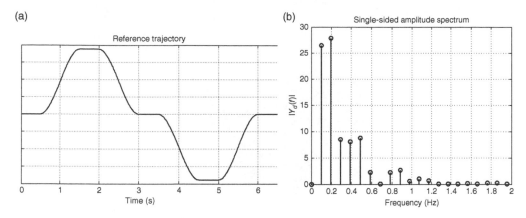

Figure 7.27 The reference trajectory (a) and its corresponding frequency spectrum (b).

because these frequencies have the most effect on the convergence rate. The second range includes those frequencies where the model accurately fits the actual system response, and hence, the control action is effective.

Executing the design procedure gives the feedback and learning controllers as follows:

$$C(z) = \frac{0.9582z^3 + 0.7857z^2 - 0.168z - 0.04756}{z^3 - 0.01955z^2 + 0.005592z + 0.01334}$$

$$L(z) = \frac{4.544z^3 + 6.16z^2 - 1.603z - 0.02429}{z^3 - 0.01955z^2 + 0.005592z + 0.01334}$$

where $\mu_1 = \sqrt{(0.8)}$ (upper bound on $|G(e^{j\omega})|$ in the first frequency range – see (7.125) and (7.126)) and $\mu_2 = \sqrt{0.95}$ in the second range. The resulting plot of $|G(e^{j\omega})|$ is given in Figure 7.28a and confirms that the design specifications are met.

The designed ILC law was experimentally tested over 10 trials. On completion of each trial, the RMS error was computed, resulting in Figure 7.28b.

Figure 7.29 shows the $RMS(e_k)$ of the tracking error plotted against the trial number and time to assess tracking performance. This last figure confirms the convergence of the error, and the largest

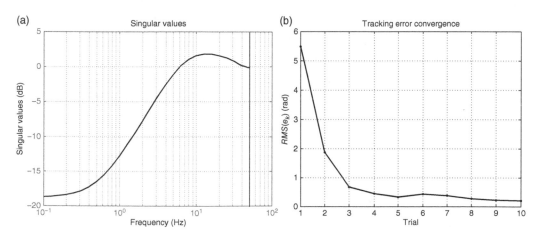

Figure 7.28 (a) Plot of $|G(e^{j\omega})|$. (b) The error convergence.

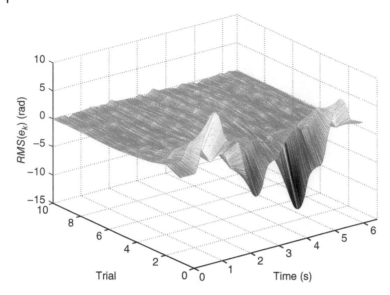

Figure 7.29 $RMS(e_k)$ error against time and trial number.

error occurs at the time instants when the system accelerated or decelerated. This feature corresponds to the low-frequency behavior of the system. Moreover, this plot demonstrates the good tracking performance of the design.

7.4 HOILC Design

Consider the HOILC law for a discrete linear system with the state, input, and output, respectively, matrices A, B, and C, see also Section 1.5. Also in the control law $u_{k+1}(p) = u_k(p) + \Delta u_{k+1}(p)$, set

$$\Delta u_{k+1}(p) = \sum_{j=1}^{M} K_{j-1} e_{k+1-j}(p+1) \tag{7.129}$$

Then proceeding as in the case of $M = 1$ results in the following state-space model of the controlled dynamics:

$$\eta_{k+1}(p+1) = A\eta_{k+1}(p) + \sum_{j=1}^{M} \hat{B}_{j-1} e_{k+1-j}(p)$$

$$e_{k+1}(p) = \hat{C}\eta_{k+1}(p) + \sum_{j=1}^{M} \hat{D}_{j-1} e_{k+1-j}(p) \tag{7.130}$$

where

$$\hat{B}_0 = BK_0, \quad \hat{B}_j = BK_j$$
$$\hat{C} = -CA, \quad \hat{D}_0 = I - CBK_0$$
$$\hat{D}_j = -CBK_j, \quad j = 1, \dots, M-1 \tag{7.131}$$

or, in equivalent unit memory form,

$$\eta_{k+1}(p+1) = A\eta_{k+1}(p) + \overline{B}\bar{e}_k(p)$$
$$\bar{e}_{k+1}(p) = \overline{C}\eta_{k+1}(p) + \overline{D}\bar{e}_k(p) \tag{7.132}$$

where

$$\bar{e}_k(p) = \begin{bmatrix} e_{k+1-M}^T(p) & \cdots & e_{k-1}^T(p) & e_k^T(p) \end{bmatrix}^T$$

$$\overline{B} = \begin{bmatrix} \hat{B}_{M-1} & \cdots & \hat{B}_1 & \hat{B}_0 \end{bmatrix}$$

$$\overline{C} = \begin{bmatrix} 0 \\ 0 \\ \vdots \\ 0 \\ \hat{C} \end{bmatrix}, \quad \overline{D} = \begin{bmatrix} 0 & I & 0 & \cdots & 0 \\ 0 & 0 & I & \ddots & 0 \\ 0 & 0 & 0 & \cdots & 0 \\ \vdots & \vdots & \vdots & \ddots & I \\ \hat{D}_{M-1} & \hat{D}_{M-2} & \hat{D}_{M-3} & \cdots & \hat{D}_0 \end{bmatrix} \tag{7.133}$$

Applying the z-transform to this last state-space model gives

$$\bar{e}_{k+1}(z) = G(z)\bar{e}_k(z) \tag{7.134}$$

where $G(z)\overline{C}(zI - \hat{A})^{-1}\overline{B} + \overline{D}$. The only term in transfer-function matrix $G(z)$ which relates to the convergence performance is the bottom row, i.e.

$$e_{k+1}(z) = \begin{bmatrix} G_{M-1}(z) & \cdots & G_1(z) & G_0(z) \end{bmatrix} \bar{e}_k(z) \tag{7.135}$$

Consider the case when the design objective requires that the H_∞ norm of $[G_{M-1}(z), \ldots, G_1(z), G_0(z)]$ is less than some specified positive constant γ. Then the following results are established in [272].

Theorem 7.15 *For a given $\gamma > 0$, the discrete linear repetitive process representing the ILC dynamics described by (7.132) is stable along the trial and*

$$\left\| \begin{bmatrix} G_{M-1}(z) & \cdots & G_1(z) & G_0(z) \end{bmatrix} \right\|_\infty < \gamma \tag{7.136}$$

if there exist compatibly dimensioned matrices $P_1 > 0$, and N_2 such that on setting $\mu = \gamma^2$ the following LMIs are feasible:

$$\begin{bmatrix} -P+Q & (\star) \\ A_1P + B_1N & -P \end{bmatrix} < 0 \tag{7.137}$$

$$\begin{bmatrix} -P & (\star) \\ A_2P + B_2N & -P \end{bmatrix} < 0 \tag{7.138}$$

where

$$A_1 = \begin{bmatrix} A & 0 & 0 & \cdots & 0 \\ -CA & 0 & 0 & \cdots & 1 \end{bmatrix}, \quad B_1 = \begin{bmatrix} 0 & B \\ 0 & -CB \end{bmatrix}$$

$$A_2 = \begin{bmatrix} A & 0 & 0 & \cdots & 0 \\ 0 & 0 & 1 & \cdots & 0 \\ \vdots & \vdots & \vdots & \ddots & \vdots \\ 0 & 0 & 0 & \cdots & 1 \\ -CA & 0 & 0 & \cdots & 1 \end{bmatrix}, \quad B_2 = \begin{bmatrix} 0 & B \\ 0 & 0 \\ \vdots & \vdots \\ 0 & 0 \\ 0 & -CB \end{bmatrix} \tag{7.139}$$

and

$$P = \text{diag}[P_1, I], \quad Q = \text{diag}[0, (1 - \mu)], \quad N = \text{diag}[0, N_2]$$

If these LMIs are feasible, stabilizing control law matrices are given by

$$\begin{bmatrix} K_{M-1} & \cdots & K_1 & K_0 \end{bmatrix} = N_2 \tag{7.140}$$

The following result shows that when γ is small, the tracking error is guaranteed to converge to zero.

Theorem 7.16 *If for* $\gamma \in [0,1/\sqrt{M})$ *the design in the previous theorem is feasible, the tracking error converges to zero as* $k \to \infty$. *Moreover, the convergence is monotonic in the sense that*

$$\max\{\|e_{k+1}\| \quad \cdots \quad \|e_{k-M+2}\|\} < \gamma \max\{\|e_k\| \quad \cdots \quad 2\|e_{k-M+1}\|\}$$

The last theorem shows that γ characterizes how quickly the tracking error converges to zero under a HOILC law. Also, optimization is possible using

$$\min_{P_1>0,\gamma<1,N_2} \mu \tag{7.141}$$

7.5 Inferential ILC Design

In some control applications, the performance variables are explicitly distinguished from those that are measured. Consequently, the former of these two variable sets are not available for use in real-time feedback control but are often available after a task is complete. This fact enables the application of ILC to performance variables. In a recent work [29], it has been shown that some ILC control laws may not be directly implementable in this setting, but, as detailed next, the linear repetitive process setting for design can be applied.

In some applications, such as motion systems, increasing performance requirements enforce an explicit distinction between measured and performance variables. Still, the latter may not be available for use in real-time feedback control. Reasons include computational constraints or physical restrictions on sensor placement, where the former is reduced by increased computational power at a cheaper cost, but the latter will continue to arise. Further discussion of this matter is given in [29].

Performance variables are often available offline, e.g. when the final product is inspected after the processing is complete, the actual performance becomes available. Hence, batch-to-batch control using these variables is possible, and ILC is well established in this general area, see, e.g. [48, 94]. However, a direct combination of ILC using the performance variables with feedback control action can lead to a conflict, where some early evidence of this problem was given in [150]. Also, [268] proposed to use observers to infer the performance variables from the real-time measurements instead of direct performance measurement. In [29], performance variables and different real-time measured variables for feedback control are termed inferential ILC.

This section demonstrates that a design in the repetitive process setting for this form of ILC avoids possible unstable behavior that would remain undetected if the lifting approach to design were used. Also, this approach includes that of [268] as a particular case. The results are motivated by a printing system application as in [29]. Such systems are examples where the performance variables cannot be measured directly in real-time.

Figure 7.30 shows the side-view of the positioning drive in a printer where the paper position z is controlled using the motor. The paper positioning derive is traditionally controlled by feedback using encoder position measurements, which is relatively cheap. However, high-tracking accuracy using the encoder measurement (y) does not imply good printing performance (z) because of mechanical deformations in the drive. The introduction of a scanner in the printhead enables line-by-line measurements of the printing performance z and the possibility of improved control.

Direct measurement of the performance of the form considered is not available for real-time feedback control in this application. Still, ILC can be applied where the feedback control variables (y) are not equal to the variables for ILC (z). This arrangement corresponds to the block diagram of Figure 7.31 (where the notation in [29] is followed).

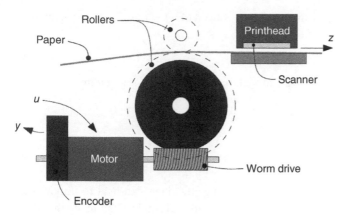

Figure 7.30 Side-view of the positioning drive in a printer. Source: [29]/J. Bolder.

Figure 7.31 Block diagram of the feedback control problem.

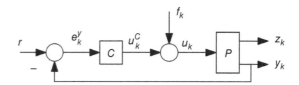

The system P in this last figure is described by

$$\begin{bmatrix} z_k \\ y_k \end{bmatrix} = Pu_k \tag{7.142}$$

where P has two outputs, i.e. the performance variable z_k and the measured variable y_k in the SISO case, which is considered for ease of presentation only. Also, the system input

$$u_k = u_k^C + f_k \tag{7.143}$$

is the sum of the feedback control signal u_k^C and f_k. In standard printing systems, it is assumed that $y_k \approx z_k$ and then a feedback controller is implemented as $u_k^C = C(r - y_k)$, where C is assumed to be fixed and designed to ensure that the controlled system is stable in the internal sense. If an ILC law is applied, the feedforward signal f_k is of the form $f_{k+1}(r, z_k, f_k)$, where one choice is

$$f_{k+1} = Q(f_k + Le_k^z) \tag{7.144}$$

where $e_k^z = r - z_k$, L is a learning filter and Q is a robustness filter. Hence,

$$f_{k+1} = Q(1 - LJ)f_k + L(1 - JC)r, \quad J = \frac{P^z}{1 + CP^y}, \quad P = \begin{bmatrix} P^z \\ P^y \end{bmatrix} \tag{7.145}$$

Consider the case when

$$P = \begin{bmatrix} P^z \\ P^y \end{bmatrix} = \begin{bmatrix} 1 \\ 3 \end{bmatrix}, \quad C = \begin{bmatrix} 1 & 1 \\ 0.5 & 0 \end{bmatrix} \tag{7.146}$$

leading to the controlled dynamics

$$\begin{bmatrix} z_k \\ y_k \end{bmatrix} = \begin{bmatrix} -0.5 & 1 & 3 \\ -05 & 0 & 1 \\ -1.5 & 0 & 3 \end{bmatrix} \begin{bmatrix} r \\ f_k \end{bmatrix} \tag{7.147}$$

where this system is stable. Next, consider the ILC law (7.144) with $Q = 1$ and L chosen such that the ILC law converges from trial-to-trial where

$$f_\infty = \lim_{k\to\infty} f_{k+1} = (1 + CP^y - CP^z)P^{z^{-1}}r \tag{7.148}$$

and results in $e_\infty^z = 0$. Also a minimal state-space model for (7.148) and e_∞^z is given by

$$\begin{bmatrix} f_\infty \\ e_\infty^z \end{bmatrix} = \begin{bmatrix} 1 & -2 \\ -0.5 & 1 \\ 0 & 0 \end{bmatrix} r \tag{7.149}$$

For this example, a bounded nonzero r gives $e_\infty^z = 0$, but f_∞ may be unbounded (a pole of $F(z)$ is on the unit circle).

The example above confirms that in the inferential setting, ILC laws can lead to unbounded control signals and that this critical aspect does not appear in analysis and design using the lifted setting. To begin the analysis using the discrete repetitive process setting, consider the following conformally partitioned state-space representations for the system P and feedback controller C:

$$P = \begin{bmatrix} A^P & B^P \\ C^{Pz} & D^{Pz} \\ C^{Py} & D^{Py} \end{bmatrix}, \quad C = \begin{bmatrix} A^C & B^C \\ C^C & 0 \end{bmatrix} \tag{7.150}$$

and for ease of presentation, the controller is assumed to be strictly proper, which guarantees well posedness. Also, let the state-space representations for L and Q in the control law be

$$Q = \begin{bmatrix} A^Q & B^Q \\ C^Q & D^Q \end{bmatrix}, \quad L = \begin{bmatrix} A^L & B^L \\ C^L & D^L \end{bmatrix} \tag{7.151}$$

Figure 7.32 shows the block diagram the system considered with both the feedback and ILC laws applied.

Introduce the following vector constructed from (7.150) and (7.151)

$$X_{k+1}(p) = \left[\left(x_{k+1}^P(p)\right)^T \ \left(x_{k+1}^C(p)\right)^T \ \left(x_{k+1}^L(p)\right)^T \ \left(x_{k+1}^Q(p)\right)^T \right]^T \tag{7.152}$$

and

$$Y_{k+1}(p) = f_{k+1}(p) \tag{7.153}$$

Then using $e_k^z = r - z_k$ and $u_k = C(r - y_k) + f_k$ results in the discrete linear repetitive process state-space model:

$$\begin{aligned} X_{k+1}(p+1) &= \mathbb{A}X_{k+1}(p) + \mathbb{B}U_{k+1}(p) + \mathbb{B}_0 Y_k(p) \\ Y_{k+1}(p) &= \mathbb{C}X_{k+1}(p) + \mathbb{D}U_{k+1}(p) + \mathbb{D}_0 Y_k(p) \end{aligned} \tag{7.154}$$

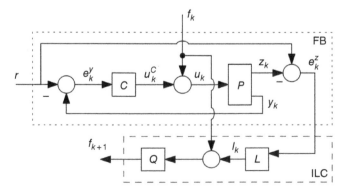

Figure 7.32 Block diagram of the combined feedback control and ILC.

where $U_{k+1}(p) = r(p)$ and

$$
\mathbb{A} = \begin{bmatrix} A^P & B^P C^C & 0 & 0 \\ -B^C C^{Py} & A^C - B^C D^{Py} C^C & 0 & 0 \\ -B^L C^{Pz} & -B^L D^{Pz} C^C & A^L & 0 \\ -B^Q D^L C^{Pz} & -B^Q D^L D^{Pz} C^C & B^Q C^L & A^Q \end{bmatrix}
\tag{7.155}
$$

$$
\mathbb{B} = \begin{bmatrix} 0 \\ B^C \\ B^L \\ B^Q D^L \end{bmatrix}, \quad \mathbb{B}_0 = \begin{bmatrix} B^P \\ -B^C D^{Py} \\ -B^L D^{Pz} \\ B^Q(I - D^L D^{Pz}) \end{bmatrix}
\tag{7.156}
$$

$$
\mathbb{C} = \begin{bmatrix} -D^Q D^L C^{Pz} & -D^Q D^L D^{Pz} C^C & D^Q C^L & C^Q \end{bmatrix}
\tag{7.157}
$$

$$
\mathbb{D} = D^Q D^L, \quad \mathbb{D}_0 = D^Q(I - D^L D^{Pz})
\tag{7.158}
$$

Asymptotic stability of the discrete linear repetitive process defined above holds if and only if $\rho(\mathbb{D}_0) < 1$. As the numerical example given above demonstrates, the resulting limit profile can be unstable, i.e. a bounded nonzero reference signal produces a limit profile that is unbounded as the temporal variable increases. Application of stability along the trial gives the following result proved as Theorem 4.10 in [29].

Theorem 7.17 *The discrete linear repetitive process described by (7.154) and (7.155)–(7.158) is stable along the trial if and only if*

(a)
$$
\rho(\overline{Q}(I - \overline{L}\,\overline{J})) < 1
$$

(b)
$$
\rho(A^{CP}) < 1
$$

(c)
$$
\rho(A^L) < 1
$$

(d)
$$
\rho(A^Q) < 1
$$

(e)
$$
\rho(Q(z)(I - LJ(z))) < 1, \quad \forall\, |z| = 1
$$

with
$$
J(z) = \frac{P^z(z)}{1 + C(z)P^y(z)}
\tag{7.159}
$$

where $\overline{Q}, \overline{L}$, and $\overline{J}$ are finite-time matrix representations of Q, L, and J, and

$$
A^{CP} = \begin{bmatrix} A^P & B^P C^C \\ -B^C C^{Py} & A^C - B^C D^{Py} C^C \end{bmatrix}
\tag{7.160}
$$

is the system matrix of the closed-loop interconnection of C and P.

This last theorem establishes several connections between stability along the trial and ILC convergence. First, the discrete linear repetitive process representation's asymptotic stability is equivalent to the finite-time convergence condition reported as Theorem 1 in [176]. Stability along the trial demands that all time-domain dynamics for a fixed trial k are stable, and in the inferential ILC setting, this requires that the L and Q filters are stable and $\rho(A^{CP}) < 1$, i.e. C is an internally stabilizing controller. Also, the third condition for stability along the trial of this last theorem is equivalent to the frequency domain convergence condition $\rho(Q(z)(1 - L(z)J(z)) < 1$ given as Theorem 6 in [176].

Other ILC design methods, such as the lifting approach, use a weaker stability property for the inferential case than in the repetitive process setting. In particular, for unstable systems, there is the preliminary step of designing a stabilizing control law and then ILC design for the resulting controlled dynamics. In the above example, a convergent ILC law is assumed without a specific L. The following result shows that this design cannot be stable along the trial for any L if C includes integral action and $Q = 1$.

Theorem 7.18 *[29] Given $Q = 1$ and*

$$P(z) = \begin{bmatrix} P^z(z) \\ P^y(z) \end{bmatrix} \tag{7.161}$$

let C have a minimal state-space representation defined by $\{A^C, B^C, C^C, D^C\}$ that is stable, i.e. $\rho(A^{CP}) < 1$ and A^C has an eigenvalue at $z = 1$. Then for any learning filter L either condition (b) or condition (c) of Theorem 7.17 cannot hold.

This last result shows that either condition (b) or condition (c) for stability along the trial can hold but not simultaneously. Hence, stability along the trial can often not be achieved for inferential ILC. Moreover, deleting the integral action from C or adding a robustness filter Q are only partial solutions. In the former of these options, feedback control action is often already present before ILC design is attempted. This action is used to attenuate trial-varying disturbances. Also, introducing a robustness filter in the design does not imply that stability along the trial can be achieved.

One solution to the above problem is to replace the parallel structure with the serial alternative shown in Figure 7.33, where the ILC signal now forms the reference for the feedback loop instead of a feedforward signal in the parallel structure.

The result of Theorem 7.18 shows that stability along the trial cannot be present for any learning filter L if $Q = 1$ and C includes integral action. Also, the serial structure (Figure 7.33) is a special case of the parallel structure (Figure 7.32) on setting $C = 0$ and $P = J_{ser}$, where $z_k = J_{ser}\eta_k$. In this setting J_{ser} is the closed-loop system within the dotted area denoted by FB in Figure 7.33. Hence, the theory for the parallel structure given above can be applied to the serial alternative with appropriate

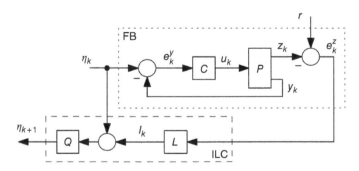

Figure 7.33 Block diagram of the serial ILC.

changes to the definitions of C and P. The following corollary to Theorem 7.17 gives stability along the trial conditions for the structure of Figure 7.33.

Corollary 7.1 *Stability along the trial holds for serial inferential ILC if and only if*

(a)
$$\rho(\overline{Q}(I - \overline{L}\,\overline{J}_{ser})) < 1$$

(b)
$$\rho(A^{J_{ser}}) < 1$$

(c)
$$\rho(A^{L}) < 1$$

(d)
$$\rho(A^{Q}) < 1$$

(e)
$$\rho(Q(z)(I - L(z)J_{ser}(z))) < 1, \quad \forall\, |z| = 1$$

where
$$J_{ser}(z) = \frac{C(z)P^z(z)}{1 + C(z)P^y(z)} \tag{7.162}$$

and J_{ser} is the system matrix corresponding to a minimal realization of (7.162) and $\overline{Q}, \overline{L},$ and $\overline{J}_{ser}$ are finite-time matrix representations of $Q, L,$ and J_{ser}.

As described above, the learning filter must contain integral action in the parallel case but for the serial structure $\lim_{z\to 1} J_{ser}(z) = \frac{P^z(z)}{P^y(z)}$. Moreover, this limit does not include C^{-1} and hence, if C contains integral action, then L does not need such action, and stability along the trial can be achieved. As an example, return to (7.146) with controller

$$C = \begin{bmatrix} 1 & 1 \\ 0.5 & 0.5 \end{bmatrix} \tag{7.163}$$

resulting in the controlled system $z_k = J_{ser}\eta_k$ or

$$z_k = \begin{bmatrix} A^{J_{ser}} & B^{J_{ser}} \\ C^{J_{ser}} & D^{J_{ser}} \end{bmatrix} \eta_k = \begin{bmatrix} 0.4 & 0.4 \\ -1 & 0 \end{bmatrix} \eta_k \tag{7.164}$$

which is asymptotically stable since $\rho(A^{J_{ser}}) < 1$. Consider also the case when $L = 5$ and $Q = 1$.

Conditions (a)–(d) of Corollary 7.1 hold for this example, and condition (e) of this result also holds since

$$\rho(Q(I - L(z)J_{ser}(z))) \le \frac{2}{3}, \quad \forall\, |z| = 1 \tag{7.165}$$

and hence the resulting limit profile must be asymptotically stable. This limit profile is described by

$$\begin{bmatrix} \eta_\infty \\ e^z_\infty \end{bmatrix} = \begin{bmatrix} 0 & 2 \\ -1 & 5 \\ 0 & 0 \end{bmatrix} r \tag{7.166}$$

which is stable and $e^z_\infty = 0$ results.

An alternative to the approach given above is for the feedback controller to use the real-time measurements together with an observer structure [268]. However, in [29], it is shown that this approach is a special case of the design in this section.

7.6 Concluding Remarks

This chapter has established that linear repetitive process stability theory forms a comprehensive setting for the design of ILC systems that, unlike designs based on other 2D systems descriptions, have been followed through to experimental validation, including robust designs. The linear repetitive process state-space models considered in this chapter do not include all possible dynamics exhibited by such processes. One example is wave linear repetitive processes [55], where some preliminary ILC results are given. See also [158] for highly promising initial results of repetitive process-based design for interconnected systems.

Recent years have seen the development of a stability theory and control law design setting for nonlinear repetitive processes, based on using the concept of dissipativity with an extension to control law design; see, e.g. [196]. This theory is used in Chapter 8 for ILC design in the presence of input saturation, with experimental validation, and in Chapter 12 for stochastic ILC design.

8

Constrained ILC Design

In some practical applications, constraints will arise due to physical limitations, e.g. actuator limits, or performance requirements, e.g. limitations on the performance along the trials. Hence, there is a need to extend iterative learning control (ILC) designs to include constraints. Previous results in this area include [49], where a novel nonlinear controller for process systems with input constraints that requires only limited knowledge of the process model is given. In [99], an ILC problem with soft constraints was considered, and Lagrange multiplier methods were used to develop a solution. Moreover, [142] used optimal quadratic design to formulate a constrained ILC problem and suggested that this setting can deal with constraints. Some other results in this area are given in [238, 277, 285].

In the first part of this chapter, a new method of ILC law design with input saturation is developed for discrete linear time-invariant systems. It is noted that an extension to differential dynamics is immediate. The analysis uses vector Lyapunov function-based stability theory for nonlinear repetitive processes [196]. A comparative study of the performance of this new design and an alternative is given. Finally, experimental validation results are given.

This chapter also considers linear time-varying (LTV) systems based on successive projection methods, with analysis and design. Supporting experimental validation results using the rack feeder of Section 2.2.3 are also given.

8.1 ILC with Saturating Inputs Design

8.1.1 Observer-Based State Control Law Design

This section follows, in the main, [201], with preliminary results in [198]. The notation in [201] is used and discrete-time systems described by the following state-space model in the ILC setting are considered:

$$
\begin{aligned}
x_k(p+1) &= Ax_k(p) + B\psi_k(p) \\
\psi_k(p) &= \mathrm{sat}(u_k(p)) \\
y_k(p) &= Cx_k(p), \quad 0 \le p \le N-1, \quad k \ge 0
\end{aligned}
\tag{8.1}
$$

where $x_k(p) \in \mathbb{R}^n$, $u_k(p) \in \mathbb{R}^m$, $y_k(p) \in \mathbb{R}^m$ are, respectively, the state, control input, and output vectors. The vector-valued saturation function $\mathrm{sat}(u_k(p)) \in \mathbb{R}^m$ has entries

$$
\psi_k(p)_i = \mathrm{sat}(u_k(p))_i
$$

Iterative Learning Control Algorithms and Experimental Benchmarking, First Edition.
Eric Rogers, Bing Chu, Christopher Freeman and Paul Lewin.
© 2023 John Wiley & Sons Ltd. Published 2023 by John Wiley & Sons Ltd.

$$= \begin{cases} U_i & \text{if } u_{k,i}(p) > U_i \\ u_{k,i}(p) & \text{if } -U_i \le u_{k,i}(p) \le U_i \\ -U_i & \text{if } u_{k,i}(p) < -U_i \end{cases} \tag{8.2}$$

for $1 \le i \le m$, $k \ge 0$, where $u_{k,i}(p)$ denotes entry i in $u_k(p)$ and the U_i are specified positive constants. It is also assumed that the pair $\{A, B\}$ is controllable, and the pair $\{C, A\}$ is observable. Also, for a relative degree $h \ge 1$, it is assumed that $\det[CA^{h-1}B] = m$ and

$$CA^i B = 0, \quad i = 1, 2, \dots, h-2$$

Under this condition, the pair $(I, CA^{h-1}B)$ is controllable and the equation that describes error dynamics below will be well defined.

As in previous chapters, let $r(p) \in \mathbb{R}^M$, $0 \le p \le N-1$, denote the supplied reference vector and then $e_k(p) = r(p) - y_k(p)$ is the error on trial k. The control design problem is to construct a control input sequence $\{u_k\}_k$, such that

$$\lim_{k \to \infty} \|e_k(p)\| = \|e_\infty(p)\|$$
$$\|e_k(p)\| \le \kappa \varrho^k + \mu, \; \kappa > 0, \; \mu \ge 0, \; 0 < \varrho < 1$$
$$\lim_{k \to \infty} \|u_k(p)\| = \|u_\infty(p)\| \tag{8.3}$$

where $\| \cdot \|$ denotes the norm on the underlying function space, $e_\infty(p)$ and $u_\infty(p)$ are bounded variables and $u_\infty(p)$ is the learned control. If there is no saturation present, the design developed in this section reduces to the case for linear dynamics and $\lim_{k \to \infty} \|e_k(p)\| = 0$ is ensured.

The structure of the ILC law is

$$\psi_{k+1}(p) = \text{sat}(u_{k+1}(p)) = \text{sat}(\psi_k(p) + \delta u_{k+1}(p)) \tag{8.4}$$

where $\delta u_{k+1}(p)$ is the control update. If all state variables are not available for measurement, an observer is required, where the analysis given includes both cases.

In this section, the state observer for (8.1) has the structure:

$$\hat{x}_k(p+1) = A\hat{x}_k(p) + B\psi_k(p) + F(y_k(p) - C\hat{x}_k(p)) \tag{8.5}$$

where $\hat{x}_k(p)$ is the estimated state vector and F is the observer gain matrix, which is guaranteed to exist by the assumption that the pair (C, A) is observable. Let $\tilde{x}_k(p) = x_k(p) - \hat{x}_k(p)$ denote the estimation error and introduce the increments of the estimate and estimation error, respectively, as

$$\hat{\xi}_{k+1}(p+1) = \hat{x}_{k+1}(p) - \hat{x}_k(p)$$
$$\tilde{\xi}_{k+1}(p+1) = \tilde{x}_{k+1}(p) - \tilde{x}_k(p)$$

Consider the case when

$$\delta u_{k+1}(p) = K_1 \hat{\xi}_{k+1}(p+1) + K_2 e_k(p+h) \tag{8.6}$$

where K_1 and K_2 are matrices of compatible dimensions to be designed. Then the augmented state-space model coupling the dynamics of (8.1), (8.5), and (8.6) can be written as follows:

$$\tilde{\xi}_{k+1}(p+1) = (A - FC)\tilde{\xi}_{k+1}(p)$$
$$\hat{\xi}_{k+1}(p+1) = FC\tilde{\xi}_{k+1}(p) + (A + BK_1)\hat{\xi}_{k+1}(p) + BK_2\bar{e}_k(p) + B\varphi_k(p)$$
$$\bar{e}_{k+1}(p) = G\tilde{\xi}_{k+1}(p) - CA^{h-1}(A + BK_1)\hat{\xi}_{k+1}(p) + (I - CA^{h-1}BK_2)\bar{e}_k(p) - CA^{h-1}B\varphi_k(p)$$
$$\tag{8.7}$$

where

$$\varphi_k(p) = \text{sat}(u_{k+1}(p-1)) - \text{sat}(u_k(p-1)) - \delta u_{k+1}(p-1)$$

$$\bar{e}_k(p) = e_k(p+h-1)$$

$$G = -C\left((A-FC)^h + \sum_{i=0}^{h-1} A^i FC(A-FC)^{h-1-i}\right) \tag{8.8}$$

Introduce the vector $\eta_{k+1}(p) = \left[\tilde{\xi}_{k+1}^T(p) \ \hat{\xi}_{k+1}^T(p)\right]^T$. Then the resulting controlled ILC dynamics can be written as follows:

$$\eta_{k+1}(p+1) = \begin{bmatrix} A-FC & 0 \\ FC & A+BK_1 \end{bmatrix} \eta_{k+1}(p) + \begin{bmatrix} 0 \\ BK_2 \end{bmatrix} \bar{e}_k(p) + \begin{bmatrix} 0 \\ B \end{bmatrix} \varphi_k(p)$$

$$\bar{e}_{k+1}(p) = \begin{bmatrix} G & -CA^{h-1}(A+BK_1) \end{bmatrix} \eta_{k+1}(p) + [I - CA^{h-1}BK_2]\bar{e}_k(p) - CA^{h-1}B\varphi_k(p) \tag{8.9}$$

and introduce the notation:

$$K = \begin{bmatrix} K_1 & K_2 \end{bmatrix}, \quad \overline{C} = \begin{bmatrix} 0 & I & 0 \\ 0 & 0 & I \end{bmatrix}, \quad \zeta_k(p) = \begin{bmatrix} \eta_{k+1}^T(p) & \bar{e}_k^T(p) \end{bmatrix}^T$$

This state-space model is in the form of a discrete nonlinear repetitive process.

It follows from (8.2) that

$$-2U_i \leq (\text{sat}(u_{k+1}(p))_i - (\text{sat}(u_k(p))_i \leq 2U_i, \quad i = 1, \ldots, m \tag{8.10}$$

and using (8.8) and (8.6) gives for $i = 1, 2, \ldots, m$,

$$-2U_i - (K\overline{C}\zeta_k(p))_i \leq (\varphi_k(p))_i \leq 2U_i - (K\overline{C}\zeta_k(p))_i \tag{8.11}$$

Given (8.11), it follows that entries of $\varphi_k(p)$ satisfy the quadratic constraints:

$$F_i[(\varphi_k(p))_i, (\zeta_k(p))_i] = \left[1 + \frac{1}{2U_i}((\varphi_k(p))_i + (K\overline{C}\zeta_k(p))_i)\right] \times \left[1 - \frac{1}{2U_i}((\varphi_k(p))_i + (K\overline{C}\zeta_k(p))_i)\right]$$

$$\geq 0, \quad i = 1, 2, \ldots, m \tag{8.12}$$

Using (8.6) and (8.8), $\varphi_k(p)$ is the difference between the saturated and nonsaturated ILC laws and is expressed in terms of the constraints (8.12), where the saturation levels are imposed on the design. In case of nonsaturated control $\varphi_k(p) = 0$ and the constraints of (8.12) are redundant.

In the presence of saturation, the ILC dynamics are nonlinear, and there has been recent work on developing a stability theory for nonlinear repetitive processes. One approach is based on the use of vector Lyapunov functions [196]. This theory forms the basis for the development of the ILC designs in this section.

Consider a vector Lyapunov function of the following form for the ILC dynamics described by (8.9)

$$V(\eta_{k+1}(p), \bar{e}_k(p)) = \begin{bmatrix} V_1(\eta_{k+1}(p)) \\ V_2(\bar{e}_k(p)) \end{bmatrix} \tag{8.13}$$

where $V_1(\eta_{k+1}(p)) > 0$, $\eta_{k+1}(p) \neq 0$, $V_2(\bar{e}_k(p)) > 0$, $\bar{e}_k(p) \neq 0$, $V_1(0) = 0$, $V_2(0) = 0$, and define the counterpart of the divergence operator along the trajectories of (8.9) as follows:

$$\mathcal{D}_d V(\eta_{k+1}(p), \bar{e}_k(p)) = V_1(\eta_{k+1}(p+1)) - V_1(\eta_{k+1}(p)) + V_2(\bar{e}_{k+1}(p)) - V_2(\bar{e}_k(p)) \tag{8.14}$$

For ease of presentation, this last property will be referred to as the divergence operator from this point onward. The following result is proved as Theorem 1 in [201].

Theorem 8.1 *Suppose that there exists a vector Lyapunov function (8.13) and positive scalars* $c_1, c_2, c_3,$ *and* γ *such that*

$$c_1 ||\eta_k(p)||^2 \le V_1(\eta_k(p)) \le c_2 ||\eta_k(p)||^2 \tag{8.15}$$

$$c_1 ||\bar{e}_k(p)||^2 \le V_2(\bar{e}_k(p)) \le c_2 ||\bar{e}_k(p)||^2 \tag{8.16}$$

$$\mathscr{D}_d V(\eta_{k+1}(p), \bar{e}_k(p)) \le \gamma - c_3(||\eta_{k+1}(p)||^2 + ||\bar{e}_k(p)||^2) \tag{8.17}$$

Then, for dynamics described by (8.9), $||e_k(p)||$ *is monotonically decreasing as* $k \to \infty$ *for all* $0 \le p \le N - 1$ *and the conditions of (8.3) hold.*

Remark 8.1 The results in this section are established using vector Lyapunov functions in explicit form, i.e. $V = [V_1 \ V_2]^T, V_1 > 0, V_2 > 0$ and (8.14). An alternative would be to use the scalar form $V_c = V_1 + V_2$ with $V_1 > 0$ and $V_2 > 0$, but the increment of this function along trajectories of (8.9) would have to be calculated in addition to the right-hand side of (8.14). The merits of one approach over the other are, if any, applications specific. However, the divergence gives a clearer physical interpretation, which, similar to the usual Lyapunov function, can be interpreted as the generalized energy of the system.

Remark 8.2 The difference between Theorem 8.1 and Theorem 2 of [196] is the presence of the parameter γ and if $\gamma > 0$, the error converges to some non-zero value as a geometric progression as $k \to \infty$. In [196] $\gamma = 0$ and convergence of the error to zero is guaranteed as $k \to \infty$ with the same rate. This previously established result cannot, however, be applied to the dynamics considered in this section due to the specific constraints (8.12) for the nonlinear term $\varphi_k(p)$ in (8.9) Setting $\gamma = 0$ recovers the case with no saturation, i.e., monotonic trial-to-trial error convergence.

To develop the ILC law design, introduce the notation:

$$\bar{A} = \begin{bmatrix} A - FC & 0 & 0 \\ FC & A & 0 \\ G & -CA^h & I \end{bmatrix}, \bar{B} = \begin{bmatrix} 0 \\ B \\ -CA^{h-1}B \end{bmatrix}, D_U = \text{diag}\left[\frac{1}{4}U_i^2\right], i = 1, 2, \dots, m, T_U = D_U^{-1} \tag{8.18}$$

Also, let the matrices $X = \text{diag}[X_1 \ X_2] > 0, \ Y = [Y_1, \ Y_2]$, and Z be such that

$$\begin{bmatrix} X & (\bar{A}X + \bar{B}Y\bar{C})^T & X & (Y\bar{C})^T \\ \bar{A}X + \bar{B}Y\bar{C} & X & 0 & 0 \\ X & 0 & Q^{-1} & 0 \\ Y\bar{C} & 0 & 0 & R^{-1} \end{bmatrix} \ge 0$$

$$X > 0, \quad \bar{C}X = Z\bar{C} \tag{8.19}$$

where $Q > 0$ and $R > 0$ are weighting matrices to be selected. Also, if (8.19) is solvable relative to these variables, then, see [196], the linear control law $\phi_k(p) = K\bar{C}\zeta_k(p)$ guarantees trial-to-trial error convergence.

The following result is proved as Theorem 2 in [201].

Theorem 8.2 *Suppose that for given constraints (8.12) and specified matrices $Q > 0$ and $R > 0$ there exist matrices $X = \text{diag}[X_1,\ X_2] > 0$, Y, and Z satisfying (8.19) and matrices $W = \text{diag}[W_1,\ W_2] > 0$, $T = \text{diag}\left[\ T_1\ \dots\ T_m\ \right] > 0$, $i = 1, \dots, m$, satisfying the linear matrix inequality (LMI):*

$$
\begin{bmatrix}
-W & -(K\overline{C}W)^T & (\overline{A}W + \overline{B}K\overline{C}W)^T \\
-K\overline{C}W & TT_U & TT_U\overline{B}^T \\
(\overline{A}W + \overline{B}K\overline{C}W) & \overline{B}TT_U & -W
\end{bmatrix} \prec 0
\tag{8.20}
$$

with $K = YZ^{-1}$ is feasible. Then the trial error sequence formed by $e_k(p)$ converges to a norm-bounded function $e_\infty(p)$ as $k \to \infty$ $\forall\ 0 \le p \le N - 1$ such that $\|\bar{e}_k(p)\|$ is monotonically decreasing and the conditions of (8.3) hold. Moreover, the control law matrix $K = [K_1\ K_2]$ is given by

$$
K = YZ^{-1}
\tag{8.21}
$$

In an analysis, intermediate variables, e.g. (8.6), have been introduced. Substituting for these in terms of the problem variables gives the following control law for implementation in the unsaturated case:

$$
u_k(p) = u_{k-1}(p) + K_1(\hat{x}_k(p) - \hat{x}_{k-1}(p)) + K_2(r(p+h) - y_{k-1}(p+h))
\tag{8.22}
$$

and in the presence of saturation

$$
u_k(p) = \text{sat}(u_{k-1}(p) + K_1(\hat{x}_k(p) - \hat{x}_{k-1}(p)) + K_2(r(p+h) - y_{k-1}(p+h)))
\tag{8.23}
$$

To highlight the application of Theorem 8.2 and the effects of saturation on the ILC design, consider the system described by

$$
G(s) = \frac{23.7356(s + 661.2)}{s^3 + 426.7s^2 + 1.744 \times 10^5 s}
\tag{8.24}
$$

This model was transformed to the state-space form and discretized using the exact method (the zero-order hold) with the sampling period 0.01 seconds. The resulting discrete-time model is controllable and observable.

The choice of the observer poles is, as in other areas, a matter of judgment based on the knowledge of the application under consideration. As one choice observer poles at $-0.1 \pm 0.05, 0.9$ results in $F = \left[\ 0.5437\ \ -0.0031\ \ 0.0113\ \right]^T$ (designed by a pole placement algorithm). For control with the saturation level $U = 2.7$, application of Theorem 8.2 with $Q = \text{diag}[\ I,\ \ I,\ \ 10^3\ ]$, $R = 5 \times 10^{-3}$ gives

$$
K = [K_1\ K_2], \quad K_1 = [-20.8 - 16.9 - 6889], \quad K_2 = 420
$$

The simulations for the ILC laws (8.22) and (8.23) are for the reference trajectory shown in Figure 8.1a and zero-boundary conditions. During the simulation, a normally distributed random variable with a standard deviation of 10^{-4} was added to the output signal to assess the influence

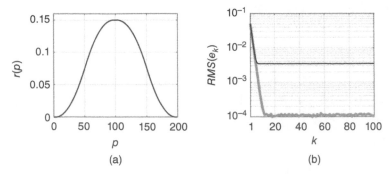

Figure 8.1 Application of the ILC laws (8.22) and (8.23) to the system described by (8.24): (a) the reference signal, (b) the $RMS(e_k)$ tracking errors for (8.22) (gray line) and (8.23) with saturation level $U = 2.7$ (black line).

of measurement noise on the control quality. The root-mean-square (RMS) error first defined in (6.103), i.e.

$$RMS(e_k) = \sqrt{\frac{1}{N} \sum_{p=0}^{N-1} ||e_k(p)||^2} \tag{8.25}$$

is used as a measure of ILC performance.

The gray line in Figure 8.1b shows $RMS(e_k)$ for the ILC law (8.22). This law needs 12 trials to obtain accurate tracking, where $RMS(e_{12})$ is close to the standard deviation of the noise signal. The black line in Figure 8.1b shows $RMS(e_k)$ for the ILC law (8.23) with saturation level $U = 2.7$. The control signal progressions for both ILC laws over the first 20 trials are given in Figure 8.2.

It is of interest to directly compare the saturated and unsaturated control signals on particular trials. Figure 8.3a,b shows the unsaturated control signal (gray line) and saturated (black line) with saturation level $U = 2.7$ for the 4th and 100th trials, respectively. These are representatives of the early trials and after many trials, which show the increasing presence of the saturation effect as the trial number k increases. These results also show that the noise signal added to the output does not significantly affect the control signal.

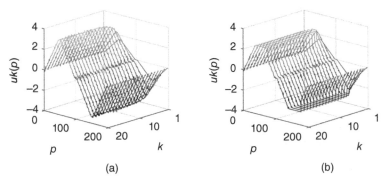

Figure 8.2 The control progression for the ILC laws: (a) without saturation and (b) with saturation level $U = 2.7$.

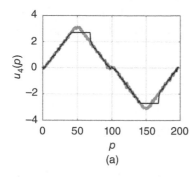 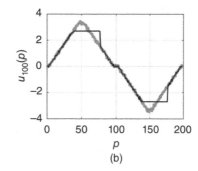

(a)

(b)

Figure 8.3 The control signals for ILC law without saturation (gray line) and with saturation level $U = 2.7$ (black line): (a) for trial $k = 4$ and (b) and $k = 100$.

8.1.2 ILC Design with Full State Feedback

The result of Theorem 8.2 is, in general, sufficient but not necessary. Hence, the interest in developing different forms of sufficient conditions that could limit the design's conservativeness. The analysis in this section develops such a design for the case when all entries in the current trial state vector can be directly measured.

The entries in the vector Lyapunov function used in Section 8.1.1 are quadratic forms. The matrices in these forms are obtained by solving the Lyapunov inequality in block-diagonal form, and this inequality depends on the choice of the weighting matrices. In the design developed in this section, the matrices of the quadratic forms are obtained by solving an algebraic Riccati equation in block-diagonal form. In this case, it is possible to introduce additional parameters that give additional freedom to the design.

Given that all entries in the state vector are available, (8.6) in this case can be taken as follows:

$$\delta u_{k+1}(p) = K_1 \xi_{k+1}(p+1) + K_2 e_k(p+h) \tag{8.26}$$

where $\xi_{k+1}(p+1) = x_{k+1}(p) - x_k(p)$ and the resulting controlled ILC dynamics can be written as follows:

$$\xi_{k+1}(p+1) = (A + BK_1)\xi_{k+1}(p) + BK_2\bar{e}_k(p) + B\varphi_k(p)$$

$$\bar{e}_{k+1}(p) = -CA^{h-1}(A + BK_1)\xi_{k+1}(p) + (I - CA^{h-1}BK_2)\bar{e}_k(p) - CA^{h-1}B\varphi_k(p) \tag{8.27}$$

Also, introduce

$$\zeta_k(p) = \begin{bmatrix} \xi_{k+1}(p) \\ \bar{e}_k(p) \end{bmatrix}, \quad \bar{A} = \begin{bmatrix} A & 0 \\ -CA^h & I \end{bmatrix}$$

$$\bar{B} = \begin{bmatrix} B \\ -CA^{h-1}B \end{bmatrix}$$

and let the matrix $P = \text{diag}[P_1 \ P_2] > 0$ be the solution of the discrete matrix Riccati inequality, where ≤ 0 denotes a symmetric negative semi-definite matrix:

$$\bar{A}^T P \bar{A} - (1 - \sigma)P - \bar{A}^T P \bar{B}(\bar{B}^T P \bar{B} + R)^{-1} \bar{B}^T P \bar{A} + Q \leq 0 \tag{8.28}$$

where $0 < \sigma < 1$ and the weighting matrices are such that $Q > 0$ and $R > 0$. This inequality can be easily reformulated as an LMI for the matrix variable $X = \text{diag}[X_1 \ X_2]$, where $X_1 = P_1^{-1}$ and $X_2 = P_2^{-1}$:

$$\begin{bmatrix} (1-\sigma)X & X\bar{A}^T & X \\ \bar{A}X & X + \bar{B}R^{-1}\bar{B}^T & 0 \\ X & 0 & Q^{-1} \end{bmatrix} \geq 0, \quad X > 0 \tag{8.29}$$

where ≥ 0 denotes a symmetric positive semi-definite matrix. If this LMI is feasible, then $P = X^{-1}$, and the following result is proved as Theorem 3 in [201].

Theorem 8.3 *Suppose that for specified weighting matrices $Q > 0$, $R > 0$, and scalar $0 < \sigma < 1$, there exists a matrix $X = \text{diag}[X_1 \ X_2] > 0$, satisfying (8.29) and*

$$K = -(\bar{B}^T P\bar{B} + R)^{-1}\bar{B}^T P\bar{A}\Theta \tag{8.30}$$

where $P = X^{-1}$ and Θ is a symmetric matrix satisfying the LMI

$$\begin{bmatrix} M - M\Theta - \Theta M - Q & \Theta\sqrt{M} \\ \sqrt{M}\Theta & -I \end{bmatrix} \leq 0 \tag{8.31}$$

where $M = \bar{A}^T P\bar{B}(\bar{B}^T P\bar{B} + R)^{-1}\bar{B}^T P\bar{A}$, and since M is positive semi-definite there exists one square root denoted by $\sqrt{M}$. Suppose also that for given constraints (8.12), the matrices $W = \text{diag}[W_1 \ W_2] > 0$, $T_U = \text{diag}[T_i] > 0$, $i = 1, \ldots, m$, satisfy the LMI

$$\begin{bmatrix} -W & -(KW)^T & (\bar{A}W + \bar{B}KW)^T \\ -KW & -TT_U & TT_U\bar{B}^T \\ \bar{A}W + \bar{B}KW & \bar{B}TT_U & -W \end{bmatrix} < 0 \tag{8.32}$$

Then the trial error sequence formed by $e_k(p)$ converges to the norm bounded function $e_\infty(p)$ as $k \to \infty$ $\forall \ 0 \leq p \leq N - 1$ such that $\|\bar{e}_k(p)\|$ is monotonically decreasing and the conditions of (8.3) hold. Moreover, the control law matrices K_1 and K_2 in (8.26) are given in (8.30), since $K = \begin{bmatrix} K_1 & K_2 \end{bmatrix}$.

The weighting matrices Q and R can, as one possibility, be chosen based on the principles of linear quadratic regulator theory, where the matrix Θ provides an additional design option. For example, it can often turn out that some entries of the control law matrix have large values, which create problems in practical implementation. By choosing the matrix Θ in diagonal form, it is possible to decrease the corresponding diagonal elements provided the inequality (8.31) is valid. By (8.30), this will make it possible to reduce the required values of the entries in the control law matrix. Moreover, the design parameter σ directly affects the relative stability margin. Moreover, Theorem 8.2 can be reformulated in terms of Theorem 8.3 and vice versa.

Remark 8.3 The form of saturation considered in this section belongs to the class of sector nonlinearity models, and controllability of the pair (A, B) plays a critical role in analysis, allowing, e.g., the use of the S-procedure [31] in the proof of Theorem 8.2. Moreover, the solvability of (8.28) guarantees stability of $\bar{A} + \bar{B}K$, which, given the structure of the matrix $\bar{A}$, is possible if and only if the matrix $A + BK_1$ is stable, i.e., controllability of (A, B) is required.

8.1.3 Comparison with an Alternative Design

As discussed earlier in this chapter, other ILC laws in the presence of saturation have been reported. This section gives a comparative study of the performance of the ILC law of Theorem 8.3

and the D-type ILC reported in [238], which also illustrates the application of the new design to multiple-input multiple-output (MIMO) examples. This study begins by designing an ILC law of the form of Theorem 8.3 and then the comparative results are given and discussed.

The example used in this previous work is velocity tracking of the joints of the two-link robot manipulator. The linearized, continuous-time, state-space model of the robot considered is

$$\dot{x}(t) = A_c x(t) + B_c u(t)$$
$$y(t) = C_c x(t) \tag{8.33}$$

where

$$A_c = \begin{bmatrix} 0 & 1 & 0 & 0 \\ 0 & -2 & 0 & -1.5 \\ 0 & 0 & 0 & 1 \\ 0 & -0.75 & 0 & 2 \end{bmatrix}, \quad B_c = \begin{bmatrix} 0 & 0 \\ 2.5 & 0.45 \\ 0 & 0 \\ 0.45 & 3 \end{bmatrix}$$

$$C_c = \begin{bmatrix} 0 & 1 & 0 & 0 \\ 0 & 0 & 0 & 1 \end{bmatrix}$$

To undertake design, the MIMO model (8.33) was discretized using the exact method (the zero-order hold) with the sampling period $T_s = 0.001$ seconds. The resulting discrete-time model and (8.33) have relative degree $h = 1$.

As in [238], it is assumed that the model inputs must be in the range $[-2.1, 2.1]$; therefore, the following saturation levels are used $U = U_1 = U_2 = 2.1$. The matrices of the ILC law (8.4) with the control update (8.26) were calculated by solving the LMI (8.29) for

$$\sigma = 0.9, \quad Q = \text{diag}[Q_1, Q_2], \quad Q_1 = 10^6 \times \text{diag}[2, 6, 2, 6]$$
$$Q_2 = 8.6 \times 10^5 I, \quad R = 5 \times 10^6 I$$

and (8.30) for $\Theta = I$, resulting in

$$K_1 = \begin{bmatrix} 0.0027 & -13.1581 & 0.0004 & -2.1087 \\ 0.0004 & -2.0848 & 0.0032 & -15.6696 \end{bmatrix}$$

$$K_2 = \begin{bmatrix} 4.1771 & 0.6814 \\ 0.6798 & 4.9284 \end{bmatrix} \tag{8.34}$$

For $K = \begin{bmatrix} K_1 & K_2 \end{bmatrix}$ in this case, it can be verified that (8.32) holds.

The simulation studies were undertaken for the same reference velocity signal

$$y_{ref}(p) = \begin{bmatrix} y_{ref1}(p) & y_{ref2}(p) \end{bmatrix}^T$$

as in [238] (and using the notation of this paper), where $y_{ref1}(p) = y_{ref2}(p)$ and their waveform (the same in each channel) is shown in Figure 8.4 (solid line). Normally distributed random variables with standard deviation 10^{-4} were added to all components of the state vector on each trial to evaluate the robustness of the control law to measurement noise.

In Figure 8.5, the black solid line shows $RMS(e_k)$, see (8.25), calculated for all performed trials. The progressions of the input vector signal and the tracking error vector signal for the first 50 trials are shown in Figure 8.6. The results show that the ILC Law needs 22 trials to obtain a very small tracking error close to the variance of the noise signals added to the state vector. Moreover, the control signal did not saturate for any of the trials. On the first trial, $RMS(e_1)$ is ≈ 0.65. Adding noise to the state vector causes little noise in the input signal vector.

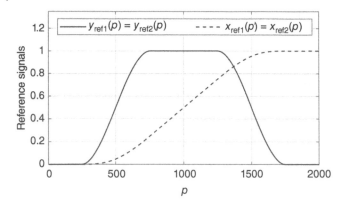

Figure 8.4 The reference signal (for one output).

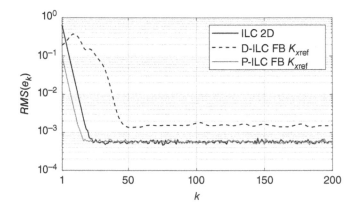

Figure 8.5 The RMS(e_k) tracking error for the considered ILC laws applied to the linearized model of the two-link robot, where ILC 2D denotes the design of this section, D-ILC FB K_{xref} the D-type design of (8.35), and P-ILC FB K_{xref} for the design of (8.36) and (8.37).

The control law of Theorem 8.3 is next compared with the D-type ILC design developed in [238] for the continuous-time model (8.33), which has the structure:

$$u_{k+1}(t) = \text{sat}\left(u_{k+1}^{\text{fb}}(t) + \text{sat}\left(u_{k+1}^{\text{ff}}(t)\right)\right) \tag{8.35}$$

where

$$u_{k+1}^{\text{fb}}(t) = K(x_{\text{ref}}(t) - x_{k+1}(t))$$
$$u_{k+1}^{\text{ff}}(t) = \text{sat}\left(u_k^{\text{ff}}(t)\right) + q\Gamma\dot{e}_k(t)$$

where K is the feedback gain matrix, $x_{\text{ref}}(t)$ is the reference state vector, q is the learning rate, and $\Gamma > 0$ is a gain matrix. The simulation of this control law uses the gain matrices and reference signals as in [238], where the feedback gain matrix is selected to place the controlled system poles at $[-10, -5, -5 \pm 3j]$, and hence,

$$K = \begin{bmatrix} 17.4313 & 4.8541 & -5.9502 & -2.1913 \\ -1.9248 & -1.6376 & 14.0213 & 4.8626 \end{bmatrix}$$

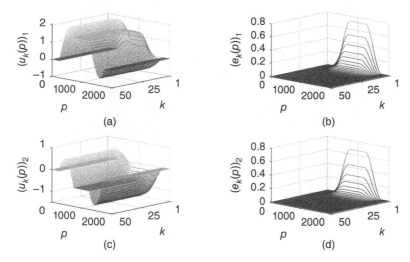

Figure 8.6 Progressions of the input vector signal (a) channel 1 and (c) channel 2 and the tracking error vector signal (b) channel 1 and (d) channel 2 generated by applying the ILC law given by (8.4) and (8.26) to the linearized model of the two-link robot model.

The learning rate and the gain matrix Γ were taken as $q = 0.95$ and $\Gamma = (C_c B_c)^{-1}$. Hence

$$q\Gamma = \begin{bmatrix} 0.3905 & -0.0586 \\ -0.0586 & 0.3255 \end{bmatrix}$$

This alternative design requires a reference signal vector

$$y_{\text{ref}}(p) = \begin{bmatrix} y_{\text{ref}1}(p) & y_{\text{ref}2}(p) \end{bmatrix}^T$$

the same as the previous design and a reference state vector

$$x_{\text{ref}}(p) = \begin{bmatrix} x_{\text{ref}1}(p) & y_{\text{ref}1}(p) & x_{\text{ref}2}(p) & y_{\text{ref}2}(p) \end{bmatrix}^T$$

whose entries are also shown in Figure 8.4. Normally distributed random variables with standard deviation 10^{-4} was added to all elements of the state vector as before. The saturation level for the D-type law (8.35) is as in [238], i.e. $U = 2.1$.

The $RMS(e_k)$ calculated for all trials is shown in Figure 8.5 (black dashed line). Figure 8.7 shows the progressions of the input vector signal and the tracking error vector signal for the first 50 trials. Compared to the new ILC law given by (8.4) and (8.26), the D-type ILC law requires significantly more trials to obtain a small tracking error. Moreover, $RMS(e_k)$ increases for the first few trials, the input signals reach their limit values, and the input signal is noisier due to the derivative action in the D-type ILC law.

A possible advantage of the control law (8.35) is a simpler implementation because there is no state vector from the previous trial $x_k(t)$ in the control update. In the numerical example considered above the $RMS(e_1)$ is small (≈ 0.19) due to the introduction of the feedforward signal $Kx_{\text{ref}}(t)$. It is, however, possible to include this last feature (if required) into the new ILC law as follows.

Routine manipulations give that the ILC law (8.4) with the control update (8.26) is equivalent to

$$u_{k+1}(p) = \text{sat}(K_1 x_{k+1}(p) + v_{k+1}(p)) \tag{8.36}$$

where

$$v_{k+1}(p) = v_k(p) + K_2 e_k(p + h) \tag{8.37}$$

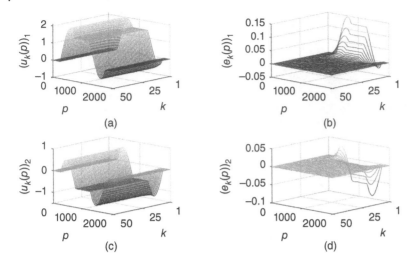

Figure 8.7 Progressions of the input vector signal and the tracking error vector signal generated by applying the D-type ILC law to the model of the linearized two-link robot model.

which also does not require the use of $x_k(p)$. Moreover, the feedback part of (8.36) can be easily extended (if required) to include the feedforward term $Kx_{\text{ref}}(p)$, analogous to the D-type ILC (8.35), i.e.

$$u_{k+1}(p) = \text{sat}(K_1x_{k+1}(p) - K_1x_{\text{ref}}(p) + v_{k+1}(p)) \tag{8.38}$$

The ILC law (8.36) and (8.37) was also applied to the linearized discrete-time two-link robot model (under the same conditions). All entries in the state vector were disturbed by noise signals with the same parameters as above, with gain matrices given by (8.34).

The gray solid line in Figure 8.5 shows the $RMS(e_k)$ plot calculated for all performed trials. Figure 8.9 shows the progressions of the input vector signal and the tracking error vector signal for the first 50 trials. The ILC law (8.37) and (8.38) needs only 17 trials to achieve accurate tracking and the input signals do not saturate.

For this design, the first trial $RMS(e_1) \approx 0.1$, which is less than the corresponding value obtained for the ILC law given by (8.4) and (8.26). This difference is caused mainly by use of the feedforward signal $-K_1x_{\text{ref}}(p)$, note also that at the first trial, the control signals $v_1(p) = 0$ due to the assumed boundary conditions. The performance of the new design exceeds that of the D-type ILC law (8.35). Since storage of $K_1x_k(p)$ is not required, simplified implementation is possible.

It is of interest to compare the ILC law (8.4) with the control update (8.26) and the P-type ILC given by (8.36) with (8.37) and also their behavior under input saturation. Simulations were undertaken for the same model, reference signal, and noise signal parameters as for previous studies and the saturation level used was $U = 1.7$.

In Figure 8.8, the gray dotted line shows the $RMS(e_k)$ plot calculated for the control system with the ILC law given by (8.4) and (8.26), the black solid line for the P-type ILC. Figure 8.9 shows the waveforms of the input vector signal and the tracking error vector signal for the first 50 trials of the control system with the P-type ILC law. These results confirm that both ILC laws are equivalent, and both control systems are stable for the adopted saturation levels. Perfect tracking is not achieved due to the control signal limitation.

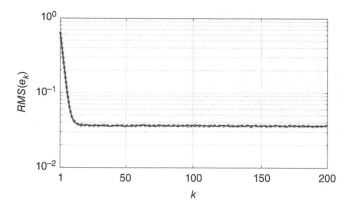

Figure 8.8 The RMS(e_k) tracking errors for ILC law (8.4) with the control update (8.26) (gray dotted line) and the P-type ILC law given by (8.36) with (8.37) (black solid line) applied to the linearized model of the two-link robot with saturation level $U = 1.7$.

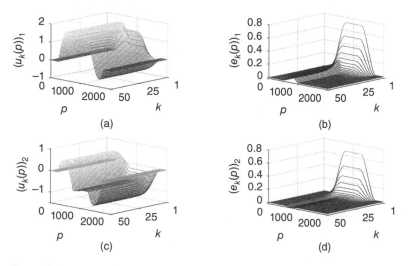

Figure 8.9 Progressions of the input vector signal and the tracking error vector signal after applying the ILC law (8.37) and (8.38) to the linearized two-link robot model.

8.1.4 Experimental Results

Experimental validation of the new design uses a testbed whose mechanical part consists of two permanent magnet synchronous motors (PMSMs) connected by a clutch, see Figure 8.10. The control goal is to track the positions of the motor shafts with an actuator formed by PMSM A, and the second motor (i.e. PMSM B) generates the load torque. Figure 8.11 gives a block diagram of the experimental setup.

The motors are powered by AKD-P006 and AKD-P012 servo amplifiers operating in the current/torque control mode. Moreover, the ILC law is implemented on a PC with Linux RTAI operating system sampled at 0.002 seconds. Communication between the PC and the servo amplifiers is via an EtherCAT bus. The control input to the both servo amplifiers is the reference current, i_A^{ref} and i_B^{ref}, respectively. From the AKD-P006, the following are read: motor shaft angular

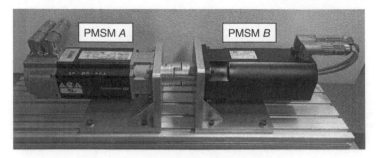

Figure 8.10 Two PMSMs connected by a clutch.

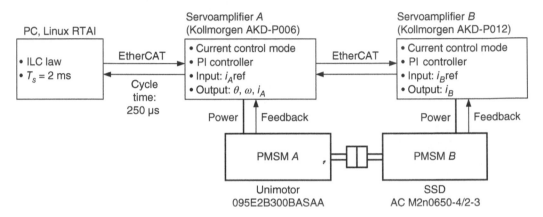

Figure 8.11 Block diagram of the experimental setup.

position θ, motor shaft angular velocity ω, and current i_A of PMSM A and from the AKD-P012 is read the current i_B of PMSM B.

The electrical part of drive A (servoamplifier A + PMSM A) is modeled (using the notation in [201]) by the first-order system:

$$i_A^{\text{ref}}(t) = T_{\text{cA}}\frac{di_A(t)}{dt} + i_A(t) \tag{8.39}$$

where T_{cA} is the time constant of the current control loop of drive A. A continuous-time model of the mechanical part of the experimental setup is

$$T_e(t) = i_A(t)k_{\text{tA}} = J\frac{d^2\theta(t)}{dt^2} + b\frac{d\theta(t)}{dt} + T_l(t) \tag{8.40}$$

where $T_e(t)$ denotes the electromagnetic torque generated by drive A, k_{tA} is the torque constant of drive A, J is the overall mass moment of inertia, b the resulting friction coefficient and $T_l(t)$ denotes the load torque generated by drive B. The numerical values of these parameters are $T_{\text{cA}} = 0.8 \times 10^{-3}$ s, $k_{\text{tA}} = 0.93$ N m/A, $J = 9.3 \times 10^{-4}$ kg m^2, $b = 2.4 \times 10^{-3}$ kg m^2/s.

To apply the new ILC design, the differential equations (8.39) and (8.40) were written in the form of state-space equations, and then both were separately discretized using the exact method (the zero-order hold) with sampling period $T_s = 0.002$ seconds. Finally, discrete-time models were connected in series to obtain

$$x_k(p + 1) = Ax_k(p) + Bu_k(p) + Ed_k(p)$$
$$y_k(p) = Cx_k(p) \tag{8.41}$$

where

$$A = \begin{bmatrix} 1 & 0.0020 & 0.0020 \\ 0 & 0.9949 & 1.9948 \\ 0 & 0 & 0.0821 \end{bmatrix}, \quad B = \begin{bmatrix} 0 \\ 0 \\ 0.9179 \end{bmatrix}$$

$$E = \begin{bmatrix} -0.0021 \\ -2.1450 \\ 0 \end{bmatrix}, \quad C = \begin{bmatrix} 1 & 0 & 0 \end{bmatrix}$$

The state vector $x_k(p)$, the control input $u_k(p)$, and the disturbance input $d_k(p)$ on trial k are

$$x_k(p) = \begin{bmatrix} \theta_k(p) \\ \omega_k(p) \\ i_{Ak}(p) \end{bmatrix}, \quad u_k(p) = i_{Ak}^{\text{ref}}(p), \quad d_k(p) = T_{lk}(p)$$

Bode plots for (8.41) between the control input and each element of the state vector coincide with the corresponding data from the frequency analysis of the experimental setup, confirming the correctness of the model. Moreover, the relative degree of the experimental setup model is $h = 2$.

The ILC law for implementation is

$$u_k(p) = \text{sat}\left(u_{k-1}(p) + K_1(x_k(p) - x_{k-1}(p)) + K_2 e_{k-1}(p+2)\right) \tag{8.42}$$

where K_1 and K_2 are calculated using (8.30) of Theorem 8.3, $P = X^{-1}$ and $X = \text{diag}[X_1 \ X_2]$ is a solution to (8.29) with the following parameters:

$$Q = \text{diag}[0.2 \times 10^6 \ 10^4 \ 10^4 \ 10^{10}], \quad R = 1.5, \quad \sigma = 0.9$$
$$\Theta = \text{diag}[1 \ 1 \ 1 \ 1.2]$$

These parameters are such that the values of the gain are relatively small on the scale of the considered system while maintaining a sufficient margin of stability and satisfaction of constraints (8.31). Also, the constraints (8.32) hold for all values of the saturation levels considered below. As a result, the following solution is obtained

$$K_1 = [-36.7873 \ -0.6336 \ -1.2859], \quad K_2 = 17.3509$$

The reference signal for the experiments is shown in Figure 8.12a. The load torque excited by the drive B is generated starting from trial 102, and its waveform is in Figure 8.12b. Moreover, the load torque was zero for trials 1 to 101. Five experiments with different saturation levels were performed for such settings, i.e. $U = 4.0, 4.1, 4.2, 4.3,$ and 4.4 A. For the first trial in each case, the

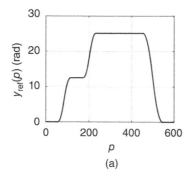

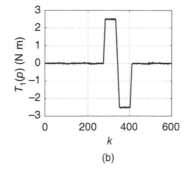

Figure 8.12 The reference signal (a) and the load torque (b).

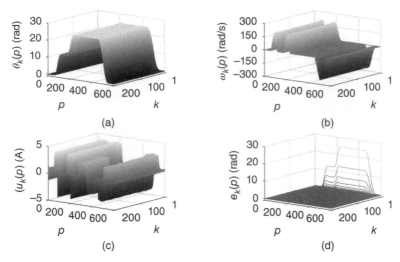

(a)
(b)
(c)
(d)

Figure 8.13 Progressions of the motor shaft position, the motor shaft velocity, the control input, and the tracking error for ILC with the control signal saturation level 4.1 A.

ILC signal $u_1(p)$ was calculated for $x_0(p) = 0$ and $e_0(p) = y_{\mathrm{ref}}(p)$. The total number of ILC trials for each saturation level was 200.

Figure 8.13 shows the progressions of the motor shaft angular position $\theta_k(p)$, the motor shaft angular velocity $\omega_k(p)$, the input $u_k(p)$, and the tracking error $e_k(p)$ for the ILC law with saturation level of 4.1 A. These display good performance. The plots for the other cases are omitted but show similar properties to those in the last figure, and below, the trial-to-trial convergence properties of these experimental results is discussed in terms of $RMS(e_k)$.

In Figure 8.14, the $RMS(e_k)$ values for all experiments are given. These results confirm that none of the saturation levels cause the errors to grow without bound. Moreover, in all cases, the load torque generated by the drive B starting from the 102nd trial is compensated by the ILC law on subsequent trials. The impact of the control signal limitation on the tracking error for the 200th trial is shown in Figure 8.15. The greater the saturation level, the more accurate the tracking achieved.

The results also highlight how the tracking error depends on the saturation level, see the results in the last figure for $U = 4.0$, 4.1, 4.2, and 4.3 A. In the case of $U = 4.4$ A, the control signal is not

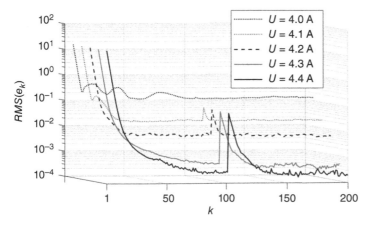

Figure 8.14 $RMS(e_k)$ for different saturation levels.

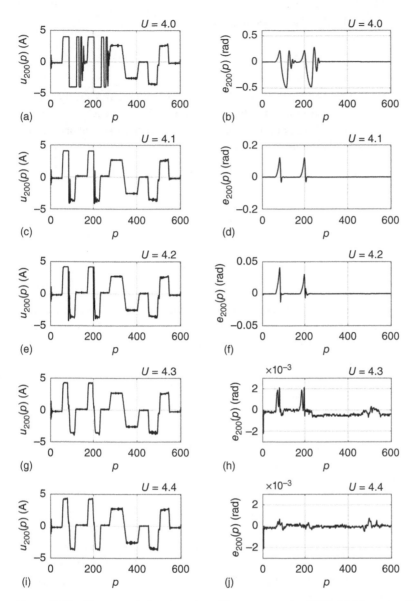

Figure 8.15 The control signals and tracking errors at the 200th trial for saturation levels U = 4.0 ((a) and (b)), 4.1 ((c) and (d)), 4.2 ((e) and (f)), 4.3 ((g) and (h)) and 4.4 A ((i) and (j))

limited, and, as expected, tracking error values close to the resolution of the motor shaft angular position signal are obtained. The position of the motor shaft is read from servo amplifier A with a resolution of $2\pi/2^{16} \approx 10^{-4}$ rad.

8.2 Constrained ILC Design for LTV Systems

8.2.1 Problem Specification

The analysis in this section is based on [54] with preliminary results in [53]. Consider a single-input single-output (SISO), with an immediate extension to the MIMO case, discrete LTV system described in the ILC setting by the state-space model:

$$x_k(p+1) = A_p x_k(p) + B_p u_k(p)$$

$$y_k(p) = C_p x_k(p) \tag{8.43}$$

where $x_k(p) \in \mathbb{R}^n$, $u_k(p) \in \mathbb{R}^l$, and $y_k(p) \in \mathbb{R}^m$ are, respectively, the state, input, and output vectors and no loss of generality arises from assuming zero state initial conditions on each trial. Defining the supervectors for the state, input, and output on each trial as in the time-invariant case (see (1.37)) leads to the lifted representation for the dynamics, i.e. $Y_k = GU_k$, where

$$G = \begin{bmatrix} C_1 B_0 & 0 & \cdots & 0 & 0 \\ C_2 A_1 B_0 & C_2 B_1 & \ddots & 0 & 0 \\ \vdots & & \ddots & \ddots & \vdots & 0 \\ C_N A_{N-1} \cdots A_1 B_0 & \cdots & \cdots & \cdots & C_N B_{N-1} \end{bmatrix} \tag{8.44}$$

and

$$E_k = R - GU_k \tag{8.45}$$

In the case of LTI systems, G is nonsingular when the system relative degree is one, i.e. $CB \neq 0$, and hence, there is a unique solution to the ILC problem. Still, for LTV systems, G can be singular depending on the product $C_{p+1} B_p$. Consequently, there may be no solution to the ILC problem. To ensure that the control design objective is achievable, it is assumed in this section that the reference trajectory r lies in the range of the matrix G, i.e. there exists an input U_∞ such that $R = GU_\infty$ is satisfied. If this condition holds but G is not invertible, then there will be an infinite number of solutions, which is in clear contrast to the LTI case.

Remark 8.4 The assumption that $R \in range(G)$ is more general and less restrictive than assuming that G is invertible as considered in, e.g. [142].

Constraints on, e.g. inputs, input rates, and states and outputs can be divided into two classes, termed, respectively, hard and soft. Examples of the former classes are constraints on magnitude(s) at each point in time, e.g. on outputs or on actuators that must be satisfied. Examples of soft constraints include total energy usage, which should be minimized in many applications.

This section only considers hard input constraints for SISO systems, but the extension to MIMO systems and state and output constraints follow by routine changes in the analysis. In particular, the input is assumed to be constrained to lie a set Ω, which is taken to be a closed convex set in a Hilbert space H. Often, the set Ω has a simple structure, e.g. the following constraints are frequently encountered (over $0 \le p \le N-1$)

(a) input saturation constraint: $\Omega = \{u \in H : |u(p)| \le M(p)\}$
(b) positive input rate constraint: $\Omega = \{u \in H : |u(p) - u(p-1)| \le \lambda(p)\}$
(c) input sign constraint: $\Omega = \{u \in H : 0 \le u(p)\}$
(d) input energy constraint:

$$\Omega = \left\{ u \in H : \sum_{p=0}^{N-1} u^2(p) \le M \right\}$$

where $M(p)$, $\lambda(p)$, and M are positive scalars defining the input constraint bounds. Solving a constrained problem is, in general, more complicated than its unconstrained counterpart.

The constrained ILC design problem is to find an ILC law that enforces trial-to-trial error convergence using a sequence of trial inputs that satisfy the imposed constraint, i.e.

$$U_k \in \Omega, \quad \forall k \geq 0 \tag{8.46}$$

This section uses successive projection to develop a solution to this problem and then describes the experimental validation. Background on successive projection in ILC for time-invariant systems is given in, e.g. [49, 52].

The key idea in successive projection is to consider the design problem as iteratively finding a point in the intersection of two closed convex sets in some Hilbert spaces, which can then be solved using the well-known projection method as represented by the following result.

Theorem 8.4 *[97, 190] Let S_1 and S_2 be two closed convex sets in a Hilbert space X and define the projection operators $P_{S_1}(\cdot)$ and $P_{S_2}(\cdot)$ as*

$$P_{S_1}(x) = \arg\min_{\hat{x} \in S_1} \|\hat{x} - x\|_X^2 \tag{8.47}$$

$$P_{S_2}(x) = \arg\min_{\hat{x} \in S_2} \|\hat{x} - x\|_X^2 \tag{8.48}$$

where $\|\cdot\|$ denotes the induced norm in X. Then, given the initial estimate $x_0 \in X$, the sequences $\{\tilde{x}_k\}_{k\geq0}$ and $\{x_k\}_{k\geq0}$ generated by

$$\tilde{x}_{k+1} = P_{S_1}(x_k), \quad x_{k+1} = P_{S_2}(\tilde{x}_{k+1}), \quad k \geq 0 \tag{8.49}$$

are uniquely defined for each $x_0 \in X$ and satisfy the monotonic convergence condition:

$$\|\tilde{x}_{k+2} - x_{k+1}\|_X^2 \leqslant \|\tilde{x}_{k+1} - x_k\|_X^2 \tag{8.50}$$

If either set is compact or finite dimensional, the sequences $\{\tilde{x}_k\}_{k\geq0}$ and $\{x_k\}_{k\geq0}$ converge to fixed points $\tilde{x}^* \in S_1$ and $x^* \in S_2$, i.e.

$$\lim_{k\to\infty} \tilde{x}_k = \tilde{x}^*, \quad \lim_{k\to\infty} x_k = x^* \tag{8.51}$$

defining the minimum distance between two sets, i.e.

$$\|\tilde{x}^* - x^*\|_X^2 = \min_{\tilde{x} \in S_1, x \in S_2} \|\tilde{x} - x\|_X^2 \tag{8.52}$$

Also if $S_1 \cap S_2 \neq \emptyset$, the following convergence condition is satisfied

$$\|x_{k+1} - x\|_X^2 \leqslant \|x_k - x\|_X^2, \quad \forall x \in S_1 \cap S_2, \quad k \geq 0 \tag{8.53}$$

The ILC design for LTV systems (and also for LTI systems) can be formulated as the iterative determination of a point in the intersection of S_1 and S_2 in the Hilbert space $H = \mathcal{Y} \times \mathcal{U}$, (in this case $\times$ denotes the Cartesian product) where $\mathcal{Y}$ contains the outputs, $\mathcal{U}$ the inputs and

$$S_1 := \{(E, U) \in H : E = R - GU, U \in \Omega\}$$
$$S_2 := \{(E, U) \in H : E = 0\} \tag{8.54}$$

or the following two sets

$$S_1 := \{(E, U) \in H : E = R - GU\}$$
$$S_2 := \{(E, U) \in H : E = 0, U \in \Omega\} \tag{8.55}$$

where the Hilbert spaces $\mathcal{Y} = R^N$ and $\mathcal{U} = R^N$ have the following inner products (and associated induced norms):

$$\langle Y, X \rangle_Q = Y^T Q X, \quad \forall\, Y, X \in \mathcal{Y}$$
$$\langle U, V \rangle_R = U^T R V, \quad \forall\, U, V \in \mathcal{U}$$

where $Q > 0$ and $R > 0$ are compatibly dimensioned matrices to be selected. This setting's use leads to two algorithms that will be detailed below after considering LTV linear systems with no constraints.

In the case of no constraints, consider any initial input U_0 with corresponding error E_0. Then by use of the successive projection setting, the input sequence $U_{k+1}, k = 0,1,2,\dots$, defined by

$$U_{k+1} = \arg\min_U \left\{ \|R - GU\|_Q^2 + \|U - U_k\|_R^2 \right\} \tag{8.56}$$

solves the ILC design problem. This solution is the unconstrained norm optimal iterative learning control (NOILC) design of Chapter 5 for LTV systems. Moreover, it can be implemented as per its LTI counterpart.

Consider any initial input U_0 and corresponding error E_0. Then it is routine to show that the ILC design with cost function (8.56) achieves monotonic trial-to-trial error convergence to zero, i.e.

$$\|E_{k+1}\|_Q \le \|E_k\|_Q, \quad \forall\, k \ge 0 \tag{8.57}$$

and

$$\lim_{k\to\infty} E_k = 0 \tag{8.58}$$

Furthermore, the input converges to the solution of the following optimization problem:

$$\lim_{k\to\infty} U_k = U_\infty = \arg\min_U \{\|U - U_0\|_R, : R = GU\}$$

and hence, the minimum energy solution occurs when $U_0 = 0$.

For LTI systems, there is only one solution to the tracking problem, but, as stated above, for LTV systems, there can be an infinite number of solutions if G is not invertible. In this case, the algorithm finds the best solution, which has particular appeal in physical applications.

Consider the constrained problem formulation (8.54) for LTV systems. Then, given any initial input U_0 satisfying the constraints with corresponding tracking error E_0, Theorem 8.4 gives that the input sequence $U_{k+1}, k = 0,1,2,\dots$, defined by

$$U_{k+1} = \arg\min_{U\in\Omega} \left\{ \|R - GU\|_Q^2 + \|U - U_k\|_R^2 \right\} \tag{8.59}$$

satisfies the constraints and iteratively solves the constrained ILC problem for LTV systems. For comparative studies, this solution is referred to as Algorithm 1.

Suppose that trial-to-trial error convergence to zero is possible, i.e. $S_1 \cap S_2 \ne \emptyset$. Then the following result shows that Algorithm 1 has the desirable property that the trial-to-trial error norm will decrease monotonically to zero and is termed perfect tracking.

Theorem 8.5 *When perfect tracking is possible, Algorithm 1 achieves monotonic trial-to-trial error convergence to zero, i.e.*

$$\|E_{k+1}\|_Q \le \|E_k\|_Q, \quad k = 0,1,\dots \tag{8.60}$$

and

$$\lim_{k\to\infty} E_k = 0, \quad \lim_{k\to\infty} U_k = U^* \tag{8.61}$$

Moreover,

$$\|U_{k+1} - U^*\|_R \le \|U_k - U^*\|_R, \quad \forall\, k \ge 0 \tag{8.62}$$

i.e. the trial inputs approach the converged solution U^* monotonically in norm.

In this result, (8.60) and (8.62) follow from Theorem 8.4 and also Theorem 1 in [190], which links this algorithm to repetitive process stability theory. Trial-to-trial error convergence to zero may not be possible due to the input constraints Ω, i.e. $S_1 \cap S_2 = \emptyset$. The following result establishes that in such cases, the algorithm computes a "best" approximation.

Theorem 8.6 *In the case when perfect tracking is not possible, Algorithm 1 converges to a solution U_s^* of the following optimization problem:*

$$U_s^* = \arg\min_{U \in \Omega} \|R - GU\|_Q \tag{8.63}$$

i.e. convergence to the minimum error norm can be achieved. Moreover, this convergence is monotonic in the sense that

$$\|E_{k+1}\|_Q \le \|E_k\|_Q, \quad k = 0,1,\dots \tag{8.64}$$

To explain this last result, Theorem 8.4 gives that the algorithm converges strongly to a fixed point of $P_{S_2} P_{S_1}$, i.e. a point $(0, U_s^*) \in S_2$ satisfying

$$(0, U_s^*) = P_{S_2} P_{S_1} (0, U_s^*) = P_{S_2} \{(E, U) = \arg\min_{U \in \Omega} \|R - GU\|_Q^2 + \|U - U_s^*\|_R^2\} = (0, U) \tag{8.65}$$

and hence,

$$U_s^* = \arg\min_{U \in \Omega} \|R - GU\|_Q \tag{8.66}$$

Moreover, monotonic trial-to-trial error convergence is established similarly to the proof of Theorem 8.5. Furthermore, this algorithm has the practically relevant property of trial-to-trial error convergence and the "best" possible solution. This property is of practical relevance since improved tracking performance can be observed from one trial to the next but at the cost of solving a constrained optimization problem.

Remark 8.5 The choice of Q and R (in the scalar case for simplicity) affects the algorithm's convergence properties. A smaller weighting R will allow more aggressive input changes and generally result in a faster trial-to-trial error convergence rate. Conversely, a smaller weighting Q on the error norm will emphasize the input change during the optimization. This choice, in turn, will result in more "cautious" updating of the input signal and, hence, slower trial-to-trial error convergence.

8.2.2 Implementation of Constrained Algorithm 1 – a Receding Horizon Approach

To apply Algorithm 1 requires the solution of the constrained quadratic programming (QP) problem (8.56) after each trial is complete. This task is an off-line computation between two consecutive trials, for which many efficient solvers are available, see, e.g. [86, 171]. Computational load difficulties could, however, still arise, e.g. the dimension of U_k is very high due to a high sampling rate and or a long trial length.

In such cases solving the QP problem will be very computationally demanding or even unmanageable. Next, an alternative, termed Algorithm 2, which uses a receding horizon method is given that draws on previous research for LTI systems [52]. The result is reduced computational complexity and increased robustness against model uncertainty.

Consider any initial input U_0 satisfying the constraints with associated tracking error E_0. The input sequence $U_{k+1}(p), k = 0, 1, 2, \ldots$, defined by the following procedure also satisfies the constraints and iteratively solves the constrained ILC design problem.

- At time instant p and for the current system state $x_{k+1}(p)$, use the LTV system model to predict the system output $y_{k+1}(p)$ over a fixed future interval $[p + 1, p + N_y]$
- Solve the following optimal control problem over a fixed future interval, say $[p, p + N_u - 1]$, subject to the constraints:

$$u_{k+1,p}^{\text{opt}} = \arg \min_{u_{k+1,p} \in \Omega} \left\{ \sum_{i=p+1}^{p+N_y} \left\| e_{k+1}(i) \right\|_Q^2 + \sum_{i=p}^{p+N_u-1} \left\| u_{k+1,p}(i) - u_k(i) \right\|_R^2 \right\} \qquad (8.67)$$

where

$$u_{k+1,p} = \begin{bmatrix} u_{k+1}(p) & \cdots & u_{k+1}(p + N_u - 1) \end{bmatrix}^T$$

- Apply only the first step input in the resulting optimal control sequence and measure or estimate the system state reached at time instant $p + 1$.
- Repeat the above steps until $p = N - 1$, starting from the (new) current state.

The selection of the control and prediction horizons N_u and N_y, respectively, can significantly affect this last algorithm's convergence properties. Selecting larger values of N_u and N_y will give more accurate solutions, but the computational effort could be a factor in some cases. In the extreme case if $N_u = N_y = N$, this algorithm is the same as Algorithm 1. Conversely, selecting values for these parameters that are too small could lead to relatively poorer error convergence. The balance between the computational power available and the obtained performance has to be based on knowledge of the particular example under consideration.

8.2.3 Constrained ILC Algorithm 3

Another constrained algorithm for ILC design can be obtained starting from (8.55), i.e.

- $S_1 = \{(E, U \in H : E = R - GU\}$
- $S_2 = \{(E, U) \in H : E = 0, U \in \Omega\}$

The resulting algorithm takes the form: Given any initial input U_0 satisfying the constraint with associated error E_0, the input sequence $U_{k+1}, k = 0, 1, 2, \ldots$, defined by the solution of the input unconstrained NOILC optimization problem:

$$\tilde{U}_k = \arg \min_U \left\{ \| R - GU \|_Q^2 + \| U - U_k \|_R^2 \right\} \qquad (8.68)$$

followed by the simple input projection:

$$U_{k+1} = \arg \min_{U \in \Omega} \| U - \tilde{U}_k \|_R \in \Omega \qquad (8.69)$$

also satisfies the constraint and iteratively solves the constrained ILC problem. For comparative studies, this solution is referred to as Algorithm 3.

Convergence properties of Algorithm 3 are given by the following two theorems for the cases when, respectively, $S_1 \cap S_2 \neq \emptyset$ and $S_1 \cap S_2 = \emptyset$. These results are established in similar manner to their counterparts for Algorithm 1.

Theorem 8.7 *When perfect tracking is possible, i.e. $S_1 \cap S_2 \neq \emptyset$, Algorithm 3 solves the constrained ILC problem in the sense that*

$$\lim_{k \to \infty} E_k = 0, \quad \lim_{k \to \infty} U_k = U_\infty \tag{8.70}$$

Moreover, convergence is monotonic with respect to the following performance index:

$$J_k = \|EE_k\|_Q^2 + \|FE_k\|_R^2 \tag{8.71}$$

where

$$
\begin{aligned}
E &= I - G\left(G^T Q G + R\right)^{-1} G^T Q \\
F &= \left(G^T Q G + R\right)^{-1} G^T Q
\end{aligned}
\tag{8.72}
$$

Additionally,

$$\|U_{k+1} - U_\infty\|_R \leq \|U_k - U_\infty\|_R, \quad \forall k \geq 0 \tag{8.73}$$

i.e. the input iterates approach the converged solution u_∞ monotonically in the norm.

This algorithm first computes the NOILC solution and then projects it onto the constraint set, which is computationally much simpler than Algorithm 1. However, this does not come for free since by Theorem 8.5, Algorithm 1 achieves monotonic trial-to-trial error convergence in the tracking error norm, but this latest algorithm can only achieve this convergence property in the "weighted" tracking error norm as shown in previous theorem.

When perfect tracking is not possible, i.e. $S_1 \cap S_2 = \emptyset$, an approximation of the "best" solution (8.63) can be achieved, as established by the next result.

Theorem 8.8 *When perfect tracking is not possible, i.e. $S_1 \cap S_2 = \emptyset$, Algorithm 3 converges to a solution U_s^* of the following optimization problem:*

$$U_s^* = \arg \min_{U \in \Omega} \left\{ \|EE_k\|_Q^2 + \|FE_k\|_R^2 \right\} \tag{8.74}$$

Moreover, this convergence is monotonic with respect to the following performance index:

$$J_k = \|EE_k\|_Q^2 + \|FE_k\|_R^2 \tag{8.75}$$

where again

$$
\begin{aligned}
E &= I - G\left(G^T Q G + R\right)^{-1} G^T Q \\
F &= \left(G^T Q G + R\right)^{-1} G^T Q
\end{aligned}
\tag{8.76}
$$

In common with Algorithm 1, the choice of weighting parameters in this last algorithm affects its convergence properties. In particular, the choices made will affect the converged solution when perfect tracking is not possible. To illustrate this, consider the simplest case of scalar weights, i.e. $Q = qI$ and $R = rI$, where $q, r > 0$. A small weighting, i.e. $r \to 0$ in (8.74) (in Algorithm 3), results in convergence to

$$U_s^* = \arg \min_{u \in \Omega} \|H_d V (U - U_{uc}^*)\|_R^2 \tag{8.77}$$

where H_d is a diagonal matrix with the first q diagonal elements all equal to one and the rest are all zero, where q denotes the rank of G, V is a unitary matrix in the singular values decomposition of G, i.e. $G = U \Sigma V$ and $U_{uc}^* = G^\dagger R$ is the unconstrained ILC minimum energy solution, where $G^\dagger$ denotes the pseudoinverse of G.

If $q = N$, i.e. when G is full rank, the above is just simple a "clipping" of the unconstrained solution U^*, with corresponding (possibly) large error. Conversely, if R is large, the convergence may be slow for both algorithms, but Algorithm 3 for $r \to \infty$ will converge to

$$U_s^* = \arg \min_{U \in \Omega} \|EE_k\|_Q \tag{8.78}$$

which is the "best" solution that can be obtained. Thus, in Algorithm 3, there is a compromise between the convergence rate and the final converged tracking error norm. This compromise, however, does not exist for Algorithm 1, which will always converge to the best solution (8.78).

To implement this algorithm, the first step requires the solution of the unconstrained NOILC problem discussed previously in this section. The second step requires the solution of (8.69). In practice, the input constraint Ω is often pointwise and the solution of (8.69) is easily computed. For example, when $\Omega = \{U \in \mathbb{R}^N : |u(t)| \le M(t)\}$, the solution is

$$u_{k+1}(p) = \begin{cases} M(p) & : \ \tilde{u}_k(p) > M(p) \\ \tilde{u}_k(p) & : \ |\tilde{u}_k(p)| \le M(p) \\ -M(p) & : \ \tilde{u}_k(t) < -M(p) \end{cases} \tag{8.79}$$

for $p = 0, \ldots, N - 1$. In the case when $\Omega = \{U \in \mathbb{R}^N : \sum_{p=0}^{N-1} u^2(p) \le M$, the solution is

$$u_{k+1}(p) = M \frac{\tilde{U}_k(p)}{\sum_{i=0}^{N-1} \tilde{U}_k^2(p)}, \quad p = 0, \ldots, N - 1$$

This algorithm only requires two simple steps, which are easy to complete even for large-scale problems.

8.3 Experimental Validation on a High-Speed Rack Feeder System

8.3.1 Simulation Case Studies

In Section 8.2, Algorithms 1–3 from the previous section will be experimentally verified on the high-speed rack feeder system of Section 2.2.3. Before this step, a simulation-based study is given for inactive and active constraints, respectively. To highlight the performance differences, a slightly modified reference signal to that for the experiments is used.

The simulation results are for a trial length of 12 *seconds* with a sampling time of 0.05 *seconds*, i.e. $N = 240$. Also, the weighting matrices are chosen as $Q = I$ and $R = 0.1I$, respectively. Zero initial input, i.e. $U_0 = 0$, is used for all the algorithms. For Algorithm 2, the control and prediction horizons are chosen as $N_y = N_u = 40$.

First, the case where the constraints on the inputs (i.e. input amplitude constraint of 7.5 m/s and input change rate constraint of 30 m/s^2) are inactive for the desired reference signal is considered, i.e. perfect tracking of the reference is possible under constraints. The tracking error norm convergence over 30 trials and outputs on the last trial (together with the reference signal) for all three algorithms are given in Figures 8.16 and 8.17, respectively. Note that the results in the figures are on top of each other and almost indistinguishable.

The results in these figures confirm that all three algorithms achieve perfect tracking of the reference signal. Furthermore, in this particular set of simulations, their convergence performance is almost identical and not distinguishable in these figures. Finally, an inspection of the obtained inputs (plots not shown) shows that they all satisfy the constraint requirements.

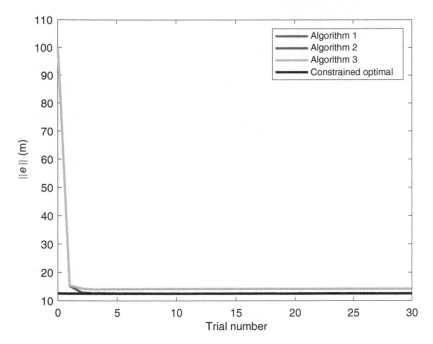

Figure 8.16 Simulation results: comparison of the error norm convergence over 30 trials for inactive constraints.

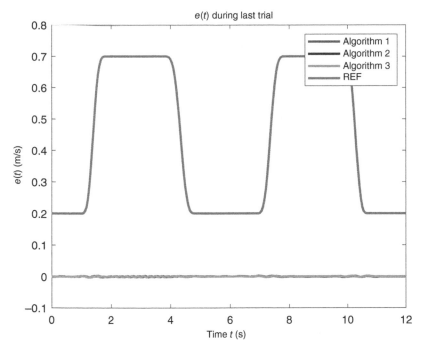

Figure 8.17 Simulation results: comparison of the output signals during the 30th trial for inactive constraints.

Consider the case where the constraints on the inputs (i.e. input amplitude constraint of 1.5 m/s and input change rate constraint of 2 m/s^2) are active for the desired reference signal, i.e. perfect tracking of the reference is not possible. The tracking error norm convergence over 30 trials, outputs on the last trial (together with the reference), and inputs on the last trial for all three algorithms are given in Figures 8.18–8.20, respectively. (Note that the results for Algorithms 1 and 2 are almost indistinguishable.)

These last three figures confirm that, as expected, Algorithm 1 (dark gray line) converges monotonically in the tracking error norm to the minimum possible value (black line in Figure 8.18, determined by Theorem 3). Algorithm 2, although with significantly reduced computational load, achieves almost identical convergence performance as Algorithm 1, which is particularly appealing for applications with moderate computational resources. Algorithm 3 has the simplest computational structure. It converges to a slightly larger final tracking error norm (compared to the previous two algorithms). This difference is also visible from the output and input signals on the last trial.

These last observations offer insight into how to choose a particular algorithm in practice. Depending on the computational resource available and the requirement of a specific practical application, the following are guidelines for selecting the algorithms: (i) Algorithm 1 should always be the first choice due to its appealing convergence properties, if the trial length is sufficiently short, or the resetting time between two consecutive trials is sufficiently long, or if the available computation hardware is powerful enough; (ii) Algorithm 2, with sufficiently large horizons, should be considered as the second preferable choice if the conditions above are not met, and finally (iii) Algorithm 3 as the last choice for scenarios where the computation power is limited (and/or a guaranteed convergence is needed).

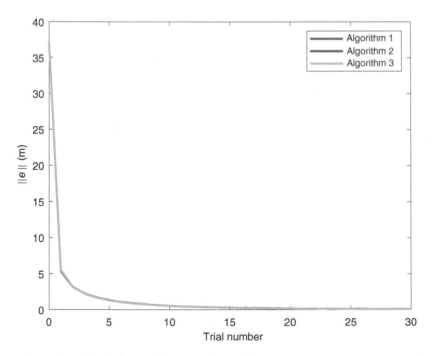

Figure 8.18 Simulation results: comparison of the error norm convergence over 30 trials for active constraints.

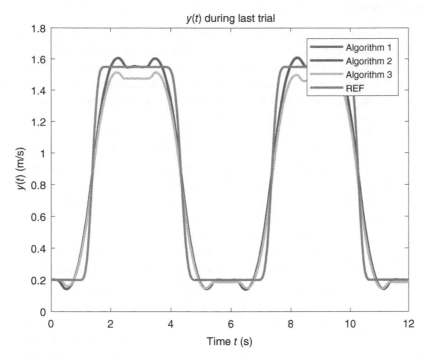

Figure 8.19 Simulation results: comparison of the output signals during the 30th trial for active constraints.

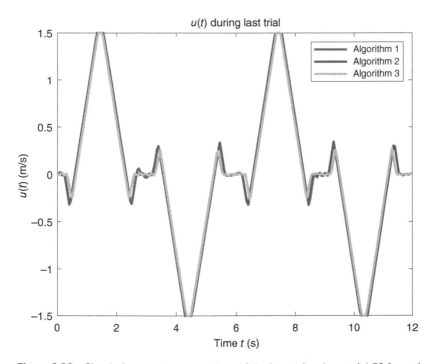

Figure 8.20 Simulation results: comparison of the input signals on trial 30 for active constraints.

8.3.2   Other Performance Issues

In an ILC application, two classes of disturbances arise, those which are repeatable and those that are not. The designs in this section can completely eliminate the effects of repeatable disturbances. However, in common with all other ILC designs, complete attenuation of nonrepeatable disturbances is not possible. The following simulation results for the rack feeder demonstrate this point. For simplicity, the case where the constraints are not active is considered (active constraints can be considered in the same way).

As a first case, consider the disturbances that are Gaussian with zero mean and standard deviation of 0.01. For each of the three algorithms, the resulting dynamics were simulated over 30 trials with the same reference as used in the experiments described below. The error norm convergence and tracking error on the final trial for the three algorithms are given, respectively, in Figures 8.21 and 8.22. These results confirm that the effects of the repeatable disturbances are completely removed and zero tracking error is achieved. (Note that the performance of the three algorithms is almost identical and indistinguishable in the figures.)

If the disturbances are nonrepeatable, the tracking error norm convergence and the error on the last trial (trial 30) are shown in, respectively, Figures 8.23 and 8.24. It can be seen from these figures that the new algorithms can still learn from past trials to reduce the tracking error norm. However, the nonrepeatable disturbances cannot be fully attenuated, and the tracking error norm cannot be reduced to zero, in common with all other ILC laws.

The model of the rack feeder used was developed from first principles, but, as in many applications, there will be inevitable model uncertainties/mismatches due to the unmodeled dynamics. If the model accuracy is good, the new algorithms work well in the presence of uncertainty. Conversely, if the model accuracy is low, there is no guarantee that the new algorithms

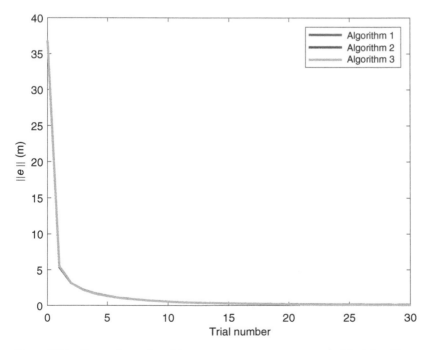

Figure 8.21 Comparison of tracking error norm convergence over 30 trials with repeatable disturbances.

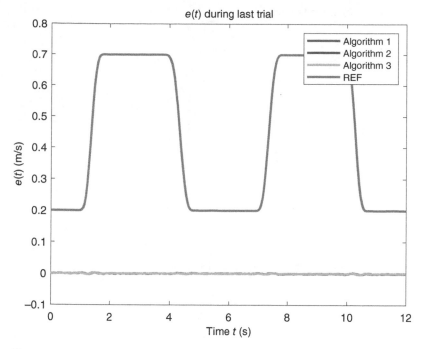

Figure 8.22 Convergence of the tracking error on trial 30 with repeatable disturbances.

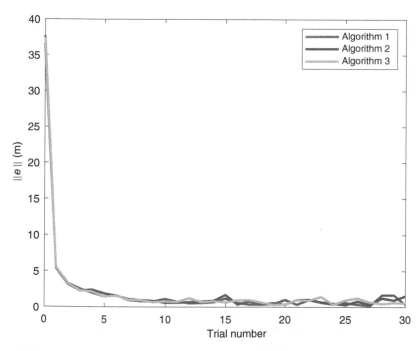

Figure 8.23 Comparison of tracking error norm convergence over 30 trials with nonrepeatable disturbances.

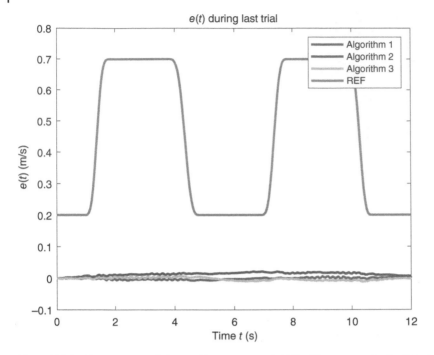

Figure 8.24 Convergence of the tracking error on trial 30 with nonrepeatable disturbances.

will work as expected. In the worst scenario, a low-quality model could lead to divergence, as the simulation results given next demonstrate.

To examine the effects of small model uncertainty, suppose that the second moment of area, I_{zB}, is estimated with a certain level of uncertainty. As above, the case considered is when perfect tracking of the reference is possible. Two scenarios are considered. The first is a 20% estimation error in the value of this parameter, representing a relatively small level of model uncertainty. The simulation results are given in Figures 8.25 and 8.26. As can be seen, for this relatively small model uncertainty, the three algorithms can still achieve perfect tracking.

Suppose that the value of I_{zB} is underestimated by 80%. The results in this case are given in Figures 8.27 and 8.28. As expected, the large uncertainty level induces divergence despite the error norm reduction on the early trials.

The frequency range of the reference signal is shown in Figure 8.29 and the new algorithms can deliver high-quality performance in response to high-frequency reference signals. For example, consider a new sinusoidal reference signal with a higher frequency (2 Hz), whose frequency spectrum is also shown in Figure 8.29 (dark gray line). The convergence performance of three new algorithms for both references is given in Figure 8.30, where the tracking error norms have been normalized (by the error norms on the first trial), to enable the convergence speeds to be directly compared.

These results confirm that the new algorithms can track higher-frequency references. The convergence speed for the higher-frequency reference is, however, slower. This conclusion is not surprising as a typical physical system has a low-pass nature (i.e. the system gain at high frequency is low). Thus, fully tracking a higher-frequency signal requires control effort, which generally takes longer (i.e. more trials) for ILC algorithms to learn.

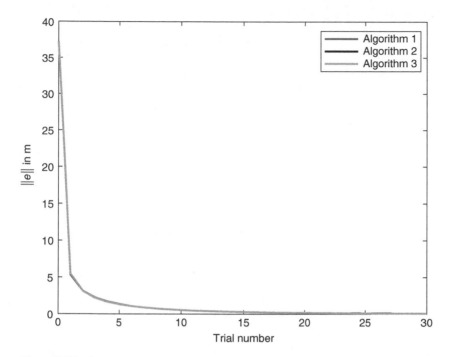

Figure 8.25 A comparison of tracking error norm convergence over 30 trials with a small level of model uncertainty.

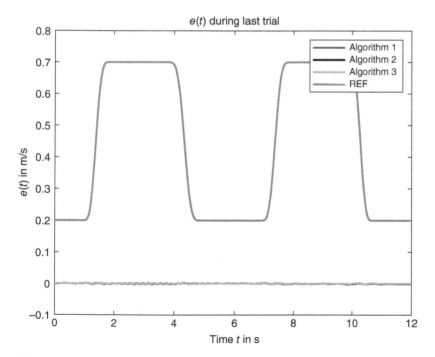

Figure 8.26 Convergence of the tracking error on trial 30 with a small level of model uncertainty.

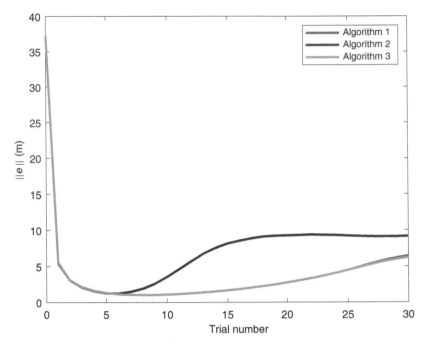

Figure 8.27 A comparison of tracking error norm convergence over 30 trials with a large level of model uncertainty.

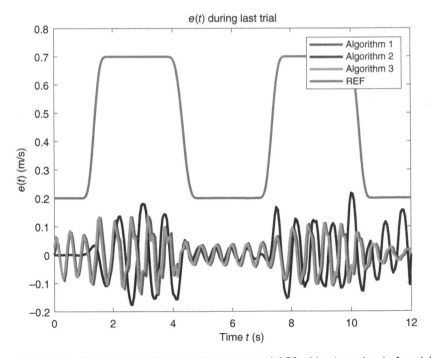

Figure 8.28 Convergence of the tracking error on trial 30 with a large level of model uncertainty.

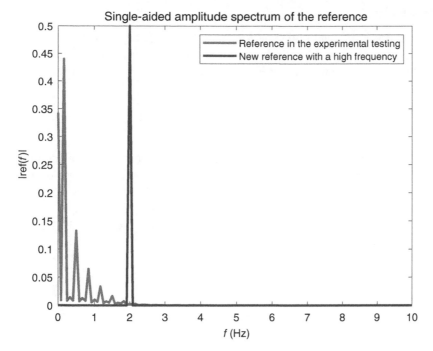

Figure 8.29 Frequency spectrums of the reference signals.

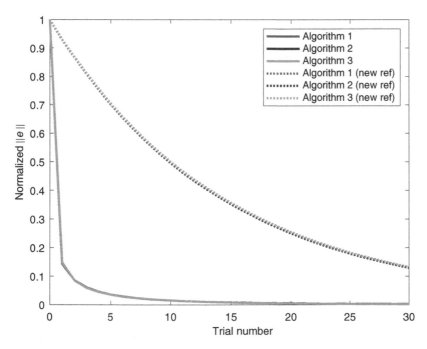

Figure 8.30 A comparison of the (normalized) tracking error norm convergence over 30 trials for the reference signal used in the experiments (solid lines) and those of a higher-frequency signal (dotted lines).

8.3.3 Experimental Results

In this section, the constrained ILC algorithms' performance is verified experimentally on the high-speed rack feeder test rig of Section 2.2.3. The computations for the rack feeder are not intensive, and hence, the focus of the experimental tests is on the convergence properties of these algorithms.

8.3.4 Algorithm 1: QP-Based Constrained ILC

This algorithm has been applied to two different scenarios. First, to the case where predefined constraints of the maximum carriage velocity and its maximum acceleration are activated during tracking the desired reference signal. The experimental results obtained are in Figure 8.31, where the control signal has been initialized with the value zero at the beginning of the experiments. Hence, the tracking error signal coincides with the desired trajectory for $k = 0$, and this ILC law rapidly decreases the tracking error after only a few trials.

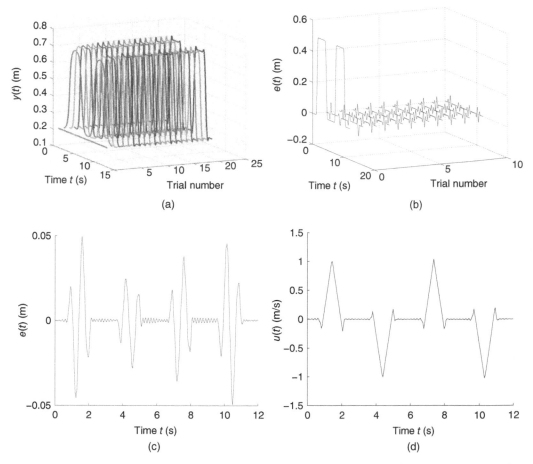

Figure 8.31 Experimental results for Algorithm 1 with active constraints (max. carriage velocity 1.5 m/s, max. acceleration 2.0 m/s^2). (a) Propagation of the system output over 20 trials. (b) Propagation of the tracking error over 20 trials. (c) Tracking error (20th trial). (d) Control signal (20th trial).

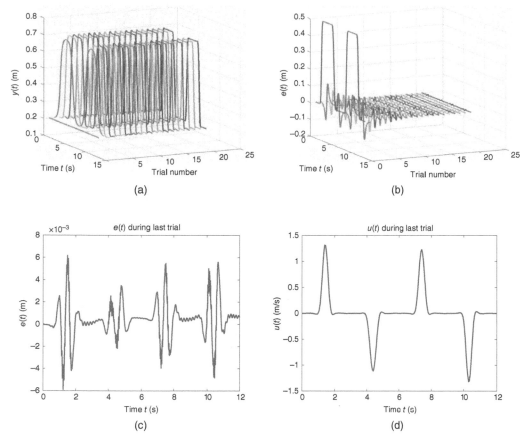

Figure 8.32 Experimental results for Algorithm 1 with inactive constraints. (a) Propagation of the system output over 20 trials. (b) Propagation of the tracking error over 20 trials. (c) Tracking error (20th trial). (d) Control signal (20th trial).

To prevent an increase in the error as the trial number increased, a feature which can often be detected in experimentally implemented ILC laws, the error signal is filtered by applying a zero-phase filter before the computation of the next trial input. In this application, a fourth-order Butterworth filter with a cutoff frequency of 5 Hz is used and no additional real-time constraints are imposed.

In the second scenario, the corresponding limits are set to values not reached during the complete sequence of experiments. As shown in Figure 8.32, the controlled error is reduced significantly from trial-to-trial in cases when none of the input and input rate constraints are active. As in the first scenario, the error is again zero-phase filtered after each trial, leading to smooth control action with excellent damping of remaining oscillations at standstill positions of the rack feeder.

8.3.5 Algorithm 2: Receding Horizon Approach-Based Constrained ILC

Figures 8.33 and 8.34 give the results of repeating the previously described experiments for Algorithm 2. Its computational requirement is less demanding ($N_u = N_y = 23$) – it only requires the solution of a QP problem with a decision variable in $\mathbb{R}^{23}$ rather than $\mathbb{R}^{240}$ as for Algorithm 1.

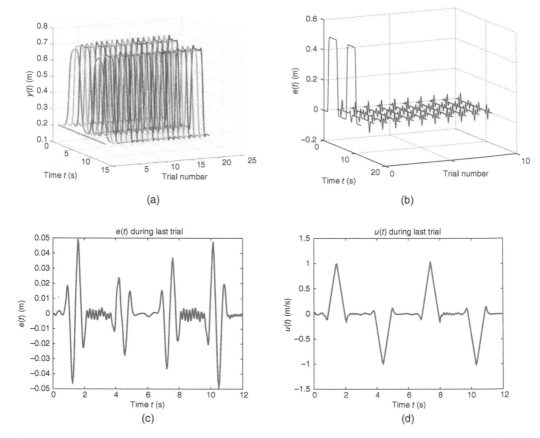

Figure 8.33 Experimental results for Algorithm 2 with active constraints (max. carriage velocity 1.5 m/s, max. acceleration 2.0 m/s^2). (a) Propagation of the system output over 20 trials. (b) Propagation of the tracking error over 20 trials. (c) Tracking error during the 20th trial. (d) Control signal during the 20th trial.

As can been seen in Figure 8.33 when the input and input rate constraints are active, both ILC laws show very similar behavior in the experimental validation. Additionally, the choice of the parameters N_u and N_y again leads to suitable damping of the elastic vibrations at a standstill. Repeating the experiment using the second algorithm with inactive constraints results in similar performance as those observed for the first algorithm (as shown in Figure 8.34).

To compare the performance of the two designs, the error norm convergence of the four experiments and the converged inputs on the last trial are given in Figures 8.35 and 8.36, respectively, showing that (i) the tracking error norms converge monotonically for Algorithm 1 (as predicted by the theory), (ii) Algorithm 2 produces similar performance as Algorithm 1 (results in the figures for both designs are in very close agreement and indistinguishable), and (iii) all kinematic constraints are satisfied and best exploited.

8.4 Concluding Remarks

This chapter has considered the constrained ILC design for both LTI and LTV systems. Overall, constrained ILC is relatively less developed than other areas but is of obvious applications interest. This topic is considered again in Chapter 11, where designs for nonlinear dynamics using the Newton method are developed.

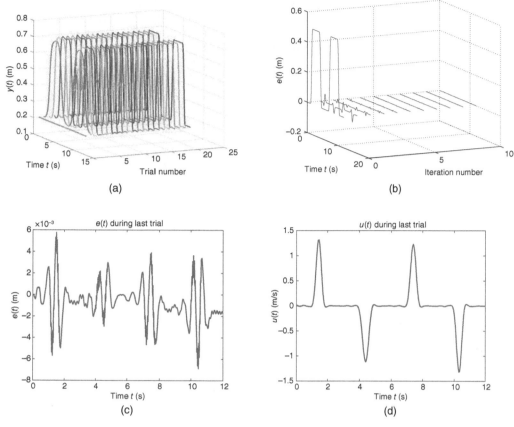

(a)

(b)

(c)

(d)

Figure 8.34 Experimental results for Algorithm 2 with inactive constraints. (a) Propagation of the system output over 20 trials. (b) Propagation of the tracking error over 20 trials. (c) Tracking error during the 20th trial. (d) Control signal during the 20th trial.

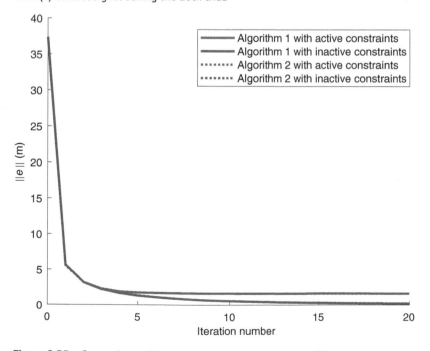

Figure 8.35 Comparison of the error norm convergence over 20 trials.

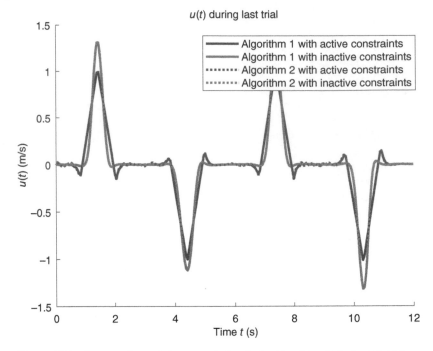

Figure 8.36 A comparison of the control signals for Algorithm 2 during trial 10.

9

ILC for Distributed Parameter Systems

Research has been testified on iterative learning control (ILC) for distributed parameter systems, see, e.g. the relevant cited references in [284] using operator theory methods. An alternative, more application-focused approach is to attempt to use finite-dimensional controllers. One way is to seek to tune fixed-structure controllers using measured data. An alternative is to approximate the dynamics by a finite-dimensional model as a basis for design.

This chapter first considers the former approach, starting with the tuning of phase-lead ILC type laws for a wind turbine application, with supporting simulation results from application to a computational fluid dynamics (CFD) representation of the dynamics. Then the design based on constructing a finite-dimensional model of the dynamics is considered, first for the wind turbine and then a heating process with supporting experimental results for the latter.

These two designs highlight that constructing the approximate finite-dimensional representation of distributed parameter dynamics cannot be based on one universal method. Finally, a design for controlling repetitive transverse loads on an elastic material is given, which highlights links with finite element modeling and sequential experimental design.

9.1 Gust Load Management for Wind Turbines

Wind turbine blades are subjected to fluctuating aerodynamic forces formed of both deterministic and stochastic elements. In the latter case, the cause is from the wind's variable nature, which varies in both frequency and magnitude, producing a variation in the aerodynamic load that passes through the turbine system (although there are other contributions to these loads). Deterministic components include the effects of the atmospheric boundary layer, stator–rotor interaction and yaw misalignment.

These forms of disturbance create loads that require management, mainly since the blades are relatively flexible in construction, and the changing loads on the blades change the blade position relative to the oncoming wind. This change, in turn, alters the aerodynamic loading and, hence, the complexity of the underpinning aeroelastic system.

Wind turbine load control involves modifying the lift on the blades. There are four main ways of achieving this: varying the blade incidence angle (variable pitch and/or predetermined blade twist), flow velocity (variable rotor speed), blade size (variable blade length), or modifying the blade section aerodynamics (active flow control) [124]. In this chapter, it is the last of these, in particular, damping fluctuations in the lift using circulation control is considered, using a model of a blade equipped with a smart rotor.

Iterative Learning Control Algorithms and Experimental Benchmarking, First Edition.
Eric Rogers, Bing Chu, Christopher Freeman and Paul Lewin.
© 2023 John Wiley & Sons Ltd. Published 2023 by John Wiley & Sons Ltd.

Smart rotors are currently an active topic of research as an advanced control concept for large-scale turbines, see [22] for a recent review, and the special issue [264] on smart rotors, to operate in conjunction with collective and individual pitch control. The aim is to improve aerodynamic performance and load control by integrating smart devices into rotor blades, mainly not only to reduce fatigue loads but also to attenuate ultimate/extreme loads and regulate power. Aerodynamic devices, or in general control terms actuators, such as trailing edge flaps, e.g. [8], microtabs, e.g. [123], and active vortex generators can be embedded into the blade structure and actively and independently controlled to meet set objectives. This form of control offers potential benefits in turbine efficiency and reliability, thereby increasing blade life and ultimately reducing the cost of energy [265].

Smart rotors can be manipulated across the operational envelope to increase energy capture. At low-wind speeds, these devices can be activated to improve the aerodynamic performance of the blade. At high-wind speeds, loads can be managed more effectively to maintain performance. A reduction in loads would also increase component lifespan and reduce maintenance. Also, a decrease in maintenance costs is significant as the offshore industry expands. Remote environments and difficult access mean a dramatic increase in the cost of maintenance operations. This requirement and increased rotor blade diameters have driven the need for significant wind turbine control advancement.

Wind turbines operate in the atmospheric boundary layer. In this layer, there is a wind shear with a nonuniform mean velocity profile and a regular variation in wind speed past the blade throughout a cycle, even in reasonably steady, nongusting conditions, see Figure 9.1. Consequently, the flow past the blade will contain an oscillatory component. This component will be more pronounced toward the blade's tip, where the speed differential will have its largest value.

Next, a CFD representation of the underlying dynamics is developed and used to evaluate the performance resulting from applying a phase-lead ILC law. These results are followed by a model-based design ILC design based on constructing a finite-dimensional, state-space model that approximates the dynamics. As a representative model-based design, norm optimal iterative learning control (NOILC) is applied. Successful design and implementation of such control laws would free up resources in an area where control is necessary but supporting technology.

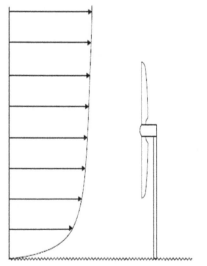

Figure 9.1 Schematic of a wind turbine in the atmospheric boundary layer (left-hand side of the figure).

Other research on ILC for wind turbines, concentrating on its use as an analysis and design tool to determine the control necessary to accurately track reference trajectories during transitions between below and above-rated wind conditions, has been reported, e.g. [138]. Also, there has been research reported on applying feedback control schemes to smart rotors. For example, [137] investigated alleviation of extreme loads using trailing edge flaps and a proportional plus integral plus derivative (PID) controller explicitly designed to compensate for the fatigue load. The flow model used in this alternative research is three-dimensional but quasi-steady, unlike the unsteady, nonlinear, two-dimensional model used in this section.

Appropriately designed smart rotor control systems offer a significant reduction of blade loads by deploying devices that implement spanwise distributed load control. Faster active load control is possible, as is active feedback based on local measurements and these features supply additional support for applying ILC designs in this area. Next, the development of the CFD model of the dynamics, i.e. the flow past the blade and the effects of disturbances, is described, based on the airfoil shown schematically in Figure 9.2.

The flow past an airfoil is described by a member of the general class of stochastic nonlinear partial differential equations. In this chapter, CFD is used to produce an approximation to the dynamics that allow the development of approximate models suitable for control law design and evaluation. The basic features required are a representation of the flow over the airfoil and modeling of the disturbances present. For the problem considered, it is assumed that the incoming flow is oscillatory but with additive disturbances in the form of vortices convected with the flow, resulting in a nonperiodic flow.

In fluid mechanics, a vortex is a region in a fluid where the flow revolves around an axis line, straight or curved. These are convected with the flow over the airfoil, resulting in a nonperiodic flow. This work uses a simple, in relative terms, CFD model based on a panel method, see below, to simulate flow past an airfoil. The incoming flow is assumed to be oscillatory but with added disturbances in the form of vortices that are convected with the flow, giving a nonperiodic flow.

The vortices interact with each other and with the airfoil, resulting in a nonlinear system. Moreover, the motion of the vortices is found by solving the Euler equations, as detailed below. Hence, this model introduces both nonperiodicity and nonlinearity into the problem in a realistic manner. It is based on a solution to the governing equations for inviscid flow with rotational disturbances convecting in a physically realistic way. (An inviscid flow has zero viscosity, and hence, the viscous forces can be neglected. As a consequence, the general Navier–Stokes equations can be simplified to the Euler form.)

Devices such as trailing edge flaps and microtabs operate by generating circulation (vorticity) in the blade's trailing edge region, thereby directly affecting the lift on the blade. Figure 9.3 shows a

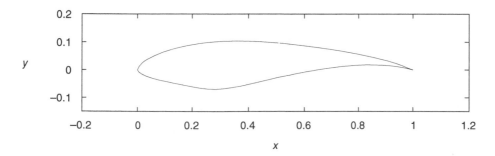

Figure 9.2 An NREL S825 airfoil.

Figure 9.3 Schematic of trailing-edge flaps operating as an actuator.

Trailing-edge flaps

schematic of trailing-edge flaps on a blade. In the CFD model, the flow at the airfoil's trailing edge is manipulated to model these devices, thereby providing the control system's actuation.

Flow past a wind turbine blade has two characteristic time scales, the first is the period of rotation of ≈ 2 seconds (for a modern turbine blade), and the second is the typical time for the flow to pass the blade section. The first of these remains relatively constant, but the second varies along the blade due to the change in the chord (length) and the blade section's velocity. This effect, in turn, results in a significant variation in the characteristic time for the flow to pass the blade. Moreover, in the current application, the interest is in the disturbances convected past the blade, where the latter is used as a reference.

Standard aerodynamic practice for cases such as that given above is to convert the problem to nondimensional form using the mean free stream velocity V_∞ on a blade section and the airfoil chord (length) H as reference values. Moreover, $v^* = V_\infty v = V_\infty(v_x, v_y)$ are the velocity components in $x^* = Hx = H(x, y)$, where the asterisk denotes a dimensional quantity. Time is nondimensionalized using H/V_∞ and, hence, $t^* = \frac{H}{V_\infty}t$. The variation of time scales along the blade is now represented by a change in the rotation's nondimensional period, which increases with the radius. Nondimensional quantities are used throughout the remainder of this section and the next.

The lift of an airfoil/wing section arises, in the main, from the pressure exerted by the fluid on the surface of the airfoil. The angle of attack (AoA) in normal operating conditions is not high enough to provoke separation. Hence, the flow remains attached, and the pressure distribution on the surface of the blade can be calculated by assuming the flow is inviscid. This assumption greatly simplifies the calculations but excludes extreme cases such as rapid direction changes or shear when separation is provoked.

If an extreme example, such as that above, arises, any simple model of the flow of the type considered may give inaccurate results. In which case, full Navier–Stokes simulations are required to provide a complete picture of the flow. Still, such simulations are too expensive computationally to allow a detailed investigation of the effects of applying control laws. As a feasible alternative, a relatively simple panel method, as one option, is used to construct an approximate but usually accurate flow model to form a basis for the development of ILC laws that could eventually be applied to full-scale simulations or experiments.

Panel methods are used to satisfy the boundary conditions on the surface of the body. A set of $2N$ discrete panels are used to enforce the boundary conditions at the surface of the body. There are N vortex panels with constant vorticity on each panel, denoted by λ_i, of varying strength. Vortex panels are placed just above the body's surface and operate by setting the tangential velocity at the surface to zero. It follows that, in principle, the kinematic boundary condition of zero normal velocity is satisfied but only within numerical error. The particular method used below is from [56].

The vortex panels in this application are augmented by a similar set of N source panels, of strength κ_i, placed just below the surface such that zero normal velocity at the surface is explicitly enforced. For a body with a smooth shape, usually, only vortex panels are used as the source panels contribute little since κ_i will be small and tends to zero as N increases, but increased computational effort is required.

A sharp trailing edge is present in the case considered, and a jump in circulation at the trailing edge is used as actuation. Both of these features will result in relatively large numerical errors using only vortex panels. Thus, source panels are also used to improve the accuracy and the numerical conditioning of the procedure. Details of the vortex and source panels and the velocity fields they generate can be found in many standard texts, e.g. [9].

In combination, the source and vortex panels produce a set of $2N$ linear equations in $2N$ unknowns, i.e. λ_i and κ_i. However, this problem is not well posed, and the matrix will be singular (within numerical error). This situation arises because there is an infinite set of solutions to the problem. For the details, see, e.g. [9].

The physically relevant solution of this problem corresponds to the case when the flow leaves the trailing edge smoothly, aligned with the trailing edge and known as the Kutta condition, see, e.g. [9]. Although the wall-normal velocity is zero, a nonzero tangential (slip) velocity exists just above the surface. Numerically, the Kutta condition implies that $v_a = v_b$, where, respectively, v_a and v_b are the tangential velocities in the direction of the trailing edge at the midpoints of the panels adjacent to the trailing edge.

Trailing edge devices used for lift control act by modifying the flow near the trailing edge, generating vorticity and thereby altering the circulation on the body and, hence, the lift, e.g. a trailing edge flap redirects the flow. This effect can be modeled simply in the current framework by allowing a jump in the tangential velocity at the trailing edge rather than applying the Kutta condition, which will directly change the way the flow leaves the airfoil at the trailing edge, i.e. by using

$$u = v_b - v_a \tag{9.1}$$

as the control input and the lift as the output of the system.

A nonzero value of u directly forces a change in the flow at the trailing edge and hence, a change in circulation and lift. This fact provides a relatively simple, relative to a full Navier–Stokes simulation, but realistic model of the flow and actuation to investigate control schemes using trailing edge devices for load control.

The base flow consists of the free-stream velocity $\mathbf{V}_0(t) = (V_{0x}(t), 0)$ and the velocity field generated by the vortex and source panels $v_p(x, t) = (v_{px}, v_{py})$, the latter including the effects of the actuation. Also, disturbances are introduced into the flow in the form of discrete vortices. The Euler equations governing two-dimensional inviscid incompressible flow can be written in vorticity form as follows:

$$\frac{D\omega}{Dt} = \frac{\partial \omega}{\partial t} + v_x \frac{\partial \omega}{\partial x} + v_y \frac{\partial \omega}{\partial y} = 0 \tag{9.2}$$

where

$$\omega = \frac{\partial v_y}{\partial x} - \frac{\partial v_x}{\partial y} \tag{9.3}$$

is the (single component of) vorticity, and $D/Dt = \partial/\partial t + v_x \partial/\partial x + v_y \partial/\partial y$ is the material derivative, i.e. the rate of change with time of a material quantity convected with the flow.

Equation (9.2) is a statement of the fact that in two-dimensional inviscid flow, vorticity is convected with the flow at the local fluid velocity [23]. Hence, the motion of an individual discrete vortex can be tracked by solving

$$\frac{dx_v}{dt} = v(x_v, t) \tag{9.4}$$

where x_v is the position of the (core of the) vortex. The complete velocity field v is given by the sum of three components, i.e.

$$v(x, t) = V_0 + v_p + v_v \tag{9.5}$$

with, for M discrete vortices,

$$v_v = \sum_{j=1}^{M} v_{vj} \tag{9.6}$$

Also, v_{vj} is the velocity field generated by an individual vortex and is given by

$$v_{vj} = \Gamma_j \frac{(-(y - y_{vj}), x - x_{vj})}{|x - x_{vj}|^2} F(|x - x_{vj}|) \tag{9.7}$$

where Γ_j is the strength of vortex j, $F(s) = \int_0^s \gamma(s)ds$, and $\gamma(s)$ is the vorticity distribution of the core of the vortex. In this case, the standard model with a Gaussian distribution is used, as in [56], where $2\pi \int_0^\infty \gamma(s)ds = 1$ and hence, Γ_j is the circulation of the vortex.

As the result of the interaction between the discrete vortices and the body, a nonlinear system of equations results, for which no closed-form solution exists. A numerical solution can be produced by applying any standard time-stepping method to (9.4) for each of the M discrete vortices, using the velocity at the center of the vortex from (9.5). Here, as in [56], a second-order Runge–Kutta method is used to move the vortices

$$\hat{x}_{vj} = x_{vj}^k + \frac{1}{2}\Delta t \, v(x_{vj}^k, t_k)$$
$$x_{vj}^{k+1} = x_{vj}^k + \Delta t \, v(\hat{x}_{vj}, t_{k+1/2}) \tag{9.8}$$

where $t_k = k\Delta t$. The aim is to damp the lift's variation due to the unsteady nature of the free stream velocity $V_{x0}(t)$, and the influence of the vortices on the airfoil as they are convected past the body. The lift, which will be used as the output to the system, is most conveniently calculated from

$$L(t) = -\frac{1}{2} \sum_{i=1}^{N} \lambda_i \Delta_i v_x(x_{ci}, t) \tag{9.9}$$

where Δ_i is the length of panel i and $v_x(x_{ci}, t)$ is the streamwise component of velocity at $x = x_{ci}$, which, as discussed above, is in the flow just above the midpoint of the panel.

The basic panel code was tested against results from standard sources. In particular, the lift coefficients and pressure distributions for inviscid flow past the airfoil obtained from the method given above were compared with those found using XFOIL [69], a well-validated code commonly used for airfoil calculations. Based on the comparison, 404 source and 404 vortex panels were used. As usual, the panels were clustered toward the trailing edge, where the curvature is largest. Moreover, as in [56], the discrete vortices' core size, ϵ, was taken as one quarter of the mean panel length.

9.1.1 Oscillatory Flow

Consider the case with no vortices, and hence, the variation in the lift comes from that in the free stream velocity, and the aim is to damp this fluctuation. The flow past the airfoil is assumed to be periodic with velocity

$$V_{0x} = 1 + A \sin(2\pi t/T) \tag{9.10}$$

where A is the amplitude of the oscillation and T its period. A time step of $\Delta t = 0.005$ is used, with amplitude $A = \frac{1}{10}$, and period $T = \frac{1}{4}$. Since there are no vortices in the flow, the problem is linear in the unknowns (the vortex and source panel strengths). At each step, the latest values are used to update the control input.

First, consider P-type ILC of the form

$$u^k = u^{k-1} + \mu E^{k-1} \tag{9.11}$$

where u^k is the control input for step k, and E^k is the error for this step given by

$$E^k = L^k - L_r \tag{9.12}$$

where L^k is the lift at step k and L_r is the target value for the lift, obtained by setting $A = 0$ in (9.10).

Figure 9.4 shows the control input u_k and the error E^k for the control law (9.11) with $\mu = 20$. Also, shown is the error with no control. A better result is obtained with a larger gain of $\mu = 50$, as shown in Figure 9.5, although early in the run, there is short term high-frequency (time step) fluctuation in the solution. Increasing the gain significantly beyond this value results in unstable controlled dynamics.

The flow considered has a forced oscillatory component, the effect of which on the lift is only partially damped by the control action. This operates over N_c steps, where $N_c = T/\Delta t$. Label the cycles as $j, j = 0, 1, \ldots$, and the step within a cycle as $k_c, k_c = 0, 1, \ldots, N_c - 1$, and hence, $k = jN_c + k_c$. The discussion that follows considers the application of the phase-lead ILC law

$$u_j^{k_c} = u_{j-1}^{k_c} + \mu E_{j-1}^{k_c+\Delta} \tag{9.13}$$

As the problem is linear, the stability of the controlled dynamics can be investigated directly. In this case, as the only vorticity in the flow field is bound to the surface in the vortex panels, the lift may be calculated directly from $L = -V_{0x}\Gamma$, where Γ is the circulation, i.e. the sum of the bound

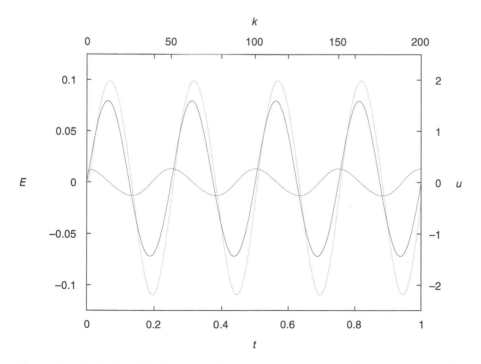

Figure 9.4 Control law (9.11) with $\mu = 20$. dark gray plot, error E^k with no control ($u^k = 0$). gray plot, error E^k with control. light gray plot, control input u^k.

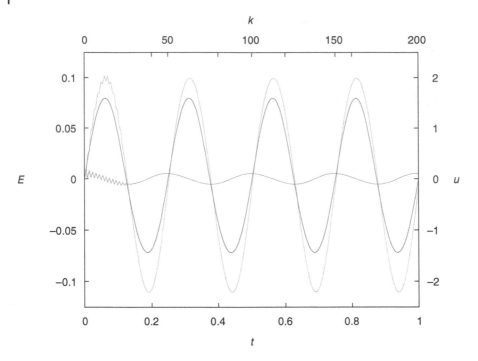

Figure 9.5 Control law (9.11) with $\mu = 50$. dark gray plot, error E^k with no control ($u^k = 0$). gray plot, error E^k with control. light gray plot, control input u^k.

vorticity on the surface and given by $\Gamma = \sum\limits_{i=1}^{N} \lambda_i \Delta_i$. Also by linearity

$$\Gamma = A \, V_{0x} + B \, u \tag{9.14}$$

and

$$L = -A \, V_{0x}^2 - B \, V_{0x} u \tag{9.15}$$

In discrete form

$$L_j^{k_c} = -A(V_{0x}^{k_c})^2 - B \, V_{0x}^{k_c} u_j^{k_c} \tag{9.16}$$

and therefore,

$$
\begin{aligned}
E_j^{k_c} &= -A \, (V_{0x}^{k_c})^2 - B \, V_{0xj} u_j^{k_c} - L_r \\
&= -A \, (V_{0x}^{k_c})^2 - B \, V_{0x} u_{j-1}^{k_c} - L_r - \mu B \, V_{0x}^{k_c} E_{j-1}^{k_c+\Delta} \\
&= E_{j-1}^{k_c} - \mu B \, V_{0x}^{k_c} E_{j-1}^{k_c+\Delta}
\end{aligned}
$$

Hence,

$$\frac{E_j^{k_c}}{E_{j-1}^{k_c}} = 1 - \mu B \, V_{0x}^{k_c} \frac{E_{j-1}^{k_c+\Delta}}{E_{j-1}^{k_c}} \tag{9.17}$$

For $\Delta = 0$, the controlled dynamics will be stable if $0 < \mu B \, V_{0x}^{k_c} < 2$, but the error will decay monotonically only if $0 < \mu B \, V_{0x}^{k_c} \leq 1$. The error changes sign throughout the cycle, and hence, stability cannot be guaranteed for any other value of Δ. With the maximum value of V_{0x} of 1.1, monotonic decay occurs when $\Delta = 0$ if $\mu < 25$, as $B = 0.03635$. Calculations were performed using this value as a guide, with the results as predicted; for $\mu = 25$, the disturbance was damped almost instantaneously, for values between 25 and 50, the error decayed but with overshoots, and for $\mu > 50$, the error increased monotonically in magnitude over the cycles.

The performance of the controlled dynamics with $\Delta = 0$ and $\mu = 10$, with the disturbance decaying to zero and the actuation taking a periodic form, is shown in Figures 9.6 and 9.7 for times early in the simulation. (See [255] for further simulation results for this case.) However, as expected from the stability analysis, this is not the case for both larger and smaller values of Δ. With $\Delta = 4$ a small, high-frequency oscillation, which is growing by $t = 3$, can be seen in the error (Figure 9.6). If the calculation is continued, this disturbance grows exponentially in magnitude. Taking $\Delta = -1$ or 1 (not shown), initially, the control appears to succeed, as shown in Figures 9.6 and 9.7. Again, however, a high-frequency component to the solution grows as the calculation proceeds.

Figure 9.7 also shows the change in the control input at each step, $u^k - u^{k-1}$ for the case with $\Delta = 0$, which operates 90° out of phase with u and at an order of magnitude lower in amplitude. A series of simulations were performed with different parameters (amplitude A, period T, and gain μ satisfying $0 < \mu B V_{0x}^{k_c} \leq 1$), and for all cases with $\Delta \neq 0$, the same pattern was found, i.e. initial decay followed by uncontrolled growth.

A comparison of the error E^k for the different control laws is shown in Figure 9.8, demonstrating the superior performance of the phase-lead ILC control law. Hence, $\mu = 10$ will be used for the ILC control law in the analysis below as it provides good attenuation of the error but with an allowance for nonlinear effects.

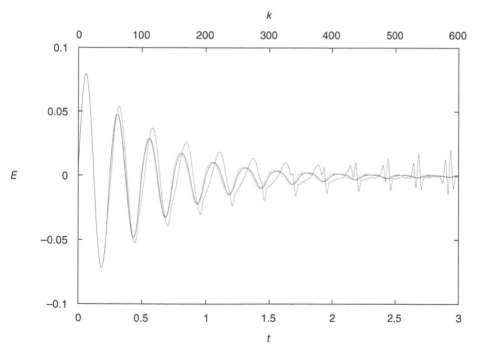

Figure 9.6 Error E^k for controller (9.13) with $\mu = 10$ and $\Delta = -1$ (light gray plot), $\Delta = 4$ (gray plot), and $\Delta = 0$ (dark gray plot).

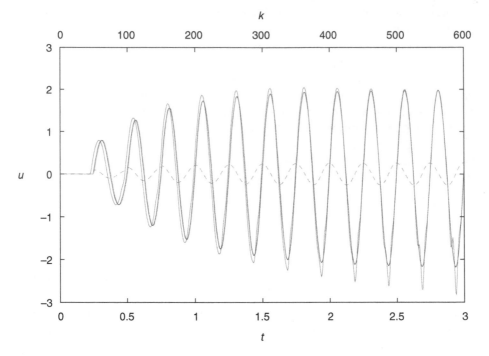

Figure 9.7 Control input u^k for controller (9.13) with $\mu = 10$ and $\Delta = -1$ (gray plot), $\Delta = 4$ (light gray plot), and $\Delta = 0$ (dark gray plot). Also shown is the change in the control signal ($u^k - u^{k-1}$) for the case with $\Delta = 0$ (gray dashed line plot).

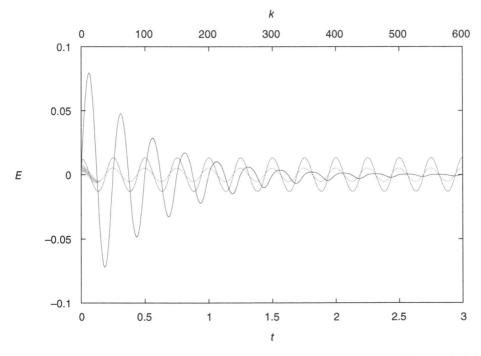

Figure 9.8 The error E_k for controller (9.11) with $\mu = 20$ (gray plot) and (9.11) with $\mu = 50$ (light gray plot), and controller (9.13) with $\mu = 10$ and $\Delta = 0$ (dark gray plot).

9.1.2 Flow with Vortical Disturbances

In this section, vortices are introduced into the flow, and it is no longer periodic, but it still has a periodic component from the free stream. The flow is now nonlinear, and the stability analysis for the phase-lead ILC law given above does not apply, as it assumes both linearity and periodicity.

Consider the flow with an oscillatory free stream specified by $A = 0.1$ and $T = 0.25$ in (9.10). with two vortices introduced into the flow upstream of the airfoil, one with strength $\Gamma_1 = \frac{1}{10}$ placed at $x_{v1} = (-15, 0.25)$ and the other also with strength $\Gamma_2 = \frac{1}{10}$, but at $x_{v2} = (-9, -0.35)$ at the start of the simulation ($t = 0$).

For these starting values, vortex 1 will pass above the airfoil and vortex 2 below. Consequently, a significant disturbance in the lift is generated, in addition to that from the oscillation in the free stream velocity. Figure 9.9 shows the error for this flow with no control for the time that the vortices are passing the airfoil. In addition to the oscillation in lift arising from the free stream, large disturbances are generated by the vortices.

The ILC law (9.13) with $\Delta = 0$ and $\mu = 10$ and a target value of the lift for undisturbed flow ($L_r = 0.379$) was applied to this case. This action suppresses most of the oscillation's effect in the free stream, but not the disturbance due to the vortices, see Figure 9.10.

The control law (9.11) reduces the magnitude of the fluctuations in the lift when applied to the oscillatory flow (recall Figures 9.4 and 9.5). In the current case with two vortices, it damps a substantial proportion of the disturbance generated by the vortices but leaves a residual oscillation, as shown for $\mu = 20$ in Figure 9.11.

To suppress the residual oscillation, an ILC law including both P- and phase-lead action is considered as one option, i.e. of the form

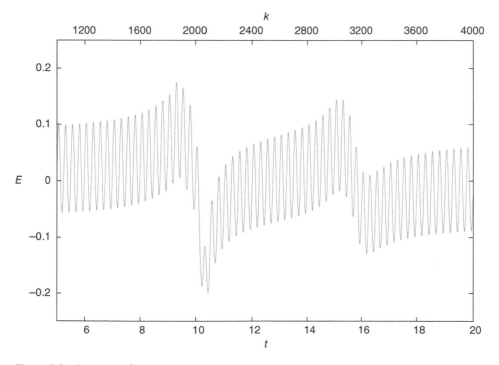

Figure 9.9 The error E^k for oscillatory flow past the airfoil with two vortices and no control applied.

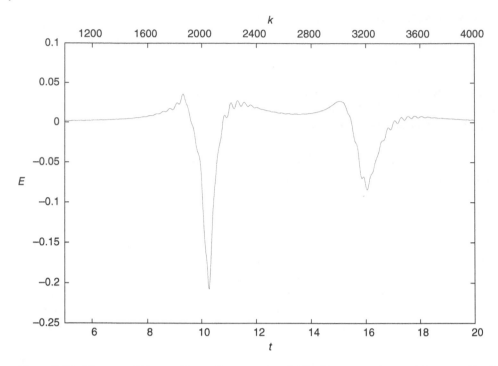

Figure 9.10 The error E^k for oscillatory flow past the airfoil with two vortices and the controller (9.13) with $\Delta = 0$ and $\mu = 10$.

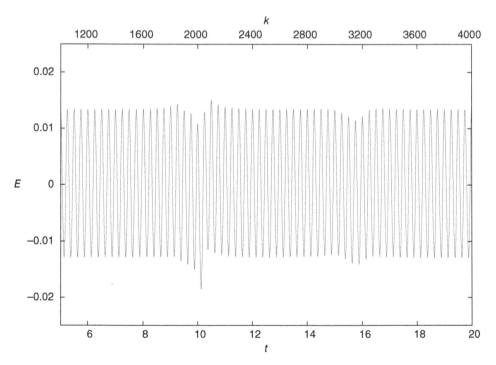

Figure 9.11 The error E^k for oscillatory flow past the airfoil with two vortices and the controller (9.11) with $\mu = 20$.

$$\hat{u}_j^{k_c} = u_{j-1}^{k_c} + \mu_0 E_{j-1}^{k_c+\Delta} \tag{9.18}$$

$$\bar{u}^k = \bar{u}^{k-1} + \mu_1 E^{k-1} \tag{9.19}$$

and

$$u_j^{k_c} = \hat{u}_j^{k_c} + \bar{u}^k \tag{9.20}$$

The resulting error for this law with $\mu_0 = 10$, $\Delta = 0$, and $\mu_1 = 20$ is shown in Figure 9.12. The oscillatory component of the fluctuation is almost eliminated. Moreover, the disturbance from the vortices has been substantially damped. The control signal u^k in this case closely tracks the lift for the uncontrolled flow (see Figure 9.13), generating a counterbalancing force to the inherent fluctuation in the lift. Increasing the values of the gains above these values did not significantly affect the controller's performance. Also, $\Delta \neq 0$ has the same problem as in the case of pure oscillatory flow.

9.1.3 Blade Conditioning Measures

In the wind turbine area, two measures are commonly used to measure performance in terms of the blades' conditioning, i.e. the fatigue and the peak loads, which can also be viewed as the degree of damping present. The 2-norm (using the notation of [255]) measures the first of these

$$\mathscr{L}_2 = \left[\frac{1}{T_1 - T_0} \int_{T_0}^{T_1} (L(t) - L_r)^2 \, dt \right]^{\frac{1}{2}} \tag{9.21}$$

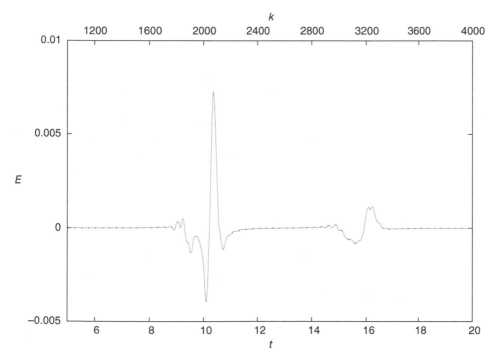

Figure 9.12 The error E^k for oscillatory flow past the airfoil with two vortices and the controller (9.18)–(9.20) with $\mu_0 = 10$, $\Delta = 0$, and $\mu_1 = 20$.

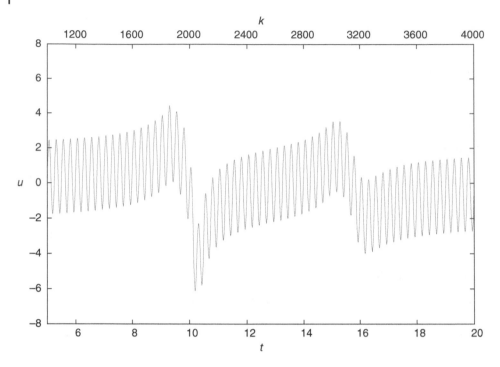

Figure 9.13 The control input u^k for oscillatory flow past the airfoil with two vortices and the controller (9.18)–(9.20) with $\mu_0 = 10$, $\Delta = 0$, and $\mu_1 = 20$.

and the second is measured by the ∞-norm

$$\mathscr{L}_\infty = \max_k \mid L^k - L_r \mid \tag{9.22}$$

The integration to evaluate the $\mathscr{L}_2$ norm was from $T_0 = 5$, which allows the controller to settle to $T_1 = 25$, when both vortices have passed the airfoil. Table 9.1 gives values of $\mathscr{L}_2$ and $\mathscr{L}_\infty$ for various cases, and also the ratio of the measures for controlled versus uncontrolled flow. In the case of the example considered, case 1 in Table 9.1, the ILC law produces a reduction of two orders of magnitude in $\mathscr{L}_2$ and a value of $\mathscr{L}_\infty$ less than 4% of that for the uncontrolled flow.

The airfoil shown in Figure 9.2 is at zero degrees AoA. Pitch control (adjusting the AoA) can be used to maintain a near-constant loading on the turbine as the mean flow rate varies. In this case, it is also possible to successfully deploy the simple structure of ILC laws of this section, see [255]. Next, the influence of actuator dynamics and trial varying ILC laws is considered.

9.1.4 Actuator Dynamics and Trial-Varying ILC

A wide range of airfoils can be used in this area. The results given next uses an airfoil generated using the Karman–Trefftz [35] transform of a circle, i.e.

$$z = n\frac{(1 + \frac{1}{\zeta})^n + (1 - \frac{1}{\zeta})^n}{(1 + \frac{1}{\zeta})^n - (1 - \frac{1}{\zeta})^n} \tag{9.23}$$

where $z = x + jy$ is a complex variable in the new space (airfoil profile) and $\zeta = \chi + j\eta$ is a variable in original space (circle). The parameter $n = 1.9$ and the coordinates of the center of the circle

Table 9.1 Error norms for selected cases.

Case/Figures	Norm	No control	Control	Ratio × 100
1:	$\mathcal{L}_2$	6.63×10^{-2}	7.45×10^{-4}	1.1
	$\mathcal{L}_\infty$	0.198	7.28×10^{-3}	3.7
2:	$\mathcal{L}_2$	0.121	2.26×10^{-3}	1.9
	$\mathcal{L}_\infty$	0.351	2.60×10^{-3}	0.7
3:	$\mathcal{L}_2$	0.149	6.25×10^{-3}	4.2
	$\mathcal{L}_\infty$	0.885	8.01×10^{-2}	9.1
4:	$\mathcal{L}_2$	5.52×10^{-2}	2.59×10^{-4}	0.5
	$\mathcal{L}_\infty$	0.122	3.26×10^{-3}	2.7
5:	$\mathcal{L}_2$	6.05×10^{-2}	3.27×10^{-5}	0.1
	$\mathcal{L}_\infty$	0.187	4.70×10^{-3}	2.5
6:	$\mathcal{L}_2$	2.71×10^{-2}	1.36×10^{-4}	0.5
	$\mathcal{L}_\infty$	4.94×10^{-2}	4.60×10^{-3}	9.3
7:	$\mathcal{L}_2$	7.27×10^{-2}	8.51×10^{-4}	1.2
	$\mathcal{L}_\infty$	0.253	7.33×10^{-3}	2.9

$x_0 = -0.05$ and $y_0 = 0.2$ are used. The profile is normalized using the chord length. This airfoil profile is not one of the standard types used in wind turbines, but it has a similar shape and properties. Moreover, conformal transformation approach greatly simplifies the calculation of the terms involved in generating the reduced-order model.

The results given so far assume perfect actuation. They, therefore, make no allowance for the time delay between the computation of the control signal and the response of the actuator to reach the required value. One possible model for this effect is of the form

$$\dot{\hat{u}}(t) = \lambda(u(t) - \hat{u}(t)) \tag{9.24}$$

where λ denotes the actuator's speed of response, u is the signal calculated by the ILC law, and $\hat{u}$ is the signal applied to the flow.

Consider the case when the phase-lead ILC control law is applied with $\mu_0 = 50$ and $\mu_1 = 1$. In this case, complete damping of the oscillatory response cannot be achieved. Figure 9.14 shows the error signals for various values of λ and the cases of no delay and no control. These show that the error converges to a particular value over the early trials in both cases. Still, the effect of the actuation delay causes no further improvement to occur over subsequent trials. Moreover, the smaller the value of λ, the more pronounced the oscillations in the transient performance.

One possible counter to these undesirable features would be to increase the value of μ_1 used, and Figure 9.15 gives the results obtained for $\lambda = 1$, and $\mu_1 = 30$ and $\mu_1 = 40$, where pronounced disturbances are evident.

A possible way of removing these difficulties is to use a trial-varying ILC law, e.g.

$$\hat{u}_i^{k_c} = u_{i-1}^{k_c} + \mu_1 \Delta t E_{i-1}^{k_c+\delta} \tag{9.25}$$

where $\mu_1 = f(i)$. Figure 9.16 shows the error for the following possible choices for the gain functions:

$$\mu_1 = 1$$

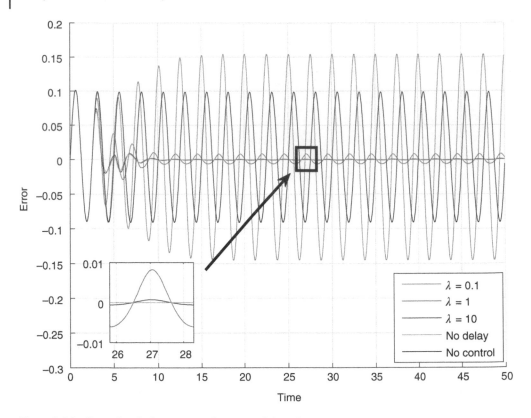

Figure 9.14 Error signals for a range of actuator delays λ.

$$\mu_{11} = \begin{cases} e^{0.1i}, & i = 1,2,\ldots,32 \\ e^{3.2} + \log 10(i-31), & i = 33,\ldots,80 \end{cases}$$

$$\mu_{12} = \begin{cases} e^{0.1i}, & i = 1,2,\ldots,32 \\ e^{3.2} + 4\log(i-31), & i = 33,\ldots,80 \end{cases}$$

$$\mu_{13} = \begin{cases} e^{0.13i}, & i = 1,2,\ldots,23 \\ e^{2.99} + \log(i-22), & i = 24,\ldots,80 \end{cases}$$

$$\mu_{14} = \begin{cases} e^{0.13i}, & i = 1,2,\ldots,23 \\ e^{2.99} + 4\log(i-22), & i = 24,\ldots,80 \end{cases} \tag{9.26}$$

A wide range of gain functions, i.e. those above and others not detailed were evaluated, where the common feature in most cases is that the gain rises exponentially for approximately 20–30 trials and then grows logarithmically to limit the value. Figure 9.16 gives the gain functions corresponding to a selection of those given in (9.26).

The error norms as a function of the trial number are given in Figures 9.17–9.19, where the results in the last plot start from the 5th trial after the error for the fixed-gain ILC law stabilizes and no further convergence can be obtained (light gray plot). After 80 trials, both error norm measures are significantly smaller than those for the standard ILC law. Also, the possibility exists that additional trials could lead to further decreases in the oscillations by the trial-varying ILC law.

Further tuning of the trial-varying ILC law will eventually lead to much more minor relative improvements in performance. Moreover, problems could arise with "high gains" if nonperiodic disturbances are present. In such cases, model-based design is the alternative, which brings

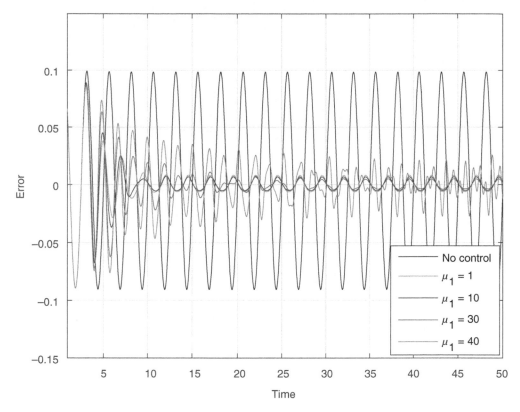

Figure 9.15 Error signals for $\lambda = 1$ and various values of μ_1.

the need to construct an approximate finite-dimensional model of the dynamics from the CFD model. Numerous approaches are possible. Of those available, a method based on constructing the finite-dimensional state-space model using proper orthogonal decompositions (PODs) is applied as detailed next.

9.1.5 Proper Orthogonal Decomposition-Based Reduced Order Model Design

In essence, POD is a method of information compression that eliminates the redundant information from the snapshots generated, either experimentally or, as in this section, numerically. The POD functions are generated using the data snapshots. Moreover, the modes' information heavily depends on the data set, as does the modes' ability to approximate the system's state. This method has been extensively used in a wide range of applications. For recent control-related research, see, e.g. [30]. In this section, the airfoil is generated using the transform (9.23).

Assume that the flow is periodic. Therefore, the total velocity field u can be decomposed as a sum of the steady mean flow component u_m, an oscillatory component, and the unsteady component $\hat{u}$. Also, the input control term κu_c is included and hence,

$$u = u_m \left(1 + A \sin(\omega t)\right) + \hat{u} + \kappa u_c \tag{9.27}$$

where u_m is the steady mean flow, $A \sin(\omega t)$ is the oscillatory component, κ is the sum of the control inputs from the time the circulation is turned on, where $\kappa = \sum_i u_i$, and u_c is the velocity field generated by an airfoil with unit circulation.

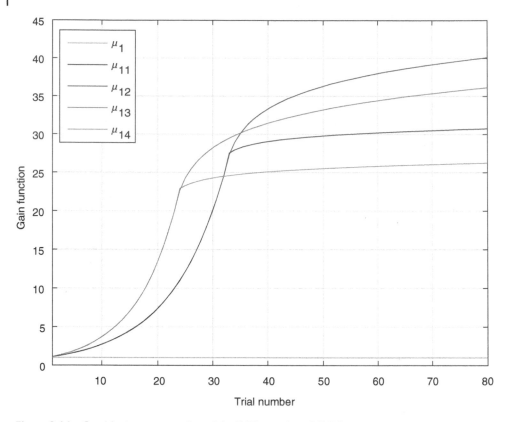

Gain function

Trial number

Figure 9.16 Graphical representation of the ILC law gains of (9.26).

The unsteady flow $\hat{u}$ can be represented by a set of POD modes $\phi_j(x)$ and their coefficients $a_j(t)$, see also, e.g. [24, 25]

$$\hat{u} = \sum_{j=1}^{H} a_j(t)\phi_j(x) \tag{9.28}$$

where H is the number of snapshots. The coefficients $a_j(t)$ and modes $\phi_j(x)$ are the solutions of the eigenvalue problem $Ca_j = \lambda_j a_j$, where $C_{ji} = \langle \hat{u}_j, \hat{u}_i \rangle$, $\lambda_1, \ldots, \lambda_H$ are the eigenvalues arranged such that $\lambda_1 > \lambda_2 > \cdots > \lambda_H$ and $a_j, \ldots, a_H$ are the eigenvectors. The inner product is defined as the integral of a dot product, i.e. $\langle a, b \rangle = \int a \cdot b \, dx$.

The Euler equations govern the flow of inviscid, incompressible fluid with constant density:

$$\begin{cases} \frac{\partial u}{\partial t} + \mathbf{u} \cdot \nabla \mathbf{u} = -\nabla p \\ \nabla \cdot \mathbf{u} = \mathbf{0} \end{cases} \tag{9.29}$$

where $\mathbf{u} = (u, v)$ is the velocity vector with components in the x and y directions and p denotes the pressure. To develop a reduced-order model of the system, the velocity field, (9.27) with the unsteady part described by (9.28) is substituted into the governing Euler equations (9.29). Moreover, the expression for the derivatives of the coefficients $\frac{\partial}{\partial t}a_i(t)$ are obtained by taking the inner product with respect to ϕ_i, giving

$$\frac{\partial}{\partial t}\sum_{j=1}^{N} a_j(t)\langle \phi_i, \phi_j \rangle = \frac{\partial}{\partial t}a_i(t)\langle \phi_i, \phi_i \rangle = \frac{\partial}{\partial t}a_i(t) \tag{9.30}$$

since the modes are orthogonal, i.e. $\langle \phi_i, \phi_i \rangle = \|\phi_i\|$ and are normalized such that $\|\phi_i\| = 1$.

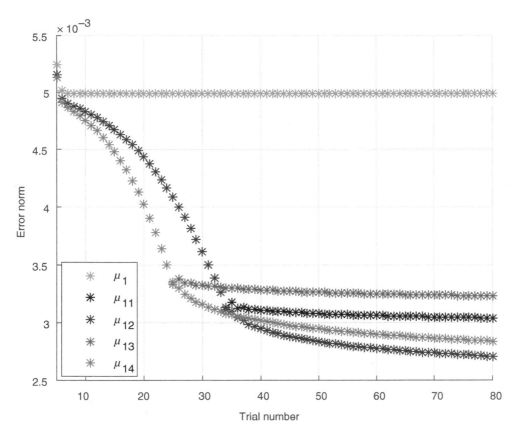

Figure 9.17 $\mathscr{L}_2$ norms for the different ILC law gains.

The nonlinear terms are small and can be neglected to obtain a linear representation. For wake flows, it is also routine to delete the pressure term $\langle \phi_i, -\nabla p \rangle$ together with $\langle \phi_i, u_m \cdot \nabla u_m \rangle$, see [24, 172] for the supporting discussion. The resulting state equation has the form:

$$\frac{\partial}{\partial t} a_i(t) = - \left[\sum_{j=1}^{H} a_j(t) \left[(1 + A \sin \omega t)(\langle \phi_i, \phi_j \cdot \nabla \mathbf{u}_m \rangle + \langle \phi_i, \mathbf{u}_m \cdot \nabla \phi_j \rangle) \right] \right.$$
$$+ \kappa \left[(1 + A \sin(\omega t))(\langle \phi_i, \mathbf{u}_m \cdot \nabla \mathbf{u}_c \rangle + \langle \phi_i, \mathbf{u}_c \cdot \nabla \mathbf{u}_m \rangle) \right]$$
$$+ A\omega \cos \omega t \langle \phi_i, \mathbf{u}_m \rangle + 2A \sin \omega t \langle \phi_i, \mathbf{u}_m \cdot \nabla \mathbf{u}_m \rangle$$
$$\left. + \frac{\partial}{\partial t} \kappa \langle \phi_i, \mathbf{u}_c \rangle \right] \tag{9.31}$$

where the inner product for each term can be calculated as, e.g.

$$\langle \phi_i, \mathbf{u}_m \cdot \nabla_m \rangle = \int \int \left[\phi_{ix} \left(\mathbf{u}_m \frac{\partial \mathbf{u}_m}{\partial x} + \mathbf{v}_m \frac{\partial \mathbf{u}_m}{\partial y} \right) \right.$$
$$\left. + \phi_{iy} \left(\mathbf{u}_m \frac{\partial \mathbf{u}_m}{\partial x} + \mathbf{u}_m \frac{\partial \mathbf{u}_m}{\partial y} \right) \right] d\mathbf{x} \, d\mathbf{y} \tag{9.32}$$

On completing all calculations, the result can be written in matrix form with the states $a_i(t)$ and input $u = \kappa$ as follows:

$$\frac{\partial}{\partial t} a(t) = A(t)a(t) + B_1(t)u(t) + B_2 \frac{\partial}{\partial t} u(t) + O(t) \tag{9.33}$$

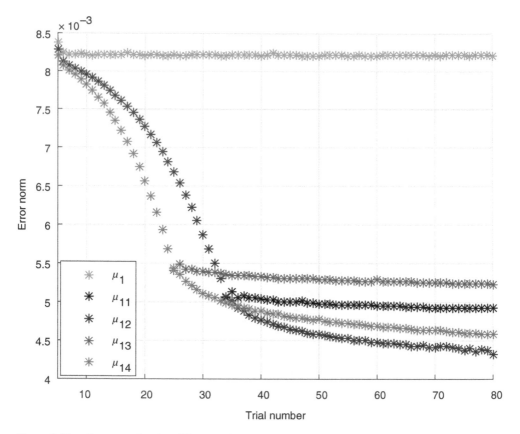

Figure 9.18 $\mathscr{L}_\infty$ norms for the different ILC law gains.

Figure 9.19 The $\mathscr{L}_2$ and $\mathscr{L}_\infty$ error norms.

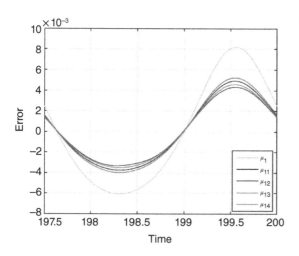

The free term $O(t)$ can be removed from the state equation and instead included in the output's calibration. To eliminate the input derivative on the right-hand side of (9.33), the state is redefined as $x = a - B_2 u$ to give

$$\frac{\partial}{\partial t} x(t) = A(t)x(t) + B(t)u(t) \tag{9.34}$$

where $B(t) = B_1(t) + A(t)B_2$. In what follows, zero-order hold discretization is applied, and the resulting state equation represents the dynamics.

The output equation is defined by estimating the lift on the blade, where the airfoil's surface is a streamline with the velocity tangential to the surface and denoted by u_t. Also, the normal velocity is zero and denote the distance along the streamline by s. Then, see, e.g. [9],

$$\frac{\partial u_t}{\partial t} + u_t \frac{\partial u_t}{\partial s} = \frac{\partial u_t}{\partial t} + \frac{1}{2}\frac{\partial u_t^2}{\partial s} = -\frac{\partial p}{\partial s} \tag{9.35}$$

Integrating this last equation from $s_0 = 0$ to s and setting $p_0 = -\frac{1}{2}u_{t0}^2$, the pressure at point s is given by

$$p_s = -\left(\int_0^s \frac{\partial u_t}{\partial t} ds + \frac{1}{2}u_{ts}^2 \right) \tag{9.36}$$

The tangential velocity at each point on the surface can be calculated using the x and y components of the velocity, where the quadratic term in the equation is linearized by rewriting the tangential velocity as follows:

$$u_t = u_r(\mathbf{x}, t) + \sum_{j=1}^{H} a_j(t)\phi_j(\mathbf{x}) + \kappa \mathbf{u}_c \tag{9.37}$$

where u_r includes the mean and oscillatory parts of the flow and approximating the quadratic term as follows:

$$u_t^2 = u_r^2(\mathbf{x}, t) + 2u_r(\mathbf{x}, t)\left(\sum_{j=1}^{H} a_j(t)\phi_j(t)(\mathbf{x}) + \kappa \mathbf{u}_c \right) \tag{9.38}$$

The integral in (9.36) is calculated by dividing the surface of the airfoil into $\hat{n}$ panels and for each panel,

$$\frac{\partial}{\partial t} \int u_{t,\hat{n}+\frac{1}{2}} ds = \frac{u_{t,\hat{n}+\frac{1}{2}}^p - u_{t,\hat{N}+\frac{1}{2}}^{p-1}}{\Delta t} l \tag{9.39}$$

where $l = \left[\Delta x^2 + \Delta y^2 \right]^{\frac{1}{2}}$ is the length of the panel and p is the time step. The lift on each panel is equal to the y component. Also, the total lift is calculated as the sum of the lift on each panel.

To calculate the lift, L, $\mathbf{u}_m$, $\mathbf{u}_c$, and the POD modes are used. This equation will contain the input and state derivative terms and hence,

$$L = C_1(t)a(t) + D_1(t)u + C_2\frac{\partial}{\partial t}a(t) + D_2\frac{\partial}{\partial t}u(t) + O_2(t) \tag{9.40}$$

and to obtain a discrete representation, the derivatives are approximated as follows:

$$\frac{\partial}{\partial t}a(t) = \frac{a(p) - a(p-1)}{\Delta t}, \quad \frac{\partial}{\partial t}u(t) = \frac{u(p) - u(p-1)}{\Delta t} \tag{9.41}$$

Introducing the state transformation $x = a - B_2 u$ gives

$$L(p) = C(p)x(p) + D(p)u(p) + \hat{C}x(p-1) + \hat{D}u(p-1) + \hat{O}(p) \tag{9.42}$$

where $C(p) = C_1(p) + C_2$, $D(p) = D_1(p) + D_2 + C_1(p)B_2 + C_2 B_2$, $\hat{C} = -C_2$, and $\hat{D} = -D_2 - C_2 B_2$.

Next, by way of an example, a reduced-order model constructed using a data set generated in the CFD simulation is validated by comparing the reconstructed lift to the lift values obtained from the CFD simulation. Subsequently, the model is used to design and evaluate an NOILC law to reduce lift fluctuations.

The data set for the construction of the POD reduced-order model below was generated using the panel code [27]. An oscillatory flow of amplitude $A = 0.1$ (for the x component only) and period $T = 2.5$ seconds ($\omega = 2.5133$) was introduced to the flow. Moreover, to obtain the modes that will perform well for different input signals, the control input used to generate the snapshot was chosen as a sum of sinusoids of different frequencies:

$$u(t) = \mu \left(\sin\left(\frac{\omega t}{3} \right) + \sin\left(\frac{\omega t}{2} \right) + \sin(\omega t) + \sin(2\omega t) + \sin(2.5\omega t) \right) \tag{9.43}$$

where $\mu = 0.000\,25$.

A time step of $\Delta t = 0.005$ seconds was used in the simulation, but the snapshots were captured every $\Delta t_s = 0.1$ seconds, resulting in 301 snapshots. Also, the flow velocity was captured at each point of the regular grid around the airfoil, which was represented by a circle in computational space using the conformal transformation (9.23) with $n = 1.9$. The coordinates of the center of the circle were taken as $x_0 = 0.05$ and $y_0 = 0.2$. Also, the profile was normalized using the cord length.

The regular grid around the airfoil was generated in computational space using $\zeta = re^{j\theta} + \zeta_c$ with lines of constant radius r, angle θ and then mapped into the physical space. In this example, 81 different values of the radius were used and 100 different values of the angle. Hence, the snapshot consists of 8100 data points in the x and y directions for each point.

Construction of the reduced-order model used $N = 6$ modes since they can reconstruct over 80% of the energy in the flow (see Figure 9.20). The energy is calculated as follows:

$$E_j = \frac{\lambda_j}{\sum_{j=1}^{H} \lambda_j} \tag{9.44}$$

The model was simulated for 100 seconds, and the comparison of lift obtained using CFD simulation and the POD model with the derivative terms neglected is shown in Figure 9.21. The error in the lift at each time step is calculated using

$$\delta L[\%] = \left| \frac{L_{\text{CFD}} - L_{\text{model}}}{L_{\text{CFD}}} \right| \times 100\% \tag{9.45}$$

and the average error in the last trial was 2.1% for the full model and 2.4% for the approximate model used below for ILC design.

As a first design, NOILC with diagonal Q and R, in each case a scalar times the identity matrix, is considered, where the target value for the lift is $L_{\text{tar}} = 0.66$. The results are in Figure 9.22, where the error norm is reduced to below 10^{-3} in less than 10 trials and 10^{-6} in less than 100 trials. Let q and r, respectively, be the diagonal values. Higher values of q not only produce faster trial-to-trial error convergence but also result in larger input signals. Conversely, increasing the value of r used produces slower trial-to-trial convergence but demonstrates the potential for better robustness.

This design delivers improved performance against simple-structure ILC laws, e.g. phase lead. The results of detailed comparative studies in this respect are given in [178]. The results given are for the case when $q = 10$ and $r = 1$, see Figure 9.23. These results show that the trial-to-trial error convergence diverges from as early as the third trial. The reason for this behavior is that the model does not contain the unsteady part of the lift arising from the derivative in velocity.

A NOILC law for the state-space model without derivative terms was first designed with diagonal weighting matrices, and good convergence was obtained for diagonal weighting matrices.

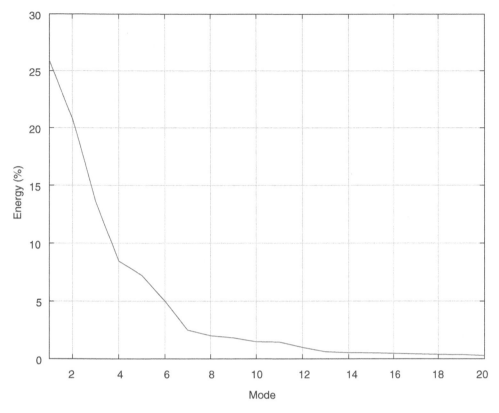

Figure 9.20   Energy in the modes.

Figure 9.24 gives the simulation results for this design with both weighting matrices equal to the identity matrix of compatible dimensions, where the target value for the lift assuming no oscillatory component is $L_{\text{tar}} = 0.66$. For this design, the lift converges monotonically to the desired value for the reduced-order model with neglected derivative terms after two trials. The same control law applied to the model with derivative terms present does not result in convergence (dark gray line in the figure).

An alternative NOILC design based on the model described by (9.34) and (9.42) is next considered. In this case, the input–output representation of the dynamics was calculated recursively for each time step p and used in the design. The results obtained for this control law are given in Figures 9.25 and 9.26, and the mean squared error for this final design converges to zero in approximately 25 trials in the trial domain (see Figure 9.26). However, the time-domain transients are not ideal since, in the early trials, peaks in lift appear for the points close to the beginning/end of the trials due to the change in the control input (high-derivative action).

This last effect can be minimized by average filtering the control input signal or slowly applying the control input signal over the first few trials. In an attempt to increase design robustness, the cost function identity-weighting matrix on the control was multiplied by three, resulting in slightly slower convergence. Still, finally the lift reaches the target value of $L_{\text{tar}} = 0.66$ at around $p = 80$, see Figure 9.25.

A more comprehensive treatment of this recent ILC application area is given in [178]. Much further profitable research is possible in this area, both in the application and the underlying theory. In this last aspect, the POD model results in a time-varying approximate model of the dynamics.

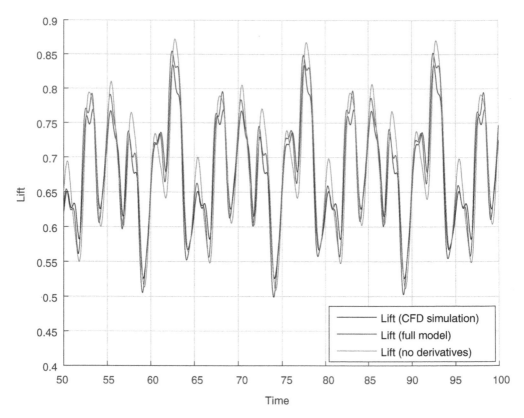

Figure 9.21 Comparison of the lift obtained in CFD simulation and that reconstructed by the reduced order model with six modes.

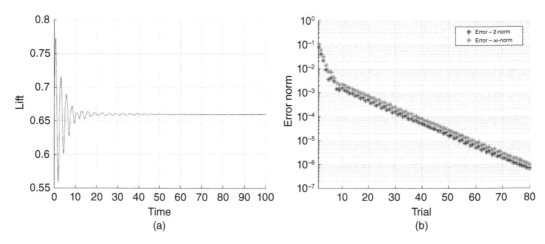

Figure 9.22 The lift and error norms NOILC based on the POD reduced order model.

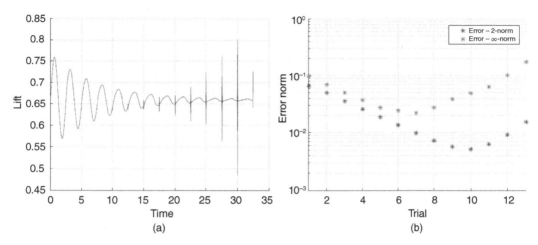

Figure 9.23 The lift (a) and error norms (b) obtained using the CFD model for design.

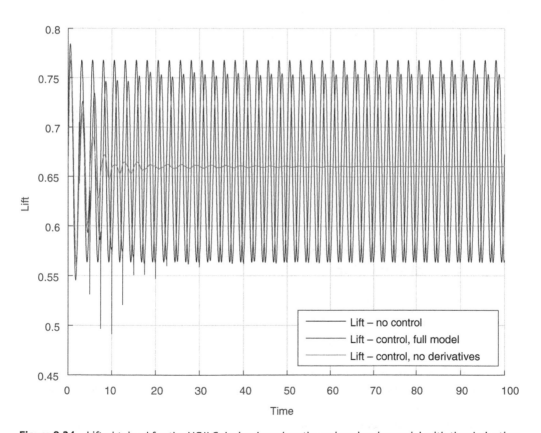

Figure 9.24 Lift obtained for the NOILC design based on the reduced-order model with the derivative terms neglected.

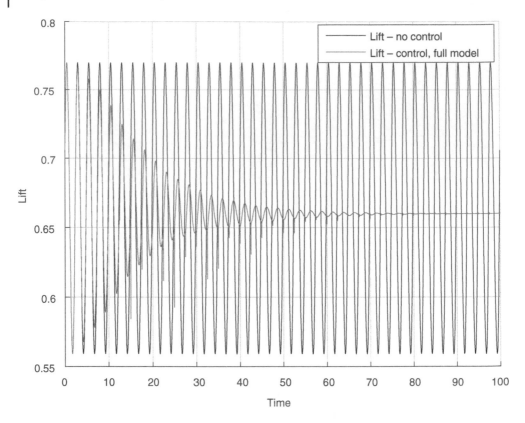

Figure 9.25 Lift obtained for the system with NOILC designed for the model with derivative terms present.

Such dynamics have been treated in Chapter 8 and will arise again in the rehabilitation application in Chapter 11.

9.2 Design Based on Finite-Dimensional Approximate Models with Experimental Validation

In this section, the analysis, design, and experimental validation are based on the system shown in Figure 9.27, for which Figure 9.28 is a schematic diagram of the metal rod that can be heated or cooled from the bottom by four independent elements, i.e. actuators, generating heat flows $\dot{Q}_i(t)$, $i = 1, \ldots, 4$. All of the side surfaces of the rod are thermally insulated. Moreover, the top surface is in direct contact with the atmosphere at a temperature of $\vartheta_a(t)$. Details of the design and development of the system can be found in [222]. The analysis in this section follows [156] and also uses the notation in this chapter.

It is assumed that measurements $\vartheta_i(z, t)$ can be made at the rod's four segments' geometric midpoints. The length l of the rod is much greater than the height h and the width b. Also, the experimental system has four actuators, but the analysis extends directly to any finite number of heated segments.

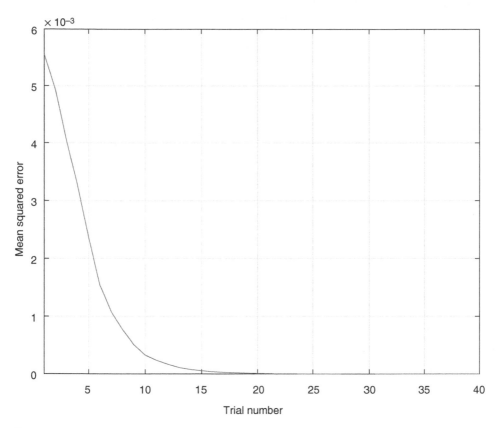

Figure 9.26 Convergence of the error for the design of Figure 9.25.

Figure 9.27 Experimental setup: an iron rod with rectangular cross-section heated or cooled from bottom by heating elements.

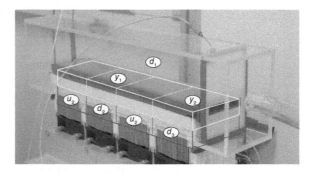

Under the assumptions made, a spatial one-dimensional temperature distribution $\vartheta(z, t)$ can be assumed and modeled by an application of the first law of thermodynamics as follows:

$$\frac{\partial q(z, t)}{\partial z} + \kappa_1 \frac{\partial \vartheta(z, t)}{\partial t} + \kappa_2 \vartheta(z, t) = \mu_c(z, t) + \kappa_2 \vartheta_a(t) \tag{9.46}$$

and by application of Fourier's heat conduction law

$$q(z, t) = -\lambda \frac{\partial \vartheta(z, t)}{\partial z} \tag{9.47}$$

where $q(z, t)$ is the heat flux density, the coefficients $\kappa_1 = \rho c_p$ and $\kappa_2 = \frac{\alpha}{h}$ depend on the density ρ, the specific heat capacity c_p, the convective heat transfer coefficient α, the height h of the rod, and λ

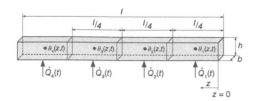

Figure 9.28 Schematic representation of the heat transfer problem in a metal rod showing the actuator and sensor locations.

is the heat conductivity. The control input $\mu_c(z, t)$ is generated by the actuators (also termed Peltier elements in some of the literature) and described by

$$\mu_c(z, t) = \sum_{i=1}^{4} a_i(z)\dot{Q}_i(t) \tag{9.48}$$

where

$$a_{c,i}(z) = \begin{cases} \frac{4}{bhl}, & \text{for } z \in [z_{i-1}, \ z_i] \\ 0, & \text{otherwise} \end{cases} \tag{9.49}$$

The positions $z_i = i\frac{l}{4}$, $i = 1,2,3,4$, along the z-axis in Figure 9.28 are the edges of the actuators. Moreover, the boundary conditions for the heat flux density $q(z, t)$ are taken as follows:

$$q(0, t) = \bar{q}_0(t), \quad q(l, t) = \bar{q}_l(t) \tag{9.50}$$

where $\bar{q}_0(t) = 0$ and $\bar{q}_l(t) = 0$ since the side surfaces are isolated. The initial temperature distribution in the rod is taken as $\vartheta(z, 0) = \bar{\vartheta}_0(z)$, or, in particular, as the initial ambient temperature, i.e. $\vartheta(z, 0) = \vartheta_a(0)$.

Substituting (9.47) into (9.46) results in the parabolic partial differential equation that describes the spatially distributed temperature in the rod. In this section, ILC design is based on approximating the dynamics by a set of ordinary differential equations. To construct this approximation, the method of integrodifferential relations combined with the projection approach is used [222].

In this method, the temperature profile in each section of the rod is approximated by

$$\tilde{\vartheta}(z, t) = \sum_{i=1}^{4} \sum_{m=0}^{M} b_{i,m,M}(z)\theta_{i,m,M}(t) \tag{9.51}$$

where the functions $b_{i,m,M}(z)$ are the Bernstein polynomials

$$b_{i,m,M}(z) = \begin{cases} b_i^{m,M}(z), & \text{for } z \in [z_{i-1}, \ z_i] \\ 0, & \text{otherwise} \end{cases}$$

$$b_i^{m,M}(z) = \binom{M}{m} \left(\frac{z - z_{i-1}}{z_i - z_{i-1}} \right)^m \left(\frac{z_i - z}{z_i - z_{i-1}} \right)^{M-m} \tag{9.52}$$

and $\theta_{i,m,M}(t)$ are the unknown time-dependent coefficients. The temperature distribution continuity between neighboring segments i and $i + 1$ is guaranteed by the assumption that

$$\theta_{i,0,M}(t) = \theta_{i-1,M,M}(t), \quad i = 2,3,4 \tag{9.53}$$

The finite-dimensional, state-space model approximation of the dynamics is obtained by following the procedure given in [222]. In application, this method requires the selection of the degree M of the Bernstein polynomials, and this is application-specific. In the considered case, the ambient temperature above the rod is homogeneous and the initial temperature distribution of the rod is equal to the ambient temperature. The heat flows transferred by the actuators are also homogeneous. Also, it is assumed that the reference temperature is smooth and hence the temperature distribution in each segment of the rod is also smooth. Based on these facts, $M = 3$ is used and a state-space model of the following form is obtained:

$$\dot{x}(t) = A_c x(t) + B_c u(t) + E_c \vartheta_a(t)$$
$$y(t) = C_c x(t) \tag{9.54}$$

where

$$x(t) = \begin{bmatrix} x_1(t) & x_2(t) & x_3(t) & x_4(t) \end{bmatrix}^T \tag{9.55}$$

and

$$x_1(t) = \begin{bmatrix} \theta_{1,0,3}(t) \\ \theta_{1,1,3}(t) \\ \theta_{1,2,3}(t) \\ \theta_{1,3,3}(t) \end{bmatrix}^T, \quad x_2(t) = \begin{bmatrix} \theta_{2,1,3}(t) \\ \theta_{2,2,3}(t) \\ \theta_{2,3,3}(t) \end{bmatrix}^T$$

$$x_3(t) = \begin{bmatrix} \theta_{3,1,3}(t) \\ \theta_{3,2,3}(t) \\ \theta_{3,3,3}(t) \end{bmatrix}^T, \quad x_4(t) = \begin{bmatrix} \theta_{4,1,3}(t) \\ \theta_{4,2,3}(t) \\ \theta_{4,3,3}(t) \end{bmatrix}^T \tag{9.56}$$

$$u_c(t) = \begin{bmatrix} \dot{Q}_1(t) & \dot{Q}_2(t) & \dot{Q}_3(t) & \dot{Q}_4(t) \end{bmatrix}^T$$

The output equation is obtained by evaluating (9.51) for the temperature sensor positions $z = \frac{3l}{8}, \frac{7l}{8}$ and hence in (9.54) for this case,

$$C_c = \begin{bmatrix} \vartheta(\frac{3l}{8}, t) & \vartheta(\frac{7l}{8}, t) \end{bmatrix}^T \tag{9.57}$$

Specification of the entries in the matrices of this state-space model will be detailed by an example later in this section. As the system model is linear, the temperatures are relative to the initial ambient values rather than their absolute counterparts. Validation of the resulting state-space model has been undertaken and is reported in [222]. Moreover, other methods of constructing a finite-dimensional approximation of infinite-dimensional dynamics are available for use, but none will be universally better than others. In the remainder of this section, the model constructed leads to a high-quality ILC design with supporting experimental results.

The experimental system enables various input–output configurations to be examined. In what follows, the problem considered is tracking the reference temperatures at the geometric midpoints of the second and fourth segments with control applied through the first and third actuators in Figure 9.28. The remaining two heating elements, located under the sectors with temperature sensors, are used to generate a disturbance to evaluate the compensation properties of the ILC design. Another form of disturbance is changes in the ambient temperature.

All these disturbances are unknown to the controller and the state-space model for design is

$$\dot{x}(t) = A_c x(t) + B_{cc} u(t) + E_{cc} d(t)$$
$$y(t) = C_c x(t) \tag{9.58}$$

where

$$u(t) = \begin{bmatrix} \dot{Q}_1(t) \\ \dot{Q}_3(t) \end{bmatrix}, \quad d(t) = \begin{bmatrix} \vartheta_a(t) \\ \dot{Q}_2(t) \\ \dot{Q}_4(t) \end{bmatrix} \tag{9.59}$$

are the control and disturbance vectors, respectively. Also,

$$B_{cc} = B_c \begin{bmatrix} 1 & 0 \\ 0 & 0 \\ 0 & 1 \\ 0 & 0 \end{bmatrix}, \quad E_{cc} = \begin{bmatrix} E_c & B_{cd} \end{bmatrix} \tag{9.60}$$

with

$$B_{cd} = B_c \begin{bmatrix} 0 & 0 \\ 1 & 0 \\ 0 & 0 \\ 0 & 1 \end{bmatrix} \tag{9.61}$$

The ILC law used in this application is shown in Figure 9.29. Design is undertaken in the discrete domain and therefore the dynamics (9.58) have been discretized using the exact method and on trial $k + 1$ the dynamics are described by

$$x_{k+1}(p + 1) = A x_{k+1}(p) + B u_{k+1}(p) + E d_{k+1}(p)$$
$$y_{k+1}(p) = C x_{k+1}(p) \tag{9.62}$$

The ILC law contains a current trial feedback loop combined with an ILC law and is detailed next.

In the design developed, the disturbance vector is treated using the equivalent input disturbance approach, see, e.g. [241] and for analysis the state-space model (9.62) is replaced by

$$x_{k+1}(p + 1) = A x_{k+1}(p) + B \left(u_{k+1}(p) + \delta_{k+1}(p) \right)$$
$$y_{k+1}(p) = C x_{k+1}(p) \tag{9.63}$$

where $\delta_{k+1}(p)$ is the equivalent disturbance vector that has the same effect on the output vector $y_{k+1}(p)$ as $d_{k+1}(p)$. Moreover, $\delta_{k+1}(p)$ is described by the disturbance model:

$$\varepsilon_{k+1}(p + 1) = A_d \varepsilon_{k+1}(p)$$
$$\delta_{k+1}(p) = C_d \varepsilon_{k+1}(p) \tag{9.64}$$

Combining (9.63) and (9.64) gives the augmented state-space model:

$$z_{k+1}(p + 1) = A_a z_{k+1}(p) + B_a u_{k+1}(p)$$
$$y_{k+1}(p) = C_a z_{k+1}(p) \tag{9.65}$$

where

$$z_{k+1}(p) = \begin{bmatrix} x_{k+1}(p) \\ \varepsilon_{k+1}(p) \end{bmatrix}, \quad A_a = \begin{bmatrix} A & BC_d \\ 0 & A_d \end{bmatrix}$$

Figure 9.29 Block diagram of the ILC law.

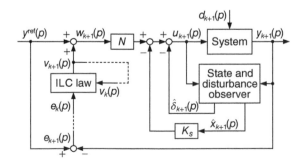

$$B_a = \begin{bmatrix} B \\ 0 \end{bmatrix}, \quad C_a = \begin{bmatrix} C & 0 \end{bmatrix} \tag{9.66}$$

where, as in [156], $y^{\text{ref}}(p)$ denotes the reference vector.

Estimates of the state vector and the equivalent disturbance $\delta_{k+1}(p)$ are obtained using the observer:

$$\hat{z}_{k+1}(p+1) = A_a \hat{z}_{k+1}(p) + B_a u_{k+1}(p) + K_L \left(y_{k+1}(p) - C_a \hat{z}_{k+1}(p) \right) \tag{9.67}$$

where $\hat{z}_{k+1}(p) = \begin{bmatrix} \hat{x}_{k+1}^T(p) & \hat{\varepsilon}_{k+1}(p) \end{bmatrix}^T$ is the observer state vector and the observer gain matrix is $K_L = \begin{bmatrix} K_{L1}^T & K_{L2}^T \end{bmatrix}^T$. Moreover, the estimated equivalent disturbance vector is obtained using

$$\hat{\delta}_{k+1}(p) = C_d \hat{\varepsilon}_{k+1}(p) \tag{9.68}$$

The current trial feedback control loop (see Figure 9.29) constructs the system input as

$$u_{k+1}(p) = N w_{k+1}(p) - K_s \hat{x}_{k+1}(p) - \hat{\delta}_{k+1}(p) \tag{9.69}$$

where N is the static feedforward gain matrix and $w_{k+1}(p)$ represents a signal by which the ILC action and the reference trajectory are applied, as specified below. Routine manipulations give the following state-space model description of the dynamics:

$$\begin{aligned} \chi_{k+1}(p+1) &= A_{cl} \chi_{k+1}(p) + B_{cl} w_{k+1}(p) + E_{cl} d_{k+1}(p) \\ y_{k+1}(p) &= C_{cl} \chi_{k+1}(p) \end{aligned} \tag{9.70}$$

where

$$A_{cl} = \begin{bmatrix} A & -BK_s & -BC_d \\ K_{L1}C & A - BK_s - K_{L1}C & 0 \\ K_{L2}C & -K_{L2}C & A_d \end{bmatrix}$$

$$B_{cl} = \begin{bmatrix} BN \\ BN \\ 0 \end{bmatrix}, \quad E_{cl} = \begin{bmatrix} E \\ 0 \\ 0 \end{bmatrix}, \quad \chi_{k+1}(p) = \begin{bmatrix} x_{k+1}(p) \\ \hat{x}_{k+1}(p) \\ \hat{\varepsilon}_{k+1}(p) \end{bmatrix}$$

$$C_{cl} = \begin{bmatrix} C & 0 & 0 \end{bmatrix} \tag{9.71}$$

Let

$$y^{\text{ref}}(p) = \begin{bmatrix} y^{\text{ref}1}(p) \\ y^{\text{ref}2}(p) \end{bmatrix} \tag{9.72}$$

and then, the tracking error vector on trial k is

$$e_k(p) = y^{\text{ref}}(p) - y_k(p) \tag{9.73}$$

The vector generated by the ILC law and the reference trajectory together form:

$$w_{k+1}(p) = y^{\text{ref}}(p) + v_{k+1}(p) \tag{9.74}$$

with

$$v_{k+1}(p) = Q(q) \left(v_k(p) + L(q) e_k(p) \right) \tag{9.75}$$

and q again denotes the forward time-shift operator. Also, the design of Q and L is detailed below in the numerical example.

Many methods to design the state feedback matrix K are available given the required controllability assumption. In [156], a method based on minimizing a quadratic cost function is used, and likewise for the observer gain matrix K_L. The static feedforward gain matrix N in (9.69) can, if required, be chosen as the inverse of the DC gain matrix of the feedback control loop to reduce the steady-state error in the feedback loop. This control action is particularly significant only on the first trial.

For the ILC design, the frequency domain description of (9.70) is used, i.e. on applying the z-transform:

$$Y_{k+1}(z) = P(z)\left(V_{k+1}(z) + Y^{\text{ref}}(z)\right) + \tilde{P}(z)D(z) \tag{9.76}$$

$$P(z) = C_{cl}(zI - A_{cl})^{-1}B_{cl} \tag{9.77}$$

$$\tilde{P}(z) = C_{cl}(zI - A_{cl})^{-1}E_{cl} \tag{9.78}$$

The ILC law (9.75) in the z-domain (where the dependence on z is suppressed when the meaning is clear) is

$$V_{k+1} = Q(V_k + LE_k) \tag{9.79}$$

and

$$E_k = Y^{\text{ref}} - Y_k \tag{9.80}$$

Stability of the system dynamics P is guaranteed by the feedback control loop and monotonic trial-to-trial error convergence under a given norm is established by the following well known result, see, e.g. [33].

Lemma 9.1 *Suppose that an ILC law of the form (9.79) is applied to an multiple-input multiple-output (MIMO) system described by (9.76). Suppose also that*

$$\|PQ(P^{-1} - L)\|_{\infty} = \gamma < 1 \tag{9.81}$$

then the ILC dynamics

$$E_{k+1} = PQ(P^{-1} - L)E_k + (I - PQP^{-1})\left((I - P)Y^{\text{ref}} - \tilde{P}D\right) \tag{9.82}$$

is stable and the trial-to-trial tracking error converges monotonically under a given norm, i.e.

$$\|E_{\infty} - E_{k+1}\|_2 \leq \gamma\|E_{\infty} - E_k\|_2 \tag{9.83}$$

to

$$E_{\infty} = \left(I - PQ(P^{-1} - L)\right)^{-1}(I - PQP^{-1})\left((I - P)Y^{\text{ref}} - \tilde{P}D\right) \tag{9.84}$$

For the application area considered, the use of an appropriately chosen Q-filter is especially important because the model used for design is an approximation of the distributed parameter dynamics.

The previous discussion in this book on how to select the Q-filter and the learning filter to achieve high-tracking accuracy over a wide-frequency range, fast and monotonic trial-to-trial error convergence under a given norm, appropriate robustness of the control system to modeling errors, and low sensitivity to measurement noise and initial conditions are equally applicable in this application. These are conflicting requirements (to some degree), and below a convex optimization problem in the frequency domain, which guarantees monotonic convergence of the tracking error vector from trial-to-trial under a given norm, is developed. This formulation assumes that the frequency responses $Q(e^{j\omega_i T_s})$ of the Q-filter have been constructed for the application under consideration.

The ILC design in this case is

$$\underset{L}{\text{minimize}} \sum_{\omega_i=\omega_{r1}}^{\omega_{r2}} \overline{\sigma}\left(M(e^{j\omega_i T_s})\right)$$

$$\text{subject to :}$$ (9.85)

$$\overline{\sigma}\left(M(e^{j\omega_i T_s})\right) \le M_{\max}, \quad \forall \omega_i \in [\omega_1 < \omega_2 < \cdots < \omega_N]$$

$$\overline{\sigma}\left(L(e^{j\omega_i T_s})\right) \le L_{\max}, \quad \forall \omega_i \in [\omega_1 < \omega_2 < \cdots < \omega_N]$$

with

$$M(e^{j\omega_i T_s}) = P(e^{j\omega_i T_s})Q(e^{j\omega_i T_s})\left(P^{-1}(e^{j\omega_i T_s}) - L(e^{j\omega_i T_s})\right)$$ (9.86)

Remark 9.1 In Section 6.3.2, a single-input single-output (SISO) design of a similar structure was given and experimentally validated for a system whose dynamics are governed by a finite-dimensional, state-space model. The core differences here are that the model used is a finite-dimensional approximation of distributed parameter dynamics and a MIMO system.

The optimization parameter $0 < M_{\max} < 1$ in this design is an upper bound on convergence rate γ (see (9.81)) and has been introduced to increase robustness against unmodeled dynamics. The second constraint of (9.85) limits the maximum gain (the maximum singular value) of the learning filter to an appropriately selected value $L_{\max}$, which again depends on the application considered. It has been added to reduce the impact of measurement noise and initial conditions on performance. In [115, 253], this problem has been solved by the introduction of a weighting function defined in the frequency domain, which complicates the design and is not required in this approach.

The constraints of (9.85) should be satisfied $\forall \omega$ in [0, ω_N], where ω_N is the Nyquist frequency. This requirement can be relaxed such that only a finite, but large, number of logarithmically spaced angular frequencies ω_i are considered (angular frequency gridding).

A consequence of limiting the maximum singular value of the learning filter could be a reduction in the trial-to-trial convergence speed, potentially over a wide range of frequencies. For a given frequency ω_i, the convergence speed depends on $\overline{\sigma}\left(M(e^{j\omega_i T_s})\right)$, where $\overline{\sigma}(\cdot)$ denotes the largest singular value of its matrix argument. In particular, reducing this value leads to a faster trial-to-trial error convergence. Moreover, in applications, it could be important to obtain a fast monotonic trial-to-trial error convergence rate of the tracking error under a specified norm over the frequency range for which the frequency spectrum of the reference signal has many components. In the optimization problem above, this is achieved by minimizing the sum of all $\overline{\sigma}\left(M(e^{j\omega_i T_s})\right)$ between the lower ω_{r1} and upper ω_{r2} limits of the angular frequency range for which the reference signal spectrum has components with significant magnitude, which is again application-dependent.

For application to the heating system, the location of the input and output signals of the approximate model (9.58) are marked in Figure 9.27 using the notation: $u_1 \equiv \dot{Q}_1(t)$ and $u_2 \equiv \dot{Q}_3(t)$ are the first and second entry of the input vector signal $u(t)$, respectively; $d_1 \equiv \vartheta_a(t)$, $d_2 \equiv \dot{Q}_2(t)$, and $d_3 \equiv \dot{Q}_4(t)$ are the entries in the disturbance vector signal $d(t)$; $y_1 \equiv \vartheta(\frac{3l}{8}, t)$ and $y_2 \equiv \vartheta(\frac{7l}{8}, t)$ are, respectively, the first and second entry of the output vector $y(t)$.

The numerical values for the system parameters of the experimental setup are as follows: $l = 0.320$ m, $b = 0.040$ m, $h = 0.012$ m, $\rho = 7800$ kg/m^3, $c_p = 420$ J/(kg K), $\alpha = 150$ W/(m^{2}K), and $\lambda = 55$ W/(m K). For these values, a minimal realization of (9.58) was constructed, resulting in a model with 10 states, which is nonminimum phase with two invariant zeros, both at $s = 0.1011$. Using the exact discretization method with a sampling period of $T_s = 1$ seconds results in a state-space model of the form (9.62), again with $x_{k+1}(p) \in \mathbb{R}^{10}$. This model is also nonminimum phase with two invariant zeros, both located at $z = 1.1065$. As in other areas of linear systems, nonminimum phase zeros have particular effects on ILC performance, see, e.g. [188].

The ambient temperature is a nonrepeatable disturbance that cannot be compensated by the ILC law. This temperature is slow-varying and therefore the disturbance observer can be used to compensate for its effect. For this reason, the disturbance model (9.64) is chosen as integral action, i.e. $A_d = C_d = I$. Also in [156], the following control law matrices were designed:

$$
K_s = \begin{bmatrix}
-0.0439 & -0.0491 \\
0.3642 & 0.1032 \\
0.2670 & 0.2436 \\
-0.2905 & -0.1920 \\
0.2442 & 0.1252 \\
0.3569 & 0.3568 \\
0.2495 & 0.3114 \\
0.1471 & 0.2493 \\
-0.0498 & -0.4204 \\
-0.1456 & 0.8236
\end{bmatrix}^{T}, \quad
K_L = \begin{bmatrix}
-0.6349 & 0.3741 \\
0.7399 & -0.4144 \\
0.9401 & -0.4358 \\
-0.6769 & 0.3089 \\
-0.1206 & 0.2423 \\
0.4635 & -0.0255 \\
-0.0629 & 0.5887 \\
0.1947 & 0.7508 \\
-0.5103 & -0.5924 \\
-0.1081 & 0.7205 \\
0.9831 & -0.4295 \\
0.4307 & 0.9887
\end{bmatrix}
$$

To achieve a small steady-state error vector on the first trial, the static feedforward gain matrix (see Figure 9.29) has been taken as the inverse of the DC gain matrix of the closed-loop feedback control system:

$$
N = \begin{bmatrix} 4.6935 & -2.4303 \\ 1.5513 & 4.6324 \end{bmatrix} \tag{9.87}
$$

Again, other choices are possible depending on the application and the performance requirements.

The particular structure of the Q and L-filters considered first have the form:

$$
Q(q) = \mathrm{diag}[Q_1(q)\ Q_2(q)] \tag{9.88}
$$

and

$$
L(q) = \mathrm{diag}[L_1(q)\ L_2(q)] \tag{9.89}
$$

Consider the matrix P for the controlled current trial feedback loop for the heating process written in the form:

$$
P = \begin{bmatrix} P_{11} & P_{12} \\ P_{21} & P_{22} \end{bmatrix}
$$

By using Bode gain plots for each of the four entries in this 2×2 frequency response matrix, the magnitudes of the off-diagonal entries are found to be smaller than those on the diagonal. Hence, (9.88) is further simplified to $Q = QI$, with the filter order n_{Qm} is chosen as 2, whereas the cut-off frequency is set to 0.1 rad/s. Above this frequency, the feedback control loop significantly damps the input vector $w_{k+1}(p)$. The maximum singular value of P and the magnitudes of P_{11} and P_{22} at this frequency are $\bar{\sigma}\left(P(e^{j0.1T_s})\right) \approx \|P_{11}(e^{j0.1T_s})\| \approx \|P_{22}(e^{j0.1T_s})\| \approx 0.05$. Given that $\|P_{11}\|$ and $\|P_{22}\|$ have similar values for frequencies from 10^{-4} rad to the Nyquist frequency; therefore, $Q_1 = Q_2$ is assumed.

The new ILC design has been applied in simulation to the heating process and the result experimentally validated. The results are given and discussed next. Then the relative performance against alternatives is described.

All optimization problems in the results that follow were solved using the MATLAB-based software CVX and SDPT3 for 1000 values of logarithmically spaced frequencies ω_i in the range $[10^{-4}, \omega_N = \frac{\pi}{T_s}]$ rad/s and $M_{max} = 0.9$ was chosen to obtain acceptable robustness.

Singular values of P below the frequency 10^{-4} rad/s and between the frequency samples ω_i do not change significantly. This guarantees that the model of the current trial feedback loop is appropriately determined and the constraints of (9.85) are satisfied below 10^{-4} rad/s and between the frequency samples ω_i.

The orders of $L_1(q)$ and $L_2(q)$ of the learning filter (9.89), relative to the order of the plant model (9.62), have been taken as $n_{L1} = n_{L2} = 10$. The model of the feedback control loop (9.70) is 22th order, where the 12 additional states are introduced by the state and disturbance observer (9.67).

One of the advantages relative to alternatives (see also the discussion at the end of this section) of the new design (9.85) is the ability to achieve a fast trial-to-trial error convergence over a frequency range of interest, i.e. ω_{r1} and ω_{r2}, respectively. In the current design, these values are selected by examining the frequency spectrum of the reference signals. Figure 9.30a shows the reference trajectories (black and gray solid line) applied during all design studies in this section together with amplitude spectrums of the reference trajectories (Figure 9.30b). These spectra have many components at low frequencies and, hence, $\omega_{r1} = 10^{-4}$ rad/s and $\omega_{r2} = 0.02$ rad/s were chosen.

Another advantage of the new design is to complement L with a limit on its maximum singular value to increase the robustness of the ILC design against measurement noise. The choice of this parameter is application-dependent, and for this application it is taken as approximately $2.5(\bar{\sigma}(P))^{-1}$ at the cut-off frequency of the Q-filter and results in $L_{max} = 50$. Completing the design gives the

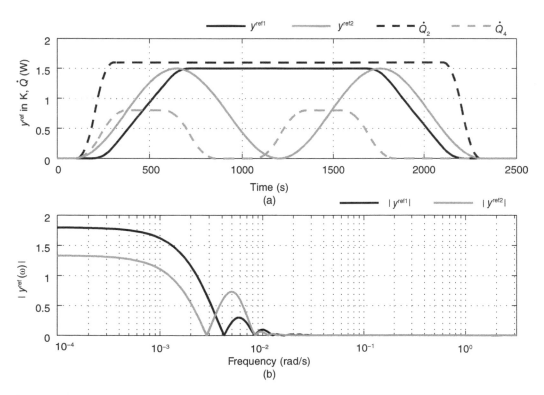

Figure 9.30 The reference and disturbance signals (a). Frequency spectrum of the reference signals (b).

entries in L of (9.89) as follows:

$$L_1(q) = 1.4050q^1 - 1.4796q^2 - 2.6064q^3 - 3.4666q^4$$
$$- 4.3478q^5 - 5.4692q^6 - 6.6882q^7 - 8.2280q^8$$
$$- 10.4490q^9 + 42.3133q^{10}$$
$$L_2(q) = 3.8432q^1 - 1.7307q^2 - 2.7839q^3 - 3.4920q^4$$
$$- 4.3099q^5 - 5.4891q^6 - 6.6589q^7 - 8.0460q^8$$
$$- 10.2179q^9 + 39.8697q^{10}$$

Figure 9.31 (black solid lines) shows the maximum singular values of M (Figure 9.31a) given by (9.86) and L (Figure 9.31b) for this design. The maximum singular values of M in the range $\omega \in [10^{-4}, 10^{-2}]$ rad/s are small, which should result in a fast convergence speed of the tracking error. The convergence rate $\gamma = \max\left(\overline{\sigma}(M)\right)$ is 0.9, i.e. equal to $M_{\max}$, due to the first constraint in (9.85). The maximum gain of the learning filter, i.e. $\max\left(\overline{\sigma}(L)\right)$, is equal to $L_{\max}$, due to the second constraint in (9.85).

All designs in this section have been validated over 10 trials with the zero boundary conditions, i.e. $v_0(p) = 0$ and $e_0(p) = 0$. The reference trajectories and the disturbance signals are shown in Figure 9.30. Also, the disturbance signals (given by black and gray dashed line in Figure 9.30) are the heat flows generated by the actuators connected to the second and fourth sections of the rod, where these sections also contain the measurements of the outputs. The ambient temperature, i.e. the entry in the disturbance vector d_{k+1} in (9.62), was set to zero for the simulations. To highlight

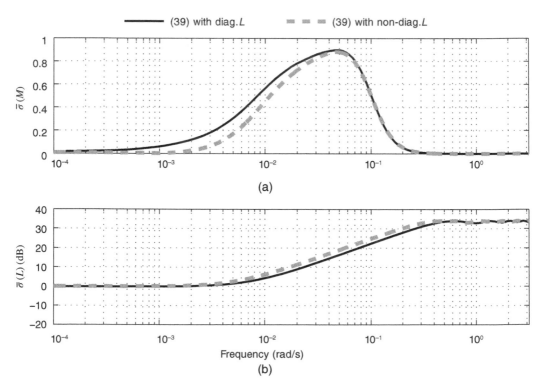

Figure 9.31 Maximum singular values of M (a, linear scale on the vertical axis) and of L (b) for the new design 9.85 for the diagonal and nondiagonal forms of the learning filter transfer-function matrix.

the compensation properties of the design, the disturbance signals were applied starting from the fifth trial, i.e. on trial $k = 5, \ldots, 10$.

The root-mean-square (RMS) value of the tracking error, abbreviated by $RMS(e_k)$, i.e. in the MIMO case

$$RMS(e_k) = \sqrt{\frac{1}{\alpha} \sum_{p=1}^{\alpha} e_k^T(p) e_k(p)}$$

is used to compare the trial-to-trial convergence of all the designs in this section, where $\alpha = 2500$ is the trial length.

The black solid line in Figure 9.32 shows the $RMS(e_k)$ resulting from the simulation of the controlled system, demonstrating the fast convergence of the tracking error. Also, the increase in the error due to the introduction of the disturbance from the fifth trial and for succeeding trials is also quickly and monotonically reduced.

To validate these simulation predictions, an experimental program was also completed. The ILC signals, input signals, output signals, and the tracking errors progressions recorded during these experiments are shown in Figures 9.33–9.36. These indicate a strong agreement with the simulation results as does the experimental $RMS(e_k)$ of the error also shown in Figure 9.32 (black dashed line), where the differences between the simulated and experimental results are due to measurement noise and modeling errors. Moreover, the temperature in the laboratory increased as the experiments progressed, i.e. the ambient temperature changed during the experiments but the simulations assumed a constant ambient temperature.

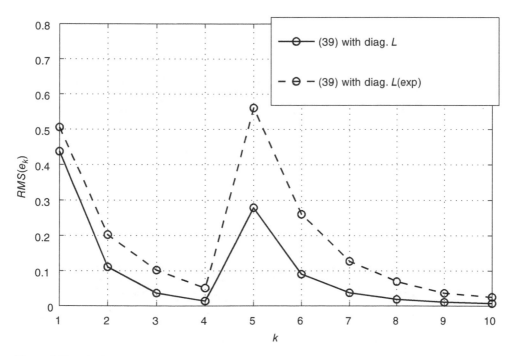

Figure 9.32 $RMS(e_k)$ values of the tracking error vector for the new design, where L(exp) denotes the experimentally measured results.

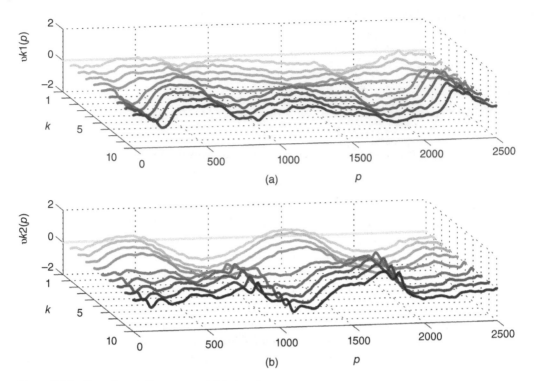

Figure 9.33 Experimentally measured ILC signals.

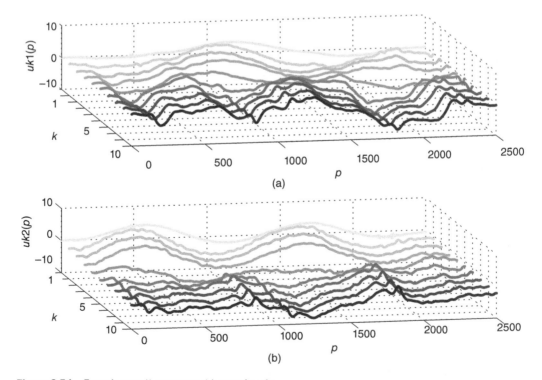

Figure 9.34 Experimentally measured input signals.

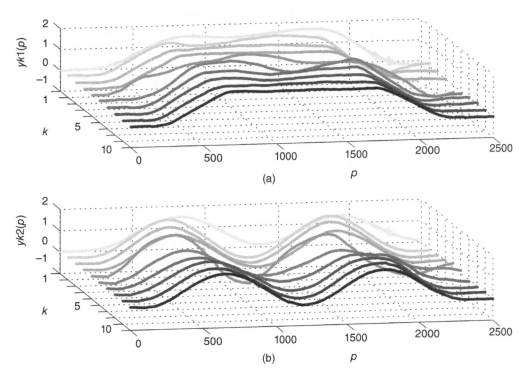

Figure 9.35 Experimentally measured output signals.

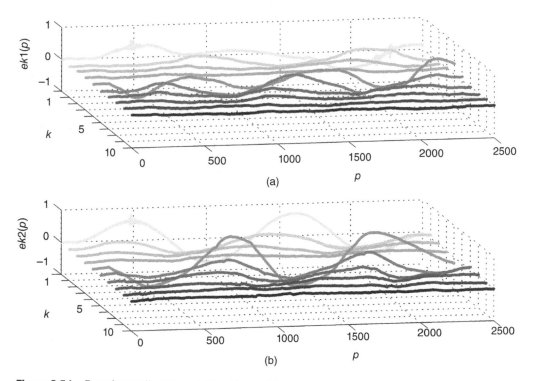

Figure 9.36 Experimentally measured tracking errors.

These results confirm that this ILC design does not require knowledge of the exact plant model and repetitive disturbance signals to achieve precision tracking. Moreover, the effects of nonrepetitive, slow-varying disturbances in the form of changes of the ambient temperature are compensated by the disturbance observer.

In Figure 9.30, $y^{\text{ref}1}(0) = y^{\text{ref}2}(0) = 0$, i.e. equal to the initial ambient temperature at the beginning of the first trial (temperatures are taken as relative to the initial ambient value). During the experiment, the ambient temperature increased and, at the beginning of the subsequent trials, $y^{\text{ref}1}(0)$ and $y^{\text{ref}2}(0)$ were lower than the ambient temperature. This means that without the current trial feedback loop, between the end of one trial and the start of the next one, the rod temperature would change due to changes in the ambient temperature and an initial tracking error would occur starting on the second trial. However, the feedback control loop together with the disturbance observer are active between the trials and compensated this unwanted effect, resulting in zero initial error on each trial.

The experimental results in this section are the first available and further development, including the use of a nondiagonal L filter, with supporting experimental validation, are given in [156]. An extensive literature search has not revealed any other reports of the experimental validation of ILC laws for distributed parameter systems.

9.3 Finite Element and Sequential Experimental Design-based ILC

This section develops an ILC design for distributed parameter systems based on discretization. The design combines the finite element method (FEM), widely used in many applications, with a sequential experiment design, allowing the collection of the most informative data on the dynamics of the distributed system on each trial. In this setting, both the process model and the control performance are improved from trial-to-trial.

The class of problems considered arises in the structural mechanics of smart structures, which apply to various engineering problems. These include smart skins for aircraft and spacecraft and ultra-precision shape-controlled positioners. One source for further background with a control content is [20]. Advances in materials science have led to many structural systems that deploy advanced sensors, and actuators and are coupled with suitable control laws. One possibility among many leads to intelligent structures that can detect impending failures.

As one example, a class of systems described by a hyperbolic PDE system is considered. Specifically, let t lie in a bounded time interval and $\Omega \subset \mathbb{R}^2$ be a bounded spatial domain with a sufficiently smooth boundary $\partial\Omega$. Then the class of systems considered have a scalar state $y(x, t)$ described by the temporal biharmonic equation [212]

$$\rho \frac{\partial^2 y(x, t)}{\partial t^2} + \kappa \nabla^4 y(x, t) = p(x, t) \tag{9.90}$$

where y is the transverse displacement, ρ is the mass density per unit area, κ is the elasticity coefficient, and p is a pressure field representing the external force distributed over the domain Ω (in general ρ and κ may be spatial functions). Also, for control design, the displacement y, a spatial curvature, is used as an additional source of output information on the system dynamics.

For analysis, (9.90) is rewritten as the following equivalent pair of the second-order equations:

$$\nabla^2 y(x, t) = w(x, t)$$
$$\rho \frac{\partial^2 y(x, t)}{\partial t^2} + \kappa \nabla^2 w(x, t) = p(x, t), \quad x \in \Omega \tag{9.91}$$

where w is the surface curvature. The first of these equations is of the Poisson type involving the Laplacian of the original unknown y, and the second gives the right-hand side of the biharmonic equation (9.90). Together they are a coupled second-order elliptic-hyperbolic description of the dynamics, to which the generalized Neumann conditions are added, i.e.

$$\vec{n} \cdot \nabla y(x, t) + q_{11}y(x, t) + q_{12}w(x, t) = g_1, \quad x \in \partial\Omega$$
$$\vec{n} \cdot \nabla w(x, t) + q_{21}y(x, t) + q_{22}w(x, t) = g_2, \quad x \in \partial\Omega \tag{9.92}$$

and the initial conditions are taken as follows:

$$y(x, 0) = y_0, \quad \dot{y}(x, 0) = y_0', \quad x \in \Omega$$
$$w(x, 0) = w_0, \quad \dot{w}(x, 0) = w_0', \quad x \in \Omega \tag{9.93}$$

where $\vec{n}$ is an outward normal vector to the boundary $\partial\Omega$.

9.3.1 Finite Element Discretization

In many applications, a critical task is to discretize a system such as (9.91) in the spatial variables to enable a control system design that is efficient and computationally tractable. In this section, the method of line semidiscretization based on finite element modeling is used, which is standard in the general subject area. The novelty in what follows is the construction of an accurate approximation of the continuous-time distributed system. Moreover, the application of an ILC law also allows trial-to-trial improvement of the model.

The weak form of the dynamics is considered first. The generalized Neumann conditions are imposed over the whole boundary without loss of generality since their Neumann counterparts can approximate the Dirichlet conditions. To simplify notation, let $y = y(x, t)$ and $w = w(x, t)$ denote the solution of the PDE system (9.91). Then multiplying these equations by arbitrary test functions $v_1 = v_1(x, t)$ and $v_2 = v_2(x, t)$ and integrating over Ω gives

$$\int_\Omega (\nabla^2 y)v_1 \, dx = \int_\Omega wv_1 \, dx$$
$$\int_\Omega \rho \frac{\partial^2 y}{\partial t^2} v_2 \, dx + \int_\Omega (\kappa \nabla^2 w)v_2 \, dx = \int_\Omega p(x, t)v_2 \, dx \tag{9.94}$$

Integrating by parts and using Green's formula (where $\cdot$ denotes the inner product) gives

$$\int_\Omega -\nabla y \cdot \nabla v_1 \, dx + \int_{\partial\Omega} \vec{n} \cdot \nabla y(x, t)ds = \int_\Omega wv_1 \, dx,$$
$$\int_\Omega \rho \frac{\partial^2 y}{\partial t^2} v_2 \, dx + \int_\Omega -\kappa\nabla^2 w \cdot \nabla v_2 \, dx + \int_{\partial\Omega} \vec{n} \cdot \nabla w(x, t) \, ds = \int_\Omega p(x, t)v_2 \, dx \tag{9.95}$$

where ds is an elementary arc length of the boundary curve $\partial\Omega$. Next, using the boundary conditions (9.92) to replace boundary integrals, the original problem can be replaced by one of finding y and w that satisfy

$$\int_\Omega (-\nabla y \cdot \nabla v_1 - wv_1)dx + \int_{\partial\Omega}(g_1 - q_{11}y - q_{12}w)v_1 \, ds = 0$$
$$\int_\Omega \rho \frac{\partial^2 y}{\partial t^2} v_2 \, dx + \int_\Omega(-\kappa\nabla w \cdot \nabla v_2 - p(x, t)v_2)dx$$
$$+ \int_{\partial\Omega}(g_2 - q_{21}y - q_{22}w)v_2 \, ds = 0 \tag{9.96}$$

This last equation is known as the variational form, and any solution of the original system (9.91) is also a solution of the variational problem. Moreover, the variational problem's solution is also termed the weak solution, and the solution pair y, w and the test functions v_1, v_2 belong to some function space $\mathcal{V}$. The next step is to project the weak system (9.96) onto the N_p-dimensional function space $\mathcal{V}_{N_p} \subset \mathcal{V}$ to obtain the finite-dimensional approximation in $\mathcal{V}_{N_p}$ that lies closest to the weak solution in terms of the energy norm. Convergence is guaranteed if the space $\mathcal{V}_{N_p}$ tends to $\mathcal{V}$ as $N_p \to \infty$.

Using an N_p-element spatial mesh on Ω and time $t \in T$ and since the differential operator is linear, it is required that the variational equation is satisfied for the N_p test functions $\phi_i \in \mathcal{V}_{N_p}$ that form a basis of this space. Further, for simplicity, the same basis is used for v_1 and v_2 and expanding the solution as a function of x in the finite element basis gives

$$y(x,t) = \sum_{j=1}^{N_p} Y_j(t)\phi_j(x), \quad w(x,t) = \sum_{j=1}^{N_p} W_j(t)\phi_j(x) \tag{9.97}$$

where Y_j and W_j are assumed to be continuously differentiable functions. On the completion of this last step, the following differential-algebraic system of equations are obtained for $i = 1, \ldots, N_p$:

$$\sum_{j=1}^{N_p} \left((K_{ij} + Q_{ij}^{11})Y_j(t) + (N_{ij} + Q_{ij}^{12})W_j(t) \right) = G_i^1$$

$$\sum_{j=1}^{N_p} M_{ij}\frac{d^2 Y_j(t)}{dt^2} + \sum_{j=1}^{N_p} \left((L_{ij} + Q_{ij}^{22})W_j(t) + Q_{ij}^{21}Y_j(t) \right) = G_i^2 + F_i \tag{9.98}$$

where

$$K_{ij} = \int_\Omega \nabla\phi_j \cdot \nabla\phi_i \, dx, \quad N_{ij} = \int_\Omega \phi_j\phi_i \, dx$$

$$L_{ij} = \int_\Omega \kappa\nabla\phi_j \cdot \nabla\phi_i \, dx, \quad Q_{ij}^{k\ell} = \int_{\partial\Omega} q_{k\ell}\phi_j \, ds$$

$$F_i = \int_\Omega p(x,t)\phi_i \, dx, \quad G_i^\ell = \int_{\partial\Omega} g_\ell\phi_i \, ds$$

$$M_{ij} = \int_\Omega \rho\phi_j\phi_i \, dx \tag{9.99}$$

In this last equation set, K, N, L, Q, M are $N_p \times N_p$ matrices and F, G, are $N_p \times 1$ vectors.

Many choices for the test-function spaces have been reported in the literature. In most cases, the software employed uses continuous functions that are linear on each element of a two-dimensional mesh. Piecewise linearity guarantees that all the integrals defining the matrices (9.99) exist. Moreover, projection onto $\mathcal{V}_{N_p}$ is linear interpolation, and the evaluation of the solution inside an element can be interpreted in terms of the nodal values of the mesh. If the mesh is uniformly refined, $\mathcal{V}_{N_p}$ approximates the set of smooth functions on Ω.

A commonly used basis for $\mathcal{V}_{N_p}$ in 2D domains is the set of so-called "tent" or "hat" functions ϕ_i. These are linear for each element and have the property that $\phi_i(x^j) = \delta_{ij}$, where δ_{ij} denotes the Kronecker delta function, i.e. this function is zero at all nodes of the mesh except for x^i, where it has value 1. Hence, the following property holds for the solution y (and similarly for w):

$$y(x^i) = \sum_{j=1}^{N_p} Y_j(t)\phi_j(x^i) = Y_i(t)$$

i.e. by solving the FEM system, the nodal values of the approximate solution are obtained. The immediate consequence is that the integrals in (9.99) only need to be computed for the elements that contain the node x^i. Also, K_{ij}, L_{ij}, and M_{ij} are zero unless x^i and x^j are vertices of the same element.

The integrals in the FEM matrices are computed by adding the contributions from each element to the corresponding entries, i.e. only if the corresponding mesh point is a vertex of the element. This operation is commonly called assembling. The so-called "local matrices" are computed for each element, and their components are added to the corresponding elements in the sparse matrices or vectors. The calculation of the FEM integrals for the triangular 2D meshes and piecewise linear test functions is somewhat technical in nature. Further details are omitted with the note that they can be obtained relatively simply by using existing FEM-based solvers, e.g. the assempde function from the MATLAB PDE Toolbox.

9.3.2 Application of ILC

In this application area, it is assumed that the state vector $\chi = \begin{bmatrix} y & w \end{bmatrix}^T$ on trial k is continuously observed over the trial length by $\hat{N}$ pointwise sensors. (The relaxation of this assumption is an open research question.) The measurement process is represented by, see, e.g. [206]

$$z_k^j(t) = C\chi(x^j, u_k, t; \theta) + \varepsilon_k(x^j, t), \quad 0 \le t \le \alpha, \quad j = 1, \dots, \hat{N} \tag{9.100}$$

where $z_k^j(t)$ is the measurement output and $x \in X$ denotes the location of the jth sensor, X denotes the region of Ω, where measurements can be made, C is the output matrix, θ is an unknown vector of system parameters to be identified based on the observations, u_k is the control input and ε_k denotes the measurement noise. As in many other applications, the noise is assumed to be a zero-mean Gaussian uncorrelated random process, i.e. $E[\varepsilon(t)] = 0$, $\text{var}(\varepsilon(t)) = \sigma^2$ [257], where E denotes the expectation operator.

It is assumed that the pressure field on trial k is a function of u_k, i.e. $p_k(x, t) = p(x, u_k(t), t)$. Hence, the control objective is to tune the input u_k over consecutive trials such that the output $z(\cdot)$ tracks a specified differentiable trajectory $r(t)$ as accurately as possible, i.e. the tracking error $e_k^j(t) = r^j(t) - z_k^j(t)$ for any $j = 1, \dots, \hat{N}$, converges from trial-to-trial, ideally to zero, but often in applications, this is replaced as to within a suitably small value as measured by an appropriate norm.

Introducing the augmented state vector $\hat{\chi}(t) = \begin{bmatrix} Y(t) & W(t) & \dot{Y}(t) & \dot{W}(t) \end{bmatrix}^T$ and using (9.98), gives the equivalent system of first-order differential–discrete equations for trial k as follows:

$$M_f \dot{\hat{\chi}}(t) = A_f \hat{\chi}(t) + B_f(u_k) + G_f \tag{9.101}$$

where

$$A_f = \begin{bmatrix} 0 & 0 & I & 0 \\ 0 & 0 & 0 & I \\ K + Q^{11} & N + Q^{12} & 0 & 0 \\ L + Q^{22} & Q^{21} & 0 & 0 \end{bmatrix}, \quad B_f(u_k) = \begin{bmatrix} 0 \\ 0 \\ 0 \\ F(u_k) \end{bmatrix}$$

$$M_f = \begin{bmatrix} I & 0 & 0 & 0 \\ 0 & I & 0 & 0 \\ 0 & 0 & 0 & 0 \\ 0 & 0 & M & 0 \end{bmatrix}, \quad G_f = \begin{bmatrix} 0 \\ 0 \\ G^1 \\ G^2 \end{bmatrix}$$

and matrix $F(u_k)$ is F calculated for $p_k(x, t)$.

Suppose that $p_k(x, t)$ depends linearly on u_k, and hence, (9.101) is linear. This lumped parameter approximation is especially suited for implementing D-type ILC laws as the solutions of (9.101) in this case can be used to approximate the derivatives of the error. The system, in this case, is ill-conditioned and sparse, and therefore, explicit time integrators are forced by the stability requirements to have very short time steps. Also, implicit solvers can be computationally expensive since they solve an elliptic problem at each time step.

9.3.3 Optimal Measurement Data Selection

To compensate for random disturbances, the control scheme is augmented by the calibration of the model. The quality of the model-estimated parameters directly influences the trial-to-trial error convergence rate. Hence, an appropriate set of measurement data on each trial is of critical importance.

Based on (9.100), the identification problem is: Given the model (9.91)–(9.93) and the measurements $z_k^j(t)$ estimate the parameter vector θ, with $\hat{\theta}$ denoting a global minimizer of the least-squares criterion. Since the covariance matrix, $\text{cov}(\hat{\theta})$, of the least-squares estimator is strongly dependent on the subset of chosen observations, a criterion, denoted by Ψ, quantifying the "goodness" of different measurements configurations is required. In this section, the route is via the Fisher information matrix, which is widely used in optimum experimental design theory, see, e.g. [16, 257], since its inverse is a good approximation of $\text{cov}(\hat{\theta})$.

The optimal measurement scheduling problem consists of choosing a subset $n < \hat{N}$ of these locations, from among all potential sites where the sensors can be placed, which provide the "best" information concerning the system dynamics from all possible sites where the sensors can be placed. The motivation for not using all the available sensors positions is reduced measurement system complexity and maintenance costs. Also, observed data with high noise-to-signal ratios must be removed to increase the estimation quality.

Introduce for each sensor i a binary variable π_i with value 1 or 0 depending on whether or not a sensor is selected for estimation. Then with $\pi = [\pi_1, \pi_2, \ldots, \pi_{\hat{N}}]$, the so-called "average" per measurement Fisher information matrix is given by

$$P(\pi) = \frac{1}{\hat{N}} \sum_{i=1}^{\hat{N}} \pi_i \frac{1}{t_f} \int_T g(t) g^T(t) dt \tag{9.102}$$

where

$$g(t) = \left[\frac{\partial \hat{\zeta}(t;\theta)}{\partial \theta_1}, \quad \ldots, \quad \frac{\partial \hat{\zeta}(t;\theta)}{\partial \theta_m} \right]_{\theta=\theta^0} \tag{9.103}$$

denotes the sensitivity vector. Also, since in the nonlinear case g depends on the estimated parameters, a preliminary estimate, denoted by θ^0, is required for its calculation. Usually, some known nominal values of the parameters θ can be used or estimates obtained from previous trials [206].

Various alternatives for the design criterion, denoted by Ψ operating on the Fisher information matrix have been reported, see, e.g. [16]. Of these, the log-determinant of this matrix, termed a *D*-optimality criterion, is most commonly used. Maximizing this quantity results in a sensor configuration corresponding to the minimum volume of the uncertainty ellipsoid for the parameter estimates. Also, the measurement selection problem can be stated as follows:

Find a vector π to maximize $\Psi(P(\pi))$, subject to

$$\sum_{i=1}^{\hat{N}} \pi_i = n \tag{9.104}$$

$$\pi_i = 0 \text{ or } 1, \quad i = 1, \ldots, \hat{N} \tag{9.105}$$

This last formulation is a 0–1 integer programming problem that requires an original and efficient solution. The development of such solutions is a longstanding research area. For small- or moderate-sized problems, the branch-and-bound scheme and extensions, see, e.g. [205, 207, 258] drastically reduces the search space, and a simple and efficient computational scheme is also available from this previous research. This scheme is not suitable for a large number of candidate observation sites, and in such cases, clusterization-free designs are effective [208, 209]. These last designs' key idea is to operate on the density of sensors per unit area instead of their positions.

The spatial sensor configuration significantly influences the estimation process's quality and, hence, the control design. One of the main difficulties associated with optimizing the measurement schedule is the dependence of the solutions of the problem considered on the actual values θ_{true} of the parameters to be estimated in addition to the control inputs. Since these values are unknown, a common approach is to construct the locally optimal designs described above for some prior estimate, θ^0, of θ_{true}. This construction improves the robustness of the design to model uncertainty. The spatial sensor configuration significantly influences the estimation process's quality and hence on the control design.

Using ILC, it is possible to use sequential experiment design [206, 257], where the block diagram representation of the overall control scheme is given in Figure 9.37. (This case can be interpreted as an extension of the control scheme developed for lumped parameter systems [133]). In particular, the identification part can be undertaken during each trial based on the measurements from sensors already collected, and the process model is iteratively tuned to higher accuracy. The control design becomes easier and faster in the next step as it benefits from the improved model. Next, a simulation-based example illustrates the new design.

Consider a repetitive transverse load applied to a thin piezoelectric membrane of thickness $h = 0.002$ m, separating the chambers of a vacuum furnace of different pressure. The membrane has an elliptic shape with normalized dimensions and a rectangular hole in its interior, see Figure 9.38. This surface distorts the out-of-plane direction due to external and internal forces arising from an applied load or the electric potential. The process has dynamics governed by (9.91) and the initial values of the system parameters are density per unit area $\rho = 5.4$ kg/m^2 and elasticity coefficient $\kappa = 50.54$ N m.

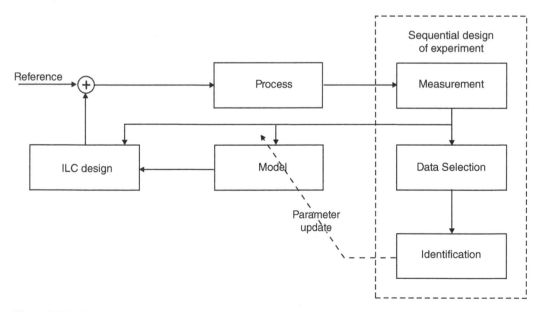

Figure 9.37 Block diagram representation of ILC with sequential experiment design.

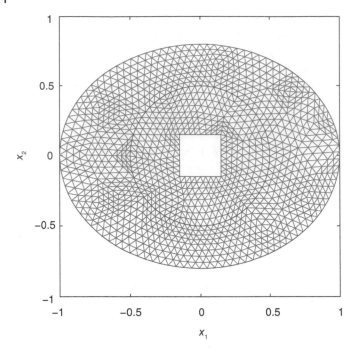

Figure 9.38 Triangular mesh of the elliptic membrane (semiaxes of lengths 1 and 0.8) with a square hole (side length 0.2) (normalized units).

Zero initial conditions are assumed for the distortion and curvature and their temporal derivatives. The external elliptic boundary of the membrane is clamped (both the distortion and its derivative are zero), and on the internal rectangular edge membrane, the following Neumann conditions hold:

$$
\begin{aligned}
y(x,t) &= 0, && \text{on the external ellipse} \\
\vec{n} \cdot \nabla y(x,t) &= 0, && \text{on the external ellipse} \\
\vec{n} \cdot \nabla y(x,t) &= 0, && \text{on the internal square} \\
\vec{n} \cdot \nabla w(x,t) &= 0, && \text{on the whole boundary}
\end{aligned}
\tag{9.106}
$$

Also, the Dirichlet boundary conditions can be embedded in the generalized Neumann conditions, e.g. setting $g_1 = q_{11}r$ and then letting $q_{11} \to \infty$ yields the Dirichlet condition since division by a very large q_{11}, results in very small normal derivative terms.

The objective is to design a temporal pressure field $p_k(x,t)$ that, when applied to the membrane over consecutive trials, achieves the reference displacement, which is the spatial profile of an elliptic paraboloid with magnitude increasing linearly for five seconds and then vanishing within the next five seconds, i.e.

$$
r = 10^{-2}(1 - \|t - 5\|/5) - (x_1^2 + (x_2/0.8)^2 - 1), \quad t \in [0,10]
$$

and illustrated in the first row of Figure 9.39.

The first step is spatial discretization of (9.101) for a spatial mesh composed of 2432 triangles and $N_p = 1284$ points, see Figure 9.38. Then the solution of resulting discrete algebraic systems was constructed for an evenly partitioned time interval (50 subintervals). It was assumed that the measurements could be made on the nodes of spatial mesh (i.e. $N = N_p$). Two scenarios were investigated, the first of which is applying a D-type ILC law with additional calibration of the model

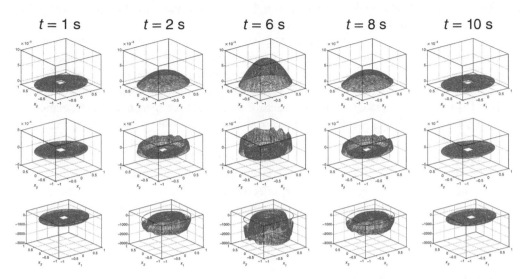

Figure 9.39 Temporal changes in the reference transverse displacement of the membrane (upper row) the final error response after 50 trials of ILC (middle row), and pressure applied (lower row).

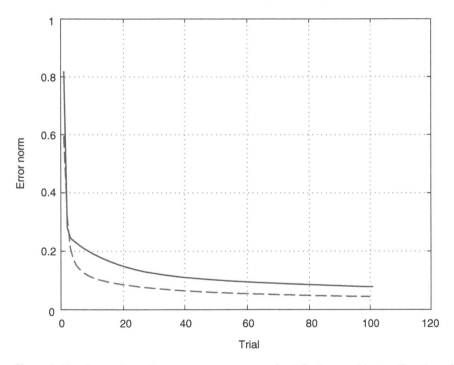

Figure 9.40 Comparison of error norm convergence for ILC scheme without calibration of the model (dark gray solid line) versus ILC with data selection for model calibration (light gray dashed line).

parameters $\theta = (\rho, \kappa)$. The second is where the number of sensor nodes used in the identification is reduced, represented by the D-optimal $n = 400$ element subset.

As a numerical example, 100 trials with the ILC law in place were performed (within approximately three minutes of computation time in the first scenario). Figure 9.39 illustrates the resulting complex process dynamics with spatial error distribution and the generated final pressure input.

Figure 9.40 shows the convergence solution produced by both ILC laws and demonstrates the potential of D-type ILC, a relatively simple structure law, to deliver acceptable performance for large-scale discretized models. Increasing the model calibration will significantly increase the trial-to-trial error convergence rate but at the cost of additional computational effort.

These are the first results on the control of PDEs based on ILC and sequential experimental design where the FEM is used to discretize the continuous dynamics. The example demonstrates that even simple structure ILC laws can be very effective in this general area and with acceptable computational cost. Despite these promising results, much further research is required to assess its potential fully.

9.4 Concluding Remarks

This chapter has considered the application of ILC to systems described by partial differential equations, including experimental validation. The focus has been on designs where a finite-dimensional approximate model of dynamics is constructed and used. For both applications areas considered, and many others, this is the most physically relevant approach. The last section establishes links to finite element models of the dynamics and sequential experimental design. This approach, in turn, could, with further research, answer the problem of where to place sensors and actuators for maximum effect.

The next chapter is one of two on nonlinear model-based ILC design.

10

Nonlinear ILC

A considerable volume of research has been reported on iterative learning control (ILC) for nonlinear systems. In the main, the analysis base is suitably modified theory for standard systems, e.g. feedback linearization, input–output linearization, backstepping, and extremum-seeking control, where one starting point for the literature is [284]. This chapter first describes an application of ILC to path following for center-articulated industrial vehicles based on feedback linearization applied to the dynamic model and for which supporting experimental results are available. Next, an input–output linearization approach for the application of ILC to stroke rehabilitation is given with supporting experimental results. The following section gives a gap metric approach to ILC design for robotic-assisted stroke rehabilitation, where for this application a particular strength is addressing an area where the difficulty in obtaining accurate models is very challenging. Finally, a general Lyapunov-based analysis is developed, and an extremum-seeking setting is briefly considered.

10.1 Feedback Linearized ILC for Center-Articulated Industrial Vehicles

The workplaces for these vehicles are often very challenging in engineering terms and coupled with requirements for very high throughput. In some industrial applications, e.g. mining, large-articulated vehicles find use to move (or haul) material between two locations, see, e.g. [159]. In many cases, the vehicles repeatedly drive the same route, where each is of finite duration. Hence, a possible application area for ILC, where the dynamics are nonlinear and design based on approximating the dynamics by a linear model, involves gross over-simplification.

The results in this section are from [62], where ILC was combined with a baseline path-following controller, and the resulting design has been experimentally validated. Considering Figure 10.1, where the notation in [62] is used throughout this section, the system configuration is defined by $q = (x, y, \theta, \phi)$, where $(x, y) \in \mathbb{R}^2$ is the position of the front component, F, $\theta \in \mathbb{S}^1$ is the heading angle of the front component, and $\phi \in \mathbb{S}^1$ is the steering (or articulation) angle, where $\mathbb{S}^1$ denotes a circle. At F the forward vehicle speed is denoted by $v \in \mathbb{R}$ and $\omega \in \mathbb{R}$ denotes the steering rate. Hence, $\omega(t) = \dot{\phi}(t)$.

Iterative Learning Control Algorithms and Experimental Benchmarking, First Edition.
Eric Rogers, Bing Chu, Christopher Freeman and Paul Lewin.
© 2023 John Wiley & Sons Ltd. Published 2023 by John Wiley & Sons Ltd.

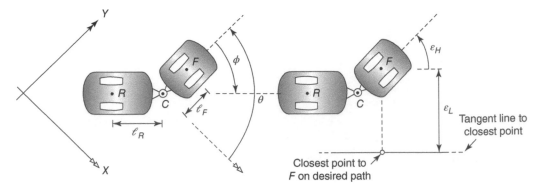

Figure 10.1 Vehicle geometric configuration (left) and path-following errors (right).

Given this notation, a kinematic model of the vehicle is

$$\dot{q} = \begin{bmatrix} \cos\theta & 0 \\ \sin\theta & 0 \\ -\dfrac{\sin\phi}{l(\phi)} & -\dfrac{l_R}{l(\phi)} \\ 0 & 1 \end{bmatrix} \begin{bmatrix} v \\ \omega \end{bmatrix} \tag{10.1}$$

where $l(\phi) = l_R + l_F \cos\phi$. In this setup, positive values of v and ω correspond, respectively, to forward movement and steering to the right.

In [62] path following is by using the errors ϵ_L and ϵ_H, where ϵ_L is the distance from F to the tangent to the desired path at the closest discrete point along the path. Also, ϵ_L is the heading error, i.e. the difference between the angle (in the configuration space) of this instantaneous target and the vehicle heading θ. In the experimental work, the error states were repeatedly computed using the unscented Kalman filter-based system described in [62]. (Stochastic ILC design is considered in Chapter 12.) This design employs feedback linearization, a well-known approach in nonlinear control systems, where one source for a detailed treatment of the basics is [116].

Assuming that the tangent to the desired path instantaneously corresponding with the X-axis in Figure 10.1 (right-hand side) incurs no loss of generality. Hence, $\epsilon_L \equiv y$, and $\epsilon_H \equiv \theta$ and the error dynamics (using (10.1)) can be written as follows:

$$\dot{\epsilon}_L = v \sin\epsilon_H$$
$$\dot{\epsilon}_H = -v\frac{\sin\phi}{l(\phi)} - \frac{l_R}{l(\phi)}\omega \tag{10.2}$$

Also introduce the nonlinear change of variables

$$z_1 = \epsilon_L$$
$$z_2 = v\sin\epsilon_H \tag{10.3}$$

to give the error dynamics

$$\begin{bmatrix} \dot{z}_1 \\ \dot{z}_2 \end{bmatrix} = \begin{bmatrix} 0 & 1 \\ 0 & 0 \end{bmatrix} \begin{bmatrix} z_1 \\ z_2 \end{bmatrix} + \begin{bmatrix} 0 \\ 1 \end{bmatrix} \eta \tag{10.4}$$

where η, the transformed control input, is

$$\eta = -\frac{v^2 \sin \phi \cos \epsilon_H}{l(\phi)} - \frac{l_R v \cos \epsilon_H}{l(\phi)} \omega \tag{10.5}$$

To stabilize this last system, a proportional plus derivative control law is one option, i.e. $\eta = \begin{bmatrix} k_P & k_D \end{bmatrix} z$, where $k_i < 0$, $i = 1, 2$.

The control input in the variable ω is obtained from (10.5) (i.e. back computed by inverting this transform) as follows:

$$\omega = -\frac{v \sin \phi}{l_R} - \frac{l(\phi)}{l_R v \cos \epsilon_H} \eta \tag{10.6}$$

Also, the heading error is limited to $\epsilon_H \in (-\frac{\pi}{2}, \frac{\pi}{2})$. At this stage, the application of ILC can begin.

In this application, various versions of ILC combined with a feedback control loop have been investigated. The base case is where ILC is used to augment the path-following controller by an additive correction to the steering angle, i.e. the steering rate. Moreover, the ILC signal is for a specific path or route and must be redesigned if it changes. Figure 10.2 shows a block diagram representation of the feedback loop combined with the ILC design, where the state vector used is defined in (10.3).

This formulation differs from previous research on the application of ILC in this area [184], where ω is directly computed as the feedforward signal. In the considered design, the ILC contribution is applied to the transformed input (10.5) before the nonlinear transformation to obtain the steering input ω is used. Consequently, ω given by (10.5) does not change with time, i.e.

$$\omega = -\frac{v \sin \phi}{l_R} - \frac{l(\phi)}{l_R v \cos \epsilon_H} (\eta + \tilde{\eta}) \tag{10.7}$$

where $\tilde{\eta}$ denotes the correction generated by the ILC law.

In this application area, the ILC law cannot be implemented over the trial length by time increments in path-following, since the trial error cannot be defined as a function of time because the vehicle will have a different pose at the same time for every subsequent trial. For discrete dynamics, let i denote the time index when the corresponding desired path point is reached

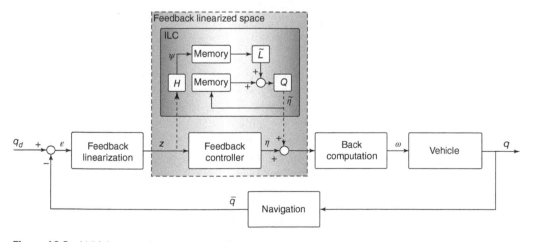

Figure 10.2 Vehicle control system block diagram.

and spatially index the system at path points in the X–Y plane (spatial ILC is also considered in Chapter 13). This approach has similarities to point-to-point ILC, see, e.g. [51]. The system state in discrete time is $q(i) = q(t = i\tilde{T})$, $i = 0, 1, 2, \dots$, where $\tilde{T} = hT$, and h denotes the number of discrete time steps between each path point.

The discrete state-space model of the linearized dynamics is

$$\begin{bmatrix} z_1(i+1) \\ z_2(i+1) \end{bmatrix} = \begin{bmatrix} 1 & \tilde{T} \\ 0 & 1 \end{bmatrix} \begin{bmatrix} z_1(i) \\ z_2(i) \end{bmatrix} + \begin{bmatrix} 0 \\ \tilde{T} \end{bmatrix} \eta(i)$$

$$\psi(i) = \begin{bmatrix} 1 & 0 \end{bmatrix} z(i) \tag{10.8}$$

where $\psi(i)$ is the output error to be learned, i.e. the lateral path-following error ϵ_L.

In this application, the ILC and feedback linearization's combined action is to learn $\tilde{\eta}(i)$ over the trials, $j \geq 1$, where this additive correction is applied to the real-time control inputs. Given $\eta = Kz$ and using (10.8), and letting (F, G, H) denote the state, input, and output matrices in (10.8), the error on trial j is

$$z_j(i+1) = Fz_j(i) + G(\eta_j(i) + \tilde{\eta}_j(i))$$

$$= \tilde{F}z_j(i) + G\tilde{\eta}_j(i), \quad \tilde{F} = F + GK \tag{10.9}$$

and it is assumed that $z_j(0) = z_0$ for all trials. Also assuming a desired system output $\psi_d = 0$, the performance error is (for an N-sample system)

$$e_j(i) = \psi_d(i) - \psi_j(i) = -\psi_j(i), \quad i = 2, 3, 4, \dots, N+1 \tag{10.10}$$

(where starting from $i = 2$ follows since the linearized model has relative degree two).

Using the lifted setting, [62] designed a phase-lead ILC law

$$\tilde{\eta}_{j+1}(i) = k_q(\tilde{\eta}_j(i) + k_p e_j(i + u)) \tag{10.11}$$

where $u \geq 1$ is the phase-lead contribution. At this stage, an applications specific aspect arises. The objective of the overall control scheme, in this case, is to reduce path-following errors. Additionally, there is the potential to improve, i.e. increase the vehicle speed when some path-following error can be tolerated. This improvement is of particular interest when productivity is critical, i.e. haul more material per unit time.

The approach to this last objective in [62] is to add a second ILC law of the form:

$$\bar{v}_{j+1} = Q_s(\bar{v}_j + L_s e_{sj}) \tag{10.12}$$

where $\bar{v}$ is the vector of desired forward vehicle speeds, Q_s is a speed learning Q-filter, L_s is a speed learning function implementing the phase-lead component of the ILC law, and e_{sj} is the speed learning error. In this form of control, learning is only directly applied to the forward vehicle speed since no controller determines the initial vehicle speed to be corrected. Instead, a constant $\bar{v}$ is chosen for the entire path, and then changes are made along the desired path to form a variable speed profile. Based on the premise of a relationship between the forward vehicle speed and path-following performance, i.e. as $\bar{v}$ increases ϵ_L increases, the error is defined as follows:

$$e_{sj} = e_t - |y_j| \tag{10.13}$$

where e_t is a tunable error threshold, and y_j is the error system output.

In this case $e_{sj} \to 0$ as $j \to \infty$. As an alternative to minimizing the path-following error, the use of (10.13) means that the desired vehicle speed $\bar{v}$ is increased when the errors are low and vice versa until e_t converges. Hence, an allowable error is introduced to ramp up the forward vehicle speed until it is considered safe, i.e. a low path-following error ϵ_L [28].

The overall design in this application combines the feedback linearization with two ILC laws to regulate the steering inputs for improved path-following performance and simultaneously allowing an increase in the desired vehicle speed. However, an increase in the path-following errors could arise, and performance, of course, depends on the design of the ILC laws.

An extensive set of field tests of these ILC laws has been undertaken using an industrial-scale autonomous vehicle, and the results reported in [62]. In common with other path-following problems, the desired path must be constructed. This last requirement is application-specific and requires detailed knowledge of the application area. The experiments reported in [62] used the desired path with two-rounded 90° corners, which represents a possible path required in an operating mine.

10.2 Input–Output Linearization-based ILC Applied to Stroke Rehabilitation

10.2.1 System Configuration and Modeling

The discussion in Section 2.4 explains why ILC can be applied in robotic-assisted upper-limb stroke rehabilitation. In Section 4.2, it is described how simple structure ILC laws can be tuned to give results where the required feature was also detected in clinical trials. This research was for planar tasks, such as reaching out over a tabletop to a cup.

For 3D tasks such as lifting the affected limb and then reaching out, e.g. to collect an object from a shelve, the dynamics are much more complicated, and nonlinear model-based ILC designs have to be used. As in the planar case, stimulation must be applied within a controlled environment to ensure safety and comfort across a broad spectrum of patient ability. This requirement may be provided by a passive/orthotic support device such as a simple sling, hinged mechanism, or orthosis, or by a robotic mechanism.

Many designs of such support systems are available. Figure 10.3 shows one such scenario where the participant's arm is in the mechanical support, and the assistive stimulation is applied through surface electrodes to the relevant muscles, in this case, the triceps and anterior deltoid. Also shown is a real-time 3D graphical environment for the patient and the therapist.

The system in this last figure applies functional electrical stimulation (FES) to two muscles in the arm and shoulder to assist the patient track a virtual reality task. The human arm is supported by a passive mechanism to provide a safe and productive environment for training across a broad spectrum of patient ability. Using systems such as this one (an ArmeoSpring, Hocoma AG), an adjustable force against gravity can be provided (in this case, by two springs). This arrangement, together with FES, has led to successful clinical trials. The development of the ILC design for this case is described next.

Each joint is aligned, as shown in Figure 10.3, in either the horizontal or vertical plane with measured joint angles $\Theta = \begin{bmatrix} \theta_1 & \theta_2 & \theta_3 & \theta_4 & \theta_5 \end{bmatrix}^T$. In this arrangement, the patient's arm is rigidly strapped to the exoskeleton support. Next, the choice of muscles to stimulate is explained.

Spasticity (velocity-dependent stiffness) in stroke typically produces resistance to arm extension due to overactivity of the biceps, wrist, and finger flexors and loss of activity in the triceps, anterior deltoid, wrist, and finger extensors, see [152] for further discussion. Triceps and anterior deltoid are, therefore, selected for stimulation to align with the clinical need to increase muscle tone and restore motor control of weakened muscles. The relationship between muscle stimulation and subsequent

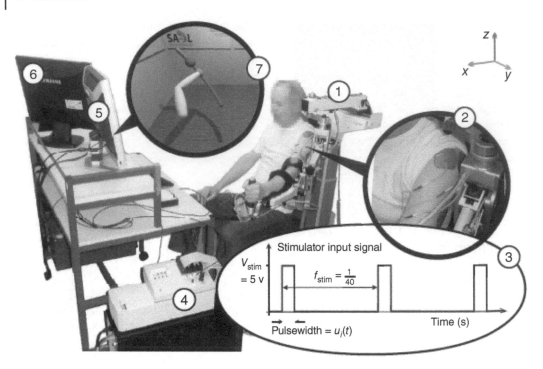

Figure 10.3 A 3D rehabilitation system, where ① is the mechanical support, ② is the surface electrodes, ③ is the FES module, ④ is the real-time processor and interface module, ⑤ is the monitor-displaying task, ⑥ therapist's monitor displaying task, and ⑦ is the real-time 3D graphics.

movement is well researched, and sophisticated muscle models exist with multiple attachment points across more than one joint and movement over complex sliding surfaces, see, e.g. [40].

In the presence of such a complexity, a simplification is an obvious approach. As described next, appropriate simplification leads to both parameter identification and control law design, which, having obtained the required ethical approval, led to clinical trials. A critical assumption in the approach described in this section is that applying FES to the triceps produces a moment about an axis orthogonal to both the forearm and the upper arm and FES to the anterior deltoid produces a moment about an axis that is fixed with respect to the shoulder.

Following on from the last assumption, the joint variables ϕ_2 and ϕ_5 are to be controlled as shown in Figure 10.4, with additional axes accounting for the remaining degrees of freedom.

A dynamic model of the support structure is

$$B_a(\Theta)\ddot{\Theta} + C_a(\Theta,\dot{\Theta})\dot{\Theta} + F_a(\Theta,\dot{\Theta}) + G_a(\Theta) + K_a(\Theta) = 0 \tag{10.14}$$

$B_a(\cdot) \in \mathbb{R}^{5\times5}$ and $C_a(\cdot) \in \mathbb{R}^{5\times5}$ are, respectively, the inertial and Coriolis matrices and $F_a(\cdot)$ and $G_a(\cdot)$, respectively, are the friction and gravity vectors. Moments produced through gravity compensation provided by each spring are represented by $K_a(\cdot) = \begin{bmatrix} 0 & 0 & k_3(\theta_3) & 0 & k_5(\theta_5) \end{bmatrix}^T$. Similarly, a dynamic model of the human arm is

$$B_h(\Phi)\ddot{\Phi} + C_h(\Phi,\dot{\Phi})\dot{\Phi} + F_h(\Phi,\dot{\Phi}) + G_h(\Phi) = \tau(u,\Phi,\dot{\Phi}) \tag{10.15}$$

where $\tau(\cdot)$ represents the moments produced by application of FES and the entries in $\Phi = \begin{bmatrix} \phi_1 & \phi_2 & \phi_3 & \phi_4 & \phi_5 \end{bmatrix}^T$ contain the anthropomorphic joint angles.

Modeling the force/moment generated by the electrically stimulated muscle is a well-researched area. There is a distinction between modeling for analysis as opposed to direct model-based control application. Detailed discussion of the last area is given in [73, 141], resulting in the use of the Hill-type model:

$$\tau_i(u_i(t), \phi_i(t), \dot{\phi}_i(t)) = h_i(u_i(t), t) \times F_{M,i}(\phi_i(t), \dot{\phi}_i(t)) \tag{10.16}$$

In this case, $i = 2$ and $i = 5$, where $u_2(t)$ and $u_5(t)$ are the FES control signals that are the pulse-width of the stimulation signal applied, respectively, to the anterior deltoid and triceps as shown in Figure 10.4.

The term $h_i(u_i(t), t)$ in (10.16) is a Hammerstein structure formed by a static nonlinearity $h_{IRC,i}(u_i(t))$ representing the isometric recruitment curve in cascade with linear activation dynamics $h_{LAD,i}(t)$. The multiplicative effect of the joint angle and joint angular velocity on the active torque developed by the muscle is represented by the term $F_{m,i}(\phi_i, \dot{\phi}_i)$. Also, let $h_{LAD,i}(t)$ have a continuous-time state-space model description with matrices (respectively, state, input, and output) $\{A_{m,i}, B_{m,i}, C_{m,i}\}$.

Setting

$$u = \begin{bmatrix} 0 & u_2 & 0 & 0 & u_5 \end{bmatrix}^T \tag{10.17}$$

gives

$$\tau(u, \Phi, \dot{\Phi}) = \begin{bmatrix} 0 & \tau_2(u_2, \phi_2, \dot{\phi}_2) & 0 & 0 & \tau_5(u_5, \phi_5, \dot{\phi}_5) \end{bmatrix}^T \tag{10.18}$$

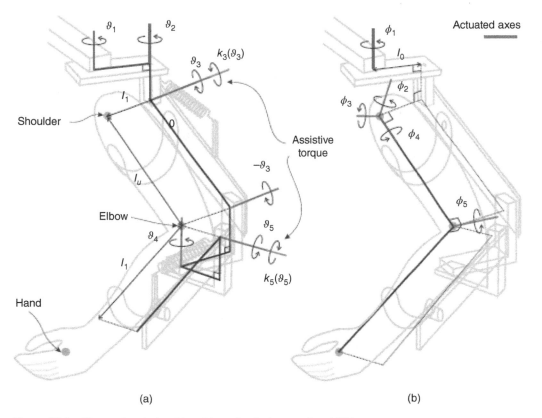

(a) (b)

Figure 10.4 Kinematic relationships: (a) mechanical support and (b) human arm.

Also within suitable ranges, it can be shown there exists a bijective transformation $\Phi = \kappa(\Theta)$ that enables the combined system to be written in the form:

$$B(\Phi)\ddot{\Phi} + C(\Phi, \dot{\Phi})\dot{\Phi} + F(\Phi, \dot{\Phi}) + G(\Phi) + K(\Phi)$$
$$= \tau(u, \Phi, \dot{\Phi}) - J_h^T(\Phi)h \tag{10.19}$$

where $J_h^T(\Phi)h$ is the system Jacobian, and h represents the externally applied force and torque. Formulas for the parameters in (10.19) and procedures for identifying them are given in [74]. The analysis that follows is also based on this publication and uses its notation.

10.2.2 Input–Output Linearization

The ILC designs for the stroke rehabilitation application covered in Section 4.2 are based on canceling each muscle nonlinearity, $h_{\mathrm{IRC},i}(\cdot)$ and assuming each muscle actuates only the corresponding joint to enable the design of separate feedback and single-input single-output (SISO) ILC laws. To improve performance and enable an extension to a much broader range of support mechanisms and choice of stimulated muscles, a combined input–output linearization and ILC design is now developed, and conditions for stability of the joint angles not actuated developed.

The linear actuation dynamics are taken to be second order with the state-space model:

$$\dot{x}_i = \begin{bmatrix} -d_{i,1} & -d_{i,2} \\ 1 & 0 \end{bmatrix} x_i + \begin{bmatrix} 1 \\ 0 \end{bmatrix} h_{\mathrm{IRC},i}(u_i)$$
$$\frac{\tau_i}{F_{m,i}(\phi_i, \dot{\phi}_i)} = \begin{bmatrix} n_{i,1} & n_{i,2} \end{bmatrix} x_i, \quad i = 2, 5 \tag{10.20}$$

with transfer-function:

$$G_{\mathrm{LAD},i}(s) = C_{m,i}(sI - A_{m,i})^{-1}B_{m,i} = \frac{n_{i,1}s + n_{i,2}}{s^2 + d_{i,1}s + d_{i,2}} \tag{10.21}$$

Let $x = \begin{bmatrix} \Phi^T & \dot{\Phi}^T & x_2^T & x_5^T \end{bmatrix}^T$. Then the controlled dynamics (10.19) can be written as follows:

$$\dot{x} = f(x) + g(x) \begin{bmatrix} h_{\mathrm{IRC},2}(u_2) \\ h_{\mathrm{IRC},5}(u_5) \end{bmatrix}$$
$$h(x) = \begin{bmatrix} \phi_2 \\ \phi_5 \end{bmatrix} \tag{10.22}$$

where

$$f(x) = \begin{bmatrix} \dot{\Phi} \\ p_1(\Phi, \dot{\Phi}) \\ p_2(\Phi, \dot{\Phi}) + (B^{-1}(\Phi))_{2,2}F_{m,2}(\phi_2, \dot{\phi}_2)(C_{m,2}x_2) \\ p_3(\Phi, \dot{\Phi}) \\ p_4(\Phi, \dot{\Phi}) \\ p_5(\Phi, \dot{\Phi}) + (B^{-1}(\Phi))_{5,5}F_{m,5}(\phi_5, \dot{\phi}_5)(C_{m,5}x_5) \\ A_{m,2}x_2 \\ A_{m,5}x_5 \end{bmatrix} \tag{10.23}$$

and row i of $p(\Phi, \dot{\Phi})$ is given by

$$p_i(\Phi, \dot{\Phi}) = -(B^{-1}(\Phi)(C(\Phi, \dot{\Phi})\dot{\Phi} + F(\Phi, \dot{\Phi}) + G(\Phi) + K(\Phi)))_i \qquad (10.24)$$

$$g(x) = \begin{bmatrix} g_1^T(x) & g_2^T(x) \end{bmatrix}^T$$
$$= \begin{bmatrix} 0 & 0 & 0 & 0 & 0 & 0 & 0 & 0 & 0 & 0 & B_{m,2}^T & 0 \\ 0 & 0 & 0 & 0 & 0 & 0 & 0 & 0 & 0 & 0 & 0 & B_{m,5}^T \end{bmatrix}^T \qquad (10.25)$$

and

$$h(x) = \begin{bmatrix} h_1(x) \\ h_2(x) \end{bmatrix} = \begin{bmatrix} \phi_2 \\ \phi_5 \end{bmatrix}$$

At this stage, geometric nonlinear control theory [116] results can be applied. Using this theory, an input–output linearizing controller for the 2×2 system in this case is

$$\begin{bmatrix} h_{\mathrm{IRC},2}(u_2) \\ h_{\mathrm{IRC},5}(u_5) \end{bmatrix} = v^{-1}(x)(v - \mu(x)) \qquad (10.26)$$

with control input $v = \begin{bmatrix} v_1 & v_2 \end{bmatrix}^T$. Also,

$$\mu_i(x) = L_f^{k_i} h_i(x), \quad v_{ij}(x) = L_{g_j} L_f^{k_i-1} h_i(x) \qquad (10.27)$$

for $i, j = 1, \ldots, m$ and k_i is the relative degree of output i.

The entries on the right-hand side of this last equation are defined in terms of Lie derivatives, where for $h_i(x)$, these are

$$L_f h_i(x) = \frac{\partial h_i}{\partial x} f(x), \quad L_{g_i} h_i(x) = \frac{\partial h_i}{\partial x} g_i(x) \qquad (10.28)$$

where $L_f^j h_i(x)$ and $L_{g_i} L_f^{j-1} h_i(x)$ are given by $L_f(L_f^{j-1} h_i(x))$ and $L_{g_i}(L_f^{j-1} h_i(x))$. Also, the relative degree k_i satisfies

$$\begin{cases} L_{g_i} L_f^{k_i-1} h_i(x) \neq 0 \\ L_{g_i} L_f^n h_i(x)) = 0, \quad n = 1, 2, \ldots, (k_i - 2) \end{cases} \qquad (10.29)$$

The standard design (10.26) in this case is

$$\begin{bmatrix} h_{\mathrm{IRC},2}(u_2) \\ h_{\mathrm{IRC},5}(u_5) \end{bmatrix} = \begin{bmatrix} L_{g_1} L_f^{k_1-1} h_1(x) & L_{g_2} L_f^{k_1-1} h_1(x) \\ L_{g_1} L_f^{k_2-1} h_2(x) & L_{g_2} L_f^{k_2-1} h_2(x) \end{bmatrix}^{-1}$$
$$\times \left(v - \begin{bmatrix} L_f^{k_1} h_1(x) \\ L_f^{k_2} h_2(x) \end{bmatrix} \right) \qquad (10.30)$$

with $k_1 = 3$ if $n_{2,1} = 0$ and $k_1 = 4$ otherwise and $k_3 = 3$ if $n_{5,1} = 0$, and $k_2 = 4$, otherwise.

Applying the above results to (10.22) gives the control signals

$$u_2 = \begin{cases} h_{\mathrm{IRC},2}^{-1} \left(\dfrac{\frac{\partial}{\partial x}\left(\frac{\partial f_7(x)}{\partial x} f(x) \right) f(x) - v_1}{(B^{-1}(\Phi))_{2,2} F_{m,2}(\phi_2, \dot{\phi}_2) n_{2,2}} \right), & \text{if } n_{2,1} = 0 \\[2em] h_{\mathrm{IRC},2}^{-1} \left(\dfrac{\frac{\partial f_7(x)}{\partial x} f(x) - v_1}{(B^{-1}(\Phi))_{2,2} F_{m,2}(\phi_2, \dot{\phi}_2) n_{2,1}} \right), & \text{otherwise} \end{cases} \qquad (10.31)$$

$$u_5 = \begin{cases} h_{IRC,5}^{-1} \left(\dfrac{\frac{\partial}{\partial x}\left(\frac{\partial f_{10}(x)}{\partial x} f(x) \right) f(x) - v_2}{(B^{-1}(\Phi))_{5,5} F_{m,5}(\phi_5,\dot{\phi}_5) n_{5,2}} \right), & \text{if } n_{5,1} = 0 \\[4mm] h_{IRC,5}^{-1} \left(\dfrac{\frac{\partial f_{10}(x)}{\partial x} f(x) - v_2}{(B^{-1}(\Phi))_{5,5} F_{m,5}(\phi_5,\dot{\phi}_5) n_{5,1}} \right), & \text{otherwise} \end{cases} \tag{10.32}$$

Also, it is assumed that $n_{i,1} \neq 0$, $i = 2, 5$, but the same analysis applies to all cases. Hence, the decoupled signals obtained are

$$\phi_2^{(4)} = v_1, \quad \phi_5^{(4)} = v_2 \tag{10.33}$$

where

$$\phi_i^{(k)} = \frac{\delta^k}{\partial t^k} \phi_i \tag{10.34}$$

To achieve baseline tracking and disturbance rejection, (10.34) must be stabilized. Of the many possible controllers, a linear quadratic regulator is used. In particular, a state feedback control law

$$v = \begin{bmatrix} \hat{\phi}_2^{(4)} & \hat{\phi}_5^{(4)} \end{bmatrix}^T - K\xi \tag{10.35}$$

is designed to result in the error state-space model:

$$\begin{bmatrix} e^{(1)} \\ e^{(2)} \\ e^{(3)} \\ e^{(4)} \end{bmatrix} = \begin{bmatrix} 0 & I & 0 & 0 \\ 0 & 0 & I & 0 \\ 0 & 0 & 0 & I \\ -A_0 & -A_1 & -A_2 & -A_3 \end{bmatrix} \begin{bmatrix} e^{(0)} \\ e^{(1)} \\ e^{(2)} \\ e^{(3)} \end{bmatrix} \tag{10.36}$$

or, in compact from,

$$\dot{\xi} = A\xi, \quad \xi = \begin{bmatrix} e^{(0)} \\ e^{(1)} \\ e^{(2)} \\ e^{(3)} \end{bmatrix} \tag{10.37}$$

where $e = \Phi_i - \hat{\Phi}_i$, $i = 2, 5$, and $K = \begin{bmatrix} A_0 & A_1 & A_2 & A_3 \end{bmatrix}$.

The dynamics of (10.37) are stabilized by choosing K to minimize the cost function:

$$J(v) = \int_0^\infty (\xi^T Q\xi + v^T Rv)dt \tag{10.38}$$

where $Q > 0$ and $R > 0$ and subject to

$$\dot{\xi} = \begin{bmatrix} 0 & I & 0 & 0 \\ 0 & 0 & I & 0 \\ 0 & 0 & 0 & I \\ 0 & 0 & 0 & 0 \end{bmatrix} \xi + \begin{bmatrix} 0 \\ 0 \\ 0 \\ I \end{bmatrix} \left(v - \begin{bmatrix} \hat{\phi}_2^{(4)} \\ \hat{\phi}_5^{(4)} \end{bmatrix} \right) \tag{10.39}$$

Figure 10.5 gives a block diagram representation of the overall control system, including the observer that provides estimates of the entries in ξ by minimizing the error covariance. The cost function (10.38) can be solved over $0 \leq t \leq \alpha < \infty$, which results in a time-varying K, but this has no effect on stability analysis, which only relies on the stability of the error dynamics (10.36).

The tracking control developed above is for the controlled joints. It is, therefore, necessary to establish the stability of the remaining joints. Consult Theorems 1 and 2 in [74] for the proof of this property.

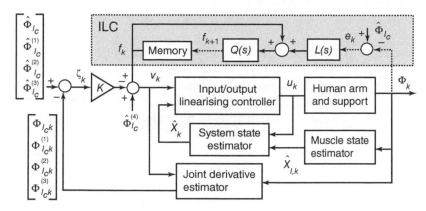

Figure 10.5 The linearization-based ILC scheme.

Given the linearized model, consider now the application of an ILC law described in Laplace transform terms as follows:

$$f_{k+1}(s) = Q(s)(f_k(s) + L(s)e_k(s)) \tag{10.40}$$

In this case, the transfer-function matrix describing the relationship between v_k and Φ_k is diagonal, i.e. of the form:

$$G(s) = \text{diag}[G_1(s), G_2(s)] \tag{10.41}$$

where

$$G_i(s) = -\frac{1}{a_{i3}s^3 + a_{i2}s^2 + a_{i1}s + a_{i0}}, \quad i = 1, 2 \tag{10.42}$$

Choosing $Q(s) = \text{diag}[Q_1(s), Q_2(s)]$ and $L(s) = \text{diag}[L_1(s), L_2(s)]$ results in the independent ILC error dynamics (where the subscript i denotes an entry in the error vector):

$$e_{i,k+1}(s) = Q_i(s)(1 - G_i(s)L_i(s))e_{i,k}(s), \quad i = 1, 2 \tag{10.43}$$

Hence, a sufficient condition for trial-to-trial error convergence is

$$|Q_i(j\omega)(1 - G_i(j\omega)L_i(j\omega))| < 1, \quad \forall \omega, \quad i = 1, 2 \tag{10.44}$$

and the design problem is reduced to two SISO control loop designs. Moreover, the voluntary effort of the patient can be treated as a trial-independent disturbance, i.e. constant over the trials.

In application, the disturbance varies from trial-to-trial. Also, there is model uncertainty. The latter leads to the input–output linearization producing a transfer-function matrix that is not equal to $G(s)$, and the resulting mismatch will propagate through the ILC design. The stability properties, in this case, can be analyzed using [3] and the other results and methods cited in this paper. Robust control design for an ILC law of the form (10.40) for linear dynamics is given in Chapter 6.

10.2.3 Experimental Results

This section reports the tests conducted with three unimpaired subjects, where this is a crucial stage before working with stroke patients. Moreover, ethical approval is required, and this was obtained from the faculty of Health Sciences, University of Southampton, UK (ETHICS-2010-30).

Next, the preliminary setup is described, and then the experimentally obtained results are given and discussed.

Each subject was seated in the ArmeoSpring, which was adjusted to their arm dimensions, and the level of support was modified such that their arm was raised 5-cm above their lap. Each subject's workspace was established, and nine reference trajectories were then calculated to form $\hat{\phi}_2$ and $\hat{\phi}_5$. In application, these correspond to lifting and extending the upper arm and forearm in three different directions (center, off-center, and far) and with three different lengths (proximal, middle, and distal).

Figure 10.6 shows the reference trajectories used, with trial lengths between 5 and 10 seconds. The surface electrodes were placed on the anterior deltoid and triceps muscles to enable the maximum appropriate movement to complete the task (if possible). Each FES channel produces a sequence of electrical pulses with a fixed frequency of 40 Hz and a fixed amplitude of 5 V.

Each pulse width is the controlled variable (0–300 μs), where u_2 corresponds to the anterior deltoid and u_5 to the triceps (as shown in Figure 10.3, ②). Each signal is then amplified by a commercial stimulator to obtain a fixed pulse amplitude of between 0 and 120 mA, set at the beginning test session to give a comfortable contraction. Further details on this critical aspect can be found in [84] and the cited references. See also [74] for the muscle identification procedure and other matters related to conducting the tests.

In the results given below, the weighting matrices in the cost function (10.38) are $Q = I$ and $R = 0.01 \times I$, chosen (as in other optimal control applications) to achieve a satisfactory balance between accuracy of tracking and control effort. The ILC design is gradient-based where in (10.40) $L_i(s) = \gamma G_i^T(s)$, $Q_i(s) = 1$ and γ is a scalar gain. Also, (10.44) holds in this case if

$$0 < \gamma < \frac{2}{|G_i(j\omega)|^2}, \quad \forall \, \omega \tag{10.45}$$

The choice of $\gamma = 0.8$ was used, as a compromise between trial-to-trial error convergence and robustness of the ILC design, to obtain the results given next.

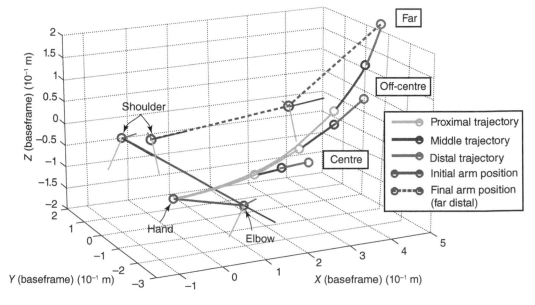

Figure 10.6 The nine reference trajectories.

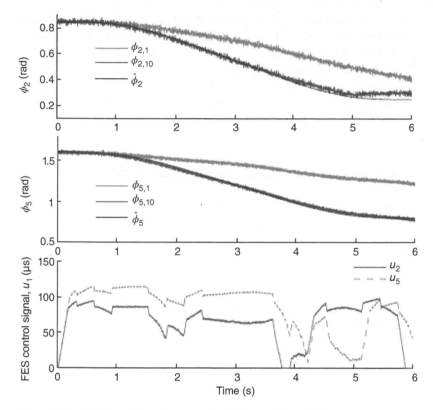

Figure 10.7 Trial $k = 10$ results using the ILC law with $\gamma = 0.8$.

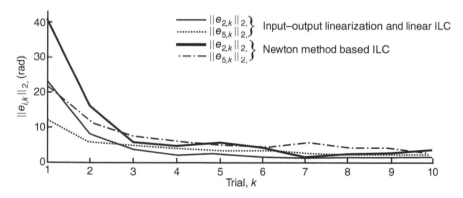

Figure 10.8 Tracking errors and control action.

Figure 10.7 gives the tracking results for ϕ_2 and ϕ_5, including the corresponding FES input signals and the tracking errors. Figure 10.8 gives the results after 10 trials for one of the subjects. In this case, the first trial ($k = 1$) is the action of the linearizing control alone. These results confirm that trial-to-trial error convergence is in a small number of trials with, of equal importance, a sequence of trial inputs that are comfortable for the subject. The next chapter returns to this design, where its merits relative to Newton-based ILC, which does not involve the preliminary step of linearizing the dynamics, are discussed. (The results of this Newton-based design are included in Figure 10.8.)

This section has described a combined input–output linearization and ILC approach that builds on control structures that have a proven track record in upper-limb stroke rehabilitation. The scheme in this section significantly extends the scope of current ILC approaches in this area and has a form that is suitable for a general application across a broad range of support structures and stimulated muscle sets. The framework provides a basis for future research on using ILC in FES-based rehabilitation, an area which is increasing in terms of the muscles which are stimulated, and the functionality of the movements controlled.

10.3 Gap Metric ILC with Application to Stroke Rehabilitation

This section describes the use of the gap metric to provide a substantial extension of the scope of existing ILC robustness analysis by addressing: (i) unstructured uncertainties, (ii) a general ILC update class, and (iii) a full generalization of the task. Also, a feedback controller is included to maximize the impact of these new results. Analysis is based on the nonlinear gap metric [91], which is applied to ILC by reformulating the within-in trial feedforward action as trial-to-trial feedback action. The resulting gap on the trial-to-trial dynamics is then translated back to the original system. (In this section, the notation follows, in the main, that used in [75].)

Along the trial dynamics are represented the mapping:

$$N : L_2^m[0, \alpha] \to L_2^p[0, \alpha] : u_k \mapsto y_k, \quad N(0) = 0 \tag{10.46}$$

which is assumed to be bounded-input bounded-output (BIBO) stable, and the control scheme is shown in Figure 10.9. In this case, the feedback controller

$$K : L_2^p[0, \alpha] \to L_2^m[0, \alpha] : (e_k + v_k) \mapsto w_k, \quad K(0) = 0 \tag{10.47}$$

is included, as in other ILC designs, to ensure baseline performance and rejection of the external disturbances $u_{0,k}$, $y_{0,k}$. Also, the controlled dynamics $[N, K]$ is assumed to be well posed.

The ILC design problem is (as before) to ensure that the output sequence $\{y_k\}$ tracks a reference vector $\hat{y} \in L_2^p[0, \alpha]$ as k increases. This tracking is to be achieved by updating the feedforward signal v_k using the tracking error:

$$e_k = \hat{y} + y_{0,k} - y_k \tag{10.48}$$

In some application areas, e.g. stroke rehabilitation, it may be more appropriate to require tracking of q subintervals, i.e. $[t_{j-1}, t_j], j = 1, \ldots, q$, where $0 = t_0 \le t_1 \le \cdots, T_q = \alpha$ are distinct points in $[0, \alpha]$. Moreover, the case of isolated time points is included by setting $t_{j-1} = t_j$.

To include these last two cases in the analysis, introduce the projection operator:

$$P : L_2^p[0, \alpha] \to L_2^{p_1}[0, t_1] \times \cdots \times L_2^{p_q}[t_{q-1}, t_q] : y \mapsto \begin{bmatrix} (Py)_1 \\ \vdots \\ (Py)_q \end{bmatrix}$$

$$((Py)_j)(t) = P_j y(t), \quad t \in [t_{j-1}, t_j], \quad j = 1, 2, \ldots, q \tag{10.49}$$

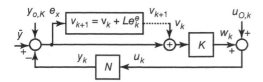

Figure 10.9 The controlled structure.

to extract the required subinterval components, where focusing on the rehabilitation application, P_j is a $p_j \times p$ matrix of full row rank specifying the output components involved in the movement stipulated over the time interval $[t_{j-1}, t_j]$. The control problem now is

$$\lim_{k \to \infty} v_k = v_\infty, \quad v_\infty := \min_{v_k} \| \hat{y}^e - y_k^e \|^2 \tag{10.50}$$

where the "extended" reference and output are $\hat{y}^e = P\hat{y}$ and the "extended" output is $y_k^e = Py_k$. The choice of $P = I$ recovers the standard case of tracking over the complete trial length (set $q = 1$, $t_1 = \alpha, P_1 = I$).

The ILC law considered is

$$v_{k+1} = v_k + Le_k^e, \quad k \geq 0, \quad v_0 = 0 \tag{10.51}$$

where $L : L_2^{p_1}[0, t_1] \times \cdots \times L_2^{p_q}[t_{q-1}, t_q] \to L_2^p[0, \alpha]$ and $L(0) = 0$. Suppose also that L is to be designed based on a linear, potentially unstable, approximation $M : L_2^m[0\ \alpha] \to L_2^p[0, \alpha]$, of the actual (or true) system dynamics N. Then the following result is Proposition 2.1 in [75], and is an extension of the ILC convergence criteria to the case of the extended task description (10.49) and multiple-input multiple-output (MIMO) dynamics whose input and output dimensions are arbitrary.

Proposition 10.1 *Let the linear operator K be designed such that the controlled dynamics $[M, K]$ is gain stable. Suppose also that L in the ILC law (10.51) is such that*

$$\|I - PGL\| = \gamma < 1 \tag{10.52}$$

where $G = (I + MK)^{-1}MK$. Then the design objective (10.50) holds with

$$v_\infty := \lim_{k \to \infty} v_k = L(PGL)^{-1}\hat{y}^e - \hat{y}, \quad y_\infty^e := \lim_{k \to \infty} y_k^e = \hat{y}^e \tag{10.53}$$

Alternatively, if L is chosen such that

$$\|I - LPG\| = \gamma < 1 \tag{10.54}$$

then

$$v_\infty := \lim_{k \to \infty} v_k = (LPG)^{-1}L\hat{y}^e - \hat{y} \tag{10.55}$$

Let Im($\cdot$) and Ker($\cdot$) denote, respectively, the image and kernel of an operator. Then the necessary and sufficient conditions for the existence of L to satisfy (10.52) and (10.54), respectively, of this last result for an arbitrary $\hat{y}^e$ are

$$\text{Im}(PG) = L_2^{p_1}[0, t_1] \times \cdots \times L_2^{p_q}[t_{q-1}, t_q] \Leftrightarrow \text{Ker}(PG)^* = \{0\} \tag{10.56}$$

$$\ker(PG) = \{0\} \Leftrightarrow \text{Im}(PG)^* = L_2^p[0, \alpha] \tag{10.57}$$

where $(PG)^* = G^*P^*$ is the adjoint of the operator PG. Next, the gap metric is used to establish robust performance when the controllers K and L designed based on the approximate model M are applied to the actual dynamics N. This analysis is undertaken by embedding the dynamics of trial k over $t \in [0, \alpha]$ as a single-time instant of the associated lifted system representation.

Write the signals in Figure 10.9 as follows:

$$v_k = v(k), \quad e_k = e(k), \quad y_k = y(k)$$
$$y_{0,k} = y_0(k) \in L_2^p[0, \alpha], \quad u_k = u(k)$$
$$u_{0,k} = u_0(k) \in L_2^m[0, \alpha] \tag{10.58}$$

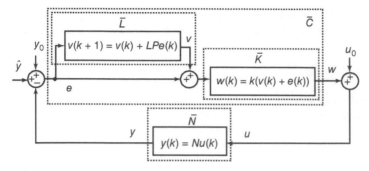

Figure 10.10 The ILC and feedback control scheme.

Also, define the corresponding lifted signal spaces

$$v, e, y \in L_2^p[0, \alpha] \times \mathbb{N}, \quad u \in L_2^m[0, \alpha] \times \mathbb{N}$$

Next, the trial-to-trial operators $\overline{N}, \overline{K}, \overline{L}$ are defined in terms of the corresponding along the trial operators as follows:

$$\overline{N} : L_2^m[0, \alpha] \times \mathbb{N} \to L_2^p[0, \alpha] \times \mathbb{N} : u \mapsto y : y(k) = Nu(k)$$
$$\overline{K} : L_2^p[0, \alpha] \times \mathbb{N} \to L_2^m[0, \alpha] \times \mathbb{N} : (v + e) \mapsto w : w(k) = K(v(k) + e(k))$$
$$\overline{L} : L_2^p[0, \alpha] \times \mathbb{N} \to L_2^p[0, \alpha] \times \mathbb{N} : e \mapsto v : v(k+1) = v(k) + LPe(k), \quad v(0) = 0 \qquad (10.59)$$

for $k > 0$, and $\overline{M}$ is obtained by replacing N by M in the first equation of (10.59). Also, introduce

$$\overline{C} : L_2^p[0, \alpha] \times \mathbb{N} \to L_2^m[0, \alpha] \times \mathbb{N} : e \mapsto w : w = \overline{K}(\overline{L} + I)e \qquad (10.60)$$

Figure 10.10 shows the trial-to-trial system representation of the dynamics, which are described by

$$\Pi_{\overline{N} \| \overline{C}} : \begin{pmatrix} u_0 \\ y_0 + \hat{y} \end{pmatrix} \mapsto \begin{pmatrix} u \\ y \end{pmatrix} \qquad (10.61)$$

with

$$\begin{pmatrix} \bar{u}_k \\ \bar{y}_k \end{pmatrix} = \Pi_{\overline{N} \| \overline{C}} \begin{pmatrix} 0 \\ \hat{y} \end{pmatrix} (k) \qquad (10.62)$$

Moreover, given (10.58)–(10.62), robust performance when the control scheme operates with the true system is given by the following result, Theorem 3.1 in [75].

Theorem 10.1 *Suppose that L is designed such that either (10.52) or (10.54) of Proposition 10.1 hold. Then the true combined feedback and ILC dynamics $[\overline{N}, \overline{C}]$ are BIBO stable if the biased gap satisfies*

$$\delta(\overline{M}, \overline{N}) < b_{\overline{M} \| \overline{C}}^{-1} \qquad (10.63)$$

where the gain bound for $[\overline{M}, \overline{C}]$ is

$$b_{\overline{M} \| \overline{C}} = \frac{\left\| \begin{pmatrix} I \\ M \end{pmatrix} K(I + MK)^{-1} \right\| \, \|L\| \, \|P(I + MK)^{-1}(-M, I)\|}{1 - \gamma} + b_{M \| K} \qquad (10.64)$$

and $b_{M,K}$ is the purely feedback component, i.e. $L = 0$, and given by

$$b_{M\|K} = b_{\overline{M}\|\overline{K}} = \left\| \begin{pmatrix} I \\ M \end{pmatrix} (I + KM)^{-1} (I, K) \right\| \tag{10.65}$$

Moreover, the convergence of the true system satisfies

$$\left\| \Pi_{\overline{N},\overline{C}} \right\| \begin{pmatrix} 0 \\ \hat{y} \end{pmatrix} \le b_{\overline{M}\|\overline{C}} \frac{1 + \delta(\overline{M}, \overline{N})}{1 - b_{\overline{M}\|\overline{C}} \delta(\overline{M}, \overline{N})} \tag{10.66}$$

It is possible to interpret (10.63) as stabilizing a "ball" of systems in the uncertainty space about the nominal system $\overline{M}$, which has radius $b_{M\|K}^{-1}$ in the case of feedback action alone, but reduces when ILC action is added (due to the additional term on the right-hand side of (10.64)). Also, the right-hand side of this last equation is finite for all cases when $\|L\|$ is bounded. Hence, the radius of this ball is always greater than zero and increases in size as $\|L\|$ reduces to zero.

This last theorem gives a method to design both the feedback and ILC controller parts to weight performance against robustness. The following result, Theorem 3.2 in [75], shows the relationship between the gap metrics for the systems defined by $\overline{M}$, $\overline{N}$, and M, N, i.e. between the lifted and original systems.

Theorem 10.2 *The gap measure of the mismatch between the lifted and original systems satisfies*

$$\delta(\overline{M}, \overline{N}) \le \delta(M, N) \le \sup_{\|u\|\ne 0, k>0} \frac{\|(N|_{\bar{u}_k} - M)u\|}{\|u\|} \tag{10.67}$$

where $\delta(M, N)$ is the biased gap between the unlifted systems.

In Theorem 10.1, $\delta(\overline{M}, \overline{N})$ can be replaced by either $\delta(M, N)$ or $\sup_{k>0}\|(N|_{\bar{u}_k} - M)u\|$. This result gives explicit conditions guaranteeing robust performance. Moreover, (10.67) enables direct computation using the standard along-the-trial operators M and N. In robustness analysis for ILC systems, this approach gives robust performance bounds for unstructured uncertainties. Moreover, the effect of the tracking objective, feedback controller, and ILC update design is characterized unlike other robust ILC designs in Chapter 6. Also, (10.66) bounds convergence with respect to (u_k, y_k), which are related to the ideal nominal signals, as explained in [75].

Figure 10.11 summarizes the control design procedure for stabilization of an unknown true system, and next, a procedure for completing the design is given.

1: Assuming the task is feasible in practice, reduce the time intervals or isolated instants over which tracking is required such that the resulting task objective operator P has minimal norm.

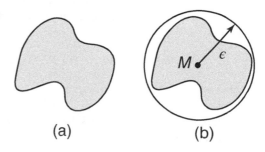

Figure 10.11 Design procedure: (a) uncertainty space known to contain the true system N and (b) stabilized ball of systems with $\epsilon = b_{\overline{M}\|\overline{C}}^{-1}$.

(a) (b)

2: Design the feedback controller K to give baseline tracking and disturbance rejection while minimizing $b_{M\|K}$ given by (10.65).

3: Compute an ILC operator L to satisfy (10.52), (10.54) of Proposition 10.1. For maximum robustness, the choice made must minimize the right-hand side of (10.64). In turn, it is required that both L and γ are minimized. Moreover, reducing both L and γ corresponds to slower convergence, and, if convergence speed is not an issue, they can be made arbitrarily small.

4: Inspect robust uncertainty by computing $b_{\overline{M\|C}}^{-1}$ using (10.64) and compare against realistic cases of model uncertainty found using the right-hand term in (10.67). If the radius of the "ball" of stabilized plants is deemed to be too small, then compute alternative feedback and ILC controllers.

5: Apply the controller and asses the performance. If necessary, redefine the task (to reduce $\|P\|$) slow learning (to reduce $\|L\|$) or detune feedback tracking performance (to reduce $b_{M\|K}$), to increase the radius of the stabilized ball of systems. Alternatively, reidentify the plant model M to reduce the mismatch, $\delta(M, N)$ between it and the true system N.

The following result, Theorem 4.1 in [75] enables the ILC operator L to be computed such that (10.52) holds if (10.57) holds, or to satisfy (10.54) if 10.57 holds.

Theorem 10.3 *Let L in the ILC law (10.51) be given by*

$$L = (PG)^*(I + PG(PG)^*)^{-1} \tag{10.68}$$

Then if (10.56) holds, (10.52) is satisfied and v_∞ is the minimum input solution to (10.50), i.e.

$$v_\infty = \min_{v_k} \|v_k\|^2 : e_k^e = 0 \tag{10.69}$$

Alternatively, if (10.57) holds, (10.54) is satisfied and v_∞ also solves (10.50).

Next, the use of this design procedure is illustrated by a stroke rehabilitation application. Relative to alternative results on robust ILC, the current approach gives robust performance bounds for unstructured uncertainties, but the results also clearly show the effect of the tracking requirement, feedback controller, and ILC update design.

One particular feature of the results in this section is that they apply to applications where the model's accurate identification is difficult. This problem applies to stroke rehabilitation, where identifying an accurate model of human muscle response to FES is problematic. Moreover, this problem is compounded by the potentially complex effects of a stroke on the muscles associated with the lost functionality, especially when more than one muscle is required. The remainder of this section illustrates the benefits of the gap metric-based approach to ILC design.

Let $u \in \mathbb{R}^m$ denote the FES signals applied to the muscles over $0 \le t \le \alpha$. Then if the ith muscle acts about a single point with angle $y_j(t)$, the standard Hill-type model gives the resulting moment as follows:

$$\tau_{j,i}(u_i(t), y_j(t), \dot{y}_j(t)) = h_i(u_i(t), t)\tilde{F}_{Mj,i}(y_j(t), \dot{y}_j(t)) \tag{10.70}$$

where (see also the previous section) $h_i(u_i(t), t)$ is a Hammerstein structure formed from a static nonlinearity $h_{\text{IRC},i}(u_i(t))$, which represents the isometric recruitment curve, in series with stable linear actuation dynamics $H_{\text{LAD},i}$. The bounded term $\tilde{F}_{Mj,i}(\cdot)$ represents the effects of the joint angle and angular velocity on the moment generated.

In some cases, multiple muscles and/or tendons may each span any subset of joints. The general expression for the total moment produced about the jth joint is

$$\tau_j(u(t), y(t), \dot{y}(t)) = \sum_{i=1}^{m} (d_{j,i}(y_j)\tau_{j,i}(u_i(t), y_j(t), \dot{y}_j(t))) \tag{10.71}$$

where $d_{j,i}(y_i) = \frac{\partial E_i(y_j)}{\partial y_j}$ is the moment arm generated by the ith muscle with respect to the jth joint, and E is a continuous function representing the associated excursion [259]. Hence, the overall moment $\tau \in L_2^p[0, \alpha]$ is generated about the p joints of the interconnected anthropomorphic and the mechanical/robotic support structure, with associated joint angle signal $y \in L_2^p[0, \alpha]$. Moreover, this support structure is modeled by the rigid-body system:

$$B(y(t))\ddot{y}(t) + H(y(t), \dot{y}(t)) = \tau(u(t), y(t), \dot{y}(t)) \tag{10.72}$$

where

$$H(y(t), \dot{y}(t)) = C(y(t), \dot{y}(t))\dot{y}(t) + F(y(t), \dot{y}(t)) + G(y(t)) + R(y(t)) \tag{10.73}$$

In these last two equations, $B(y(t))$ and $C(y(t), \dot{y}(t))$ are, respectively, the $p \times p$ inertial and Coriolis matrices of the augmented structure, $G(y(t))$ is the $p \times 1$ combined gravity vector, and $R(y(t))$ is the $p \times 1$ assistive moment arising from the mechanical passive/robotic support. Lastly, $F(y(t), \dot{y}(t))$ is the $p \times 1$ anthropomorphic stiffness and damping vector, which represents the joint stiffness, friction, and spasticity and is modeled by

$$F(y(t), \dot{y}(t)) = F_e(y(t)) + F_v(\dot{y}(t)) \tag{10.74}$$

Consider the case when

$$F_{e,i}(y(t)) = F_{e,i}(y_i(t)), \quad F_{v,i}(\dot{y}(t)) = F_{v,i}(\dot{y}_i(t)), \quad i = 1, 2, \ldots, m$$

Finally, suppose that the support $R(y)$ is adjusted such that the passivity condition (well known in the general area) holds, i.e.

$$(y - \bar{y})^T(F_e(y) + G(y) + R(y) - \bar{\tau}) \ge 0 \tag{10.75}$$

for some pair $(\bar{\tau}, \bar{y})$.

The stimulated arm model given above can be represented by the map

$$N = H_{RB}F_m(y, \dot{y})H_{LAD}h_{IRC} \tag{10.76}$$

where

$$h_{IRC} : L_2^m[0, \alpha] \to L_2^m[0, \alpha] : u \mapsto y : y_i = h_{IRC,i}(u_i)$$
$$H_{LAD} : L_2^m[0, \alpha] \to L_2^m[0, \alpha] : y \mapsto w : w_i = h_{LAD,i}y_i$$
$$F_m : L_2^m[0, \alpha] \to L_2^p[0, \alpha] : w \mapsto \tau : \tau_i = \sum_j F_{M,i,j}w_j \tag{10.77}$$

where

$$F_{M,i,j} = d_{i,j}(y_i)\tilde{F}_{M,i,j}(y_i(t), \dot{y}_i(t)), \quad j = 1, 2, \ldots, m$$
$$H_{RB} : L_2^p[0, \alpha] \to L_2^p[0, \alpha] : \tau \mapsto y : \ddot{y} = B^{-1}(y)(\tau - H(y, \dot{y})) \tag{10.78}$$

Also, (10.76) is BIBO stable, and the last two theorems provide bounds on specific sources of modeling errors. The known structure of the system can also be used to give more detailed, explicit bounds on common sources of uncertainty within the simulated system. In stroke rehabilitation, muscle fatigue is a critical feature, and for this application, the following result, Proposition 4.3 in [75], highlights this last feature.

Proposition 10.2 *Suppose that the linear approximations to the dynamics H_{RB}, tendon function F_m and muscle curve h_{IRC} denoted, respectively, by $\overline{H_{RB}F_m}$ and $\overline{h}_{IRC}$ are used to construct a nominal model $M = \overline{H_{RB}F_m}H_{LAD}\overline{h}_{IRC}$. Suppose also that this nominal model is used to design a feedback controller K and an ILC operator L that satisfy (10.64) in Theorem 10.1. Then the system has a robust stability margin and, in particular, is stable if*

$$\Delta_{IRC} < \sup_{k>0} \frac{b_{\frac{1}{M\|C}}^{-1} - \Delta_{RB}\|\overline{h}_{IRC}\|}{\|H_{RB}F_m|_{\overline{w}(k)}\|} \tag{10.79}$$

where

$$\Delta_{IRC} = \max_i \sup_{\|u\|\neq 0,k>0} \frac{\|(h_{IRC,i}|_{\overline{u}(k)_i} - \overline{h}_{IRC,i})u\|}{\|u\|} \tag{10.80}$$

is the model mismatch arising from muscle fatigue, and

$$\Delta_{RB} = \sup_{\|u\|\neq 0,k>0} \frac{\|(H_{RB}F_m(y,\dot{y})|_{\overline{w}(k)} - \overline{H_{RB}F_m})u\|}{\|u\|} \tag{10.81}$$

is the model mismatch arising from the linearization.

This last result places a bound on the allowable muscle fatigue, characterized by the term Δ_{IRC}. Also, this mismatch increases as the patient experiences fatigue. The bound is given in terms of the modeling error due to linearization. Moreover, this bound can be assumed not to vary. The feedback and/or ILC design can always be modified to ensure that (10.79) is satisfied, e.g. reducing the convergence speed and or the range of frequencies over which convergence occurs. Within the stroke rehabilitation application area, the significance of this last result is that it allows ILC design in the presence of stipulated muscle fatigue bounds.

Remark 10.1 Another method of compensating for muscle fatigue in the application of ILC in robotic-assisted stroke rehabilitation can be found in [281]. The approach taken was to add a feedback control loop around the muscle model. It was shown that simple structure controllers in this loop could be beneficial. However, this early research considered 2D tasks, see Section 4.2, and was not tested beyond simulation studies using a system model constructed from patient trials data. Moreover, extension to 3D tasks was not achieved.

People who suffer a stroke also have problems with hand functionality and rehabilitation of tasks requiring hand action, such as switching on or off a light or opening/closing a drawer. These tasks require the use of an electrode array to stimulate the required muscles, and one way of facilitating rehabilitation is to use an interactive table display and noncontact sensors. Figure 10.12 shows an experimental system where the gap-metric ILC design of this section has been tested with 10 unimpaired participants (an essential step to obtaining ethical approval to work with stroke patients). The details of this system can be found in [75], and the remainder of the section focuses on the ILC design and the results obtained.

Each participant was asked to perform two tasks involving a light switch and closing a drawer. Speed of completion is also relevant, and two speeds were used in each case, 5 and 10 seconds. The first step is to construct the approximate model M, which was estimated at the beginning of the test session, by applying FES sequences in an isometric identification test [141].

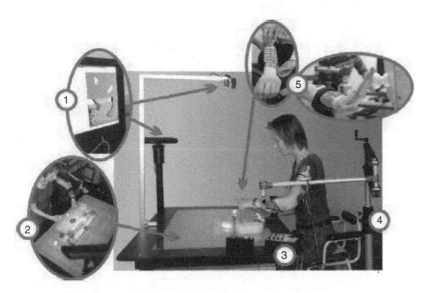

Figure 10.12 Rehabilitation system, ① motion-tracking hardware, ② interactive touch table display, ③ FES controller and multiplexer hardware, ④ SaeoMAS and perching seat, and ⑤ electrode-array. Christopher Thomas Freeman et al. (2017), ELSEVIER/CC BY 4.0.

For these two tasks, the model M was augmented to include both joint velocity and position, and the projection operator was defined using $q = 2, t_1 = t_2 = \alpha$, with $P_1 = 0, P_2 = I$ to obtain the position and velocity of the joints at $t = \alpha$ only. Reference $\hat{y}^e$ supplies the position of the light switch or drawer. For each participant, the feedback controller K was selected as a proportional plus integral plus derivative (PID) controller designed to give baseline tracking and disturbance rejection and also minimizing $b_{M\|K}$ of (10.65).

The next stage is to design the operator L to satisfy either (10.52) or (10.54) of Proposition 10.1. The system robustness is maximized by minimizing the right-hand side of (10.65). In turn, both L and γ are minimized, reducing both of these corresponds to slower convergence. Moreover, if convergence speed is not an issue (for stroke patients completing the task is arguably more important than the speed at which this is achieved), this can be made arbitrarily small.

Examining robust uncertainty by computing $b_{M\|\overline{C}}^{-1}$ using (10.64) is the next stage, which is followed by comparing against realistic cases of model uncertainty found using the right-hand side of (10.67). If the ball radius of stabilized systems is too small, compute alternative feedback and ILC controllers. After this, proceed to implement the control system and assess the performance. The gap metric setting provides a natural setting to proceed if the particular design does not give acceptable performance. For example, redefine the task (to reduce $\|G\|$) to slow learning (reduce $\|L\|$) or de-tune feedback tracking performance (to reduce $b_{M\|K}$ and, hence, increase the radius of the stabilized ball of systems). Alternatively, reidentify M to reduce the mismatch ($\delta(M, N)$) with the true system N.

Figure 10.13a shows the tracking results obtained using the system of the previous figure for the light switch task. These results are for the case when K was chosen such that $b_{M\|K} = 2.35$ (small value), and the bound on $\|L\|$ is 3.72. These results give rapid trial-to-trial error convergence but at the expense of more oscillatory learning transients. Figure 10.13b shows the case where K was tuned with an increased $b_{M\|K}$ value of 4.12, and the ILC convergence speed was reduced (via reduction of γ) to give a gain bound of 1.89.

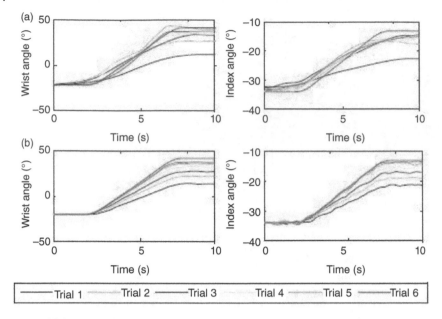

Figure 10.13 Tracking results for light switch task using (a) $b_{M\|K} = 2.35$ and (b) $b_{M\|K} = 4.12$. The reference signal is shown in light gray.

Further analysis of the results obtained is given in [75] and [162], where the latter publication gives the results of analysis of the results collected from a healthcare professional's standpoint. All tasks were achieved accurately by allowing the designer to manipulate robustness transparently. Moreover, they can also modify the task (through P) and the effort required to identify an accurate model (through the stipulated gap bound).

In summary, this section has developed a framework for robust ILC design, where, by considering unstructured uncertainties, considerably enlarges the scope of existing analysis in this field. Moreover, it embeds separate feedback control and ILC loops, with each component utilizing the other's available robustness margins to maximize overall performance. It thereby provides a mechanism for transparent control design which can balance performance and convergence speed.

A substantial generalization is the inclusion of projection operator P, which significantly enlarges the class of tasks that are supported to quantify the effect of the task on performance. This feature illustrates the framework's potential widespread applicability across ILC applications, especially in the case of stroke rehabilitation, where the difficulty in obtaining accurate models is particularly challenging.

The projection (P) is also critical to support natural motor control motions required in clinical practice. Further development should aim to extend the classes of feedback and ILC operators to nonlinear forms. The combined feedback and ILC framework could also be employed within further clinical rehabilitation programs with more functional movements, e.g. eating and brushing hair. In Section 11.4, another nonlinear ILC design will be applied to the stroke rehabilitation application.

10.4 Nonlinear ILC – an Adaptive Lyapunov Approach

Adaptive control based on Lyapunov methods has progressed very rapidly in the last decade, based, in turn, on earlier work. This fact poses the question of what can be achieved by considering ILC from an adaptive control perspective? Results on progress toward answering this general question are given in this section.

10.4.1 Motivation and Background Results

In the analysis that follows, the system dynamics are assumed to be the same on each trial but may be subject to disturbances. A particular feature is that ultimately, after a sufficiently large number of trials, the learning can be "switched off," which is critical in at least some cases. For example, convergence to within an application-specific tolerance after a finite number of trials is a specification with much industrial relevance, e.g. a pick-and-place system will only ever execute a finite number of trials. In other cases, only a finite number of trials can be completed. The question arising now is: what trial-to-trial error convergence can be achieved in this number of trials?

The ILC designs considered in this section are, in essence, adaptive control designs but used in the ILC setting where state resets occur after each trial is complete. Compared to most of the other ILC laws considered in this book, learning occurs during the trials instead of a control update between successive trials. Later in this section, the links with other forms of adaptive ILC, such as [185], are considered. Learning along the trials has also been considered in Section 5.5 for linear dynamics.

Let the set $\mathcal{U}$, consisting of elements $u : \mathbb{R}_+ \to \mathbb{R}^m$, denote the inputs and let $\mathcal{Y}$, consisting of elements $y : \mathbb{R}_+ \to \mathbb{R}^n$, denote the set of outputs. Also, let $\mathcal{C}(\mathcal{U}, \mathcal{O})$ denote all causal maps from $\mathcal{U}$ to $\mathcal{O}$. Suppose also that $H_k \in \mathcal{C}(\mathcal{U}, \mathcal{O})$ is an operator that describes the input–output behavior of a system on trial k :

$$y_k = H_k u_k, \quad k \geq 0 \tag{10.82}$$

It is also necessary to model two types of uncertainty that can occur in ILC problems. The first of these is uncertainty due to disturbances that act on the system and vary from trial-to-trial, and under a worst-case perspective, it is assumed that such disturbances are $\ell^p/\mathcal{D}$ bounded. In particular,

$$d = (d_1|d_2| \ldots |d_k| \ldots), \quad d_k \in \mathcal{D}, \quad k \geq 1 \tag{10.83}$$

where the notation | denotes concatenation, i.e.

$$d(t) = d_i(t) \quad \text{if} \quad (i-1)\alpha \leq t < i\alpha \tag{10.84}$$

and if $\mathcal{D}$ is a normed space of admissible signals defined on $0 \leq t \leq \alpha$, then d is such that

$$\|d\|_{\ell^p/\mathcal{D}} = \left(\sum_{k=1}^{\infty} \|d_k\|_{\mathcal{D}}^p \right)^{\frac{1}{p}} \leq \delta_d^{\ell^p/\mathcal{D}} = \delta_d < \infty \tag{10.85}$$

The operator H_k is of the form $H_k = G(d_k)$ and describes the input/output behavior of the system due to the disturbances alone. This representation is used to model exogenous influences on the system or disturbances to the initial conditions due to inaccurate reset after a trial. The second type of uncertainty represents the prior knowledge of the system available to the control design concerning G at the start of the first trial. This uncertainty is denoted by Δ and corresponds to the knowledge of the set membership $G \in \Delta$. In this section, the notation of [85] is followed (in the main), and hence, y_{ref} is used to denote the reference signal.

The feedforward ILC design task is (in general terms): given a reference signal, construct a sequence of inputs such that the corresponding sequence of outputs converges from trial-to-trial, i.e. ideally to zero or to within some neighborhood of the origin under an appropriate norm. Convergence is understood in the topology of $\mathcal{O}$ over the trial length. Also, the possibility of high-gain solutions is to be avoided, where one way to achieve this is to include a penalty on the control and aim to minimize a suitably chosen cost function. In the case of nonlinear dynamics, however, the result is an intractable optimal control problem.

Let $y_{\text{ref}} : \mathbb{R}_+ \mapsto \mathbb{R}$ be a specified constant reference signal or, equivalently, a reference vector $Y^{\text{ref}} : \mathbb{R}_+ \mapsto \mathbb{R}^n$, where $Y^{\text{ref}} = (y_{\text{ref}}, y_{\text{ref}}^{(1)}, \dots, y_{\text{ref}}^{(n-1)})$. Then the design problem is to construct a sequence of inputs $\{u_k\}_{k \geq 1}$, dependent only on Δ, m_k, where m_k is the memory of trials $1, \dots, k$, defined below and causally dependent on y_k, such that the corresponding outputs produced by (10.82) converge to a closed ball $B(y_{\text{ref}}, \epsilon)$ of fixed radius ϵ centered on the specified y_{ref}. The convergence is in the topology of $\mathcal{O}$, restricted to the time interval (the trial length) $0 \leq t \leq \alpha$. Still, other convergence measures, e.g. pointwise convergence, could be considered. Also, let $\mathscr{P}_k \geq 0$ denote a cost function that reflects this convergence, e.g.

$$\mathscr{P}_k = \|y_k - B(y_{\text{ref}}, \epsilon)\|_{L^2[0,\alpha]}^2 \tag{10.86}$$

To avoid the possibility of high-gain solutions, the most straightforward alternative is to assume that the reference signals are uniformly bounded, i.e. there exists γ_1 and $\gamma_2 > 0$ such that

$$\mathscr{Y}^{\text{ref}} \subset \{Y^{\text{ref}} \in C(\mathbb{R}_+, \mathbb{R}^n) \; : \; \|Y^{\text{ref}}\| \leq \gamma_1, \; |y_{\text{ref}}^{(n)}| \leq \gamma_2\} \tag{10.87}$$

(C denotes continuous functions (or a vector with entries that have this property)) and also require that all trial signals are uniformly bounded over the domain $\mathbb{N} \times [0, \alpha]$, i.e. uniformly pointwise bounded for all trials $k \geq 1$ over the trial length.

The signals on trial k, $\mathscr{K}_k$, are defined to be all the inputs u_k, outputs y_k, and internal states x_k, of the system and those of the controller, where the latter are denoted by $\hat{\theta}_k$, i.e.

$$\mathscr{K}_k = (u_k, y_k, x_k, \hat{\theta}_k) \tag{10.88}$$

and the uniform boundedness criterion requires that

$$\|\mathscr{K}(\cdot, \cdot)\|_{L^\infty(\mathbb{N} \times [0,\alpha])} < \infty \tag{10.89}$$

where

$$\mathscr{K}(k, t) = \mathscr{K}_k(t) \tag{10.90}$$

To propagate information from trial-to-trial, the memory on trial k is defined as follows:

$$m_k = F(k, m_{k-1}, \mathscr{K}_k) \in \mathscr{M} \tag{10.91}$$

for some suitably defined operator

$$F : \mathbb{N} \times \mathscr{M} \times \mathscr{K}_k \to \mathscr{M} \tag{10.92}$$

where $\mathscr{M}$ can be taken as the space of all compactly supported sequences (finite memory) or a Euclidian space (uniformly bounded memory) or (complete memory)

$$m_k = (\mathscr{K}_1, \mathscr{K}_2, \dots, \mathscr{K}_k) \tag{10.93}$$

Many feedforward ILC designs store the entire error trajectory in the memory, i.e. $\mathscr{M}_k = y_k - y_{\text{ref}_k}$. In most cases, this continuous signal is sampled to obtain a finite but large memory requirement. The designs given below store parameter estimates only and, hence, a much lower memory requirement.

The following is the (formal) definition of the feedforward ILC design considered.

Definition 10.1 Given

$$y_{\text{ref}_k} = y_{\text{ref}} \in \mathscr{Y}^{\text{ref}}, \quad d_k \in \mathscr{D}, \quad k \geq 1 \tag{10.94}$$

the feedforward ILC design problem is to construct an operator $\mathscr{B} : \mathscr{M} \to \mathscr{U}$ such that $\forall\, G \in \Delta$:

$$\mathscr{P}_k \to 0 \quad \text{as} \quad k \to \infty, \quad \|\mathscr{K}\|_{L^\infty(\mathbb{N}\times[0,\alpha])} < \infty \tag{10.95}$$

where H_k is of the form $H_k = G(d_k)$ and $u_k = \mathscr{B}m_k$.

The feedback ILC problem is defined as follows:

Definition 10.2 Given

$$y_{\mathrm{ref}_k} \in \mathscr{Y}^{\mathrm{ref}}, \quad d_k \in \mathscr{D}, \, k \geq 1 \tag{10.96}$$

the ILC feedback problem is to find an operator $\mathscr{A} : \mathscr{M} \times \mathscr{Y}^{\mathrm{ref}} \to C(\mathcal{O}, \mathscr{U})$, such that $\forall G \in \Delta$

$$\mathscr{P}_k \to 0 \quad \text{as} \quad k \to \infty, \quad \|\mathscr{K}\|_{L^\infty(\mathbb{N}\times[0,\alpha])} < \infty \tag{10.97}$$

where H_k and Ξ_k are of the form $H_k = G(d_k)$, $\Xi_k(m_k, y_{\mathrm{ref}_k}) = \mathscr{A}(m_k, y_{\mathrm{ref}_k})$ and satisfy the controlled system equations:

$$(H_k, \Xi_k(m_k, y_{\mathrm{ref}_k})) : y_k = H_k u_k, \; u_k = \Xi_k(m_k, y_{\mathrm{ref}_k})(y_k) \tag{10.98}$$

Proposition 1 in [85], given next, is critical to establishing the existence, boundedness, and convergence of solutions to the two problems defined above (where C^1 denotes the class of all differentiable functions whose derivative is continuous).

Proposition 10.3 *Let $f, h \in C^1(\mathbb{R}^n, \mathbb{R}_+)$, $g \in C^1(\mathbb{R}^p, \mathbb{R}_+)$ and suppose that f, g, h are positive definite. Also, let $[0, \tau_k)$ be the maximal interval of existence for $\mathscr{K}_k = (u_k, y_k, \xi_k, \pi_k)$ for signals $\xi_k \in C^1([0, \tau_k), \mathbb{R}^m)$, $\pi_k \in C^1([0, \tau_k), \mathbb{R}^p)$, $k \geq 1$. Also, assume that $\zeta_k \in C[0, \alpha]$ for $k \geq 1$. Suppose that $V : \mathbb{R}^{n+p} \to \mathbb{R}_+$ satisfies:*

(i) $V(\xi_k, \pi_k) = f(\zeta_k) + g(\pi_k)$
(ii) $|\mathscr{K}_k(t)| \leq v(V(\xi_k(t), \pi_k(t)))$, $\forall\, t \in [0, \tau_k)$ for some continuous function $v : \mathbb{R}_+ \to \mathbb{R}_+$
(iii) $\dot{V}(\xi_k(t), \pi_k(t)) \leq -h(\xi_k(t)) + \zeta_k(t)$, $\forall\, t \in [0, \tau_k)$
(iv) There exists $\lambda > 0$ such that

$$\lambda \int_0^\alpha h(\xi_k(t)) dt \geq \mathscr{P}_k \tag{10.99}$$

whenever the left-hand side is defined.

If $\xi_k(0) = 0, \pi_k(0) = \pi_{k-1}(\alpha)$, then the trial signals $\mathscr{K}$ are defined on $\mathbb{N} \times [0, \alpha]$ and $\forall\, K \geq 1$ and $\forall\, t \in [0, \alpha]$

$$V(\xi_K(t), \pi_K(t)) \leq V(\xi_1(0), \pi_k(0)) + \sum_{k=1}^{K-1} \int_0^\alpha \zeta_k(t) dt + \int_0^t \zeta_K(t) dt \tag{10.100}$$

$$\sum_{k=1}^{K} \mathscr{P}_k \leq \lambda g(\pi_1(0)) + \lambda \sum_{k=1}^{K} \int_0^\alpha \zeta_k(t) dt \tag{10.101}$$

If

$$\sum_{k=1}^{\infty} \int_0^\alpha \zeta_k(t) dt < \infty$$

then $\displaystyle\sum_{k=1}^{\infty} \mathscr{P}_k < \infty$ and $\|\mathscr{K}\|_{L^\infty(\mathbb{N}\times[0,\alpha])} < \infty$.

It is not assumed that solutions exist, hence, all assumptions made are in terms of $[0, \tau_k)$ – the maximal interval of existence of the solutions on trial k. In ILC applications, the trial length is finite and hence, it suffices to check that $\tau_k > \alpha$. The requirement of condition (ii) in this last result can be verified by bounding $|\xi_k|$, $|\pi_k|$ by V and then, in turn, bounding $|u_k|$, $|y_k|$ in terms of ξ_k, π_k, and y_{ref}.

To illustrate this last result, consider when a chain of integrators describes the dynamics with matched uncertainty. In particular,

$$\Delta = \{G(d) : \mathcal{U} \to \mathcal{Y} \mid G(d)(u(t)) = x(t)$$

$$\dot{x} = Ax + B(\theta^T \phi(x) + u + d'), \; x(0) = d'', \theta \in \mathbb{R}^p, d = (d', d'')^T\} \tag{10.102}$$

where the nonlinearity $\phi : \mathbb{R}^n \to \mathbb{R}^p$ is known and also

$$A = \begin{bmatrix} 0 & 1 & 0 & \dots & 0 \\ 0 & 0 & 1 & \dots & 0 \\ \vdots & \vdots & \vdots & \ddots & \vdots \\ 0 & 0 & 0 & \dots & 1 \\ 0 & 0 & 0 & \dots & 0 \end{bmatrix}, \quad B = \begin{bmatrix} 0 \\ 0 \\ \vdots \\ 0 \\ 1 \end{bmatrix} \tag{10.103}$$

In the case of no disturbances, corresponding to $d = 0$, the standard adaptive tracking design is as follows: Suppose $y_{\text{ref}} : \mathbb{R}_+ \to \mathbb{R}$ is a given reference trajectory and let

$$Y^{\text{ref}} = \begin{bmatrix} y_{\text{ref}} & y_{\text{ref}}^{(1)} & \dots & y_{\text{ref}}^{(n-1)} \end{bmatrix}^T \tag{10.104}$$

Also let a be a vector such that $A_* = A - Ba^T$ is Hurwitz (all eigenvalues of this matrix have strictly negative real parts) and if $Q > 0$ take $P > 0$ as the solution of the Lyapunov equation $A_*^T P + PA_* = -Q$. Finally, set $b = (P^T + P)B$ and consider a positive adaption gain β.

Consider the control law

$$\Xi(\hat{\theta}_0, \beta)(y_{\text{ref}}) : \; u = -\hat{\theta}^T \phi(x) - a^T(x - Y^{\text{ref}}) + y_{\text{ref}}^{(n)}$$

$$\dot{\hat{\theta}} = \beta(x - Y^{\text{ref}})^T b\phi(x), \; \hat{\theta}(0) = \hat{\theta}_0 \tag{10.105}$$

resulting in the controlled dynamics:

$$\dot{x} - \dot{Y}^{\text{ref}} = A_*(x - Y^{\text{ref}}) + B(\theta - \hat{\theta})^T \phi(x)$$

$$\dot{\hat{\theta}} = \beta(x - Y^{\text{ref}})^T b\phi(x), \quad x(0) = x_0, \quad \hat{\theta}(0) = \hat{\theta}_0 \tag{10.106}$$

and it can be shown that the Lyapunov function

$$V(x, \hat{\theta}) = (x - Y^{\text{ref}})^T P(x - Y^{\text{ref}}) + \frac{1}{2\beta}(\theta - \hat{\theta})^T(\theta - \hat{\theta}) \tag{10.107}$$

satisfies

$$\dot{V} = -(x - Y^{\text{ref}})^T Q(x - Y^{\text{ref}}) \tag{10.108}$$

and, hence, by La-Salles theorem $x - Y^{\text{ref}} \to 0$ as $t \to \infty$.

To apply this last design to the feedback ILC problem, consider $\mathcal{M} = \mathbb{R}^p$ and define the ILC design by

$$m_k = \theta_{k-1}(\alpha), \; m_1 = 0$$

$$\mathcal{A}m = \Xi(m, \beta)(\cdot) \tag{10.109}$$

and the cost function

$$\mathcal{P}_k = \int_0^\alpha (x_k - Y^{\text{ref}})^T Q(x_k - Y^{\text{ref}}) dt$$

$$\geq \sqrt{\underline{\lambda}(Q)} \|x_k - y_{\text{ref}_k}\|^2_{L^2[0,\alpha]} \tag{10.110}$$

where $\underline{\lambda}(Q)$ denotes the minimum eigenvalue of Q. A further assumption is required, i.e. all reference trajectories are compatible in the sense that

$$Y_k^{\text{ref}}(0) = 0, \quad k \geq 1 \tag{10.111}$$

since it is assumed that $x_0 = 0$ (below x_0 will be perturbed, but its nominal value will still be zero), and the following result is Proposition 2 in [85].

Proposition 10.4 *Consider Δ defined by (10.102) with $d = (d', d'')^T = 0$. Then the ILC design problem (10.109) based on the adaptive controller (10.105) solves the feedback ILC problem with cost function given by (10.110).*

In many ILC designs, it is assumed that the system resets to the same initial condition on each trial. The first case of the effects of disturbances on the controlled dynamics considered is when the initial condition reset is not perfect but subject to ℓ^1 perturbations, i.e.

$$\|d''\|_{\ell_1} = \sum_{k=1}^\infty \|x_k(0)\|_2 < \infty \tag{10.112}$$

The following result is Proposition 3 in [85] and solves the ILC feedback design problem in this case.

Proposition 10.5 *Let Δ be given by the first line in (10.102) with $d = (0, d'')^T$ and such that $\|d''\|_{\ell^1} \leq \delta_{d''}$. Suppose also that $\|Y_k^{\text{ref}}\|$ is uniformly bounded for $k \geq 1$. Then the ILC design given by (10.109) based on the adaptive controller of (10.105) is a solution to the feedback ILC design, with cost function given by (10.110).*

Perturbations that are nondecaying require modifications to the basic design. For example, if the initial condition is perturbed in ℓ^∞, the following example shows that unstable behavior can result.

Consider the case when $n = p = 1$, $\alpha = 1$, $\phi(x) = x$, $\theta = 0$, $y_{\text{ref}} = 0$ and $x_0 = 1$. This choice of initial condition corresponds to $x_k(0) = d_k'' = 1$ and also $d'' \in \ell^\infty$. Also, set $a = 1$, $\beta = 1$, resulting in controlled dynamics described by

$$\dot{x}_k = u_k, \quad x_k(0) = 1$$

$$u_k = -\hat{\theta}_k x_k - x_k$$

$$\dot{\hat{\theta}}_k = x_k^2, \quad \hat{\theta}_k(0) = \hat{\theta}_{k-1}(\alpha), \quad \hat{\theta}_1(0) = 0 \tag{10.113}$$

Two cases are considered, starting with that when $\int_0^\alpha x_k^2(t) dt \to 0$ as $k \to \infty$. Then

$$\alpha_k = \inf \left\{ t \in [0, \alpha] \ : \ x_k(t) = \frac{1}{2} \right\} \tag{10.114}$$

is defined for large k. Hence, by the Cauchy–Schwarz inequality

$$\left(\int_0^{\alpha_k} x_k^2 \, dt \right) \left(\int_0^{\alpha_k} u_k^2 \, dt \right) \geq \left(\int_0^{\alpha_k} x_k u_k \, dt \right)^2$$

$$= \left(\int_0^{\alpha_k} x_k \dot{x}_k \, dt \right)^2 = \frac{3}{8} \tag{10.115}$$

Hence, $\int_0^\alpha u_k^2 \, dt \to \infty$ as $k \to \infty$ and therefore,

$$\|u_k\|_{L^\infty[0,\alpha]} \to \infty, \quad \text{as } k \to \infty \tag{10.116}$$

As a second case, suppose that $\int_0^\alpha x_k^2(t)dt \nrightarrow 0$ as $k \to \infty$. Then there exists an $\epsilon > 0$ and a sequence $k_1, \ldots, k_n, \ldots$, such that $\int_0^\alpha x_{k_i}^2 \, dt \geq \epsilon$ for $i \geq 1$. Hence,

$$\hat{\theta}_{k_i+1}(0) = \hat{\theta}_{k_i}(\alpha)$$

$$= \hat{\theta}_{k_i}(0) + \int_0^\alpha x_{k_i}^2 \, dt \geq \hat{\theta}_{k_i}(0) + \epsilon$$

$$\geq \hat{\theta}_1(0) + (i-1) \to \infty, \quad \text{as } i \to \infty \tag{10.117}$$

The most straightforward modification to the adaptive design to ensure stability in the presence of ℓ^∞ / L^∞ disturbances is to introduce a dead-zone. To implement this modification, replace the second entry in (10.105) by

$$\dot{\hat{\theta}} = \beta D(\Omega_0, x - Y^{\text{ref}})(x - Y^{\text{ref}})^T b \phi(x), \quad \hat{\theta}(0) = \hat{\theta}_0 \tag{10.118}$$

where $D(\Omega_0, e) = 0, 1$, respectively, if $e \in \Omega_0, e \notin \Omega_0$. Moreover, Ω_0 is a neighborhood of zero of the form:

$$\Omega_0 = \{ e \in \mathbb{R}^n \ : \ e^T P e \leq \eta_0 \} \tag{10.119}$$

It can be shown that the adaptive design only achieves convergence of $x - Y^{\text{ref}} \to \Omega_0$. Consequently, the performance of the corresponding ILC design is measured by the cost function:

$$\mathscr{P}_k = \int_{\alpha_k} (x_k - Y_k^{\text{ref}})^T Q(x_k - Y_k^{\text{ref}}) dt \tag{10.120}$$

where α_k, $k \geq 1$, is defined to be the time set

$$\alpha_k = \{ t \in [0, \alpha] \ : \ (x_k - Y_k^{\text{ref}})(t) \notin \Omega_0 \} \tag{10.121}$$

The following is Proposition 4 in [85].

Proposition 10.6 *Let Δ be given by (10.102) with $d = (d', 0)^T$ such that $\|d'\|_{\ell^\infty / L^\infty} \leq \delta_{d'}$. Then for sufficiently small $\delta_{d'}$, the ILC design given by (10.109) based on the adaptive controller given by (10.105) and (10.118) is a solution to the feedback ILC design with the cost function given by (10.120).*

In the previous research, establishing good trial-to-trial error convergence rates has been extensively considered, see, e.g. [4]. In what follows, it is shown that the best controller over the first k trials has performance inversely proportional to k. To establish this result, the following general result is required (Lemma 1 in [85]).

Lemma 10.1 *Under the conditions of Proposition 10.3, suppose that $\zeta_k = 0$, $\forall k \geq 1$. Then*

$$\inf_{1 \leq j \leq k} \mathscr{P}_j \leq \lambda \frac{g(\pi_1(0))}{k} \tag{10.122}$$

The primary result on the convergence rate is as follows (Proposition 5 in [85]).

Proposition 10.7 *For Δ given in the first line of (10.102) and $d = 0$, the ILC design given by (10.109) based on the adaptive controller (10.105) and (10.118) solves the stopping feedforward ILC problem with a linear trial-to-trial convergence rate when the associated cost functions are of the form (10.110).*

Previously in this section, strict feedback systems have been considered. Such systems are the simplest scenario in nonlinear control, i.e. when the uncertainty is matched. In which case, the Lyapunov analysis is completed in the same coordinate system as the original state-space equations. Of course, this is not the generic situation in nonlinear control and a coordinate transformation is required before a Lyapunov analysis design can be completed.

The need for a coordinate transformation complicates the treatment of the initial condition resetting. In particular, the initial state may be reset identically in the original state-space, but this property does not automatically hold in the transformed state-space and this aspect requires careful consideration. An example of how this feature can cause unstable behavior is given next.

Consider the strict feedback system described by

$$
\begin{aligned}
\dot{x}_1 &= x_2 + \phi_1(x_1), \quad x(0) = 0 \\
\dot{x}_2 &= \theta\phi_2(x_1, x_2) + u, \quad \theta \in \mathbb{R} \\
y &= x_1
\end{aligned}
\tag{10.123}
$$

where the aim is stabilization, i.e. to track the reference trajectory $y_{\text{ref}} = 0$. A standard backstepping design [135] for this system in the case when ϕ_1 and ϕ_2 are known, and θ is an a priori unknown scalar is now developed.

Introduce for this last system

$$
\begin{aligned}
z_1 &= x_1 \\
z_2 &= x_2 - \gamma_1
\end{aligned}
\tag{10.124}
$$

where

$$
\gamma_1 = -z_1 - \phi_1(x_1)
\tag{10.125}
$$

and consider the control law

$$
\begin{aligned}
u &= \gamma_2 = -z_1 - z_2 - \hat{\theta}\phi_2 - \left(1 + \frac{\partial\phi_1}{\partial x_1}\right)(x_2 + \phi_1) \\
\dot{\hat{\theta}} &= z_2\phi_2
\end{aligned}
\tag{10.126}
$$

resulting in the system

$$
\begin{aligned}
\dot{z}_1 &= -z_1 + z_2 \\
\dot{z}_2 &= -z_1 - z_2 + (\theta - \hat{\theta})\phi_2 \\
\dot{\hat{\theta}} &= z_2\phi_2
\end{aligned}
\tag{10.127}
$$

Using the Lyapunov function,

$$
V(z, \hat{\theta}) = \frac{1}{2}\left(z_1^2 + z_2^2 + (\theta - \hat{\theta})^2\right)
\tag{10.128}
$$

gives

$$
\dot{V} \le -(z_1^2 + z_2^2)
\tag{10.129}
$$

Hence, $y \to 0$ as $t \to \infty$.

In the ILC case, start with the control law initialization

$$\hat{\theta}_1(0) = 0, \quad \hat{\theta}_k(0) = \hat{\theta}_{k-1}(\alpha), \quad k \geq 2 \tag{10.130}$$

As established below, a particular choice of ϕ_1 and ϕ_2 results in unstable control action, i.e.

$$\|u_k\|_{L^\infty} \to \infty, \quad k \to \infty \tag{10.131}$$

Take $\phi_1(0) > 0$ and introduce

$$\phi_2 = z_2 = x_1 + x_2 + \phi(x_1) \tag{10.132}$$

Then the critical fact is that the transformed state vector entries in the definition of V are not reset to zero at the start of each trial. Therefore, in contrast to the analysis above, the original state vector's resetting does not imply that the transformed state vector has zero entries. This fact is established next, based on two cases.

In the first case, suppose that $\int_0^\alpha z_{k2}^2 \, dt \not\to 0$ as $k \to \infty$ (where the subscript $k2$ denotes z_2 on trial k). Hence,

$$\hat{\theta}_k(0) = \sum_{i=1}^{k-1} \int_0^\alpha \dot{\hat{\theta}}_i \, dt = \sum_{i=1}^{k-1} \int_0^\alpha z_{i2}^2 \, dt \to \infty, \quad k \to \infty \tag{10.133}$$

Then

$$\|u_k\|_{L^\infty} \geq |u_k(0)| \geq \hat{\theta}_k(0)\phi_2(0) - \lambda \to \infty, \quad k \to \infty \tag{10.134}$$

for some $\lambda \in \mathbb{R}$ since $\phi_2(0) = \phi_1(0) > 0$.

As the second case, suppose that $\int_0^\alpha z_{k2}^2 \, dt \to 0$ as $k \to \infty$. Then, since $z_{k2}(0) = \phi_1(0) > 0$, it follows that $\dot{z}_{k2}(0) \to \infty$ as $k \to \infty$. Also,

$$\dot{z}_{k2} = u_k(0) + \theta\phi_2(0) + \left(1 + \frac{\partial \phi_1}{\partial x_1}\big|_{x_1=0}\right)(x_2(0) + \phi_1(0)) \tag{10.135}$$

and hence,

$$|\dot{z}_{k2}| = |u_k(0)| + |\theta\phi_2(0) + \left(1 + \frac{\partial \phi_1}{\partial x_1}\big|_{x_1=0}\right)(x_2(0) + \phi_1(0))| \tag{10.136}$$

As the last term in (10.136) is constant for all $k \geq 1$, it follows that $|u_k(0)| \to \infty$ as $k \to \infty$ and hence, $\|u_k\|_{L^\infty} \to \infty$ as $k \to \infty$. This example demonstrates that adaptive control design must be used with care in ILC since the state reset can be destabilizing.

To develop the adaptive design, start with

$$\begin{aligned}
\Delta = \big\{ H : \mathcal{U} \to \mathcal{Y} \ : \ &H(u(t)) = x(t), \quad \theta \in \mathbb{R}^p \\
&\dot{x}_i = x_{i+1} + \phi_1(x_1, \ldots, x_i)^T \theta, \quad x(0) = 0, \quad 1 \leq i \leq n-1 \\
&\dot{x}_n = \phi_n(x_1, \ldots, x_n)^T \theta + u \\
&\phi_1(0) = \cdots = \phi_n(0) = 0 \big\}
\end{aligned} \tag{10.137}$$

A coordinate transform was used in the previous analysis under the assumption that $x_0 = 0$, where this choice incurred no loss of generality. In the analysis below, it is also required to assume that the origin is an equilibrium point, where no loss of generality arises in assuming that $\phi_1(0) = \cdots = \phi_n(0) = 0$ if $x = 0$ is an equilibrium point. This assumption is required to avoid an instability phenomenon, as in the second example above.

To avoid this instability, the trajectory initialization method [135] when the reference signal sequence is persistently exciting could be used. An alternative is the tuning functions design, also from [135], which is outlined next in the non-ILC case.

Set $x_{n+1} = 0$, $\beta_0 = 0$ and

$$z_i = x_i - y_{\text{ref}}^{(i-1)} - \beta_{i-1}$$

$$\beta_i = -z_{i-1} - c_i z_i - w_i^T \hat{\theta}$$

$$+ \sum_{k=1}^{i-1} \left(\frac{\partial \beta_{i-1}}{\partial x_k} x_{k+1} + \frac{\partial \beta_{i-1}}{\partial y_{\text{ref}}^{(k-1)}} y_{\text{ref}}^{(k-1)} \right)$$

$$- \kappa_i |w_i|^2 z_i + \frac{\partial \beta_{i-1}}{\partial \hat{\theta}} \Gamma \tau_i + \sum_{k=2}^{i-1} \frac{\partial \beta_{i-1}}{\partial \hat{\theta}} \Gamma w_i z_k \tag{10.138}$$

$$\tau_i = \tau_{i-1} + w_i z_i$$

$$w_i = \phi_i - \sum_{k=1}^{i-1} \frac{\partial \beta_{i-1}}{\partial x_k} \phi_k \tag{10.139}$$

where it is assumed that $c_1, \ldots, c_n > 0$. The adaptive controller is

$$\Xi(\hat{\theta}_0, \Gamma)(y_{\text{ref}}) : u = \beta_n + y_{\text{ref}}^{(n)}$$

$$\dot{\hat{\theta}} = \Gamma \tau_n, \quad \hat{\theta}(0) = \hat{\theta}_0 \tag{10.140}$$

Also, define the coordinate transformation $T : \mathbb{R}^n \times \mathbb{R}^p \to \mathbb{R}^n \times \mathbb{R}^p$ by $T(x, \hat{\theta}) = (z, \hat{\theta})$. Then since there is an equilibrium point at the origin, this transformation has the properties that (i) $T(0, \hat{\theta}) = (0, \hat{\theta})$, $\forall \hat{\theta} \in \mathbb{R}^p$ since it is assumed that $Y_k^{\text{ref}}(0) = x(0) = 0$, and (ii) $z_1 = y - y_{\text{ref}}$ since $\beta_0 = 0$.

The following is Proposition 6 in [85].

Proposition 10.8 *Consider the adaptive design $\Xi(y_{\text{ref}}, \hat{\theta}_0)$ defined above applied to any system $G \in \Delta$, where Δ is given by (10.137). Consider also the following choice of Lyapunov function for this system:*

$$V(z, \hat{\theta}) = \frac{1}{2} z^T z + \frac{1}{2}(\theta - \hat{\theta})^T(\theta - \hat{\theta}) \tag{10.141}$$

Then

$$\dot{V} \leq -\sum_{i=1}^n c_i z_i^2 \tag{10.142}$$

The similarities between the Lyapunov functions for the two designs do not mean that it is relatively easy to ensure trial-to-trial error convergence for the ILC design. This problem arises because resetting the state x does not correspond to resetting the transformed state z; Proposition 10.3 cannot be directly applied. Under the extra assumption that 0 is an equilibrium point, $x = 0$ implies that $z = 0$ and, therefore, Proposition 10.3 applies. (See the second example above.)

Returning to the ILC design, introduce

$$m_k = \theta_{k-1}(\alpha), \quad m_1 = 0$$

$$\mathscr{A} m = \Xi(m, \Gamma)(\cdot) \tag{10.143}$$

and associated L^2 cost function:

$$\mathscr{P}_k = \|z_k\|_{L_2^2[0,\alpha]}^2 \geq c_1 \|x_{1k} - y_{\text{ref}_k}\|_{L_2^2[0,\alpha]}^2 \tag{10.144}$$

and again, it is assumed that $Y_k^{\text{ref}}(0) = x(0) = 0$. The following is Proposition 7 in [85].

Proposition 10.9 *Suppose that Δ is defined by (10.137). Then the ILC design (10.143) based on the adaptive control law (10.140) solves the feedback ILC problem with the L^2 cost function (10.144).*

The corresponding solution to the stopping feedforward ILC problem has a linear mean square trial-to-trial convergence rate by Lemma 10.1.

The analysis above extends to output feedback of ILC as detailed in [85], which also discusses the special case of linear adaptive ILC. Overall, the results in this section demonstrate that adaptive control designs can be employed in ILC, where the principal drawback of such designs is the requirement for online updating, rather than all computations occurring at the end of each trial. This fact, in turn, means that the implementing of such designs are more complex than many other ILC alternatives.

In contrast to most ILC designs, however, these control schemes are universal to parameter variations, involve no growth constraints on nonlinearities, are valid for any known relative degree, and achieve quantifiable convergence rates. Convergence rate analysis is straightforward. In particular, a linear rate of convergence of the mean square error can be achieved for nonminimum phase linear systems. The ILC problem of learning unseen trajectories (generalization) is easily achieved within this framework.

Although the designs require an uncertain parametrization, the existence of such a parametrization is assured in the case of finite-dimensional linear systems and can with appropriate treatment of modeling errors also be assured in the nonlinear case by using appropriate function approximators in the manner of, e.g. [236], where such approximators are widely used in ILC, e.g. [93], to parameterize input vectors.

10.5 Extremum-Seeking ILC

Data-driven, or nonmodel-based, ILC has received considerable attention in recent years. In this section, an approach based on extremum seeking is described. The analysis is for SISO time-varying nonlinear ILC of discrete-time systems, where the extension to MIMO systems is immediate.

A critical feature of extremum seeking is its ability to locate an optimum with respect to some measure without assuming knowledge of the models that govern the dynamics of the systems considered. The results in this section are based on [131], which, in turn, gives references to the basics [13]. Examples of application in particular areas include biochemical reactors [96] and gas-turbine combustors [165]. An ILC related application is in [223] and the notation in this section is (in the main) that used in [131].

The systems considered are defined over the trial length $\mathbb{T} = \{0, 1, \ldots, \alpha\}$ on trial $k + 1$ by

$$x_{k+1}(p + 1) = f(x_{k+1}(p), u_{k+1}(p), p), \quad x(0) = \bar{x}_k$$
$$y_{k+1}(p) = h(x_{k+1}(p), u_{k+1}(p), p) \tag{10.145}$$

where $f : \mathbb{R}^n \times \mathbb{R} \times \mathbb{T} \to \mathbb{R}^n$ and $h : \mathbb{R}^n \times \mathbb{R} \times \mathbb{T} \to \mathbb{R}$ are assumed to be locally Lipschitz functions in each argument. Moreover, the dependence of both the state and output equations on p accounts for repeated disturbances on each trial's state and output dynamics. Also, the error on trial k (where r again denotes the trajectory signal) is $e_k = r - y_k$ and in [131] minimization of this quantity is in terms of the ℓ_2 norm.

Remark 10.2 If the uncontrolled system is unstable, a stabilizing controller must regulate the transient response along the trials. Additionally, optimizing the system's initial condition requires its dimension to be known. If this information is not available, one option is to take it as constant for all trials.

In the extremum-seeking control setting, the first step is to transform the ILC problem to one of static optimization. Let $\mathcal{U}$ denote the set of functions $u : \mathbb{T} \to \mathbb{R}$ and for a vector $v \in \mathbb{R}^{n+\alpha+1}$ introduce (following the terminology in [131]) the demultiplexer $D : \mathbb{R}^{n+\alpha+1} \to \mathbb{R}^n \times \mathcal{U}$ as

$$D(v) = (w, z)$$
$$w = \begin{bmatrix} v_1 & \cdots & v_n \end{bmatrix}^T$$
$$z(p) = v_{n+1+p}, \quad p = 0, 1, \ldots, \alpha \tag{10.146}$$

Also, given a vector $w \in \mathbb{R}^n$ and $z \in \mathcal{U}$, define the multiplexer $M : \mathbb{R}^n \times \mathcal{U} \to \mathbb{R}^{n+\alpha+1}$ by

$$v = M(w, z) \tag{10.147}$$

$$v_i = w_i, \quad i = 1, 2, \ldots, n \tag{10.148}$$

$$v_p = z(p), \quad p = n+1, n+2, \ldots, n+\alpha+1 \tag{10.149}$$

These definitions are the link between the system response and optimization methods as detailed below.

Consider the case when $\bar{x}_k \in \mathbb{R}^n$, $u_k \in \mathcal{U}$, and $\theta_k := M(\bar{x}_k, u_k)$ are given and define, using (10.145), $\mathbb{Q} : \mathbb{R}^{n+\mathbb{T}+1} \to \mathbb{R}$ as

$$\mathbb{Q}(\theta_k) = \|r - \Sigma(D(\theta_k)\|_2 = \|r - \Sigma(\bar{x}_k, u_k)\|_2$$
$$= \|r - y_k\|_2 = \bar{e}_k \tag{10.150}$$

where $\mathbb{Q}$ is locally Lipschitz continuous by the assumptions already introduced, and for linear time-invariant systems $\mathbb{Q}$ is a convex function. This formulation transforms the ILC problem to one where static optimization can be applied, and hence, amenable to a wide range of local and global optimization algorithms. Consequently, θ_{k+1} can be designed using

$$\theta_{k+1} = \Gamma(\bar{e}_k, \ldots, \bar{e}_{k-s}, \theta_k, \ldots, \theta_{k-s}) \tag{10.151}$$

applied to Σ using the demultiplexer D and is an higher-order iterative learning control (HOILC) law unless $s = 1$.

Extremum-seeking ILC design can be summarized as in Figure 10.14. In this setting, the demultiplexer D has an input vector of real numbers from the optimization method and outputs the initial condition and control input to the system for the subsequent trial. One strength of this method is the option to use local, gradient-based, and nonconvex global sampling-based optimization methods.

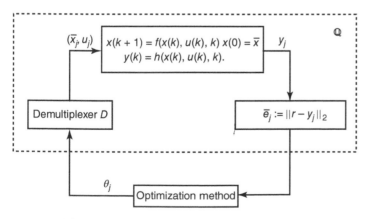

Figure 10.14 Extremum-seeking ILC application.

Local optimization methods generally use estimates of the first (or higher-order) cost function derivatives in computing the control signals.

Global-based optimization methods can be employed in cases where significant variations in the initial condition and control input can be tolerated as the trials progress. These algorithms, which do not use the information on the gradients of the cost function, can be used to find global minima in the presence of local minima in ILC design. Moreover, the framework extends to allow saturation constraints on the inputs. These aspects are further detailed in [131].

10.6 Concluding Remarks

This chapter has developed ILC designs for nonlinear dynamics, where two of them have been demonstrated in two application areas. The theory that is used in these areas is undergoing continual development in the control systems research area. In the stroke rehabilitation application, it has been shown that the gap metric approach is particularly relevant. Supporting experimental results have also been given for this gap metric application. Extremum seeking, covered briefly in the last section, has attributes that make it a potentially powerful method in more recent applications for ILC, especially based on data-driven modeling, one of which is covered in the final Chapter 13 of this book. The next chapter develops Newton ILC, using well-known numerical solution algorithms for nonlinear algebraic equations.

11

Newton Method Based ILC

This chapter considers Newton-based iterative learning control (ILC) design, where the first step is to rewrite the nonlinear dynamics as a set of static nonlinear multivariable equations akin to extremum-seeking ILC considered in Chapter 10. Once in this equivalent setting, longstanding results for such multivariable equations can be employed. As one application, this method of ILC design is applied to the stroke rehabilitation application area, where the analysis is supported by experimental results.

11.1 Background

The Newton method is a longstanding technique in numerical analysis [183]. As this chapter demonstrates, it is possible to formulate some nonlinear ILC design problems as a Newton iteration and used well-established results from this area of numerical analysis to develop efficient and reliable solutions. The starting point for analysis in the ILC setting is the discrete-time state-space model, single-input single-output (SISO) case is considered for ease of presentation

$$
\begin{aligned}
x_k(p+1) &= f(x_k(p), u_k(p), p), \quad 0 \le p \le N-1 \\
y_k(p) &= h(x_k(p), u_k(p), p), \quad x_k(0) = x_0
\end{aligned}
\tag{11.1}
$$

where $x_k(p) \in \mathbb{R}^n$, $r(p)$ again denotes the specified reference signal, and it is assumed without loss of generality that the state initial vector is the same on all trials.

Early results on this setting for nonlinear ILC design, includes [269], which considered the design of an ILC law of the form:

$$
u_{k+1}(p) = u_k(p) + L_k e_k(p)
\tag{11.2}
$$

where L_k is the trial-varying learning gain, using a DC motor model as a specific case. Trial-to-trial error convergence was established by applying the contraction mapping theory. A more strict proof of trial-to-trial error convergence in the sup-norm is in [280].

In [46], P-type ILC, with constant learning gain, $L_k = L$, $\forall\, k$ in (11.2) was used in a cascade control system with an inner proportional plus integral plus derivative (PID) controller for the same model as in [269]. Moreover, delayed signals were used in the ILC law to compensate for the system delay. Experimental application of the design to a proportional-valve-controlled pneumatic X–Y (2D plane) table was also reported.

Iterative Learning Control Algorithms and Experimental Benchmarking, First Edition.
Eric Rogers, Bing Chu, Christopher Freeman and Paul Lewin.
© 2023 John Wiley & Sons Ltd. Published 2023 by John Wiley & Sons Ltd.

Remark 11.1   Recall also Section 9.1.2 for a more recent consideration of trial-varying ILC laws with application in the use of smart rotors for wind turbines.

Consider the system

$$y = F(u) \tag{11.3}$$

where F is a nonlinear static function mapping one Banach space into another, and y and u are sampled sequences. In [279], numerical linearization was applied to approximate the nonlinear ILC problem whose dynamics can be described in abstract form, see below, by this last equation, and the control law (11.2) was applied with learning gain optimization.

In the case of the P-type ILC law (11.2), [269] used approximate Newton directions and also an automatic step-length update to obtain linear-rate, local trial-to-trial convergence starting from representing the dynamics in the form (11.3). This work set the learning gain on trial k equal to the Newton index, see below, but did not address the related question of how to estimate this index for the general case. Instead, local convergence of the control input was established, and an illustrative simulation example based on a multilink robotic manipulator model was given.

Consider the differential version of (11.1). Then [112] developed a Newton-type ILC law for such systems, where it was assumed that F was either known or identified. The ILC law was derived using the Taylor series expansion and applied in simulation to an electrostrictive actuator.

This early work focused on general nonlinear systems where the dynamics are reformulated as nonlinear algebraic equations that, in turn, avoid the unwanted effects arising from the calculation of nonlinear inverse systems and approximating Newton directions. It is also the case that the nonlinear ILC law can be decomposed into a sequence of linear time-varying (LTV) ILC problems. Moreover, for accurate implementation, the local linear problems are solved precisely, and arbitrary globally convergent linear ILC laws ensure this property. The remainder of this chapter details the development of several versions of this general law and, as an application, its use in stroke rehabilitation, see Section 2.4 for the background on this application area, is described.

11.2 Algorithm Development

This section and the next is based on [145–147] and starts from (11.1). The SISO case is considered for ease of presentation, and it is assumed without loss of generality that the initial state vector is the same on all trials. Moreover, the lifted system is used.

To write the dynamics of (11.1) as a set of algebraic equations in $\mathbb{R}^N$, introduce the following supervectors:

$$U_k = \begin{bmatrix} u_k(0) & u_k(1) & \dots & u_k(N-1) \end{bmatrix}^T$$
$$Y_k = \begin{bmatrix} y_k(1) & y_k(2) & \dots & y_k(N) \end{bmatrix}^T \tag{11.4}$$

Then the systems dynamics can be described by the following algebraic functions:

$$y_k(1) = h[x_k(1)] = h[f(x_k(0), u_k(0))] = g_1[x_k(0), u_k(0)]$$
$$y_k(2) = h[x_k(2)] = h[f(x_k(1), u_k(1))] = g_2[x_k(0), u_k(0), u_k(1)]$$
$$\vdots$$
$$y_k(N) = h[x_k(N)] = h[f(x_k(N-1), u_k(N-1))]$$
$$= g_N[x_k(0), u_k(0), u_k(1), \dots, u_k(N-1)] \tag{11.5}$$

Moreover, since the initial state vector is the same on each trial, the dynamics of (11.1) can be represented by the following algebraic function in $\mathbb{R}^N$

$$Y_k = g(U_k)$$
$$g(\cdot) = [g_1(\cdot), g_2(\cdot), \dots, g_N(\cdot)]^T \tag{11.6}$$

and the ILC problem is equivalent to solving the N-dimensional nonlinear algebraic equation:

$$F(U_k) = g(U_k) - R = 0 \tag{11.7}$$

for U_k, where the ideal sequence $\{U_k\}_k$ has a limit U_∞ such that $F(U_\infty) = 0$, and R is the supervector corresponding to the reference trajectory $(r(p))$.

There is a strong connection between ILC laws and iterative numerical solutions of algebraic equations. Both use the results from previous trials to force the solution closer to the desired. Hence, it is possible to use iterative algorithms for solving nonlinear equations to solve nonlinear ILC design problems. The Newton method is one choice from the class of iterative algorithms for solving nonlinear equations and, as will become apparent below, is especially useful in ILC design. Hence, no others are considered in what follows.

Consider again the P-type ILC law (11.2) where for each trial $L_k = L$. Then the Newton method is an iterative numerical method to solve (11.7) and is a special case of the parallel-cord method, see, e.g. [183], for solving nonlinear multivariable equations. This fact, in turn, means that the Newton method's well-established convergence properties apply to the ILC design problem.

Consider the nonlinear multivariable equation $F(b) = 0$ with $F : \mathbb{R}^N \to \mathbb{R}^N$. Then the parallel-cord method is defined by

$$b_{k+1} = b_k - H^{-1}F(b_k), \quad k = 0, 1, \dots \tag{11.8}$$

where the coefficient matrix H is assumed to have constant entries and is invertible. Suppose that b^* is a solution of $F(b) = 0$ and $F'(b^*)$ exists. Then it is known [183] that this algorithm is locally convergent if and only if

$$\sigma = \rho(I - H^{-1}F'(b^*)) < 1 \tag{11.9}$$

and $F'(b^*)$ denotes the Jacobian matrix. Also, the local convergence property guarantees the existence of a $\delta > 0$ such that, when $|b - b^*| < \delta$, the algorithm converges in norm to b^*. Moreover, the smaller the σ is the faster the convergence, and this property can be influenced by appropriate selection of the matrix H.

In application to ILC, rewrite (11.7) as $F(U_k) = -E_k$, and take the particular choice of $H = F'(U_k)^{-1} = g'(U_k)^{-1}$ to obtain the Newton's method ILC law:

$$U_{k+1} = U_k + g'(U_k)^{-1}E_k, \quad k = 1, 2, \dots \tag{11.10}$$

Under certain conditions [183], the Newton method (11.8) can exhibit local quadratic convergence, i.e. the convergence of U_k to the learned control U_∞ satisfies

$$\|U_{k+1} - U_\infty\| \le c\|U_k - U_\infty\|^2, \quad c > 0 \tag{11.11}$$

The remaining problem is to find an efficient method to implement this ILC law without numerical difficulties that could arise due to one or more of the following: (i) the calculation of the derivative $g'(\cdot)$, (ii) the calculation of the inverse $g'(\cdot)^{-1}$, and (iii) the dimension N (number of samples along the trial).

Given the requirement for ILC design to inherit advantageous features from the Newton method, any form of approximation is not acceptable. One way to address this issue is to select a particular form of approximation to the Newton direction, but this may degrade the method's properties. For example, the design in [18] only achieves linear-local rate trial-to-trial error convergence due to the approximation used.

11.2.1 Computation of Newton-Based ILC

Write (11.10) in the form

$$U_{k+1} = U_k + z_{k+1} \tag{11.12}$$

and compute $z_{k+1} = g'(U_k)^{-1} E_k$ by solving

$$g'(U_k) z_{k+1} = E_k \tag{11.13}$$

and the calculation of the inverse is avoided. In ILC terms, g represents the nonlinear system dynamics. Hence, the derivative $g'(u_k)$ is equivalent to the linearization of (11.1) in k at $(u_k(p), x_k(p))$. Moreover, solving (11.13) is equivalent to finding z_{k+1} that produces the linearized system $g'(U_k)$ generating the desired E_k. In particular, for each k, the following linear system has to be solved:

$$\hat{Y} = g'(U_k)\hat{U} \tag{11.14}$$

where $\tilde{Y}$ is required to track E_k, and $\tilde{U}$ represents z_{k+1} in the Newton ILC law.

The linearization of the system dynamics at (u_k, x_k) is

$$\tilde{x}(p+1) = A(p)\tilde{x}(p) + B(p)\tilde{u}(p)$$
$$\tilde{y}(p) = C(p)\tilde{x}(p) \tag{11.15}$$

where

$$\tilde{x} = \delta x_{k+1} = x_{k+1} - x_k, \quad A(t) = \left(\frac{\partial f}{\partial x}\right)_{u_k, x_k}$$

$$\tilde{u} = \delta u_{k+1} = u_{k+1} - u_k, \quad B(t) = \left(\frac{\partial f}{\partial u}\right)_{u_k, x_k}$$

$$\tilde{y} = \delta y_{k+1} = y_{k+1} - y_k, \quad C(t) = \left(\frac{\partial h}{\partial x}\right)_{u_k, x_k} \tag{11.16}$$

The initial condition for (11.16) is zero by the assumption of the same initial condition on each trial. Also, the input $\tilde{U}$ of the linearized system gives the update for U_k and the difficulty of calculating the derivative can be avoided by simulating the linearized system.

At this stage, the design can be completed by any ILC law that results in global convergence for an arbitrary LTV system. Possible choices include norm optimal iterative learning control (NOILC) (see Chapter 5), its predictive extension, or an adaptive adjoint law. By considering the linear algebraic equation (11.15) as a LTV ILC problem, the N-dimensional equation is transformed to an n-dimensional dynamic system. Hence, avoiding the difficulty of solving a potentially large-dimensional equation. Two numerical examples to illustrate the basics of Newton-based ILC design are given in [146, 147].

11.2.2 Convergence Analysis

The Newton ILC law is a special case of the Newton method and, hence, it inherits the convergence properties of this method. Semilocal quadratic convergence is established by the Newton–Kantorovich theorem stated next, see, e.g. [183] for a detailed treatment.

Theorem 11.1 *Consider the nonlinear equation $F(x) = 0$ and suppose that $F : D \subset \mathbb{R}^N \to \mathbb{R}^N$ is Frechet-differentiable on a convex set $D_0 \subset D$ and also*

$$\|F'(a) - F'(b)\| \leq \gamma \|a - b\| \tag{11.17}$$

Suppose also that there exists $x_0 \in D_0$ satisfying $\|F'(x_0)\| < \beta$, and that $\zeta = \beta\gamma\eta \leq \frac{1}{2}$, where $\eta \leq \|(F'x_0)^{-1}F(x_0)\|$. Set $q^* = \frac{1-(1-2\zeta)^{\frac{1}{2}}}{\beta\gamma}$, $q^{**} = \frac{1+(1-2\zeta)^{\frac{1}{2}}}{\beta\gamma}$, and suppose that the closed ball $\overline{S}(x_0, q^*) \subset D_0$. Then the Newton-type iteration

$$x_{k+1} = x_k - F'(x_k)^{-1}F(x_k), \quad k = 0, 1, \ldots \tag{11.18}$$

is well defined, remains in $\overline{S}(x_0, q^*)$, and converges to a solution x^* of $F(x) = 0$ that is unique in $S(x_0, q^{**}) \cap D_0$. Moreover,

$$\|x^* - x_k\| \leq \frac{(2\zeta)^{2^k}}{2^k \beta\gamma} \tag{11.19}$$

The bounds β, γ, and η, subject to the constraint $\zeta = \beta\gamma\eta \leq \frac{1}{2}$, on the nonlinear function F and the initial value x_0 in this last result have the following interpretations for Newton ILC:

1. $\|F'(x_0)\| < \beta$ bounds the gradient of F at x_0, which must be finite for the Newton update on the first trial to be nonzero. In particular, this assumption results in $\|g'(U_1)\| < \beta$ and requires that the output of the example under consideration does not vary infinitely under a finite change in the initial control input. Consequently, the change in the initial control input can be calculated.
2. $\|F'(a) - F'(b)\| \leq \gamma \|b - a\|$ defines the magnitude of the nonlinearity F. If $F'(a) = F'(b)$ for an arbitrary a and b, the system is linear and nonlinear otherwise. This inequality restrains the growth and amplitude of the nonlinearity. In the ILC setting, this condition becomes $\|g'(a) - g'(b)\| \leq \gamma \|b - a\|$ and limits the magnitude of the nonlinearity.
3. $\eta \leq \|F'(x_0)^{-1}F(x_0)\|$ is a measurement of the first update $F'(x_0)^{-1}F(x_0)$. Also, since $F'(x_0)$ is finite, $\eta = 0$ only when $x_0 = x^*$. Hence, there exists a neighborhood of x^* in which η increases as the distance between x_0 and the solution x^* increases. Therefore, this constraint can also be interpreted as measuring how close the initial value is to the solution. In the ILC setting, this condition becomes $\eta \leq \|g'(u_1)^{-1}g(u_1)\|$ and is a measure of whether or not the initial control input is a good guess or estimation of the ideal one.

These conclusions are qualitative only since β, γ, and η cannot often be computed for an example. The condition $\zeta \leq \frac{1}{2}$ gives local convergence of the Newton method. Moreover, when $x_0 = x^*$, $\eta \leq \|F'(x_0)^{-1}F(x_0)\| = 0$ and $\zeta = 0 \leq \frac{1}{2}$. Hence, there exists a $\Delta > 0$ such that $\|x_0 - x^*\| \leq \Delta, \zeta \leq \frac{1}{2}$, independent of the maximum value of γ. Also, for a nonlinearity of arbitrary magnitude, the Newton method can converge for an initial control input "close enough" to the solution. If F is linear, i.e. $\gamma = 0$, then $\zeta = 0$ and an arbitrary initial value in D_0 can be used.

In the case of Newton ILC, the convergence property can be expressed as follows: there always exists an initial control input u_1 close enough to the ideal control input such that $\eta \leq \|g'(u_1)^{-1}g(u_1)\|$ is sufficiently small to ensure $\zeta = \beta\gamma\eta \leq \frac{1}{2}$, and the algorithm converges to the ideal control input. Consequently, the Newton method can converge with an initial value close to the solution for an arbitrary nonlinearity. If the nonlinearity is very strong, the initial value needs to be closer to the solution.

The quantity q^* provides a measure of how close the initial control needs to be to the solution since, when all the conditions are satisfied, x_k, $k = 0, 1, \ldots$ will remain in $\overline{S}(x_0, q^*)$ and will converge to x^*, which is also in this ball. Hence, $\|x_0 - x^*\| \leq q^*$ indicates how close the initial value is to the solution. If the nonlinearity is stronger and, hence, γ is large, the maximum radius of convergence is $q*_m = \frac{1}{\beta\gamma}$ (obtained by selecting x_0 such that $\zeta = \frac{1}{2}$), and the initial value x_0 must be selected closer to the solution x^*. Conversely, if the initial value x_0 is selected very close to the solution x^* $F(x_0)$ can be close to zero and η becomes very small.

For the last case, γ can be larger subject to $\zeta = \beta\gamma\eta \leq \frac{1}{2}$, i.e. less limitations on the nonlinearity. Also for Newton ILC, the maximum distance between the initial control input and the ideal is defined by $q_m^* = \frac{1}{\beta\gamma}$ and $\overline{S}(x^*, q_m^*)$ decreases with the strength of the nonlinearity.

In the Newton method, ζ determines the convergence speed, and (11.19) shows how close the result is from the solution on trial k. Also, since $\zeta \leq \frac{1}{2}$ and both β and γ are finite, $(2^k)^{-1}$ and $(2\zeta)^{2k}$ converge to zero rapidly, and the difference between x_k and x^* reduces quickly. Also, the smaller the ζ, the faster the convergence. Moreover, $\zeta = \beta\gamma\eta$ is a measure of convergence speed, and the weaker the nonlinearity, the smaller are γ and ζ and the faster the convergence. Fast convergence will also result if the initial control input is close to the ideal since γ and ζ will be smaller.

The condition $\zeta \leq \frac{1}{2}$, leads [183] to the error estimate:

$$\|x^* - x_k\| \leq c_1 \|x_k - x_{k-1}\|^2, \quad k \geq 1 \tag{11.20}$$

for some $c_1 > 0$, i.e. subsequent entries in the sequence become closer with increasing k. Also, under certain conditions on F, the pair x_{k-1}, x_k satisfies, as shown in [183],

$$\|x_k - x^*\| \leq c_2 \|x_{k-1} - x^*\|^2 \tag{11.21}$$

for some $c_2 > 0$, when x_k is close to x^*. This property explains the rapid terminal convergence speed of the algorithm. In summary, Newton ILC is locally convergent with a quadratic convergence speed on the terminal trials.

11.3 Monotonic Trial-to-Trial Error Convergence

The analysis in the previous sections proceeded by converting an ILC problem for nonlinear system dynamics into the solution of a set of nonlinear algebraic equations, which are solved by Newton method-based ILC. The resulting ILC laws have a local convergence property. Still, it is known, see the numerical examples in [147], that nonmonotonic convergence can occur if the initial control input is not selected appropriately. This problem arises from how the law evaluates the system's nonlinearity. In particular, the linearized model on the current trial can differ significantly from one trial to the next if the control input is changed too much.

This underestimation of the system dynamics, in turn, leads to temporarily or even permanently diverging updates for the control input, resulting in overshooting or divergence behavior. This non-monotonicity may have a critical effect on practical applications. Other results on the impact of the initial control input on an ILC design can be found in [76, 77] with experimental validation on the gantry robot of Section 2.1.1.

11.3.1   Monotonic Convergence with Parameter Optimization

The system description and ILC problem of the previous section is augmented by the requirement of monotonic trial-to-trial error convergence in the 2-norm, i.e.

$$\|e_{k+1}\|_2 \le \|e_k\|_2 \tag{11.22}$$

is required. Moreover, the nonmonotonicity of the design in the previous section is caused by nonadjustable learning gain. In particular, the Newton update term indicates the direction in which an appropriate change of control input may lead to a smaller tracking error. The learning step length in the Newton direction decides the updated control input and, hence, whether or not the tracking performance improves on the subsequent trial.

One-step convergence can be achieved for linear dynamics by a unit step length in the Newton direction. For nonlinear systems, the Newton direction may not represent the one-step convergence direction. An improper step in this direction could result in the update departing from the convergence direction. Hence, a variable step length is required in the form

$$U_{k+1} = U_k + \gamma_{k+1} g'(U_k)^{-1} E_k, \quad \gamma_{k+1} > 0 \tag{11.23}$$

where γ_{k+1} is the trial-varying learning gain.

In implementation, the input for trial $k+1$ is computed as follows:

$$\begin{cases} U_{k+1} = U_k + \gamma_{k+1} Z_{k+1} \\ \left(\dfrac{\partial g}{\partial U_k}\right)_{X_k, U_k} Z_{k+1} = E_k \end{cases} \tag{11.24}$$

and the tracking performance on trial $k+1$ varies with the choice of γ_{k+1}. As detailed next, there exists a suitable value for γ_{k+1} for which (11.24) achieves monotonic trial-to-trial error convergence. (Trial-varying ILC laws have been applied (in simulation) to the wind turbine example (see Section 9.1.2).)

The basis of the relaxed method is to adjust the step length in the Newton direction by adding a relaxation index. In ILC, the relaxed Newton method for a function $F : D \subset \mathbb{R}^N \to \mathbb{R}^N$ is

$$X_{k+1} = X_k - \gamma_{k+1} F'(X_k)^{-1} F(X_k), \quad k = 0, 1, \ldots \tag{11.25}$$

and the following result from [183] is used in the subsequent development.

Lemma 11.1   *Assume that $J : D \subset \mathbb{R}^q \to \mathbb{R}$ is Gateaux-differentiable at $x \in \text{int}(D)$, i.e. the interior of D, and there exists an $h \in \mathbb{R}^q$ such that $J'(x)h > 0$. Then there exists a $\delta > 0$ such that*

$$J(x - \gamma h) < J(x), \quad \forall \gamma \in (0, \delta) \tag{11.26}$$

To apply this result to (11.25), set

$$J_k = \frac{1}{2}\|E_k\|^2 = \frac{1}{2}\|F(X_k)\|^2, \quad h_{k+1} = F'(X_k)^{-1} F(X_k) \tag{11.27}$$

where also $J'(X_k)h_{k+1} = \|F(X_k)\|^2 > 0$. Hence, by the last result, there exists a γ_{k+1} that satisfies $\|E_{k+1}\| = \|F(X_k + \gamma_{k+1} h_{k+1})\| < \|E_k\|$, i.e. monotonic trial-to-trial error convergence. Moreover, under mild conditions, there exists a γ_{k+1} for which the relaxed Newton ILC will monotonically converge from trial-to-trial. The following result from [183] describes the influence of this parameter on convergence and gives a computation formula.

Lemma 11.2 *Suppose that* $J : D \subset \mathbb{R}^q \to \mathbb{R}^1$ *is continuously differentiable on* $D_0 \subset D$ *and that, for some* $\lambda \in (0, 1]$, J' *satisfies*

$$\|J'(a) - J'(b)\| \leq f \|a - b\|^\lambda, \quad \forall a, b \in D_0 \tag{11.28}$$

Then, for all γ such that $[x - \gamma h] \subset D_0$,

$$J(x - \gamma h) \leq J(x) - \gamma J'(x)h + \left[\frac{f}{1 + \lambda}\right](\|h\|\gamma)^{1+\lambda} \tag{11.29}$$

Given this last result, convergence occurs if γ is chosen to satisfy

$$\gamma \leq \frac{1}{\|h\|}\left[(1 + \lambda)\frac{J'(x)h}{\|h\|f}\right]^{\frac{1}{\lambda}} \tag{11.30}$$

or, on setting $\lambda = 1$,

$$\gamma \leq \frac{2J'(x)h}{f\|h\|^2} \tag{11.31}$$

For the relaxed Newton method of (11.25), $h_{k+1} = F'(X)_k^{-1}F(X_k)$, and

$$J_k(X_k) = \frac{1}{2}\|E_k\|^2 = \frac{1}{2}\|F(X_k)\|^2 \tag{11.32}$$

and hence, γ_{k+1} must satisfy

$$\gamma_{k+1} \leq \frac{2\|E_k\|^2}{h\|Z_{k+1}\|^2} \tag{11.33}$$

For the ILC application, $FU_k = E_k = R - g(U_k)$, $h_{k+1} = -g'(U_k)^{-1}E_k = -Z_{k+1}$, and hence, γ_{k+1} must be selected such that

$$\gamma_{k+1} \leq \frac{2\|E_k\|^2}{f\|Z_{k+1}\|^2} \tag{11.34}$$

Moreover, f can be treated as a measure of the system's nonlinearity. Hence, as the nonlinearity's strength increases, the relaxation to the learning step length for Newton ILC must be reduced. Also, this parameter is usually unknown, and therefore, it is necessary to develop a numerical method to select a suitable relaxation index. A method for calculating this index is given next.

11.3.2 Parameter Optimization for Monotonic and Fast Trial-to-Trial Error Convergence

In NOILC for linear systems, see Chapter 5 and [4], monotonically convergent ILC is achieved by solving the optimization problem:

$$\min_{U_{k+1}} J_{k+1}(U_{k+1}) = \|E_{k+1}\|_Q^2 + \|U_{k+1} - U_k\|_R^2 \tag{11.35}$$

where $Q > 0$ and $R > 0$ and

$$\|E_{k+1}\|_Q^2 = E_{k+1}^T Q E_{k+1}, \quad \|U_{k+1} - U_k\|_R^2 = (U_{k+1} - U_k)^T R(U_{k+1} - U_k) \tag{11.36}$$

Also, let U_m be the minimum of (11.35). Then

$$J_{k+1}(U_m) = \|E_m\|_Q^2 + \|U_m - U_k\|_R^2 \leq \|E_k\|_Q^2 + \|U_{k+1} - U_k\|_R^2 = J_{k+1}(U_{k+1}) \tag{11.37}$$

and hence,

$$\|E_m\|_Q^2 \leq \|E_k\|_Q^2 - \|U_m - U_k\|_R^2 < \|E_k\|_Q^2 \tag{11.38}$$

where $E_m = E_{k+1}(U_m)$. Hence, monotonic error reduction can always be achieved away from the solution.

For the relaxed Newton method-based ILC, monotonic trial-to-trial error convergence can be guaranteed by optimizing γ_{k+1} using the cost function:

$$\min_{\gamma_{k+1}} J_{k+1}(\gamma_{k+1}) = \|E_{k+1}\|_Q^2 + \|U_{k+1} - U_k\|_R^2$$

$$= \|E_{k+1}\|_Q^2 + \|\gamma_{k+1} Z_{k+1}\|_R^2 \qquad (11.39)$$

Using Lemma 11.1, it can be shown that there exists a $\gamma_{k+1} > 0$ generating a new control input U_{k+1} such that $\|E_{k+1}\|_Q^2 < \|E_k\|_Q^2$ and [160] the control input can converge to the ideal one if the error norm decreases sufficiently over each trial. This also follows from (11.38), which gives $\|E_k\|_Q^2 - -\|E_{k+1}\|_Q^2 > \|U_{k+1} - U_k\|_R^2$. Hence, and with a small R (SISO case), sufficient change in the control input is allowed to cause a sufficient decrease in the error.

The control update term $\|U_{k+1} - U_k\|_R^2$, and especially the control weighting R, is the key to monotonic and fast trial-to-trial error convergence as detailed next:

(i) The two reasons that lead to overshoots, large changes in control inputs and large terminating errors, are restrained by the two terms in the cost function. In particular, the term $\|U_{k+1} - U_k\|_R^2$ limits the change of the control input between two trials and the term $\|E_{k+1}\|_Q^2$ is current trial feedback. Also, Q can be adjusted to restrain the terminating errors.

(ii) Fast convergence speed results by adjusting R. Also, from (11.39), with monotonicity ensured, the larger the change in the control input, the larger the reduction in the error and the faster the convergence.

Numerical algorithms are required for the application of Newton ILC. This topic is covered in [145–147], with supporting numerical examples. A critical issue is using numerical search algorithms to find the optimal γ_{k+1}, the Levenberg–Marquardt method [237] has attractive properties in this respect.

The coverage so far in this chapter has started from a discrete-time representation of the dynamics. Newton ILC can also be applied when sampling continuous-time dynamics is first used to obtain this representation. This step is required in the next section that considers the application of Newton ILC to 3D stroke rehabilitation.

11.4 Newton ILC for 3D Stroke Rehabilitation

The analysis in this section follows, in the main [73]. The setup considered is the same as that in Section 10.3, i.e. Figure 10.3. Next, the basic equations are given using the notation of [73], and at the end of this section, a discussion is given of the relative merits of the two approaches.

Assuming rigid links, a general dynamic model of the support structure is

$$B_a(\Theta)\ddot{\Theta} + C_a(\Theta, \dot{\Theta})\dot{\Theta} + F_a(\Theta, \dot{\Theta}) + G_a(\Theta) + K_a(\Theta) = J_a^T(\Theta)h \qquad (11.40)$$

where $\theta = \begin{bmatrix} \theta_1 & \dots & \theta_q \end{bmatrix}^T \in \mathbb{R}^q$ contains the joint angles, $h \in \mathbb{R}^q$ is the externally applied force vector, and $B_a(\cdot) \in \mathbb{R}^{q \times q}$ and $C_a(\cdot) \in \mathbb{R}^{q \times q}$ are the inertial and Coriolis matrices, respectively, $J_a(\cdot)$ is the system Jacobean matrix, $F_a(\cdot) \in \mathbb{R}^q$ and $G_a(\cdot) \in \mathbb{R}^q$ are, respectively, the friction and gravitational vectors and the entries in the $K_a(\cdot) \in \mathbb{R}^q$ are the moments produced by the assistive action of the support mechanism.

A dynamic model of the human arm is

$$B_h(\Phi)\ddot{\Phi} + C_h(\Phi, \dot{\Phi})\dot{\Phi} + F_h(\Phi, \dot{\Phi}) + G_h(\Phi) = \tau(u, \Phi, \dot{\Phi}) \tag{11.41}$$

where $\Phi = \begin{bmatrix} \phi_1 & \cdots & \phi_p \end{bmatrix}^T$, $B_h(\cdot)$ and $C_h(\cdot)$ are, respectively, the compatibly dimensioned inertial and Coriolis matrices, $F_h \in \mathbb{R}^p$ and $G_h \in \mathbb{R}^p$ are, respectively, the friction and gravitational vectors. Also, if effects such as spasticity are sufficiently mild, biomechanical coupling between joints can be omitted and hence,

$$F_h(\Phi, \dot{\Phi}) = \text{diag}[F_{v,1}(\dot{\phi}_1) + F_{s,1}(\phi_1)_1, \dots, F_{v,p}(\dot{\phi}_p) + F_{s,p}(\phi_p)] \tag{11.42}$$

for which experimental verification is reported in [81]. The term $\tau(u, \Phi, \dot{\Phi})$ represents the moments generated through application of FES. Hence, with m muscles actuated, $u(t) = \begin{bmatrix} u_1(t) & \cdots & u_m(t) \end{bmatrix}^T$.

A Hill-type model is use to represent the moment generated about the jth muscle of the form:

$$\tau_{i,j}(u_j(t), \theta_i(t), \dot{\theta}_i(t)) = h_j(u_j(t), t) \times F_{M,i,j}(\theta_i(t), \dot{\theta}_i(t)) \tag{11.43}$$

and $h_j(u_j(t), t)$ is a Hammerstein structure, where a static nonlinearity $h_{\text{IRC}j}(u_j(t))$, representing the isometric recruitment curve, is cascaded with linear activation dynamics $h_{\text{LAD}j}(t)$. The multiplicative effect of the joint angle and joint angular velocity on the active torque developed by the muscle is represented by $F_{M,i,j}(\phi_i(t), \dot{\phi}_i(t))$. Also, the total moment about joint i is obtained by summing the individual contributions from j to m and hence,

$$\tau_i = \begin{bmatrix} F_{M,i,1}(\theta_i(t), \dot{\theta}_i(t)) & \cdots & F_{M,i,m}(\theta_i(t), \dot{\theta}_i(t)) \end{bmatrix} \begin{bmatrix} h_1(u_1(t)) \\ \vdots \\ h_m(u_m(t), t) \end{bmatrix} \tag{11.44}$$

Suppose a unique bijective transformation exists within the necessary joint ranges of the form $\Theta = k(\Phi)$, between coordinate systems that allow the mechanical support and human arm models to be combined. This transformation explicitly holds for all exoskeletal robots (where $q = p$) and can be extended to all end effector robot devices currently available for rehabilitation. The Lagrangian equation in one variable can be written in terms of the other through the application of the chain rule and the outcomes added to give

$$B(\Phi)\ddot{\Phi} + C(\Phi, \dot{\Phi})\dot{\Phi} + F(\Phi, \dot{\Phi}) + G(\Phi) + K(\Phi) = \tau(u, \Phi, \dot{\Phi}) - J^T(\Phi)h \tag{11.45}$$

and the expressions for the entries in this equation are given in equation six of [73]. Next, the Newton ILC analysis can commence.

Let $h_{\text{LAD}j}(t)$ have a continuous-time state-space model with matrices (respectively, state, input, and output) $\{M_{Aj}, M_{Bj}, M_{Cj}\}$, and states $x_j(t)$. Then the system of (11.45) is described over the finite time interval $0 \leq t \leq \alpha$ by the state-space model:

$$\dot{x}_s(t) = f_s(x_s(t)) + g_s(u(t))$$
$$\Phi(t) = h_s(x_s(t)), \quad \Phi(0) = \Phi_0, \quad 0 \leq t \leq \alpha \tag{11.46}$$

where

$$f_s(x_s(t)) = \begin{bmatrix} \dot{\Phi}(t) \\ B(\Phi(t))^{-1}X(\Phi(t), \dot{\Phi}(t)) \\ M_{A,1}x_1 \\ \vdots \\ M_{A,m}x_m \end{bmatrix}, \quad g_s(u(t)) = \begin{bmatrix} 0 \\ 0 \\ M_{B,1}h_{\text{IRC},1}(u_1(t)) \\ \vdots \\ M_{B,m}h_{\text{IRC},m}(u_m(t)) \end{bmatrix}$$

$$\Phi(t) = h_s(x_s(t)) = \begin{bmatrix} I & 0 & \dots & 0 \end{bmatrix} x_s(t) \tag{11.47}$$

and

$$x_s(t) = \begin{bmatrix} \Phi^T(t) & \dot{\Phi}^T(t) & x_1^T(t) & \dots & x_m^T(t) \end{bmatrix}^T \tag{11.48}$$

The elements of $X(\Phi(t), \dot{\Phi}(t))$ are

$$X_i(\Phi(t), \dot{\Phi}(t)) = \sum_j^m (M_{Cj} x_j F_{M,i,j}(\phi_i(t), \dot{\phi}_i(t))) + (J^T(\Phi(t)))_i h(t)$$
$$- C_i(\Phi(t), \dot{\Phi}(t))\dot{\Phi}(t) - F_i(\Phi(t), \dot{\Phi}(t)) - G_i(\Phi(t)) \tag{11.49}$$

A consequence of using anthropomorphic joint angles in the combined model (11.45) is their use to realize partial input–output decoupling since the muscles to be stimulated, e.g. anterior deltoid, triceps, primarily generate torque about a single axis. However, this axis may not be dominant in a particular functional movement since there may equally be joints that may not be controlled (due to the presence of subluxations, stiffness, or limited angular range of movement). To embed this flexibility in controlled joint selection, together with the clinical requirement that only a subset of muscles are stimulated, define the set $\mathscr{P}$ containing the controlled joint indices, with elements $\mathscr{P} = \{p_1, \dots, p_{n_p}\}$, $n_p \leq p$, and controlled variables $\phi_{p_1}, \dots, \phi_{n_p}$. Also, let $\mathscr{M}$ contain the indices of the input channels that are stimulated by functional electrical stimulation (FES), with elements $\mathscr{M} = \{m_1, \dots, m_{n_m}\}$, $n_m \leq m$.

The tasks presented to a patient during a treatment session are in the form of repeated tracking movements for the affected upper limb, with a rest period between the end of one attempt and the beginning of the next. During this period, the affected limb is returned to the starting position. As first discussed in Section 2.4, this fits exactly into the ILC setting. In Section 10.2.1, full or partial linearization was required before applying SISO loops. The rest of this section demonstrates that Newton ILC removes the need for the preliminary linearization step.

Figure 11.1 shows the overall control scheme, where $\hat{\Phi}(t)$ denotes the reference trajectory and $e_k(t) = \hat{\Phi}(t) - \Phi(t)$ is the error on trial k. To apply ILC in this application, the design of a stabilizing feedback control loop of the form

$$\dot{x}_{c,k}(t) = f_c(x_{c,k}(t), e_k(t))$$
$$z_k(t) = h_c(x_{c,k}(t), e_k(t)) \tag{11.50}$$

is required, where $f_c(\cdot)$ and $h_c(\cdot)$ are continuously differentiable. The controlled system dynamics are obtained by combining the stimulated arm model (11.47) and the feedback controller, linking the signals $v_k(t)$ and Φ_k shown in Figure 11.1, where $v_k(t) \in L_2^m[0, \alpha]$ and $\Phi_k(t) \in L_2^p[0, \alpha]$ are, respectively, the ILC input and output vectors.

The state-space model of the controlled dynamics is

$$\dot{x}_{k+1}(t) = f(x_k(t), v_k(t))$$
$$\Phi_k(t) = h_s(x_{s,k}(t)) = h(x_k(t)) \tag{11.51}$$

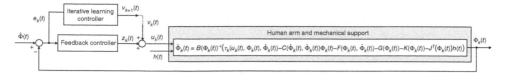

Figure 11.1 Block diagram of the 3D rehabilitation system.

and

$$\dot{x}_{k+1}(t) = \begin{bmatrix} \dot{x}_{s,k}(t) \\ \dot{x}_{c,k}(t) \end{bmatrix} \tag{11.52}$$

$$f(x_k(t), v_k(t)) = \begin{bmatrix} f_s(x_{s,k}(t)) + g_s(h_c(x_{c,k}(t), \hat{\Phi}(t) - h_s(x_{s,k}(t))) + v_k(t)) \\ f_c(x_{c,k}(t), \hat{\Phi}(t) - h_s(x_{s,k}(t))) \end{bmatrix} \tag{11.53}$$

If only a subset of the inputs and outputs is used, then for any $a \in L_2^p[0, \alpha]$ and $b \in L_2^m[0, \alpha]$

$$a^{\mathscr{P}} = \begin{bmatrix} a_{p_1} \\ \vdots \\ a_{p_{n_p}} \end{bmatrix} \in L_2^{n_p}[0, \alpha], \quad b^{\mathscr{M}} = \begin{bmatrix} b_{m_1} \\ \vdots \\ b_{m_{n_m}} \end{bmatrix} \in L_2^{n_m}[0, \alpha] \tag{11.54}$$

Moreover, the continuity property of $f(\cdot)$ and $h(\cdot)$ in (11.51) ensures that the input and output variables used can be explicitly represented by the nonlinear mapping:

$$G : L_2^{n_m}[0, \alpha] \to L_2^{n_p}[0, \alpha] : v^{\mathscr{M}} \mapsto \Phi^{\mathscr{P}} \tag{11.55}$$

or, in component form,

$$(Gv^{\mathscr{M}})(t) = \begin{bmatrix} G_{p_1} v^{\mathscr{M}} \\ \vdots \\ G_{p_{n_p}} v^{\mathscr{M}} \end{bmatrix}, \quad G_{p_i} : L_2^{n_m}[0, \alpha] \to L_2[0, \alpha] : v^{\mathscr{M}} \mapsto \phi_{p_i} \tag{11.56}$$

which represents the relationship between the input and output signals in an arbitrary close region about $v^{\mathscr{M}}(t)$. Also, it is routine to establish that this mapping has the properties of uniqueness and continuity.

Consider an operating point $(v, x) = (\bar{v}, \bar{x})$. Then the dynamics $\Phi^{\mathscr{P}} = Gv^{\mathscr{M}}$ can be locally represented by the LTV state-space model:

$$\dot{\tilde{x}}(t) = A(t)\tilde{x}(t) + B(t)\tilde{v}^{\mathscr{M}}(t)$$
$$\tilde{\Phi}^{\mathscr{P}}(t) = C(t)\tilde{x}(t), \quad \tilde{x}(0) = \bar{x}(0) \tag{11.57}$$

where

$$A(t) = \frac{\partial}{\partial x}(f(x(t), v(t))_{\bar{x}(t),\bar{v}(t)}, \quad B(t) = \frac{\partial}{\partial v^{\mathscr{M}}}(f(x(t), v(t))_{\bar{x}(t),\bar{v}(t)}$$
$$C(t) = \frac{d}{dx} h^{\mathscr{P}}(x(t))_{\bar{x}(t)} \tag{11.58}$$

The corresponding linear mapping is $\overline{G} : L_2^{n_m}[0, \alpha] \to L_2^{n_p}[0, \alpha] : \tilde{v}^{\mathscr{M}} \mapsto \tilde{\Phi}^{\mathscr{P}}$ given by

$$(\overline{G}\tilde{v}^{\mathscr{M}})(t) = \begin{bmatrix} \overline{G}_{p_1} \tilde{v}^{\mathscr{M}} \\ \vdots \\ \overline{G}_{p_{n_p}} \tilde{v}^{\mathscr{M}} \end{bmatrix}, \quad G_{p_i} : L_2^{n_m}[0, \alpha] \to L_2[0, \alpha] :$$

$$\tilde{v}^{\mathscr{M}} \mapsto \tilde{\phi}_{p_i}(\overline{G}_i \tilde{v}^{\mathscr{M}}(t)) = \int_0^t C_i(t)\Phi(t, \tau)B(\tau)\tilde{v}^{\mathscr{M}}(\tau)d\tau \tag{11.59}$$

where $\Phi(t, \tau)$ denotes the transition matrix for (11.57).

Denote the reference trajectory components corresponding to the controlled joints by $\hat{\phi}_i \in \mathscr{P}$. Then the general joint angle tracking problem can be stated as computing an ILC input v that minimizes the norm of the tracking error of the controlled joints, i.e.

$$\min_{v^{\mathscr{M}}} J(v), \quad J(v) = \|\hat{\Phi}^{\mathscr{P}} - Gv^{\mathscr{M}}\|^2, \quad v^{\notin \mathscr{M}} = v_0^{\notin \mathscr{M}} \tag{11.60}$$

where $v_0 = 0$ and $\hat{\Phi}^{\mathscr{P}} = \begin{bmatrix} \hat{\phi}_{p_1} & \cdots & \hat{\phi}_{p_{n_p}} \end{bmatrix}^T$. Also, depending on the system dynamics and inputs and outputs, there may be a single or an infinite number of solutions. To determine if perfect tracking is possible, let the set of achievable joint motions be

$$\mathrm{Im}(G) = \{y = Gv^{\mathscr{M}} \;:\; v^{\mathscr{M}} \in L_2^{n_m}[0, \alpha], \; v^{\notin\mathscr{M}} = v_0^{\notin\mathscr{M}}\} \tag{11.61}$$

and following result is Proposition 1 in [73].

Lemma 11.3 *Suppose that in the ILC setting the solution of (11.60) is a set of control inputs* $\{v\}_k, \; k \geq 1$ *such that if* $\hat{\Phi}^{\mathscr{P}} \in \mathrm{Im}(G)$, *then*

$$\lim_{k\to\infty} \|v_k^{\mathscr{M}} - v_\infty^{\mathscr{M}}\| = 0, \quad v_\infty^{\mathscr{M}} = \min_{v^{\mathscr{M}}} \|v\|^2 \;:\; \hat{\Phi}^{\mathscr{P}} = Gv^{\mathscr{M}} \tag{11.62}$$

Alternatively, if $\hat{\Phi}^{\mathscr{P}} \notin \mathrm{Im}(G)$, *(11.62) is replaced by*

$$\lim_{k\to\infty} \|v_k^{\mathscr{M}} - v_\infty^{\mathscr{M}}\| = 0, \quad v_\infty^{\mathscr{M}} = \min_{v^{\mathscr{M}}} \|\hat{\Phi}^{\mathscr{P}} - Gv^{\mathscr{M}}\|^2 \tag{11.63}$$

where in both cases $v_0 = 0$.

The problem defined in this last result gives that the ILC input must minimize the tracking error norm using minimal control effort. It can be solved using iterative optimization methods based on the linear approximation (11.57), and the following is Theorem 1 in [73].

Theorem 11.2 *The ILC problem given in the last result is iteratively solved in the ILC setting by first linearizing G about each new* (v_k, x_k). *Then an update step calculated using the linearized model* $\overline{G}_k$ *is applied, i.e.*

$$v_{k+1}^{\mathscr{M}} = v_k^{\mathscr{M}} + L_k e_k^{\mathscr{P}}, \quad k \geq 0, \quad v_0 = 0 \tag{11.64}$$

where $e_k^{\mathscr{P}} = \hat{\Phi}^{\mathscr{P}} - \Phi_k^{\mathscr{P}}$ *is the experiment error and* Φ_k *is the experimentally determined output. Also, if the learning operator* $L_k : L_2^{n_p}[0, \alpha] \to L_2^{n_m}[0, \alpha]$ *satisfies*

$$\|I - \overline{G}_\infty L_k\| < 1, \quad \forall k \tag{11.65}$$

where the operator norm is induced by the inner product $\langle \cdot, \cdot \rangle$, *then local convergence to* $e_k^{P} = 0$ *is guaranteed, where* $\overline{G}_k$ *is obtained by linearizing G about* (v_k, x_k) *as detailed previously in this section.*

Lemma 3 in [73] gives the ILC update calculation as follows:

Lemma 11.4 *The term* $L_k e_k^{\mathscr{P}} = v_{k+1}^{\mathscr{M}} - v_k^{\mathscr{M}} = \Delta v_k$ *in (11.64) can be computed by solving*

$$\min_{\Delta v_k^{\mathscr{M}}} \|\overline{G}\Delta v_k - e_k^{\mathscr{P}}\|^2, \quad v_0 = 0 \tag{11.66}$$

using gradient-based ILC, which gives the update

$$\Delta v_k^{j+1} = \Delta v_k^j + \beta \overline{G}_k^* \left(e_k^{\mathscr{P}} - \overline{G}_k \Delta v_k^j \right), \quad \Delta v_k^0 = 0 \tag{11.67}$$

for $j = 1, 2, \ldots, J$ *inter-trial updates, where* β *is a positive real scalar. Also,* $\overline{G}^* : L_2^{n_p}[0, \alpha] \to L_2^{n_m}[0, \alpha]$ *is given by*

$$(\overline{G}^* v)(t) = \int_0^t B^T(\tau)\Phi(t, \tau)C^T(\tau)v(\tau)d\tau \tag{11.68}$$

where $\Phi(t, \tau)$ *is the transition matrix for the state-space triple* $\{A^T(t), B^T(t), C^T(t)\}$.

Each update can be efficiently computed using the adjoint system formed from the linearized dynamics (11.57):

$$\dot{z}(t) = -A^T(t)z(t) - C^T(t)\left(e_k^{\mathscr{P}}(t) - \overline{G}_k \Delta v_k^j(t)\right)$$

$$\Delta v_k^{j+1}(t) = \Delta v_k^j(t) + \beta B^T(t)z(t) \tag{11.69}$$

where $0 < \beta < \frac{2}{\overline{\sigma}}$ and $\overline{\sigma} \geq 0$ is the smallest value satisfying $\langle e, \overline{G}_k \overline{G}_k^* e \rangle \leq \overline{\sigma}^2 ||e||^2$, $\forall\, e$. The number of updates determines the accuracy with which the solution approximates $L_k e_k^{\mathscr{P}}$, with the limiting solution $\Delta v_k^\infty = L_k e_k^{\mathscr{P}}$. Moreover, the input to be applied on the next trial is

$$v_{k+1}^{\mathscr{M}} = v_k^{\mathscr{M}} + \Delta_k^j \tag{11.70}$$

Theorem 11.2 establishes the existence of an ILC law that satisfies Lemma 11.3 at each step. Moreover, convergence in a single step occurs when G has linear and time-varying dynamics. When applied to the underlying nonlinear system G, the update inherits all the Newton method's convergence properties. These provide an upper bound on the convergence rate as a function of the linearized system's magnitude, nonlinearity, and the proximity of the initial input to the solution.

The ILC law (11.64) enforces convergence of the controlled joints, ϕ_i, $i \in \mathscr{P}$, for the cases considered in Lemma 11.3. Theorem 2 in [73] provides conditions for stability of the uncontrolled joints ϕ_i, $i \notin \mathscr{P}$, and this property is not considered further in this chapter.

11.4.1 Experimental Results

The results given in this section are for the problem considered in Section 10.2.3, where, in particular, the reference trajectories used are those in Figure 10.6. Consider first the section of the feedback controller, where the following choice has previously produced satisfactory performance in clinical trials with stroke patients [80]. In the current case, this control law implementation is

$$f_c(x_{c,k}(t), e_k(t)) = \begin{bmatrix} -\frac{1}{c} & 0 \\ 1 & 0 \end{bmatrix} x_{c,k}(t) + \begin{bmatrix} 1 \\ 0 \end{bmatrix} e_k(t)$$

$$h_c(x_{c,k}(t), e_k(t)) = \left[(k_i - \frac{k_d}{c^2}) \quad \frac{k_i}{c}\right] x_{c,k}(t) + \left(k_p + \frac{k_d}{c}\right) e_k(t) \tag{11.71}$$

where k_p, k_c, and k_d are the (proportional, integral, and derivative) controller gains and c is a small positive scalar. Combing the control law dynamics with the identified arm model (11.46) gives the controlled system dynamics in the form of (11.51).

The ILC input is given by (11.64) with $L_k = \overline{G}_k^*(\overline{G}_k \overline{G}_k^*)^{-1}$ (and hence the convergence condition (11.65) holds). This input is computed after each trial, where the state-space matrices $A(t), B(t), C(t)$ are given by (11.58) and are directly obtained from the identified model. The system (11.69) is run for $J = 100$ times in between trials to approximate the descent term $L_k e_k^{\mathscr{P}} = \Delta v_k$, using a gain β (satisfying the bound given after (11.69)). No additional damping is necessary to satisfy (11.41) due to the decoupling effect resulting from choosing $\mathscr{M} = \mathscr{P}$.

This control scheme has been tested with six unimpaired subjects, who all undertook the far center and far distal tasks shown in Figure 10.6, the first of which moves their arm out in front and the second moves it out and to their side. Figure 11.2 gives representative tracking performance results

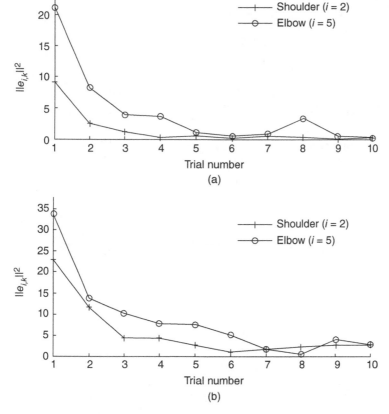

Figure 11.2 Far center error norm plots: participant 1 (a) and participant 2 (b).

for two participants. The control law gains used with participant 1 were $k_p = 20, k_i = 0.1, k_d = 2$, and $c = 0.01$ and for participant 2 $k_p = 5, k_i = 0.2, k_d = 1$, and $c = 0.001$.

In these results, the tracking error reduces quickly and maintains a low level over later trials. In some cases, the error norm increases slightly in later trials because the participant's triceps started to suffer from fatigue. However, ILC was quickly able to modify the stimulation to maintain a low error. Figures 11.3 and 11.4 show tracking performance on trial 10 for the same two participants. These exhibit close reference tracking for both controlled angles.

These results confirm that the Newton ILC scheme can yield high performance when applied to unimpaired subjects. In a small-scale clinical trial, a translation to statistically significant improvement in clinically relevant outcome measures was found [163].

11.5 Constrained Newton ILC Design

As discussed in Chapter 8, the need for constrained ILC design is motivated by application areas. In this section, a Newton ILC design in the presence of input constraints is given, based on a penalty function and an iterative method for solving an unconstrained nonlinear optimization problem.

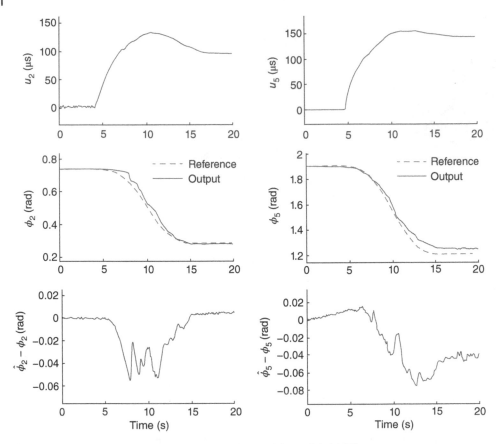

Figure 11.3 Far center tracking performance for participant 1 (trial 10).

The resulting algorithm also has monotonic and super linear convergence characteristics. This new design transforms the input inequality constraints into equality form by adding auxiliary variables. A cost function is minimized to produce the new ILC design. Finally, a simulation-based case study illustrates the performance of the new design. This section follows, in the main, [250].

Suppose that the function g in (11.6) is twice differentiable, and hence, it can be approximated as

$$g(U_k) \approx g(U_{k+1}) - G_{k+1}(U_{k+1})\Delta U_k + \frac{1}{2}\mathfrak{I}(\Delta U_k) \tag{11.72}$$

where $\Delta U_k = U_{k+1} - U_k$ and the detailed structures of the entries on the right-hand side of this last equation are given in [250]. Moreover, the matrix $G_{k+1}(U_{k+1})$ is the gradient matrix of $g(U_k)$ and let its Hessian matrix be denoted by $H_{k+1}(U_{k+1})$. Also, the derivative of the gradient matrix of (11.72) with respect to U_k is (where for ease of presentation, the arguments of matrix-valued functions are dropped from this point onward and also approximation is replaced by equality in further analysis of (11.72))

$$G_k = G_{k+1} - H_{k+1}\Delta U_k \tag{11.73}$$

or, on setting $\Delta G_k = G_{k+1} - G_k$,

$$\Delta G_k = H_{k+1}\Delta U_k \tag{11.74}$$

It may be the case that the Hessian H_{k+1} cannot be accurately obtained and instead an approximation, Θ_{k+1}, to H_{k+1} is used, where the objectives are (i) to ensure that $\Theta_k \approx H_k$ for each trial to obtain

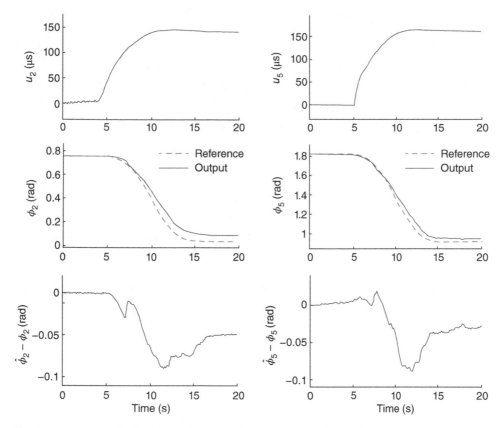

Figure 11.4 Far center tracking performance for participant 2 (trial 10).

a fast convergence rate and (ii), the calculation direction of the iterative algorithm approximates to the Newton direction. Also, if $\Theta_k > 0$, then the calculation direction of the algorithm is the descent direction of the function $g(\cdot)$ for U_k over the trial number.

Based on the above, (11.74) can be rewritten as follows:

$$\Delta G_k \approx \Theta_{k+1} \Delta U_k \tag{11.75}$$

and, on introducing (and again replacing approximation by equality in further analysis of (11.75)),

$$\Theta_{k+1} = \begin{bmatrix} \Theta_{k+1,1}^T & \Theta_{k+1,2}^T & \cdots & \Theta_{k+1,N}^T \end{bmatrix}^T, \quad \Theta_{k+1,j} \in \mathbb{R}^{N \times N}$$

$$\Delta G_k = \begin{bmatrix} \Delta G_{k,1}^T & \Delta G_{k,2}^T & \cdots & \Delta G_{k,N}^T \end{bmatrix}^T, \quad \Delta G_{k,j} \in \mathbb{R}^N, \quad j = 1, 2, \dots, N \tag{11.76}$$

gives

$$\Delta G_{k,j} = \Theta_{k+1,j} \Delta U_k \tag{11.77}$$

It is not possible to obtain $\Theta_{k+1,j}$ directly, and this term varies with k. Consequently, introduce the correction term $E_{k,j}$ such that $\Theta_{k+1,j} = \Theta_{k,j} + E_{k,j}$ and hence (11.75) can be written as follows:

$$\Delta G_{k,j} = (\Theta_{k,j} + E_{k,j}) \Delta U_k \tag{11.78}$$

The problem of ensuring that $\Theta_{k+1,j} > 0$ as k increases has been the subject of previous research, see, e.g. [214]. Moreover, the correction term $E_{k,j}$ in (11.78) must ensure that $\Theta_{k+1,j} > 0$ and one method is to choose $E_{k,j}$ to be of the form:

$$E_{k,j} = \alpha_j \mu_{k,j} \mu_{k,j}^T + \beta_j v_{k,j} v_{k,j}^T \tag{11.79}$$

where α_j and β_j are real numbers and $\mu_{k,j}$ and $v_{k,j}$ are vectors to be determined.

To solve this last problem, routine analysis, see [250], enables (11.79) to be written as follows:

$$E_{k,j} = -\frac{\Theta_{k,j} \Delta U_k (\Theta_{k,j} \Delta U_k)^T}{(\Theta_{k,j} \Delta U_k)^T \Delta U_k} + \frac{\Delta G_{k,j}(\Delta G_{k,j})^T}{(\Delta G_{k,j})^T \Delta U_k} \tag{11.80}$$

Hence, the correction formula for the Hessian matrix $\Theta_{k+1,j}$ in (11.75) can be written as follows:

$$\Theta_{k+1,j} = \Theta_{k,j} - \frac{\Theta_{k,j} \Delta U_k (\Theta_{k,j} \Delta U_k)^T}{(\Theta_{k,j} \Delta U_k)^T \Delta U_k} + \frac{\Delta G_{k,j}(\Delta G_{k,j})^T}{(\Delta G_{k,j})^T \Delta U_k}, \quad j = 1, 2, \dots, N \tag{11.81}$$

The following result is Theorem 1 in [250], which is based on a more general result for the positive-definite properties of approximate Hessian matrices (see [151] for further details) to the ILC problem considered, and gives conditions under which $\Theta_{k+1,j} \succ 0$.

Theorem 11.3 *Suppose that the matrix $\Theta_{k,j}$ satisfies $\Theta_{k,j} \succ 0$ for any $k \geq 1$ and $j = 1, 2 \dots, N$. Then $\Theta_{k+1,j} \succ 0$ if and only if $\Delta G_{k,j}^T \Delta U_k > 0$.*

Given Theorem 11.3, if $\Theta_{0,j} \succ 0$, the matrices generated by (11.81) will be such that $\Theta_{k,j} \succ 0, k \geq 1$, if $\Delta G_{k,j}^T \Delta U_k > 0$. Otherwise, setting $\Theta_{k+1,j} = \Theta_{k,j}$ will ensure that $\Theta_{k+1,j} \succ 0$, i.e. the control direction is unchanged, and the latest $\Theta_{k,j}$ is used.

The structure of Θ_{k+1} could lead to a very complicated ILC law design. As an alternative, introduce $M_{k+1} = \begin{bmatrix} M_{k+1,1}^T & M_{k+1,2}^T & \cdots & M_{k+1,N}^T \end{bmatrix}^T \in \mathbb{R}^{N \times N \times N}$ and write $\hat{\Theta}_{k+1} = M_{k+1} \times \Theta_{k+1} = \sum_{j=1}^{N} M_{k+1,j} \Theta_{k+1,j}$, where $M_{k+1,j} \in \mathbb{R}^{N \times N}$ are adjustable weighting matrices. Then using these matrices, a candidate ILC law is

$$\begin{cases} U_{k+1} = U_k + \varepsilon_k W \delta_k \\ \hat{\Theta}_k \delta_k = E_k \\ \hat{\Theta}_k = M_k \times \Theta_k \\ \Theta_k = [\Theta_{k,1}^T \quad \Theta_{k,2}^T \quad \cdots \quad \Theta_{k,N}^T]^T \\ \Theta_{k+1,j} = \Theta_{k,j} - \frac{\Theta_{k,j} \Delta U_k (\Theta_{k,j} \Delta U_k)^T}{(\Theta_{k,j} \Delta U_k)^T \Delta U_k} + \frac{\Delta G_{k,j}(\Delta G_{k,j})^T}{(\Delta G_{k,j})^T \Delta U_k} \end{cases} \tag{11.82}$$

where $\varepsilon_k \in \mathbb{R}$ is the updating step factor, $W \in \mathbb{R}^{N \times N}$ is a relaxation matrix, $\delta_k \in \mathbb{R}^N$ is the search direction, and $\hat{\Theta}_k \in \mathbb{R}^{N \times N}$ is the parameter matrix to be determined.

The role of the matrix W is to attach weights to the entries in δ_k, leading to the construction of a new search direction to speed up the convergence of the correlation elements in the search process. Also, if ILC convergence holds, then as $k \to \infty$, $\Delta U_k = U_{k+1} - U_k \to 0$ and hence, a singularity would occur in computation. In such a case, sufficiently small positive numbers are added to the denominator terms.

Theorem 11.3 assumes that it is possible to obtain a sufficiently accurate estimation of ΔG_k. Alternatively, this term, i.e. the difference between two gradient matrices for a nonlinear system, in the last equation entry in (11.82) can be replaced by an approximate representation, e.g. in terms of the tracking error. Also, potentially difficult computations in constructing the Hessian matrix are avoided. Moreover, in analysis, no loss of generality arises from selecting $W = I$. An open research question is the effects of $W \neq I$.

Constrained ILC for linear dynamics was considered in Chapter 8. Another form is vector inequality constraints of the form:

$$\Lambda U_{k+1} \leq \tilde{B} \tag{11.83}$$

where $\Lambda = [I \ -I]^T \in \mathbb{R}^{2N \times N}, \tilde{B} \in \mathbb{R}^{2N}$. One method of design for an ILC law of the form (11.82) in the presence of such constraints is by minimizing the cost function:

$$\min J_{k+1}(\varepsilon_k) = \|E_{k+1}\|^2 + \gamma \varepsilon_k^2 \tag{11.84}$$

where $\gamma > 0$ is an adjustable parameter and the inclusion of the ε_k term in the cost function is to prevent excessive change in the control input from one trial to the next. This formulation is a parameter optimal ILC cost function; see Section 5.4.

Setting $\Phi(U_{k+1}) = \Lambda U_{k+1} - \tilde{B}$, the corresponding rows in (11.83) can be written as follows:

$$\Phi_j(U_{k+1}) = \tilde{\alpha}_j^T U_{k+1} - \tilde{\beta}_j \leq 0$$

where $\tilde{\alpha}_j^T$ and $\tilde{\beta}_j$, respectively, are the jth rows of Λ and $\tilde{B}$. Moreover, the inequality constraint can be transformed into an equality by introducing the auxiliary variable vector $\eta_{k+1} \in \mathbb{R}^N$, where $\eta_{k+1,j}$ denotes the jth entry of η_{k+1} on trial $k + 1$. Hence, the problem to be solved is

$$\begin{cases} \min J_{k+1}(\varepsilon_k) = \|E_{k+1}\|^2 + \gamma \varepsilon_k^2 \\ \Phi_j(U_{k+1}) + \eta_{k+1,j}^2 = 0 \end{cases} \tag{11.85}$$

The generalized augmented Lagrange performance function based on the penalty function and multiplier method for (11.85) is

$$\tilde{J}_{k+1}(\varepsilon_k, \eta_{k+1}, \lambda_k, \sigma) = \|E_{k+1}\|^2 + \gamma \varepsilon_k^2 - \sum_{j=1}^N \lambda_{k,j}(\Phi_j(U_{k+1}) + \eta_{k+1,j}^2)$$

$$+ \sum_{j=1}^N \frac{\sigma}{2}(\Phi_j(U_{k+1}) + \eta_{k+1,j}^2)^2 \tag{11.86}$$

where λ_k denotes the Lagrange multiplier and $\lambda_{k,j}$ denotes the jth entry of λ_k on trial k, $\sigma > 0$ is the penalty factor and the updating algorithm for λ_k is $\lambda_{k+1,j} = \lambda_{k,j} - \sigma \Phi_j(U_k)$. To eliminate $\eta_{k+1,j}$ from $\tilde{J}_{k+1}(U_{k+1}, \eta_{k+1}, \lambda_k, \sigma)$, the first-order partial derivative with respect to $\eta_{k+1,j}$ is computed and set equal to zero for a minimum, resulting in

$$\eta_{k+1,j}\left(-\lambda_{k,j} + \sigma\left(\Phi_j(U_{k+1}) + \eta_{k+1,j}^2\right)\right) = 0 \tag{11.87}$$

Suppose also that $\lambda_{k,j} - \sigma \Phi_j(u_{k+1}) > 0$ and $\eta_{k+1,j}^2 = \frac{\lambda_{k,j}}{\sigma} - \Phi_j(U_{k+1})$, then

$$-\lambda_{k,j}\left(\Phi_j(U_{k+1}) + \eta_{k+1,j}^2\right) + \frac{\sigma}{2}\left(\Phi_j(U_{k+1}) + \eta_{k+1,j}^2\right)^2 = -\frac{\lambda_{k,j}^2}{\sigma} + \frac{\lambda_{k,j}^2}{2\sigma} = \frac{1}{2\sigma}(-\lambda_{k,j}^2) \tag{11.88}$$

otherwise, if $\lambda_{k,j} - \sigma \Phi_j(U_{k+1}) \leq 0$, then $\eta_{k+1,j}^2 = 0$. Hence,

$$-\lambda_{k,j}(\Phi_j(U_{k+1}) + \eta_{k+1,j}^2) + \frac{\sigma}{2}(\Phi_j(U_{k+1}) + \eta_{k+1,j}^2)^2$$

$$= -\lambda_{k,j}\Phi_j(U_{k+1}) + \frac{\sigma}{2}(\Phi_j(U_{k+1}))^2$$

$$= \frac{1}{2\sigma}\left((\sigma\Phi_j(U_{k+1}) - \lambda_{k,j})^2 - \lambda_{k,j}^2\right) \tag{11.89}$$

and, on combining (11.88) and (11.89),

$$
\begin{aligned}
&-\lambda_{k,j}(\Phi_j(U_{k+1}) + \eta^2_{k+1,j}) + \tfrac{\sigma}{2}(\Phi_j(U_{k+1}) + \eta^2_{k+1,j})^2 \\
&= \tfrac{1}{2\sigma} \left\{ \min\left[0, (\sigma\Phi_j(U_{k+1}) - \lambda_{k,j})\right]^2 - \lambda^2_{k,j} \right\}
\end{aligned} \tag{11.90}
$$

Substituting (11.90) into (11.86) gives

$$
\tilde{J}_{k+1}(\varepsilon_k, \lambda_k, \sigma) = \|E_{k+1}\|^2 + \gamma\varepsilon^2_k + \frac{1}{2\sigma}\sum_{j=1}^{N}\left\{ \left[\min\left\{0, (\sigma\Phi_j(U_{k+1}) - \lambda_{k,j})\right\}\right]^2 - \lambda^2_{k,j} \right\} \tag{11.91}
$$

The updating step factor ε_k in (11.82) can be determined by minimizing the performance function (11.91). Hence, substitute (11.82) into (11.91) and set

$$
\frac{\partial\tilde{J}_{k+1}(\varepsilon_k, \lambda_k, \sigma)}{\partial\varepsilon_k} = 0
$$

Also, it is immediate on subtracting R (the reference trajectory supervector) from both sides of (11.72) that

$$
Y_k - R \approx Y_{k+1} - R - G_{k+1}(U_{k+1})\Delta U_k + \frac{1}{2}\Im(\Delta U_k)
$$

and introducing

$$
\begin{aligned}
\tilde{G}_k &= G_{k+1} - \frac{1}{2}\Delta U^T_k H_{k+1} = G_k + H_{k+1}\Delta U_k - \frac{1}{2}\Delta U^T_k H_{k+1} \\
&\approx G_k + \Theta_{k+1}\Delta U_k - \frac{1}{2}\Delta U^T_k\Theta_{k+1}
\end{aligned} \tag{11.92}
$$

gives

$$
E_{k+1} = E_k - \tilde{G}_k(U_{k+1} - U_k)
$$

Moreover, using $U_{k+1} = U_k + \varepsilon_k W\delta_k$ gives

$$
E_{k+1} = \left(I - \tilde{G}_k\varepsilon_k W\hat{\Theta}^{-1}_k\right)E_k \tag{11.93}
$$

The following two cases arise from the analysis above.

Case 1: If $\lambda_{k,j} - \sigma\Phi_j(U_{k+1}) > 0$, then

$$
\frac{\partial\tilde{J}_{k+1}(\varepsilon_k, \lambda_k, \sigma)}{\partial\varepsilon_k} = -2E^T_k(I - \tilde{G}_k\varepsilon_k W\hat{\Theta}^{-1}_k)(\tilde{G}_k W\hat{\Theta}^{-1}_k)E_k + 2\gamma\varepsilon_k = 0
$$

or

$$
\varepsilon^*_k = \frac{E^T_k\left(\tilde{G}_k W\hat{\Theta}^{-1}_k + \left(\tilde{G}_k W\hat{\Theta}^{-1}_k\right)^T\right)E_k}{2\left(\gamma + \left\|\tilde{G}_k W\hat{\Theta}^{-1}_k E_k\right\|^2\right)} \tag{11.94}
$$

Case 2: If $\lambda_{k,j} - \sigma\Phi_j(U_{k+1}) \leq 0$, then consideration of $\frac{\partial\tilde{J}_{k+1}(\varepsilon_k, \lambda_k, \sigma)}{\partial\varepsilon_k}$ gives, after extensive but routine manipulations,

$$
\varepsilon^*_k = \frac{E^T_k N_k E_k + (\sigma B) + \lambda_k)^T\Lambda W\delta_k - \sigma(\Lambda U_k)^T\Lambda W\delta_k}{2\left(\gamma + \left\|\tilde{G}_k W\hat{\Theta}^{-1}_k e_k\right\|^2\right) + \sigma(\Lambda W\delta_k)^T(\Lambda W\delta_k)} \tag{11.95}
$$

where $N_k = \tilde{G}_k W\hat{\Theta}^{-1}_k + (\tilde{G}_k W\hat{\Theta}^{-1}_k)^T$, and ϵ^*_k denotes the optimal value of ϵ_k.

Substituting ε_k^* from (11.94) or (11.95) into (11.82) gives the ILC law

$$U_{k+1} = U_k + \varepsilon_k^* W \delta_k \tag{11.96}$$

Consider the property of monotonic trial-to-trial error norm reduction in the system obtained by applying the ILC law developed above, i.e.

$$\|E_{k+1}\| < \|E_k\|, \quad \forall k > 0 \tag{11.97}$$

To obtain a condition for this property, consider the system resulting from application of (11.96), and hence,

$$E_{k+1} = \left(I - \tilde{G}_k \varepsilon_k^* W \hat{\Theta}_k^{-1} \right) E_k \tag{11.98}$$

Then the following result, Theorem 2 in [250], establishes how the updating step factor ε_k affects trial-to-trial error convergence.

Theorem 11.4 *Consider a discrete nonlinear system described by (11.6) in the presence of input constraints defined by (11.83). Suppose also that the ILC law of the form (11.96) computed using (11.82) and (11.91) is applied. Then the resulting controlled system has monotonic trial-to-trial error convergence, i.e. the condition (11.97) is satisfied, if*

$$\lambda_{k,j} - \sigma \Phi_j(U_{k+1}) > 0 \tag{11.99}$$

holds. If this condition is not satisfied, monotonic trial-to-trial error convergence occurs if both the Lagrange multiplier $\lambda_{k,j}$ and the penalty factor σ satisfy the following constraint:

$$\tilde{\alpha}_j^T U_{k+1} - \tilde{\beta}_j + \frac{\lambda_{k,j}}{\sigma} = 0, \quad \forall k > 0, \quad j = 1, 2, \dots, N \tag{11.100}$$

Optimization of other parameters in the ILC law is possible, e.g. the matrix W, but the interest is trial-to-trial error convergence in this section. These additional cases are areas for possible future research.

Theorem 11.4 establishes that the developed ILC law considered has monotonic trial-to-trial error convergence. The analysis given next shows that this ILC law also exhibits super-linear trial-to-trial convergence, i.e. if the sequence of trial control inputs converges to the learned control U_∞, then $\lim\limits_{k \to \infty} \frac{\|U_{k+h+1} - U_\infty\|}{\|U_k - U_\infty\|} = 0, h \geq 0$. Hence, this modified Newton method-based ILC law can be considered as a special case of the Broyden–Fletcher–Goldfarb–Shanno (BFGS) algorithm. Therefore, it inherits the properties of this last algorithm, including the super-linear convergence property. In particular, suppose that the inputs U_k are generated by (11.82), and the following conditions hold:

(i) the input sequence $\{U_k\}_k$ does not terminate, but remains in a closed, bounded, convex set, where the function $g(U_k)$ is twice continuously differentiable and $g(U_k)$ has a unique stationary point U_∞;

(ii) the sequence of Hessian matrices $\{H_{\infty,j}(U_\infty) > 0, j = 1, 2, \dots, N\}$ and $\{H_{k,j}(U_k), j = 1, 2, \dots, N\}$ is Lipschitz continuous in a neighborhood of U_∞, i.e. $\forall U_k \in \Upsilon(u_\infty, \delta)$;

(iii) the sequence of matrices $\{\Theta_{k,j}\}$ is bounded in norm;

(iv) the condition $\left\| \Delta U_k^T \left(\Delta G_{k,j} - \Theta_{k,j} \Delta U_k \right) \right\| \geq r \|\Delta U_k\| \left\| \Delta G_{k,j} - \Theta_{k,j} \Delta U_k \right\|$ holds for each trial, where r is some small constant in $(0, 1)$.

Then $\lim\limits_{k\to\infty} U_k = U_\infty$ and

$$\lim_{k\to\infty} \frac{\|U_{k+h+1} - U_\infty\|}{\|U_k - U_\infty\|} = 0 \tag{11.101}$$

Hence, the control input sequence $\{U_k\}$ converges to U_∞ at an $(h+1)$-step super-linear convergence rate, for a detailed treatment (in the non-ILC case) see [173]. Super-linear convergence means that the new ILC law can deliver, provided some additional conditions are satisfied, a faster trial-to-trial error convergence rate over an alternative design that ensures monotonic trial-to-trial error convergence.

The ILC law developed in this section can be applied using the following procedure:

Step 1: For the given application, select the initial state $x_k(0)$, the initial input $U_0(p)$, reference trajectory r, the number of samples along a trial N, the penalty factor σ, and the initial Lagrange multiplier λ_0.

Step 2: Specify the maximum output tracking error ξ_{max} and a maximum number of trials to be completed k_{max}.

Step 3: Given Y_k and U_k, use (11.91) to compute ε_k^* and hence, given W and γ, the control input for the next trial.

Step 4: If the tracking error performance on trial $k+1$ is less than ξ_{max}, the ILC law has achieved the performance specification. If not, return to Step 3 and repeat. If the trial number is k_{max}, the procedure ends.

As an example to demonstrate the properties and performance of the new ILC law, a simulation case study is given based on a model developed in [153] for a chemical process. After sampling, the

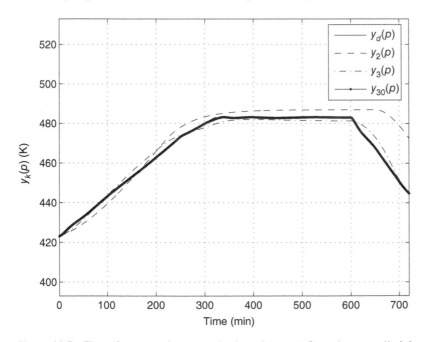

Figure 11.5 The reference trajectory and selected outputs from the controlled dynamics.

dynamics have the following discrete nonlinear state-space model in the ILC setting:

$$x_k^1(p+1) = \left(0.2 - 7.2 \times 10^{10} e^{\frac{-10^4}{x_k^2(p)}} \right) x_k^1(p) + 0.8$$

$$x_k^2(p+1) = \left(-0.8 - 10^{-7} \right) x_k^2(p) + \left(1.44 \times 10^{12} e^{\frac{-10^4}{x_k^2(p)}} \right) x_k^1(p)$$

$$+ \left(1 - 10^{-7} \right) u_k(p) + 280$$

$$y_k(p) = x_k^2(p) \tag{11.102}$$

(where the superscripts denote the entries in the state vector). The reference trajectory, see Figure 11.5, is

$$r(p) = \begin{cases} 0.2p + 423, & 0 \le p \le 250 \\ \frac{1}{7}p + \frac{3061}{7}, & 250 \le p \le 320 \\ 483, & 320 \le p \le 600 \\ -\frac{19}{60}p + 673, & 600 \le p \le 720 \end{cases} \tag{11.103}$$

which is a piecewise function.

Also, $x_k^1(0) = 1$, $x_k^2(0) = 423$, the initial control input is $u_0(p) = 293$, $0 \le p \le 720$, the penalty factor $\sigma = 0.1$, and the initial Lagrange multiplier

$$\lambda_0 = [\ 20 \quad 20 \quad \cdots \quad 20\ ]^T$$

The control input constraint is defined by

$$B = [\ 378,\ 378,\ \cdots\ 378,\ 273,\ 273,\ \cdots\ 273\ ]^T$$

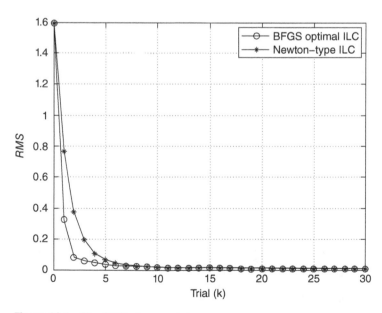

Figure 11.6 The *RMS(e_k)* error plot.

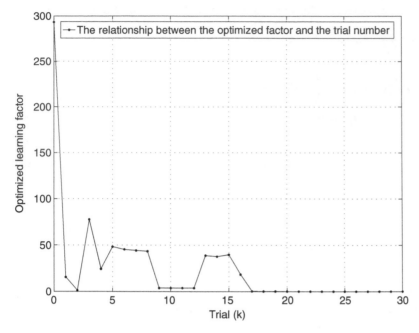

Figure 11.7 The optimized updating step factor ε_k^* along the trial axis.

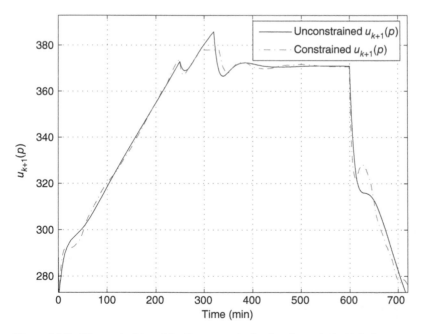

Figure 11.8 The control input for the unconstrained and constrained designs.

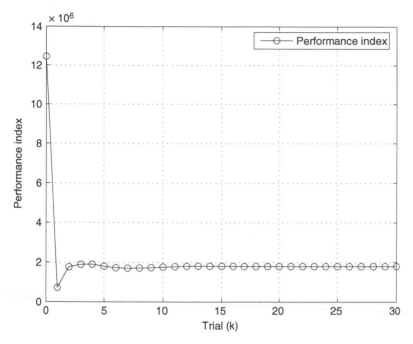

Figure 11.9 The performance index for the constrained design.

and $\Lambda = [I, -I]^T$. Moreover, to evaluate tracking performance from trial-to-trial, the convergence performance index is the $RMS(e_k)$ along each trial, i.e.

$$RMS(e_k) = \sqrt{\frac{1}{721} \sum_{p=0}^{721} e_k^2(p)}, \quad k = 0, 1, \ldots, 30 \tag{11.104}$$

Setting the $RMS(e_k)$ error at $\xi_{\max} = 5.1 \times 10^{-3}$, the maximum number of trials $k_{\max} = 30$, $W = I$, $\gamma = 0.0636$, the system temperature output tracking and $RMS(e_k)$ error plots generated with the ILC law (11.82) applied are shown, respectively, in Figures 11.5 and 11.6.

To compare the trial-to-trial error convergence speed and accuracy of the new ILC law, the Newton-type ILC algorithm used in [73] was applied to this example, giving

$$U_{k+1} = U_k + G_k^{-1} e_k, \quad G_k^{-1} = (1 - 10^{-7})^{-1}$$

The results obtained with this law are also shown in Figure 11.6 and confirm that the new ILC law outperforms the basic Newton-type ILC law. These plots also show that the tracking error is monotonically convergent from trial-to-trial even under the input constraint. Figure 11.7 shows that the optimized learning factor ε_k gradually approaches zero. Figure 11.8 gives the control inputs for both designs, and Figure 11.9 the performance function for the constrained design.

11.6 Concluding Remarks

This chapter has covered the Newton method that lends itself to ILC design in an algebraic setting and permits a direct extension of the lifting approach from linear to nonlinear dynamics. Moreover, for applications where the trial-to-trial error convergence rate is critical, it offers the prospect of super-linear convergence. Also, robotic-assisted stroke rehabilitation provides recent evidence of experimental application. In the next chapter, the subject is stochastic ILC analysis and design.

12

Stochastic ILC

This chapter considers iterative learning control (ILC) design for cases where noise cannot be ignored, even for initial control-related studies. First, a brief survey of early results is given, including a design with experimental validation. Next, frequency domain-based design of simple structure ILC laws is considered with experimental validation. Finally, design in a repetitive process setting is considered.

12.1 Background and Early Results

Disturbance rejection is, as in other areas, an essential topic in ILC analysis and design. In the deterministic case, early work includes [100, 164], where the latter contains both simulation and experimental studies. This early work led to consideration of the performance degrading effects of stochastic trial-dependent disturbances. One example of such work is [106], which developed the well-known idea of a forgetting factor for D-type ILC, with further work in [11] for P-type. The basic idea underlying this work is that as the trials progress, information from the "older" trials will be "forgotten." Hence, the contribution to the stochastic disturbances will have a more negligible effect on the current trial input.

The idea of using a learning gain that decreases with each trial has also been investigated, and early results are in, e.g. [177, 249], with emphasis on measurement noise effects, and in the form of a stochastic approximation algorithm. A trial decreasing gain was also used in [177] for a repetitive disturbance in the presence of measurement noise. Trial-varying filters in stochastic ILC laws have been investigated in, among others, [232–234], where the approach in this last reference is considered again later in this chapter. Other work proposed filtering the reference signal to reduce the influence of noise on the error, see, e.g. [175], and this topic is considered later in this chapter.

In [36], the laws referred to above are compared by statistical analysis and then by experimental application to a linear motor system. The results of this previous work are considered next, where the analysis is for the single-input single-output (SISO) case with additive load and noise disturbances.

Let $z_k(p)$ denote the controlled output on trial k, for $0 \leq p \leq N-1$, for a linear system with dynamics described in the ILC setting by

$$z_k(p) = G(q)u_k(p) + d_k(p) \tag{12.1}$$

where $u_k(p)$ is the input, $d_k(p)$ is the trial-varying load disturbance, and q again denotes the forward shift time-domain operator. Also, the measured output is given by

$$y_k(p) = z_k(p) + n_k(p) \tag{12.2}$$

Iterative Learning Control Algorithms and Experimental Benchmarking, First Edition.
Eric Rogers, Bing Chu, Christopher Freeman and Paul Lewin.
© 2023 John Wiley & Sons Ltd. Published 2023 by John Wiley & Sons Ltd.

where $n_k(p)$ is the trial-varying measurement disturbance. The input on trial $k + 1$ is (as before)

$$u_{k+1}(p) = Q(q)(u_k(t) + L(q)e_k(p)) \tag{12.3}$$

where $Q(q)$ and $L(q)$ are linear discrete and possibly noncausal filters. The measured error signal is

$$e_k(p) = r(p) - y_k(p) \tag{12.4}$$

and $r(p)$ again denotes the (deterministic) reference trajectory. Output tracking in this case is best studied for an error signal based on the measured output, i.e. the controlled error:

$$\epsilon_k(p) = r(p) - z_k(p) \tag{12.5}$$

where, as in [36], this latter signal will be referred to as the error from this point onward.

Combining (12.1)–(12.5) gives

$$\begin{aligned} \epsilon_{k+1}(p) = {}& Q(q)(1 - G(q)L(q))\epsilon_k(p) \\ & + (1 - Q(q))r(p) + Q(q)d_k(p) - d_{k+1}(p) \\ & + G(q)Q(q)L(q)n_k(p) \end{aligned} \tag{12.6}$$

Hence, even if $L(q)$ is chosen as the inverse of $G(q)$ (if possible), the presence of time-varying disturbances and the use of a filter $Q(q) \neq 1$ preclude convergence to zero error. If there are no disturbances present, the fastest trial-to-trial error convergence occurs when $L(q) = G^{-1}(q)$, which can only ever be approximately achieved.

The following sufficient condition for monotonic trial-to-trial error convergence in the sense of the 2-norm is used

$$\max_{\omega \in [0,\omega_N]} |Q(e^{j\omega T})(1 - G(e^{j\omega T})L(e^{j\omega T}))| < 1 \tag{12.7}$$

where ω_N is the Nyquist frequency, and T is the sampling period. Also, as discussed previously in this book, multiple choices for $L(q)$ are possible, e.g.

$$L(q) = \hat{G}^{-1}(q) = (1 + \Delta(q))G^{-1}(q) \tag{12.8}$$

where $\hat{G}(q)$ is a model of the system dynamics and $\Delta(q)$ represents the multiplicative uncertainty due to unmodeled dynamics, i.e.

$$G(q) = (1 + \Delta(q))\hat{G}(q) \tag{12.9}$$

In ILC, the load and measurement disturbances can be the same on each trial or vary with the trial number. The former can be learned gradually similarly to the reference trajectory. Hence, they are not considered below, and the analysis in this section requires the assumptions given next.

Assumption 12.1 The load, $d_k(p)$, and the measurement, $n_k(p)$, disturbances are trial-varying.

Assumption 12.2 The disturbances $d_k(p)$ and $n_k(p)$ are assumed to be zero-mean weak stationary sequences that are white in the trial domain with respective variances σ_d^2 and σ_n^2, i.e. with E denoting the expectation operator, $E[d_k(p)] = 0$ and $E[n_k(p)] = 0$ and

$$E[d_k(p)d_{k+h}(p)] = \begin{cases} \sigma_d^2, & h = 0 \\ 0, & \text{otherwise} \end{cases} \tag{12.10}$$

and

$$E[n_k(p)n_{k+h}(p)] = \begin{cases} \sigma_n^2, & h = 0 \\ 0, & \text{otherwise} \end{cases} \tag{12.11}$$

where h denotes the lag in the trial-domain.

Assumption 12.3 The load and noise disturbances are assumed to be uncorrelated with the system input $u_k(p)$ and with each other, i.e.

$$E[u_k(p)d_k(p+h)] = 0, \quad \forall\, h \tag{12.12}$$

$$E[u_k(p)n_k(p+h)] = 0, \quad \forall\, h \tag{12.13}$$

$$E[d_k(p)n_k(p+h)] = 0, \quad \forall\, h \tag{12.14}$$

The following results are Theorems 1 and 2, respectively, in [36] and establish how the presence of noise affects the achievable error, where the variance of the error signal is

$$E[\tilde{e}_k^2(p)] = E[[\epsilon_k(p) - E[\epsilon_k(p)]]^2 \tag{12.15}$$

Theorem 12.1 *Consider the ILC dynamics described by (12.1)–(12.3). Then under Assumptions 12.1–12.3*

$$\lim_{k \to \infty} E[\epsilon_k(p)] = \frac{1 - Q(q)}{1 - Q(q)(1 - G(q)L(q))} r(p) \tag{12.16}$$

Theorem 12.2 *Consider the ILC dynamics described by (12.1)–(12.3). Then under Assumptions 12.1–12.3, the variance of the error signal is given by*

$$E[\tilde{e}_{k+1}^2(p)] = E[[B(q)\tilde{e}_k(p)]^2] + E[[Q(q)d_k(p)]^2] + E[d_{k+1}^2(p)]$$
$$+ E[[G(q)Q(q)L(q)n_k(p)]^2] - 2E[B(q)d_k(p)Q(q)d_k(p)] \tag{12.17}$$

where $B(q) = Q(q)(1 - G(q)L(q))$.

Theorem 12.1 shows that zero error is not possible with a filter $Q(q) \neq 1$. Using both of these theorems, the 2-norm for the achieved error is available. Specifically, writing the variance as follows:

$$E[\tilde{e}_k^2(p)] = E[\epsilon_k^2(p)] - E[\epsilon_k(p)]^2 \tag{12.18}$$

then

$$E[\epsilon_k^2(p)] = E[\tilde{e}_k^2(p)] + E[\epsilon_k(p)]^2 \tag{12.19}$$

Hence, the 2-norm of the error is

$$\|\epsilon_k\|_2 = \left(\frac{1}{N} \sum_{p=0}^{N-1} E[\epsilon_k^2(p)] \right)^{\frac{1}{2}} = \left(\frac{1}{N} \sum_{p=0}^{N-1} E[\tilde{e}_k^2(p)] + E[\epsilon_k(t)]^2 \right)^{\frac{1}{2}} \tag{12.20}$$

and therefore, a small 2 norm value can only be achieved when both the variance and the error value are small.

Suppose that the ILC law is designed without consideration of stochastic disturbances, i.e. $Q(q) = 1$ and $L(q) = (1 + \Delta(q))G^{-1}(q)$. Then by Theorem 12.1

$$\lim_{k \to \infty} E[\epsilon_k(p)] = 0 \tag{12.21}$$

and by Theorem 12.2

$$E[\tilde{e}_{k+1}^2(p)] = E[[\Delta(q)\tilde{e}_k(p)]^2] + E[d_k^2(p)] + E[d_{k+1}^2(p)]$$
$$+ E[[(1 + \Delta(q))n_k(p)]^2] + 2E[d_k(p)\Delta(q)d_k(p)] \tag{12.22}$$

Hence, the expected value of the error converges to zero as desired. In the case of perfect knowledge of the system, i.e. $\Delta(q) = 0$, which is required for rapid deterministic trial-to-trial convergence, the variance of the error signal is given by

$$E[\tilde{\epsilon}_{k+1}^2(p)] = E[d_k^2(p)] + E[d_{k+1}^2(p)] + E[n_k^2(p)]$$
$$= 2\sigma_d^2 + \sigma_n^2 \tag{12.23}$$

where the stationary assumption on $d_k(p)$ has been used, and hence, the variances of $d_k(p)$ and $d_{k+1}(p)$ are equal.

This last formula shows that the error variance is twice the variance of the load disturbance plus the measurement error variance. Moreover, this fact corresponds with a result in [249] (even though only a single disturbance was considered in the latter case). More generally, these results demonstrate how the presence of nonrepetitive disturbances can have a powerful effect on the achievable tracking performance of ILC designs since they are fed back from the previous trial.

Consider the case of $Q(q) = 1$ but $L(q) = \beta(1 + \Delta(q))G^{-1}(q)$, where $\beta \in (0,1)$. Then the expected value of the error still converges to zero in the limit provided (12.7) holds, but the variance of the error from the last result is

$$E[\tilde{\epsilon}_{k+1}^2(p)] = E[(1 - \beta(1 + \Delta(q))\tilde{\epsilon}_k(p))^2] + E[d_k^2(p)]$$
$$+ E[d_{k+1}^2(p)] + E[(\beta(1 + \Delta(q))n_k(p))^2]$$
$$- 2E[d_k(p)(1 - \beta(1 + \Delta(q)))d_k(p)] \tag{12.24}$$

and for the case of no uncertainty,

$$E[\tilde{\epsilon}_{k+1}^2(p)] = (1 - \beta)^2 E[\tilde{\epsilon}_k^2(p)] - (1 - 2\beta)E[d_k^2(p)]$$
$$+ E[d_{k+1}^2(p)] + \beta^2 E[n_k^2(p)] \tag{12.25}$$

Hence, since $\beta \in (0,1)$,

$$\lim_{k \to \infty} E[\tilde{\epsilon}_{k+1}^2(p)] = E[\tilde{\epsilon}_k^2(p)] = E[\tilde{\epsilon}_\infty^2(p)] \tag{12.26}$$

where

$$E[\tilde{\epsilon}_\infty^2(p)] = \frac{2}{2 - \beta}\sigma_d^2 + \frac{\beta}{2 - \beta}\sigma_n^2 \tag{12.27}$$

This last equation shows that a reduction in the error variance is possible compared to the standard case when $\beta = 1$. However, this reduction results in a lower rate of deterministic trial-to-trial error convergence, and a compromise is required. Also, the component of the error variance due to the load disturbances agrees with that in [100]. Next, the use of a forgetting factor is considered.

An ILC law with a forgetting factor has the form:

$$u_{k+1}(p) = (1 - \gamma)u_k(p) + L'(q)e_k(p) \tag{12.28}$$

where $\gamma \in [0,1)$ is the forgetting factor and $L'(q)$ is defined below. The objective of introducing the forgetting factor is to increase the robustness of ILC laws to initialization errors, fluctuations of the dynamics, and random disturbances. On trial k, the premise is that the previous inputs are multiplied by $(1 - \gamma)^k$. Hence, the choice of γ such that $1 - \gamma < 1$ will reduce the influence of the previous trial contributions. This reduction is also the case for the effects of disturbances fed back to the input.

The analysis of this last law is the same as that for (12.3) on setting $Q(q) = 1 - \gamma$ and $L(q) = \frac{L'(q)}{1-\gamma}$. Also if $L'(q) = (1 + \Delta(q))G^{-1}(q)$, Theorem 12.1 gives that

$$\lim_{k \to \infty} E[\epsilon_k(p)] = \frac{\gamma}{\gamma + (1 + \Delta(q))}r(p) \tag{12.29}$$

Hence, if $\gamma \neq 0$, the expected value of the error cannot converge to zero.

Applying Theorem 12.2 gives the following formula for the variance in this case:

$$E[\tilde{\epsilon}_{k+1}^2(p)] = E[((\gamma + \Delta(q))\tilde{\epsilon}_k(p))^2] + (1 - \gamma^2)E[d_k^2(p)]$$
$$+ E[d_{k+1}^2(p)] + E[((1 + \Delta(q))n_k(p))^2]$$
$$+ 2(1 - \gamma)E[d_k(p)(\gamma + \Delta(q))d_k(p)] \tag{12.30}$$

If $\Delta(q) = 0$, i.e. perfect knowledge of the system is available, it follows that

$$E[\tilde{\epsilon}_\infty^2(p)] = \sigma_d^2 + \frac{1}{1 - \gamma^2}(\sigma_d^2 + \sigma_n^2) \tag{12.31}$$

Hence, the error variance is minimized when $\gamma = 0$, i.e. the standard law should be used. A different analysis setting is used in [232–234] and a similar result obtained (this analysis concludes that for multiple-input multiple-output (MIMO) systems, the best forgetting matrix is zero when the trace of the input error covariance matrix is minimized).

Consider the case when $L'(q) = L(q) = (1 + \Delta(q))G^{-1}(q)$, and $Q(q)$ as before, and hence, the forgetting factor affects the complete law. Theorem 12.1 now gives

$$\lim_{k \to \infty} E[\epsilon_k(p)] = \frac{\gamma}{1 + (1 - \gamma)\Delta(q)}r(p) \tag{12.32}$$

and again, $\gamma = 0$ is the optimum choice. Theorem 12.2 gives

$$E[\tilde{\epsilon}_{k+1}^2(p)] = (1 - \gamma)^2 E[(\Delta(q)\tilde{\epsilon}_k(p))^2] + (1 - \gamma)^2 E[d_k^2(p)]$$
$$+ E[d_{k+1}^2(p)] + (1 - \gamma)^2 E[((1 + \Delta(q))n_k(p))^2]$$
$$+ 2(1 - \gamma)^2 E[d_k(p)\Delta(q)d_k(p)] \tag{12.33}$$

and when $\Delta(q) = 0$

$$E[\tilde{\epsilon}_{k+1}^2(p)] = E[d_{k+1}^2(p)] + (1 - \gamma)^2 (E[d_k^2(p)] + E[n_k^2(p)])$$
$$= \sigma_d^2 + (1 - \gamma)^2 (\sigma_d^2 + \sigma_n^2) \tag{12.34}$$

Hence, in this case, the minimum variance of the error occurs when $\gamma = 1$ and under this condition $E[\tilde{\epsilon}_{k+1}^2(t)] = \sigma_d^2$. This value is consistent because the previous input is not fed back and only the load disturbances during the current trial influence the error. Therefore, a compromise is required between minimizing the variance of the error and ensuring that the expected value of its converged value is small.

Adaptive ILC can be applied to the along the trial dynamics, or trial-to-trial dynamics, or both. A simple structure of this form of control is to have a decreasing trial-to-trial gain, such as

$$u_{k+1}(p) = u_k(p) + \frac{L'(q)}{k+1}e_k(p) \tag{12.35}$$

In this case, $Q(q) = 1$ and $L(q) = \frac{L'(q)}{k+1}$ and by Theorem 12.1

$$\lim_{k \to \infty} E[\epsilon_k(p)] = 0 \tag{12.36}$$

and (omitting the detailed calculations for brevity) by Theorem 12.2

$$E[\tilde{\epsilon}_{k+1}^2(p)] = \left(1 - \frac{1}{k+1}\right)^2 E[\tilde{\epsilon}_k^2(p)] + \frac{1}{(k+1)^2}E[n_k^2(p)] + \frac{2}{k+1}E[d_k^2(p)] \tag{12.37}$$

As $k \to \infty$, $E[\tilde{\epsilon}_{k+1}^2(p)] = E[\tilde{\epsilon}_k^2(p)] = E[\tilde{\epsilon}_\infty^2(p)]$, and hence,

$$E[\tilde{\epsilon}_\infty^2(p)] = \frac{1}{2(k+1)}\left(E[\tilde{\epsilon}_\infty^2(p)] + E[n_k^2(p)]\right) + E[d_k^2(p)] \tag{12.38}$$

In the limit, therefore,

$$\lim_{k \to \infty} E[\tilde{\epsilon}_\infty^2(p)] = E[d_k^2(p)] = \sigma_d^2 \tag{12.39}$$

Hence, if a decreasing learning gain is used from trial-to-trial, the expected value of the error converges to zero, and its variance approaches that of the load disturbance. This outcome is the best possible. The disadvantage of this ILC law is that the error reduction rate reduces with the trial number, and eventually, the learning practically ceases. Consequently, it cannot react to changes in the desired output or the repetitive disturbances present.

For the control law (12.3), Theorem 12.1 applied in the case when $L(q) = (1 + \Delta(q))G^{-1}(q)$ gives

$$\lim_{k \to \infty} E[\epsilon_k(p)] = \frac{1 - Q(q)}{1 + Q(q)\Delta(q)} r(p) \tag{12.40}$$

Suppose also that frequency domain analysis and filter design are to be used. Then as in the previous analysis, $N \to \infty$ must be assumed. In which case, the finite-time Fourier transform can be used to undertake a "reasonably accurate" analysis. Using $\mathscr{F}$ to denote the Fourier transform, the magnitude response of the Fourier transform of (12.40) is

$$\left| \mathscr{F} \left(\lim_{k \to \infty} E[\epsilon_k(p)] \right) \right|^2 = \frac{|1 - Q(e^{j\omega T})|^2}{|1 + Q(e^{j\omega T})\Delta(e^{j\omega T})|^2} |r(\omega)|^2 \tag{12.41}$$

To converge to zero expected error requires a filter with unity magnitude and zero phase shift at frequencies where $r(\omega)$ is nonzero. By Theorem 12.2, the error variance in the absence of uncertainty in the system model is

$$E[\tilde{\epsilon}_{k+1}^2(p)] = \sigma_d^2 + E[(Q(q)d_k(p))^2] + E[(Q(q)n_k(p))^2] \tag{12.42}$$

Also, the noise terms are stationary. Hence, $\epsilon_{k+1}(p)$ also has this property and, therefore,

$$E[\tilde{\epsilon}_{k+1}^2(p)] = \sigma_d^2 + \frac{T}{2\pi} \int_{-\frac{\pi}{T}}^{\frac{\pi}{T}} |Q(e^{j\omega T})|^2 (\Phi_d(\omega) + \Phi_n(\omega)) d\omega \tag{12.43}$$

where $\Phi_d(\omega)$ and $\Phi_n(\omega)$ are, respectively, the power spectra of $d_k(p)$ and $n_k(p)$. Hence, the error variance can be made smaller than that obtained with standard law by selecting $Q(q)$ to have magnitude less than unity at frequencies where the disturbance power spectra are large. Hence, the error variance can be made smaller than that obtained with a standard law by selecting $Q(q)$ to have magnitude less than unity at frequencies where the disturbance power spectra are large. As before, a compromise to reduce the error variance without filtering at frequencies important to $r(p)$ is required to allow an acceptable converged error to be obtained.

In many cases, $r(p)$ has low-frequency content. Moreover, $\Phi_d(\omega)$ and $\Phi_n(\omega)$ often are large at high frequencies. Hence, minimizing the trial and error variances is often not a conflicting aim, and $Q(q)$ can be chosen as a low-pass filter with an appropriately chosen cut-off frequency.

As a numerical example, [36] considered the case when the "real" system has transfer-function $G(s)$ and approximate transfer-function $\hat{G}(s)$, where

$$G(s) = \frac{20}{(s + 1)(s + 20)}, \quad \hat{G}(s) = \frac{1}{s + 1} \tag{12.44}$$

with

$$r(t) = \begin{cases} 1 - \cos(0.1\pi t), & 0 \le t \le 20 \text{ s} \\ 0, & 20.1 \le t \le 30 \text{ s} \end{cases} \tag{12.45}$$

Both systems were sampled using a zero-order-hold with sampling period $T = 0.1$ seconds and $N = 301$ was chosen. Also, the load disturbance $d_k(p)$ is taken as a normally distributed zero mean random sequence with $\sigma_d^2 = 0.0025$, and the measurement $n_k(p)$ is a zero-mean, normally distributed, random sequence with variance 0.0025. Moreover, this sequence was filtered using a fifth order Butterworth high-pass filter, with a cut-off frequency of 2 Hz, to simulate high-frequency measurement noise.

The various designs arising from the analysis in [36] were simulated over 10 trials to obtain an estimate of the expected value and variance of the error at a specific time, with each simulation repeated 200 times. The expected value and variance at $p = 15$ and the 2-norm were then calculated for $k = 10$ and 200 simulations. Also, the expectation operator was taken as the mean of the simulations, and hence,

$$E[v_j(p)] = \frac{1}{N_p} \sum_{i=1}^{N_p} v_k^i(p) \tag{12.46}$$

where N_p denotes the total number of simulations completed and $v_k^i(p)$ represents an arbitrary signal at sample p on trial k and the ith simulation. Also, N_p was chosen to allow the design to converge to a stage where the errors caused by the disturbances were dominant relative to the deterministic errors. Using the same disturbance signal enabled a direct comparison of the performance of each design (even though the disturbances acting on the system were different for the 200 simulations).

In the case of the designs using forgetting factors, different ways of implementing these factors were tested. The first is when this factor only affects the previous trial term, and the second is where it affects the complete control law ($L = \frac{L'}{1-\gamma}$ in the first case and $L = L'$ for the second). A fifth-order Butterworth low-pass filter with a cut-off frequency of 0.3 Hz implemented as a zero-phase filter was used for the filtered version of the control law. The cut-off frequency is greater than the highest component of the Fourier transformed representation of $r(p)$, which occurs at 0.05 Hz.

The numerical data from this simulation study are given in [36] (Table 1), and following are the main conclusions: (i) the trade-off necessary between minimizing the error variance and keeping its expected value small is present, (ii) the standard law has the second smallest expected error but the second-largest error variance, (iii) the laws with forgetting factors included exhibit the trends predicted by the analysis, (iv) in the case of decreasing gain, the variance is much smaller than for the standard law, but the expected value of the error is larger in relative terms, and (v) the lowest error and also variance is achieved by the law with the filtering included, which also has the lowest 2-norm. This last feature arises because the reference trajectory has only low-frequency content, and hence, a low-pass filter can effectively remove the high-frequency disturbances.

In [36], the results of experimentally applying these stochastic ILC laws to a linear, permanent magnet synchronous motor are given. Motors of this type are very stiff and have no mechanical transmission components. Consequently, they do not suffer problems with backlash, resulting in very high-positioning accuracy. The experiments were conducted when the motor's position was controlled by a two-degree-of-freedom controller sampled at 2 kHz and tuned for robust control.

The experiments were based on an approximate model of the motor dynamics constructed by system identification, see [36] for the details, resulting in the model:

$$G(q) = \frac{0.002008q^3 + 0.0009185q^2 + 0.01972q - 0.009375}{q^4 - 3.104q^3 + 3.739q^2 - 2.058q + 0.4364} \tag{12.47}$$

This model was used to construct a phase-lead structure as an approximation to $\hat{G}^{-1}(q)$ as $\hat{G}^{-1}(q) = 0.85q^6$, for which (12.7) holds up to 424 Hz with $Q(q) = 1$. This filter was implemented in zero-phase format, i.e. off-line between the end of one trial and the start of the next.

All of the stochastic ILC laws given above were experimentally tested over 100 trials, and each one repeated 4 times. Also, $\epsilon_k(p)$ cannot be measured in experiments, and hence, the measured error $e_k(p)$ was used instead. In each case, the initial trial input ($k = 0$) was used as the reference trajectory. For the laws with a forgetting factor, the choice of $\beta = 10^{-6}$ was used to obtain a "reasonable value" of the converged measured error 2-norm $\|e_\infty\|_2$. Also (12.22), (12.31), and (12.34) can be used in the noise-free and zero uncertainty case with $e_k(p) = \epsilon_k(p)$ to obtain upper bounds on $\|e(p)\|_2$. Finally, the desired value of $\|e_\infty\|_2$ was taken as the value obtained with the low-pass filtered law, and the value of β is such that the two laws for this case are essentially the same.

The experiments resulted, respectively, in $\|e_{100}\|_2 = 7.4257 \times 10^{-5}$ for the low-pass filter law, 3.8603×10^{-4} for the forgetting factor law and 1.0673×10^{-4} for the decreasing gain law. In summary, the last law error is, relative to the first, approximately 1.4 times larger and the forgetting factor law approximately 5 times larger. In [36], the mean tracking performance of the forgetting factor law after 100 trials is analyzed where even with the small value of β used, a constant error still occurs and hence, the large (relative to the other two laws) value of $\|e_{100}\|_2$.

There are slightly more pronounced oscillations in the transient response along the trials for the decreasing gain law because it has slower learning of the deterministic errors, which is most likely the cause of its more significant 2-norm value. However, it has less oscillatory performance in the steady state because the noise exists at frequencies below the filter cut-off frequency. The decreasing gain assists in reducing the adverse effects of the noise. Moreover, reducing the filter cut-off frequency would reduce the sensitivity to noise but at the cost of an increased expected error.

In summary, there is a trade-off between minimizing the expected value and the variance of the error. When the noise and desired output spectra are in different frequency regions, a filtered ILC law can result in good-quality tracking performance. Suppose there is a significant overlap in the spectra. In such a case, an ILC law with a decreasing learning gain from trial-to-trial can give good robustness to noise and small tracking errors, but at the expense of slower trial-to-trial error convergence.

12.2 Frequency Domain-Based Stochastic ILC Design

The analysis in this section is based on [32, 37], where experimental results using the gantry robot of Section 2.1.1 are also given. Without loss of generality, SISO discrete linear time-invariant systems are considered. Since the analysis is in the frequency domain, it is necessary to assume an infinite time horizon (trial length).

A convenient starting point is the relationship between the error and input on a given trial, written as follows:

$$e_k(p) = -G(q)u_k(p) + w_k(p) + d(p) \tag{12.48}$$

where the extra notation relative to the previous section is that $d(p)$ is a deterministic signal and $w_k(p)$ is a stationary random disturbance. Moreover, $G(q)$ is assumed to be stable. Also, the error measurement $\hat{e}_k(p)$ is considered to be corrupted by noise, i.e.

$$\hat{e}_k(p) = e_k(p) + v_k(p) \tag{12.49}$$

where $v_k(p)$ is a stationary random noise signal.

Suppose that the ILC law has the form:

$$u_{k+1}(p) = Q(q)(u_k(p) + L(q)\hat{e}_k(p)) \tag{12.50}$$

and Figure 12.1 gives a block diagram representation of the controlled system.

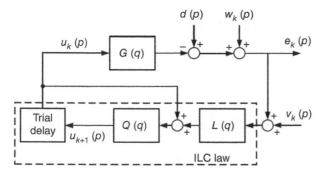

Figure 12.1 Block diagram of the controlled system formed by (12.48)–(12.50).

It is assumed that the initial trial input $u_0(p) = 0$, and that $d(p)$ is known and bounded. The following assumptions are also made.

Assumption 12.4

(i) $E[w_{j_1}(p_1)w_{j_2}(p_2)] = 0, \quad \forall j_1, j_2, p_1, p_2$
(ii)

$$E[w_{j_1}(p_1)w_{j_2}(p_2)] = 0$$
$$E[v_{j_1}(p_1)v_{j_2}(p_2)] = 0$$
$$E[w_{j_1}(p_1)d(p_2)] = 0$$
$$E[v_{j_1}(p_1)d(p_2)] = 0$$

$\forall j_1 \neq j_2$ and $\forall p_1, p_2$.

A further assumption is that $G(q)$ has relative degree zero. If the relative degree is $h > 1$ then the modification developed previously, see Section 7.3, can be used. The analysis of this section uses, in the main, signal spectra, where the spectrum of a signal, say $s(p)$, is

$$\Phi_s(\omega) = \sum_{\tau=-\infty}^{\infty} R_s(\tau)e^{-j\omega\tau}, \quad R_s(\tau) = \sum_{p=0}^{\infty} E[s(p)s(p+\tau)] \tag{12.51}$$

where $R_s(\tau)$ is the autocorrelation function of $s(p)$. Routine algebraic manipulations give

$$\begin{aligned} e_k(p) &= Q(q)(1 - L(q)G(q))e_{k-1}(p) \\ &+ (1 - Q(q))d(p) + w_k(p) - Q(q)w_{k-1}(p) \\ &- Q(q)L(q)G(q)v_{k-1}(p) \end{aligned} \tag{12.52}$$

It is not possible to find the power spectrum of e_k from this last equation because e_{k-1} is correlated with w_{k-1}. Instead, the nonrecursive solution of (12.52) is

$$e_k(p) = X_k(q)d(p) + \sum_{i=0}^{k-1} Y_i(q)(w_{k-1-i}(p) + v_{k-1-i}(p)) + w_k(p), \quad k \geq 1 \tag{12.53}$$

where

$$X_k(q) = (Q(q)(1 - L(q)G(q)))^k$$
$$+ \sum_{i=0}^{k-1}(Q(q)(1 - L(q)G(q)))^i(1 - Q(q))$$
$$Y_i(q) = -(Q(q)(1 - L(q)G(q)))^i Q(q)L(q)G(q) \tag{12.54}$$

Starting from (12.53) and (12.54), the power spectrum for the error on trial k is

$$\Phi_{e_k}(\omega) = |X_k(e^{j\omega})|^2\Phi_d(\omega) + \sum_{i=0}^{k-1}|Y_i(e^{j\omega})|^2(\Phi_w(\omega) + \Phi_v(\omega)) + \Phi_w(\omega) \tag{12.55}$$

The following result is Theorem 1 in [32] and details the convergence properties of the power spectrum in this case.

Theorem 12.3 *If*

$$\max_{\omega \in [-\pi,\pi]} |Q(e^{j\omega})(1 - L(e^{j\omega})G(e^{j\omega}))| < 1 \tag{12.56}$$

then the error power spectrum converges in k. Also, if Φ_{e_∞} denotes $\lim_{k\to\infty}\Phi_{e_k}(\omega)$, then

$$\Phi_{e_\infty} = \frac{|1 - Q(e^{j\omega})|^2}{|1 - Q(e^{j\omega})(1 - L(e^{j\omega})G(e^{j\omega}))|^2}\Phi_d(\omega) + \Phi_w(\omega)$$
$$+ \frac{|Q(e^{j\omega})L(e^{j\omega})G(e^{j\omega})|^2}{|1 - Q(e^{j\omega})(1 - L(e^{j\omega})G(e^{j\omega}))|^2}(\Phi_w(\omega) + \Phi_v(\omega)) \tag{12.57}$$

The frequency-domain condition (12.56) is also related to the convergence rate of the spectrum. See [32] for a detailed treatment of this aspect of the general problem area.

Consider the particular case of the inverse ILC law with

$$L(e^{j\omega}) = \eta(\omega)G^{-1}(e^{j\omega}) \tag{12.58}$$

where $\eta(\omega)$ is the real-valued inversion gain. Also write $Q(e^{j\omega}) = \xi(\omega)e^{j\phi(\omega)}$, where ξ and ϕ are real, $\xi \geq 0$, and $-\pi \leq \xi \leq \pi$. Suppose that a desired trial-to-trial error convergence rate, say $\overline{\gamma}$, is required. Then the optimal design problem is to find $\hat{\eta}(\omega), \hat{\xi}(\omega), \hat{\phi}(\omega)$ that solve

$$\min_{\eta(\omega),\xi(\omega),\phi(\omega)} \Phi_{e_\infty}(\omega), \quad : \quad \gamma \leq \overline{\gamma} < 1 \tag{12.59}$$

where

$$\gamma = \max_{\omega \in [-\pi,\pi]} |Q(e^{j\omega})(1 - L(e^{j\omega})G(e^{j\omega}))|^2 \tag{12.60}$$

The following results are Theorems 2 and 3, respectively, in [32], and solve this problem for the cases when $\Phi_d(\omega) \neq 0$ and $\Phi_d(\omega) = 0$.

Theorem 12.4 *If $\Phi_d(\omega) \neq 0$,*

$$\min_{\eta(\omega),\xi(\omega),\phi(\omega),\gamma(\omega)\leq\overline{\gamma}<1} \Phi_{e_\infty}(\omega) = \hat{\Phi}(\omega) \tag{12.61}$$

where

$$\hat{\Phi}(\omega) = \frac{(1 - \sqrt{\overline{\gamma}})\Phi_d(\omega)(\Phi_w(\omega) + \Phi_v(\omega))}{(1 + \sqrt{\overline{\gamma}})\Phi_d(\omega) + (1 - \sqrt{\overline{\gamma}})(\Phi_w(\omega) + \Phi_v(\omega))} + \Phi_w(\omega) \tag{12.62}$$

This condition is achieved when

$$\hat{\eta}(\omega) = \frac{(1 - \overline{\gamma})\Phi_d(\omega)}{(1 + \sqrt{\overline{\gamma}})\Phi_d(\omega) + \sqrt{\overline{\gamma}}(1 - \sqrt{\overline{\gamma}})(\Phi_w(\omega) + \Phi_v(\omega))} \tag{12.63}$$

$$\hat{\xi}(\omega) = \frac{(1 + \sqrt{\overline{\gamma}})\Phi_d(\omega) + \sqrt{\overline{\gamma}}(1 - \sqrt{\overline{\gamma}})(\Phi_w(\omega) + \Phi_v(\omega))}{(1 + \sqrt{\overline{\gamma}})\Phi_d(\omega) + (1 - \sqrt{\overline{\gamma}})(\Phi_w(\omega) + \Phi_v(\omega))} \tag{12.64}$$

$$\hat{\phi}(\omega) = 0 \tag{12.65}$$

and with convergence rate $\overline{\gamma}$.

This last result shows that the minimum spectrum occurs when the trial-to-trial error convergence approaches unity, i.e. very slow convergence. The best-possible solution is to make an ILC design insensitive to random disturbances and noise from previous trials through the $Y_i(q)$ terms. Moreover, this result is a proof (see [32] for the details) that a zero-phase filter provides the best performance in an ILC design as it provides proof of the optimality of such filters.

In the case, when $\Phi_d(\omega) = 0$ the corresponding result to (12.54) of Theorem is as follows:

Theorem 12.5 *If $\Phi_d(\omega) = 0$,*

$$\min_{\eta(\omega),\xi(\omega),\phi(\omega),\gamma(\omega)\leq\overline{\gamma}<1} \Phi_{e_\infty}(\omega) = \Phi_w(\omega) \tag{12.66}$$

which is achieved when $\hat{\xi}(\omega) = 0$ and zero convergence rate.

Suppose that the input signal has no repeated component, and then best performance occurs when the ILC law is disabled. Given these last results, a noise sensitivity plot can be constructed, as illustrated in Figure 12.2.

As discussed above, the noise sensitivity of an optimal ILC law decreases with increasing convergence rate (slow learning). Also, the spectrum for the initial trial, i.e. $k = 0$, without ILC, is $\Phi_d(\omega) + \Phi_v(\omega)$ and, hence, performance continuously improves using the optimal gains, but the performance improvement depends on the signal-to-noise ratio and desired convergence rate. Suppose, therefore, that the desired performance level is available, i.e. $\Phi_{e_\infty}^{\text{desired}}(\omega) > \Phi(\omega)$. Then this

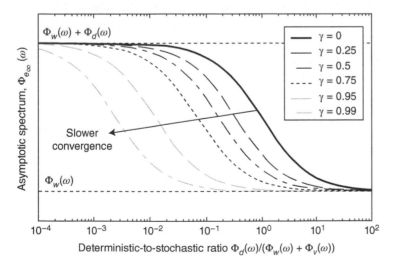

Figure 12.2 Noise sensitivity plot for the optimal ILC law.

figure shows that a sufficiently slow trial-to-trial convergence rate can always be found that meets or exceeds this value.

The results above extend to robust control with frequency-domain bounded uncertainty, i.e.

$$|G(e^{j\omega})| < |\overline{G}(e^{j\omega})| \tag{12.67}$$

where $\hat{G}(e^{j\omega})$ is a model of the bounded uncertain system. Expressing the stability condition of Theorem 12.3 in terms of the optimal results from Theorem 12.4 gives that stability holds provided

$$|\hat{\xi}(\omega)| < \frac{1}{|1 - \hat{\eta}(\omega)\hat{G}^{-1}(e^{j\omega})G(e^{j\omega})|}, \quad \omega \in [-\pi, \pi] \tag{12.68}$$

or

$$|\hat{\xi}(\omega)| < \frac{1}{1 + |\hat{\eta}(\omega)||\hat{G}^{-1}(e^{j\omega})||\overline{G}(e^{j\omega})|}, \quad \omega \in [-\pi, \pi] \tag{12.69}$$

Hence, a new robust Q-filter can be defined with

$$\hat{\xi}_r(\omega) = \min \left\{ \hat{\eta}(\omega), \frac{1}{1 + |\hat{\eta}(\omega)||\hat{G}^{-1}(e^{j\omega})||\overline{G}(e^{j\omega})|} \right\} \tag{12.70}$$

As the trial length is finite, the spectrum of the disturbance d is approximated by its Fourier transform and scaling by H as follows:

$$\Phi_d^H(\omega) = \frac{1}{H} \left| \sum_{p=0}^{H-1} d(p)e^{-\frac{j\omega p}{H}} \right|^2 \tag{12.71}$$

This term is used instead of $\Phi_d(\omega)$ in the optimal design problem.

Constructing the Q and L filters to exactly meet the optimal specifications (12.63) and (12.64) is, in general, a nontrivial task, where the problems are the construction of a zero-phase filter and optimal selection of a filter whose spectrum is the best approximation to the optimal. In the case of the former, the so-called "filtfilt" approach [101] can be used, and a numerical example is given in [32]. The latter problem is the optimal model fitting problem, where extensive results are available in the system identification literature, see, e.g. [148].

One option that may be applicable in some cases is to use the approximate or assumed deterministic-to-stochastic ratio (DSR) $\frac{\Phi_d(\omega)}{\Phi_w(\omega)+\Phi_v(\omega)}$. At frequencies where the stochastic noise is minimal, no penalty is imposed on fast learning. In this case, therefore, the choice of $\hat{\xi}(\omega) = 1$ and $\hat{\eta}(\omega) = 1$ should be made.

At frequencies where the deterministic error is minimal, there is little advantage in learning and hence, $\hat{\xi}(\omega) \approx 0$ or $\hat{\eta}(\omega) \approx 0$ should be used. The advantage of selecting $\hat{\eta}(\omega) \approx 0$ and $\hat{\xi}(\omega) \approx 1$ is that the deterministic error will be eventually be learned, but very slowly. Moreover, the advantage of setting $\hat{\xi}(\omega) \approx 0$ is improved robustness. The design guidelines above are frequency-dependent, and for a given application, $\hat{\xi}(\omega)$ and $\hat{\eta}(\omega)$ can be shaped as the DSR changes with frequency [32].

To obtain the disturbance model from an application, one route is to feed a number of zero input signals into the system and measure the corresponding output and error signals. If the system input is zero, then

$$e_k(p) = d(p) + w_k(p) \tag{12.72}$$

for M trials,

$$\sum_{k=1}^{M} e_k(p) = \sum_{k=1}^{M} d(p) + \sum_{k=1}^{M} w_k(p) \tag{12.73}$$

and as $M \rightarrow \infty$, $\sum_{k=1}^{M} w_k(p) = 0$. Hence,

$$\sum_{k=1}^{M} e_k(p) \approx Md(p), \quad d(p) \approx \frac{1}{M}\sum_{k=1}^{M} e_k(p) \tag{12.74}$$

and, therefore, on each trial

$$w_k(p) \approx e_k(p) - d(p) \tag{12.75}$$

Moreover, the spectrum can be computed as follows:

$$\Phi_w(\omega) \approx \frac{1}{M}\sum_{k=1}^{M}|FFT[w_k(p)]|^2 \tag{12.76}$$

where FFT denotes the Fast Fourier Transform.

At this stage, a transfer-function (or a constant for simplicity) can be fitted, respectively, to $\Phi_d(\omega)$ and $\Phi_w(\omega)$. Also, in implementation, the random noise term $\Phi_v(\omega)$ can be assumed to be zero or a small positive constant offset. As one example, Figure 12.3 gives the resulting Bode plots of $\Phi_d(\omega)$ and $\Phi_w(\omega)$ for the X-axis of the gantry robot of Section 2.1.1.

The filters that approximate the optimal $\hat{\eta}(\omega)$ and $\hat{\xi}(\omega)$ must have the zero-phase property, and applying the filtfilt technique from Matlab results in the fourth-order low-pass filter:

$$H(z) = \frac{0.0002 + 0.0007z^{-1} + 0.0011z^{-2} + 0.0007z^{-3} + 0.0002z^{-4}}{1 - 3.5328z^{-1} + 4.7819z^{-2} - 2.9328z^{-3} + 0.6868z^{-4}} \tag{12.77}$$

Figure 12.4 shows the Bode gain plots on trials 0, 5, 10, 15, and 20. For comparative purposes, the trial-varying filter

$$L_k(z) = \frac{1}{k+1}G^{-1}(z) \tag{12.78}$$

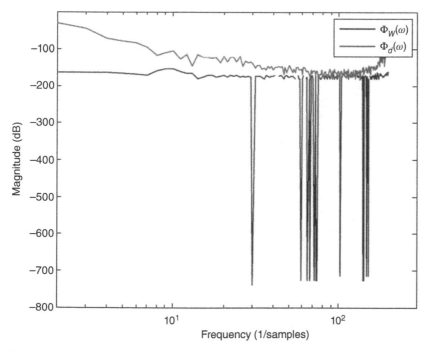

Figure 12.3 DSR and the deterministic and stochastic spectra for the X-axis of the gantry robot of Section 2.1.1.

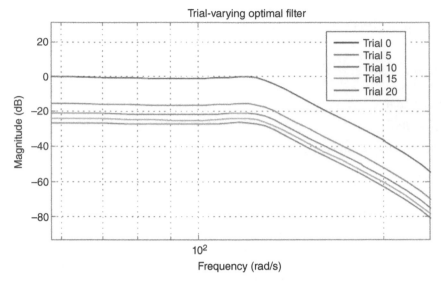

Figure 12.4 Frequency response of the optimal filters for trials 0,5, 10,15, and 20.

is used, which has a similar gain at low frequency, but at higher frequencies, the optimal filters can minimize the effect of amplifying the noise.

In the first set of experiments, on the gantry robot of Section 2.1.1, the ILC law was directly applied, i.e. the prestabilizing proportional plus integral plus derivative (PID) loop is not used. The results are in Figure 12.5, which shows that the number of trials increases, the tracking error

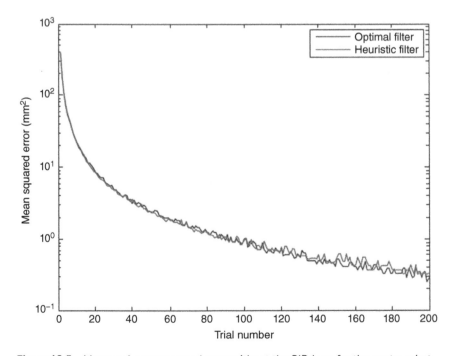

Figure 12.5 Measured mean squared error without the PID loop for the gantry robot.

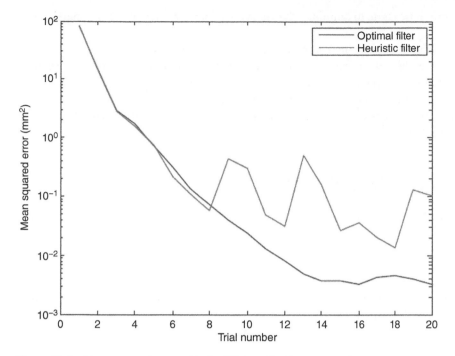

Figure 12.6   Mean squared error with the PID loop for the gantry robot.

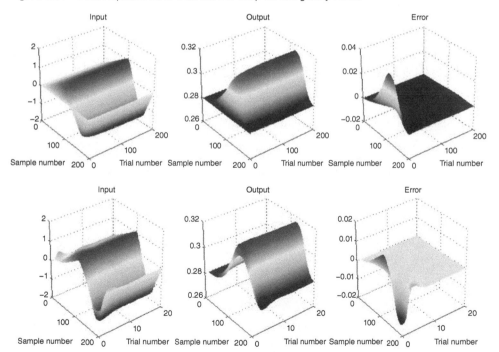

Figure 12.7   Performance of the ILC design with the PID loop.

is reduced, and the noise starts to dominate the signal. The performance of the optimal filters is superior.

Figure 12.6 gives the results obtained with the PID loop included. In this case, the noise starts to dominate the signal early in the learning process, and the optimal filters give better performance. Figure 12.7 shows the corresponding input, output, and error signals (for the case with the PID controller) and confirms that excellent performance is possible.

Next, the results of a further experimental comparison of simple structure stochastic ILC laws are given and discussed.

12.3 Experimental Comparison of ILC Laws

The laws considered in this section are from [232–234] for discrete linear time-varying (LTV) systems described in the ILC setting by

$$
\begin{aligned}
x_k(p+1) &= A(p)x_k(p) + B(p)u_k(p) + \omega_k(p) \\
y_k(p) &= C(p)x_k(p) + v_k(p)
\end{aligned}
\tag{12.79}
$$

where the notation not defined previously is the state noise vector $\omega_k(p) \in \mathbb{R}^n$ and the output measurement error $v_k(p) \in \mathbb{R}^m$. Attention will focus on the following particular cases from [232–234] for time-invariant dynamics (where again e_k denotes the error on trial k).

P-type ILC: The control law is (in fact, this is phase-lead ILC, but the terminology used in the original research is retained in this section)

$$
u_{k+1}(p) = u_k(p) + K_k e_k(p+1)
\tag{12.80}
$$

D-type ILC: This law uses the approximate error differentiation instead of the error derivative considered in [12] for continuous-time systems

$$
u_{k+1}(p) = u_k(p) + K_k(e_k(p+1) - e_k(p))
\tag{12.81}
$$

where for both laws, K_k is the $l \times m$ learning control gain matrix.

In each case, K_k is given by

$$
\begin{aligned}
K_k &= P_{u,k}(CB)^T \left((CB)P_{u,k}(CB)^T + (C - CA)P_{x,k}(C - CA)^T + CQ_tC^T + R_p \right)^{-1} \\
P_{u,k+1} &= (I - K_k CB)P_{u,k} \\
P_{x,k+1} &= AP_{x,k}A^T + BP_{u,k}B^T + Q_p
\end{aligned}
$$

where the symmetric positive definite matrices Q_p and R_p and the symmetric positive semi-definite matrices $P_{x,0}$ a $P_{u,0}$ are given by

$$
\begin{aligned}
Q_p &= E(\omega(p,k)\omega^T(p,k)) \geqslant 0, \quad \mathbf{P}_{x,0} = E_k(\delta x(0,k)\delta x^T(0,k)) \geqslant 0 \\
R_p &= E_k(v(0,k)v^T(0,k)) > 0, \quad P_{u,0} = E(\delta u(p,0)\delta u^T(p,0)) > 0
\end{aligned}
$$

and E_k denotes the expectation operator with respect to the trial domains. Also, $\delta x(0,k), \delta u(p,0)$ are, respectively, the initial state and input errors.

Figure 12.8 shows a block diagram of the control configuration considered in the continuous-time domain. This configuration is a parallel arrangement of a PID feedback controller and an ILC control law.

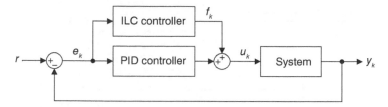

Figure 12.8 Block diagram of the controller structure and arrangement considered.

Consider the case when in continuous time a PID controller is present, and let $f_k(t)$ be the feed-forward signal from the ILC controller. Then for the P-type law,

$$f_{k+1}(t) = f_k(t) + K_k e_k(t+1) \tag{12.82}$$

and for the D-type,

$$f_{k+1}(t) = f_k(t) + K_k(e_k(t+1) - e_k(t)) \tag{12.83}$$

Also, for ease of notation, introduce $PID[e_k(t)] := K_p e_k(t) + K_i \int e_k(t)dt + K_d \dot{e}_k(t)$, where for discrete implementation, the integral term must be replaced by a discrete approximation. Then for the P-type law,

$$u_{k+1}(t) = u_k(t) + PID[e_{k+1}(t) - e_k(t)] + K_k e_k(t+1) \tag{12.84}$$

and for the D-type,

$$u_k(t) = f_k(t) + PID[e_k(t)]$$
$$u_{k+1}(t) = f_{k+1}(t) + PID[e_{k+1}(t)]$$
$$= u_k(t) + PID[e_{k+1}(t) - e_k(t)] + K_k(e_k(t+1) - e_k(t)) \tag{12.85}$$

After zero-order hold sampling with period T, $PID[e_k(t)]$ is approximated by

$$PID[e_k(p)] = K_p e_k(p) + K_i \sum_{\tau=1}^{p} \frac{(e_k(\tau) + e_k(\tau - 1))T}{2} + K_d \frac{e_k(p) - e_k(p-1)}{T} \tag{12.86}$$

Before experimental implementation, a series of designs were completed for the X-axis of the gantry robot and their performance evaluated in simulation. Figure 12.9 shows the MSE error (as defined in the source referenced for this section) results obtained using the D-type law with a variety of PID controller parameters, where $PID = \{0,0,0\}$ corresponds to removing this control loop from the scheme. These results confirm that higher PID gains can give improved performance only on the first trial and do not assist in obtaining error reduction on succeeding trials.

Figures 12.10 and 12.11 show the X-axis MSE results obtained by varying $P_{u,0}$ (the initial value of $P_{u,k}$) and Q_p, where

$$Q_p = q_p I, \quad P_{x,k} = p_{x,k} I \tag{12.87}$$

In this last equation, q_p and $p_{x,k}$ are scalars, and R_p is set to the mean value of the white noise, where this value is calculated using the error vector after each completed trial. These results indicate that selecting a larger $P_{u,0}$ and a smaller q_p gives better tracking performance, both in terms of convergence speed and the final error.

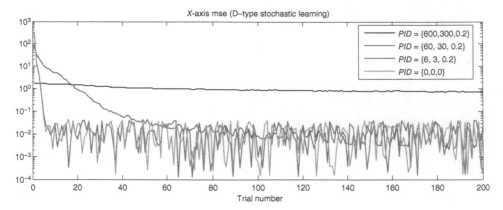

Figure 12.9 Simulation results with various PID gains ($P_{u,0} = 1000$, $Q_p = 0.001$, $P_{x,0} = 0.1$, Disturbance: 0.0002).

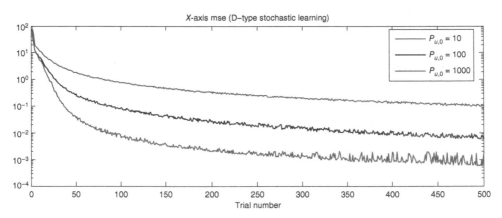

Figure 12.10 Simulation results with varying $P_{u,0}$ ($PID = \{60, 30, 0.2\}$, $Q_p = 0.001$, $P_{x,0} = 0.1$, Disturbance: 0.00005).

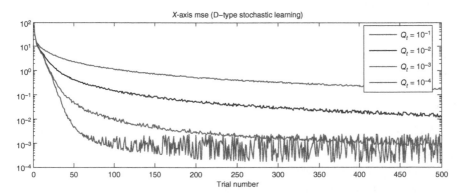

Figure 12.11 Simulation results with varying Q_p ($PID = \{60, 30, 0.2\}$, $P_{u,0} = 800$, $P_{x,0} = 0.1$, Disturbance: 0.00005).

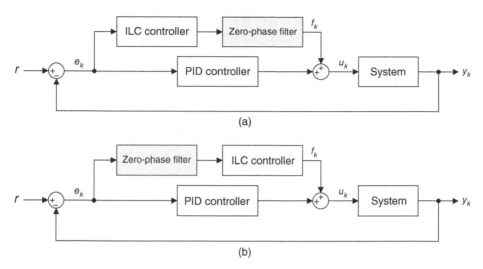

Figure 12.12 Filter arrangements. (a) Filtering the output of the ILC controller. (b) Filtering the error vector before the ILC update.

To mitigate the effects of high-frequency noise, a low-pass, zero-phase filter was used. In particular a third-order, low-pass, zero-phase Chebyshev filter with 20 dB attenuation at 15 Hz (94.25 rad/s) has been used with transfer-function

$$H(z) = \frac{0.102693 + 0.002934z^{-1} + 0.002934z^{-2} + 0.102693z^{-3}}{1 - 1.644597z^{-1} + 1.091881z^{-2} - 0.236029z^{-3}} \qquad (12.88)$$

Moreover, since the filter placement in the overall scheme is critical for convergence and tracking performance, two arrangements have been considered. In Figure 12.12a, the filter is applied to the feedforward signal, and in Figure 12.12b to the error signal before computation of the control input for the next trial. Experiments have shown that the latter arrangement provides superior performance and is therefore used in the tests whose results are given and discussed below. An experimental program, see also [38], has been undertaken for both the P- and D-type designs, with the results for the latter given first and then those for the former.

The ILC law was applied to all three axes of the gantry robot and the zero-phase filter given by (12.88) has been used in all experiments. The reference trajectories are the same as those used in all previously reported results for ILC laws implemented on the gantry robot (to enable the broadest possible comparison to be made). Figure 12.13 shows the resulting errors for all axes without the PID feedback controller for various values of $P_{u,0}$. In contrast to the simulation results, the use of larger values of $P_{u,0}$ does not lead to appreciable differences in the error values produced.

Figure 12.14 shows the errors for all axes without a PID feedback controller, using various values of q_p. In this case, the smaller values of q_p give improved performance, especially for the Y and Z-axes. However, continuing to reduce q_p gives progressively less advantage in terms of performance. From the MSE plots, it is evident that the value of q_p significantly influences learning speed.

As discussed above, including the PID feedback controller provides a higher-tracking performance level over the initial trials. This feature is illustrated in Figure 12.15, where the error plots are given for all axes, demonstrating that, with small PID gains, the ILC controller can cooperate more effectively with the PID controller. If the PID controller is disabled, the convergence rates for all axes are higher, but the performance in terms of the final level of error is diminished, especially for the Y-axis.

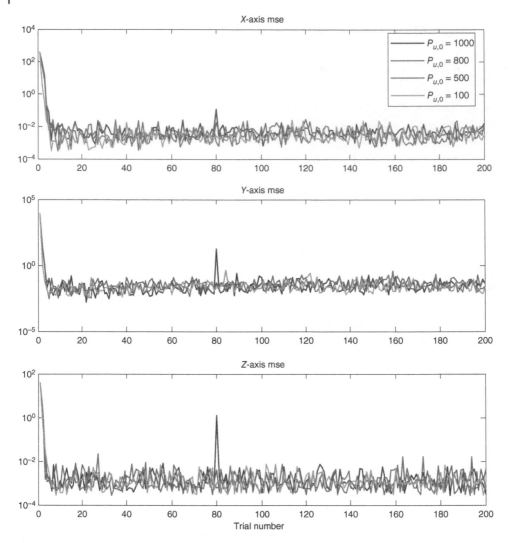

Figure 12.13 Experimental results (*PID* = {0,0,0}, Q_p = 0.001, $P_{x,0}$ = 0.1).

Keeping the PID feedback controller in place, further experiments to compare the effects of varying $P_{u,0}$ were undertaken. The results in Figure 12.16 confirm that the performance for different initial values of $P_{u,0}$ is generally quite similar. Exceptions occur in the case of the Y-axis, where the initial values of $P_{u,0}$ = 500 or $P_{u,0}$ = 800 clearly offer improvement relative to $P_{u,0}$ = 100.

Overall, these experimental results suggest that this design has significant robustness and disturbance rejection potential. Note also that substantial unexpected errors can arise on some trials (see, e.g. Figure 12.16, around trial 100 with $P_{u,0}$ = 100, and around trial 120 with $P_{u,0}$ = 200), but overall, they do not lead to long-lasting negative effects.

In the case of the P-controller, the controller arrangement and filtering method for D-type ILC were the same, but smaller values of initial parameters were used. However, this did not lead to a high level of performance. The results are in Figure 12.17, where the upper plot shows the tracking error over all trials. Ten trials only were possible due to instability over a narrow frequency band.

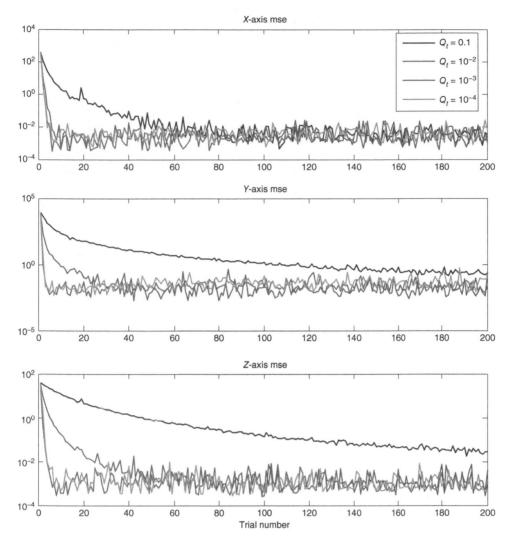

Figure 12.14 Experimental results ($PID = \{0, 0, 0\}$, $P_{u,0} = 100$, $P_{x,0} = 0.1$).

Frequency domain analysis of the error signal (see Figure 12.17 lower figure), shows that a sharp rise in magnitude is present at $\approx$11–12 Hz. Another fourth-order Chebychev filter (see (12.88)) which has a smaller cutoff frequency of $\approx$5 Hz has, therefore, been designed to solve this problem and is given by

$$H(z) = \frac{0.0002 + 0.0007z^{-1} + 0.0011z^{-2} + 0.0007z^{-3} + 0.0002z^{-4}}{1 - 3.5328z^{-1} + 4.7819z^{-2} - 2.9328z^{-3} + 0.6868z^{-4}} \tag{12.89}$$

Figure 12.18 shows a series of experimental results for all axes obtained with the modified filter in place. These results confirm that larger initial values of $P_{u,k}$, $P_{u,0}$ provide superior performance, especially for the Y and Z-axes.

Figure 12.19 shows a series of experimental results obtained using various values of Q_p. Although simulation studies indicated that smaller values of Q_p lead to improved performance, experimental results for the X-axis show little difference. Furthermore, the smaller values of this parameter could not provide superior results over the initial trials. For the Y and Z axes, the smaller values of Q_p

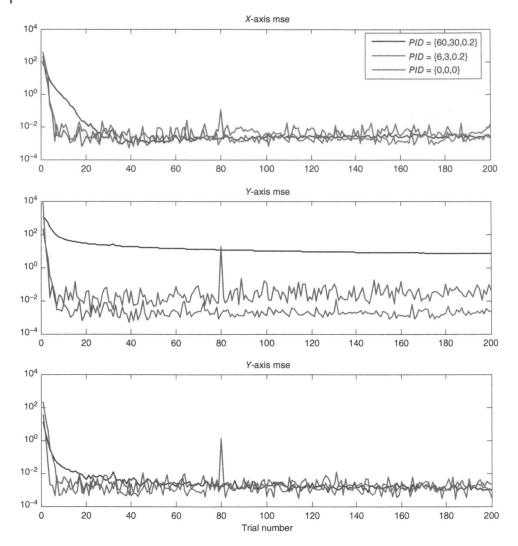

Figure 12.15 Experimental results ($P_{u,0} = 1000$, $Q_p = 0.001$, $P_{x,0} = 0.1$).

lead to a reduced level of final error. However, as with the X-axis results, too small a value of Q_p produces poorer performance over the initial trials.

Figure 12.20 gives the resulting errors for all axes without the PID feedback controller, for various values of $P_{u,0}$. In contrast to the simulation results, the use of larger values of $P_{u,0}$ does not lead to appreciable differences in the levels of error.

Figure 12.21 shows the errors for all axes without a PID feedback controller for various choices of Q_p. For smaller values of Q_p, the performance is improved, especially for the Y and Z axes. However, reducing Q_p further gives progressively less advantage in terms of performance. From the mse plots, the value of Q_p significantly influences the learning speed.

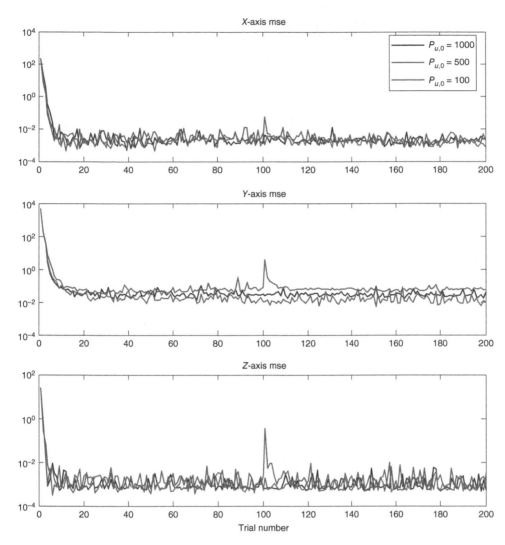

Figure 12.16 Experimental results ($PID = \{6, 3, 0.2\}$, $Q_t = 0.001$, $P_{x,0} = 0.1$).

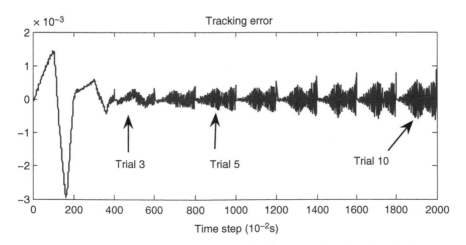

Figure 12.17 The frequency spectrum of tracking error for some trials.

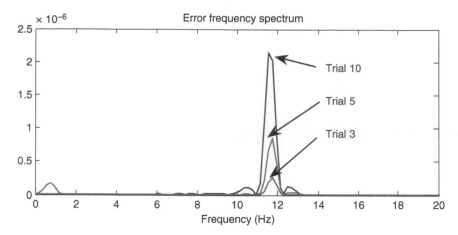

Figure 12.17 (*Continued*)

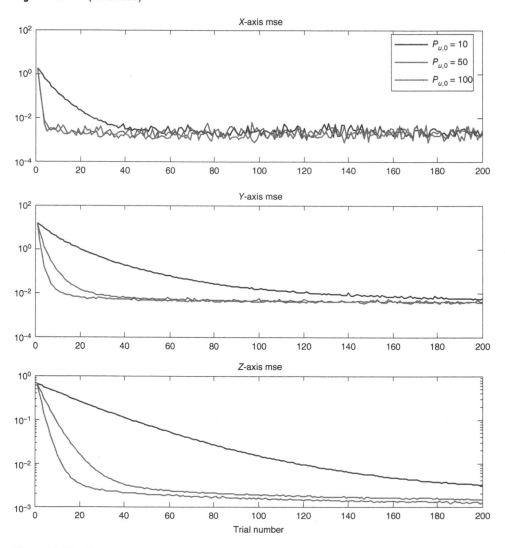

Figure 12.18 Experimental results (*PID* = {600, 300, 0.2}, $Q_t = 0.1$, $P_{x,0} = 0.1$).

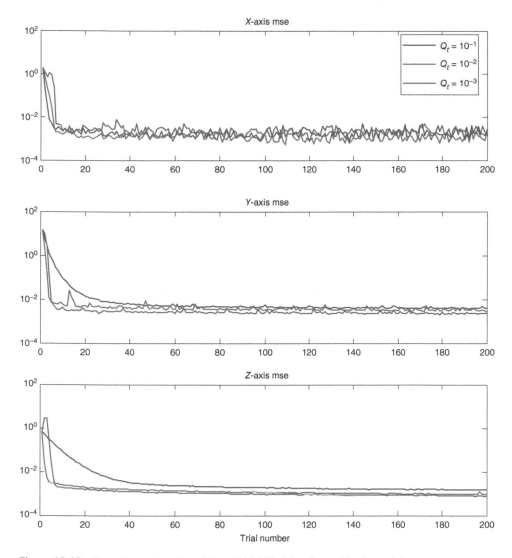

Figure 12.19 Experimental results ($PID = \{600, 300, 0.2\}$, $P_{u,0} = 50$, $P_{x,0} = 0.1$).

As detailed above, the inclusion of the PID feedback controller provides a higher level of tracking performance over initial trials. This feature is seen in Figure 12.22 where error plots for all axes are shown. These results show that the ILC controller can cooperate more effectively with the PID controller when the latter has small gains. If the PID controller is omitted, the convergence rates for all axes are higher, but the final error performance is poorer, especially for the Y-axis.

Including the PID feedback controller, further experiments were conducted to compare the effects of varying $P_{u,0}$. The results are in Figure 12.23, and it follows that the performance for different initial values of $P_{u,0}$ is generally quite similar. Exceptions occur in the case of the Y-axis, where the initial values of $P_{u,0} = 500$ or $P_{u,0} = 800$ clearly improve over those for $P_{u,0} = 100$.

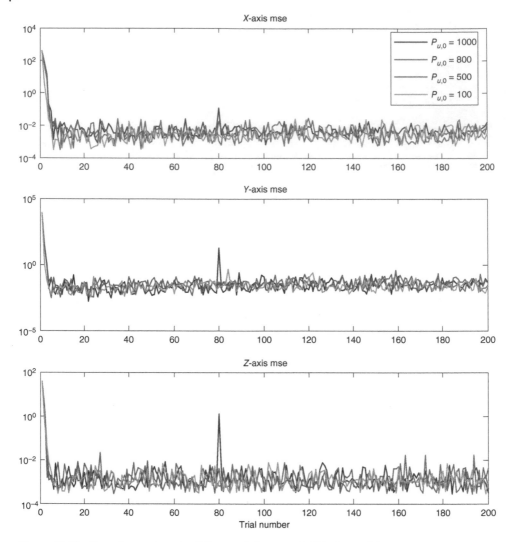

Figure 12.20 Experimental results ($PID = \{0,0,0\}$, $Q_p = 0.001$, $P_{x,0} = 0.1$).

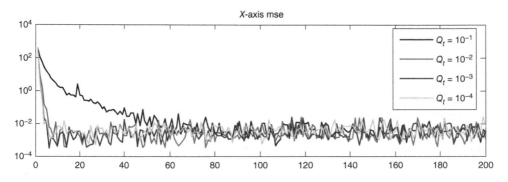

Figure 12.21 Experimental results ($PID = \{0,0,0\}$, $P_{u,0} = 100$, $P_{x,0} = 0.1$).

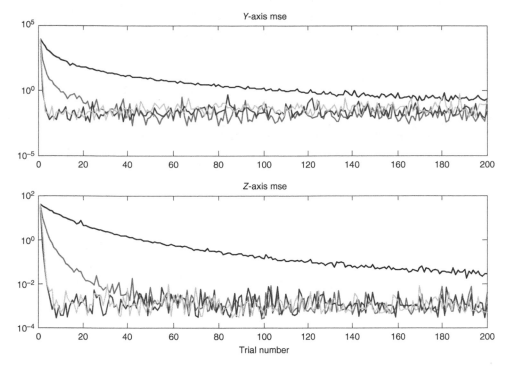

Figure 12.21 (*Continued*)

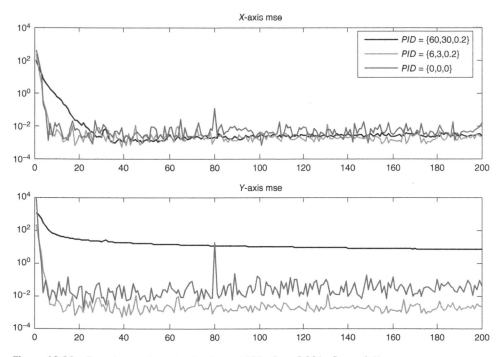

Figure 12.22 Experimental results for $P_{u,0} = 1000$, $Q_p = 0.001$, $P_{x,0} = 0.1$).

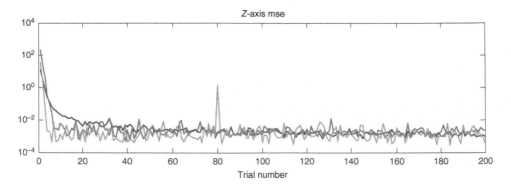

Figure 12.22 (*Continued*)

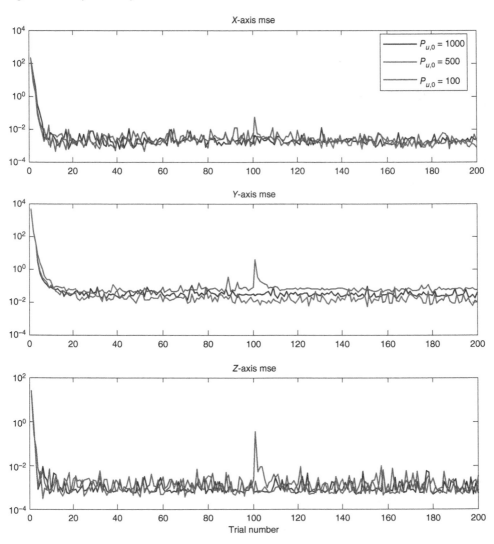

Figure 12.23 Experimental results ($PID = \{6, 3, 0.2\}$, $Q_p = 0.001$, $P_{x,0} = 0.1$).

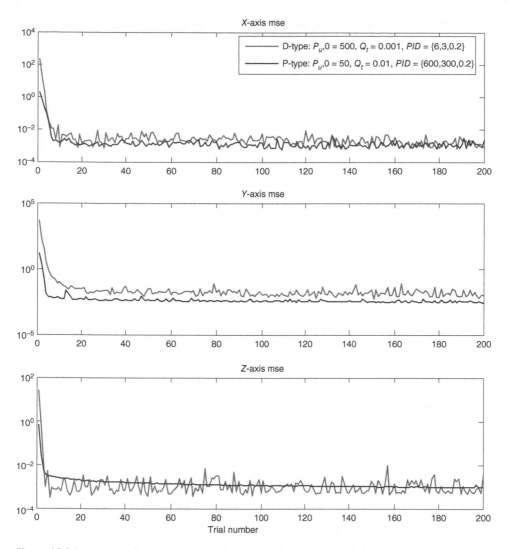

Figure 12.24 A comparison between the D-type and P-type stochastic ILC laws.

In general, these experimental results suggest that the ILC la used has the potential to deliver significant robustness and disturbance rejection potential. Note, however, that substantial unexpected errors can arise on some trials (see, e.g. Figure 12.16, around trial 100 with $P_{u,0} = 100$, and around trial 120 with $P_{u,0} = 200$), but overall, they do not lead to long-lasting negative effects.

The best-performing experimental results obtained for the two ILC laws considered are compared in Figure 12.24. These show that the P-type law gives superior performance for the Y-axis compared with the D-type law. The P-type law slightly improves on the results of the D-type law for the X-axis. However, the performance is approximately equal in the Z-axis, but the P-type law has a converged error with less oscillatory behavior along the trial.

In the next section, a state-space model-based design is developed and compared with another design in this general class.

12.4 Repetitive Process-Based Analysis and Design

The discrete linear systems considered are described in the ILC setting on trial k as follows:

$$x_k(p+1) = Ax_k(p) + Bu_k(p) + D\mu_k(p)$$
$$y_k(p) = Cx_k(p)$$
$$y_{\omega k}(p) = y_k(p) + G\omega_k(p), \quad 0 \le p \le N-1 \tag{12.90}$$

where the notation and dimensions of the state, input, and output (or trial) profile vectors are as given previously in this chapter, $\mu_k(p) \in \mathbb{R}^s$ is the state noise vector, and $\omega_k(p) \in \mathbb{R}^o$ is the measurement noise vector. It assumed that $\mu_k(p)$ and $\omega_k(p)$ are zero mean independent Gaussian white noise vectors such that $E[\mu_k(p)\mu_k^T(p)] = S_\mu$, $E[\omega_k(p)\omega_k^T(p)] = S_\omega$. Also, it is assumed that $\mu_k(p)$ is not dependent on the initial state vector on any trial.

Given a reference trajectory $r(p)$, $0 \le p \le N-1$, the ILC design problem is to construct a sequence of trial inputs $\{u_k\}_k$ such that the following conditions on the error and input $u_k(p)$ hold:

$$\lim_{k \to \infty} E[\|e_k(p)\|] = \|e_\infty(p)\|, \quad E[\|e_{k+1}(p)\|] \le E[\|e_k(p)\|], \quad \forall k$$
$$\lim_{k \to \infty} E[\|u_k(p) - u_\infty(p)\|] = 0 \tag{12.91}$$

and $\lim_{k \to \infty} E[\|e_k(p)\|^2]$ and $\lim_{k \to \infty} E[\|u_k(p) - u_\infty(p)\|^2]$ are bounded. Again, $u_\infty(p)$ is again termed the learned control.

In applications, only a finite number of trials will be completed. One design-oriented measure arising from the property $E[\|e_{k+1}(p)\|] \le E[\|e_k\|]$ is to specify a tolerance, say $\delta > 0$, and then allow the controlled dynamics to operate until this tolerance is achieved. This option is considered again in the case study given later in this section.

The ILC law on the current trial is again that used on the previous trial plus a correction, i.e. of the form

$$u_{k+1}(p) = u_k(p) + \Delta u_{k+1}(p) \tag{12.92}$$

where $\Delta u_{k+1}(p)$ is correction (update) term and the problem is to design this term to ensure (12.91), boundedness of $\lim_{k \to \infty} E[\|e_k(p)\|^2]$ and $\lim_{k \to \infty} E[\|u_k(p) - u_\infty(p)\|^2]$.

This paper follows, in the main, Pakshin et al. [199] and uses the mean square estimate, $\hat{x}_k(p)$, of the state vector $x_k(p)$, obtained by using the Kalman filter:

$$\hat{x}_k(p+1) = A\hat{x}_k(p) + Bu_k(p) + F(y_{\omega k}(p) - C\hat{x}_k(p)), \quad \hat{x}_k(0) = Fy_{\omega k}(0) \tag{12.93}$$

and $\hat{y}_k(p) = C\hat{x}_k(p)$, where $F = ASC^T(CSC^T + S_\omega)^{-1}$ and S is the solution of the algebraic Riccati equation:

$$S = ASA^T - ASC^T(CSC^T + S_\omega)^{-1}CSA^T + S_\mu \tag{12.94}$$

Introduce the estimation error as $\tilde{x}_k(p) = x_k(p) - \hat{x}_k(p)$ and the following auxiliary vectors as increments, respectively, in k of the estimated and error estimation vectors:

$$\hat{\eta}_{k+1}(p+1) = \hat{x}_{k+1}(p) - \hat{x}_k(p), \quad \tilde{\eta}_{k+1}(p+1) = \tilde{x}_{k+1}(p) - \tilde{x}_k(p) \tag{12.95}$$

Moreover, since $x_k(p)$ has to be estimated, $e_k(p) = r(p) - y_k(p) = r(p) - Cx_k(p)$ is not available for use in the control law and $\hat{e}_k(p) = r(p) - C\hat{x}_k(p)$ must be used instead. Then by routine calculations:

$$\overline{\eta}_{k+1}(p+1) = A_{11}\overline{\eta}_{k+1}(p) + A_{12}\hat{e}_k(p) + B_1 v_{k+1}(p) + D_1 w_{k+1}(p)$$
$$\hat{e}_{k+1}(p) = A_{21}\overline{\eta}_{k+1}(p) + A_{22}\hat{e}_k(p) + B_2 v_{k+1}(p) + D_2 w_{k+1}(p) \tag{12.96}$$

where

$$\bar{\eta}_{k+1}(p) = \left[\tilde{\eta}_{k+1}^T(p) \quad \hat{\eta}_{k+1}^T(p)\right]^T$$

$$v_{k+1}(p) = \Delta u_{k+1}(p-1)$$

$$w_{k+1}(p) = \left[\Delta\mu_{k+1}^T(p-1) \quad \Delta\omega_{k+1}^T(p-1)\right]^T$$

$$\Delta\mu_{k+1}(p-1) = \mu_{k+1}(p-1) - \mu_k(p-1)$$

$$\Delta\omega_{k+1}(p-1) = \omega_{k+1}(p-1) - \omega_k(p-1)$$

$$A_{11} = \begin{bmatrix} A - FC & 0 \\ FC & A \end{bmatrix}, \quad B_1 = \begin{bmatrix} 0 \\ B \end{bmatrix}, \quad D_1 = \begin{bmatrix} D & -FG \\ 0 & FG \end{bmatrix}$$

$$A_{12} = 0, \quad A_{22} = I, \quad A_{21} = [-CFC - CA], \quad B_2 = -CB, \quad D_2 = [0 - CFG]$$

The state-space model (12.96) describes a stochastic discrete linear repetitive process with additive disturbances. Moreover, the stability theory for linear repetitive processes has been extended to nonlinear stochastic processes using dissipativity theory [202]. The analysis that follows applies this theory to ILC design for discrete linear stochastic systems where a general nonlinear control law of the form:

$$v_{k+1}(p) = \varphi(\bar{\eta}_{k+1}(p), \hat{e}_k(p)), \quad \varphi(0,0) = 0 \tag{12.97}$$

is applied. As demonstrated in the case study given later in this chapter, particular cases of this control law provide extra freedom to tune the control law for performance (for both nonlinear and linear dynamics).

Definition 12.1 [202] Consider the discrete nonlinear repetitive process formed by applying (12.97) to (12.96). Then this process said to be stable along the trial in the second moment if

$$\lim_{k,p\to\infty} E[||\hat{\eta}_k(p)||^2 + ||\hat{e}_k(p)||^2] \leq \Gamma < \infty \tag{12.98}$$

where Γ is independent of N.

Further analysis makes use of a vector Lyapunov function of the form:

$$V(\xi, \epsilon) = \begin{bmatrix} V_1(\xi) \\ V_2(\epsilon) \end{bmatrix}, \quad \xi \in \mathbb{R}^n, \quad \epsilon \in \mathbb{R}^m \tag{12.99}$$

where $V_1(\xi) > 0$, $\xi \neq 0$, $V_2(\epsilon) > 0$, $\epsilon \neq 0$, $V_1(0) = 0$, $V_2(0) = 0$. (This function is the stochastic counterpart of the function used in Section 8.1 for deterministic dynamics.)

The analysis that follows uses results that are the stochastic counterparts of those used in Section 8.1. Hence, only the main results are given. Starting with the stochastic counterpart of the divergence operator of (12.99) along the trajectories of (12.96), referred to as the divergence operator for ease of presentation from this point onward,

$$DV(\xi, \epsilon) = E[V_1(\hat{\eta}_{k+1}(p+1))|\hat{\eta}_{k+1}(p) = \xi, \hat{e}_k(t) = \epsilon] - V_1(\xi)$$
$$+ E[V_2(\hat{e}_{k+1}(t))|\hat{\eta}_{k+1}(p) = \xi, \hat{e}_k(p) = \epsilon] - V_2(\epsilon) \tag{12.100}$$

Remark 12.1 $DV(\xi, \epsilon)$ on any trial $k \geq 0$ is the sum of the mean partial increments in both the trial and along the trial directions under the conditions that the arguments in the previous steps are fixed and take values ξ and ϵ. It is the natural counterpart of the divergence operator for deterministic systems, and this applies to all cases where $DV(\xi, \epsilon)$ arises in the remainder of this section.

The following result is Theorem 1 in [199].

Theorem 12.6 *[196, 202] Consider the ILC dynamics resulting from the application of (12.97) to (12.96). Suppose that there exists a function $V(\xi, \epsilon)$ of the form (12.99) and positive constants c_1, c_2, c_3, and γ satisfying the following inequalities along the trajectories of the controlled process:*

$$c_1\|\xi\|^2 \le V_1(\xi) \le c_2\|\xi\|^2 \tag{12.101}$$

$$c_1\|\epsilon\|^2 \le V_2(\epsilon) \le c_2\|\epsilon\|^2 \tag{12.102}$$

$$\mathscr{D}V(\xi, \epsilon) \le \gamma - c_3(\|\xi\|^2 + \|\epsilon\|^2) \tag{12.103}$$

Then this representation of the ILC dynamics is stable along the trial in the second moment.

In the Kalman filtering part of this last result, $\hat{x}_k(0) = Fy_{\omega k}(0)$, where $y_{\omega k}(0)$ is the initial measurement vector available to the control law on each trial k. As $y_{\omega k}(0)$ has bounded variation the same property holds for $\hat{\eta}_k(0)$. Also, for analysis and control law design purposes only, consider an auxiliary vector $z_{k+1}(p) \in \mathbb{R}^f$ given by

$$z_{k+1}(p) = C_1\hat{\eta}_{k+1}(p) + C_2\hat{e}_k(p) + C_3v_{k+1}(p) \tag{12.104}$$

for (12.96), where C_1, C_2, and C_3 are constant matrices of compatible dimensions and define the dissipativity property as follows.

Definition 12.2 A discrete linear repetitive process described by (12.96) is said to be dissipative along the trial in the second moment if there exists a vector function of form (12.99), a scalar function $S(v_{k+1}(p), z_{k+1}(p))$, a vector $z_{k+1}(p)$ of form (12.104), positive scalars c_1, c_2, c_3, and γ satisfying (12.101) and (12.102) and

$$DV(\xi, \epsilon) \le S(v_{k+1}(p), z_{k+1}(p)) : (\overline{\eta}_{k+1}(p) = \xi, \hat{e}_k(p))$$
$$+ \gamma - c_3(\|\xi\|^2 + \|\epsilon\|^2) \tag{12.105}$$

It is immediate that if S, $u_{k+1}(p)$, and $z_{k+1}(p)$ are such that $S(v(z_{k+1}(p)), z_{k+1}(p)) \le 0$, then the control law $v(z_{k+1}(p)) = \varphi(\hat{\eta}_{k+1}(p), e_k(p))$ ensures stability along the trial in the second moment of the process formed by applying (12.97) to (12.96).

Returning to the ILC design problem, introduce $\overline{\xi}_{k+1}(p) = \begin{bmatrix} \overline{\eta}_{k+1}^T(p) & \hat{e}_k^T(p) \end{bmatrix}^T$ and introduce

$$\bar{A} = \begin{bmatrix} A_{11} & A_{12} \\ A_{21} & A_{22} \end{bmatrix}, \quad \bar{B} = \begin{bmatrix} B_1 \\ B_2 \end{bmatrix} \tag{12.106}$$

Also, let the matrix $P = \text{diag}[P_1\ P_2] \succ 0$ be the solution of the discrete matrix Riccati inequality:

$$\bar{A}^T P\bar{A} - (1-\sigma)P - \bar{A}^T P\bar{B}[\bar{B}^T P\bar{B} + R]^{-1}\bar{B}^T P\bar{A} + Q \preccurlyeq 0 \tag{12.107}$$

where $0 < \sigma < 1$ is a positive scalar and $Q \succ 0$ and $R \succ 0$ are weighting matrices to be selected. Moreover, this last inequality is easily reformulated as the following linear matrix inequality (LMI) with respect to $X = \text{diag}[X_1\ X_2]$, where $X_1 = P_1^{-1}$ and $X_2 = P_2^{-1}$,

$$\begin{bmatrix} (1-\sigma)X & X\bar{A}^T & X \\ \bar{A}X & X + \bar{B}R^{-1}\bar{B}^T & 0 \\ X & 0 & Q^{-1} \end{bmatrix} \succcurlyeq 0, \quad X \succ 0 \tag{12.108}$$

If this LMI is feasible, then $P = X^{-1}$.

The following result establishes a set of stabilizing control laws for a system described by (12.96).

Theorem 12.7 *[197] A discrete linear repetitive process described by (12.96) is dissipative along the trial in the second moment with the supply function:*

$$S(v_{k+1}(p), z_{k+1}(p)) = z_{k+1}^T(p)(\bar{B}^T P \bar{B} + R)^{-1} z_{k+1}(p) + 2z_{k+1}^T(p)v_{k+1}(p)$$
$$+ v_{k+1}(p)^T \bar{B}^T P \bar{B} v_{k+1}(p) \tag{12.109}$$

with respect to input $v_{k+1}(p)$ and

$$z_{k+1}(p) = \bar{B}^T P \bar{A} \bar{\xi}_{k+1}(p) \tag{12.110}$$

where $P = X^{-1}$ and $X > 0$ is a solution of the LMI (12.108). Moreover, a set of control laws that provides stability along the trial in the second moment for (12.96) is given by

$$v_{k+1}(p) = -[\bar{B}^T P \bar{B} + R]^{-1} \bar{B}^T P \bar{A} \Theta(\bar{\xi}_{k+1}(p)) \bar{\xi}_{k+1}(p) \tag{12.111}$$

where $\Theta(\bar{\xi}_{k+1}(p))$ is a symmetric matrix function to be chosen such that

$$M - 2M\Theta(\bar{\xi}_{k+1}(p)) + \Theta(\bar{\xi})M\Theta(\bar{\xi}_{k+1}(p)) - Q < 0 \tag{12.112}$$

$\forall \bar{\xi}_{k+1}(p) \in \mathbb{R}^{n+m}$, where $M = \bar{A}^T P \bar{B}[\bar{B}^T P \bar{B} + R]^{-1} \bar{B}^T P \bar{A}$. Also

$$\lim_{k,p \to \infty} E[||e_k(p)||^2] \leq 4\text{tr}[P_1 S_1] + \text{tr}[P_2 S_2][\lambda_{\min}(P)(1 - \sigma \lambda_{\min}(P)/\lambda_{\max}(P))]^{-1} \tag{12.113}$$

where

$$S_1 = \begin{bmatrix} DS_\mu D^T & FGS_\omega G^T F^T \\ 0 & FGS_\omega G^T F^T \end{bmatrix}, \quad S_2 = CFGS_\omega G^T F^T C^T$$

Remark 12.2 As discussed above, the error $e_k(p)$ will not be available for use in implementing the control law and $\hat{e}_k(p) = r(p) - C\hat{x}_k(p)$ must be used instead. Since $\hat{e}_k(p) = e_k(p) + C\tilde{x}_k(p)$, it follows from the proof of this last result that the rate of trial-to-trial error convergence remains the same, but the variance of the error will grow from trial-to-trial.

Remark 12.3 In general, (12.112) can be difficult to check because the form of Θ is unknown. However, this function is piecewise constant, corresponding to state-dependent piecewise linear control often used in practice. For a P obtained from (12.108) and (12.112) reduces to an LMI on applying the Schur's complement formula.

Suppose that the measurements made in the application to a physical example are corrupted by noise on the control signal applied and the measured output. Also, suppose that the state-space model describing these effects is

$$\dot{x}(t) = A_0 x(t) + B_0(u(t) + \mu_k(t))$$
$$y_\rho(t) = Cx(t) + \omega_k(t) \tag{12.114}$$

where matrices A_0, B_0, and C define a minimal realization of the system dynamics and $\mu_k(t)$ and $\rho_k(t)$ are continuous-time Gaussian white noise signals with constant intensities Q_n and R_n representing, respectively, the actuator and sensor noise present. Also, sampling the dynamics of (12.114), with sampling period T, results in discrete dynamics in the ILC setting with state-space model:

$$x_k(p+1) = Ax_k(p) + Bu_k(p) + \mu_k(p) \tag{12.115}$$
$$y_{\omega k}(p) = Cx_k(p) + \omega_k(p)$$

where $A = e^{A_0 T}$, $B = \int_0^T e^{A_0 \tau} B_0 \, d\tau$ and also it is easily verified that $CB \neq 0$. Also, $\mu_k(p)$ and $\omega_k(p)$ are independent Gaussian discrete white noises with covariances:

$$S_\mu = \int_0^{T_s} e^{A_0^\tau} B_0 Q_n B_0^T e^{A_0^T} \tau \, d\tau, \quad S_\omega = R_n/T_s \tag{12.116}$$

The numerical examples in this section use the transfer-function representing the Z-axis of the gantry robot system of Section 2.2.1, sampled at 0.01 seconds and $Q_n = 5 \times 10^{-5}$, $R_n = 5 \times 10^{-8}$.

Applying Theorem 12.7 to (12.96) for this case, gives the following set, by varying Θ, of ILC laws:

$$u_k(p) = u_{k-1}(p) + K\Theta(\overline{\zeta}_k(p))\overline{\zeta}_k(p) \tag{12.117}$$

where all the variables and parameters are defined in Theorem 12.7.

As a measure to assess the performance of this ILC law and the alternative considered below, introduce on each trial k the sample mean square deviation:

$$E(k) = \sqrt{\frac{1}{N} \sum_{p=0}^{N} \|\hat{e}_k(p)\|^2} \tag{12.118}$$

Choose the matrix Θ in block diagonal form, i.e. $\Theta = \text{diag}[\Theta_1 \ \Theta_2 \Theta_3]$, where the dimensions of diagonal blocks correspond to the dimensions of $\tilde{\eta}$, $\hat{\eta}$, and $\hat{e}$. Then the ILC law (12.117) can be written as follows:

$$u_k(p) = u_{k-1}(p) + K_1\Theta_1(\tilde{x}_k(p) - \tilde{x}_{k-1}(p)) + K_2\Theta_2(\hat{x}_k(p) - \hat{x}_{k-1}(p))$$
$$+ K_3\Theta_3 \hat{e}_{k-1}(p+1) \tag{12.119}$$

where the last term can be viewed as ILC phase-lead.

Given that $\tilde{x}$ is not available for use in the control law, set $\Theta_1 = 0$ and to examine the performance of the nonlinear control law, consider the simplest case of piecewise linear control. Choose $\Theta_2 = I$ and assume, as one choice, that the required mean square tolerance along the trial length is 2.5×10^{-3}, hence,

$$\Theta_3 = \begin{cases} 1.5 & \text{if } E(k-1) > 2.5 \times 10^{-3}, \\ 0.25 & \text{if } E(k-1) \leq 2.5 \times 10^{-3} \end{cases} \tag{12.120}$$

The motivation for this choice is based on reasoning that a higher gain should provide a higher rate of error convergence over the early trials and then a switch is made to a lower gain for more effective energy consumption. In the simulation results below, the following numerical data were used: $N = 200$, $Q_1 = 0.1 \times I_3$, $Q_2 = 10$, $Q_3 = 10^2$, $Q = \text{diag}[Q_1 \ Q_2 \ Q_3]$, and $R = 10$.

Figures 12.25 and 12.26 have been generated by repeating the computations on each trial 120 times, which demonstrate that $E(k)$ decreases by a factor of 10 over 5–7 trials and the convergence is monotonic from trial-to-trial (allowing for the fact that the expected value is calculated over a finite interval). Figures 12.27 and 12.28 show how the control inputs and the outputs progress with the trial number, which are consistent with the fast trial-to-trial error convergence. Also, the control input is not excessive and rapidly settles to the same value on each successive trial.

Figure 12.25 also shows a comparison in terms of the error with a linear design, where $K_1 = 0$, $K_2 = \begin{bmatrix} -17 & -24 & -11 \ 702 \end{bmatrix}$, $K_3 = 1007$ and use has been made of the property of Remark 12.2 for such designs, with the feature that the new design gives faster convergence over early trials.

In [233], an alternative ILC law for systems described by (12.115) is developed under the assumption that the reference trajectory $r(p)$ is the output of the system:

$$x_{\text{ref}}(p + 1) = Ax_{\text{ref}}(p) + Bu_{\text{ref}}(p)$$
$$r(p) = Cx_{\text{ref}}(p) \tag{12.121}$$

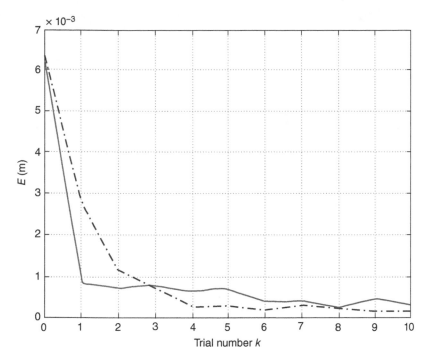

Figure 12.25 Progression of $E(k)$ against trial number (light gray line new design, dark gray line linear control law).

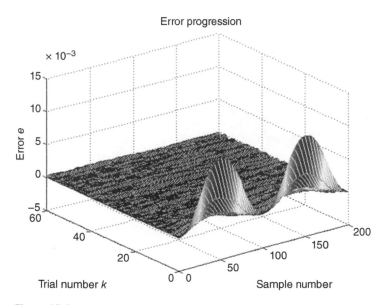

Figure 12.26 Error progression against trial number.

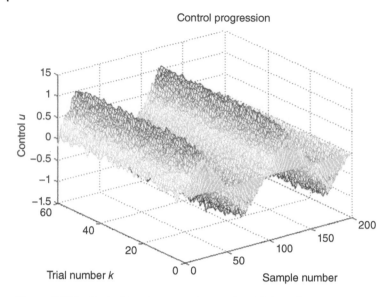

Figure 12.27 Control input progression against trial number.

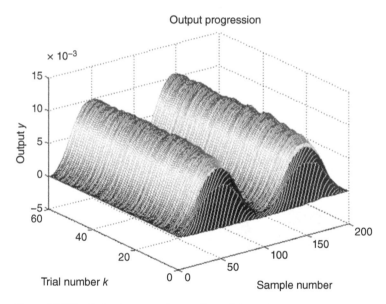

Figure 12.28 Output progression against trial number.

Consequently, there exists a unique input $u_{\text{ref}}(p)$ for given output $r(p)$, i.e. the output is realizable. Introduce $\delta x_k(p) = x_{\text{ref}}(p) - x_k(p)$, $\delta u_k(p) = u_{\text{ref}}(p) - u_k(p)$ with corresponding covariance matrices:

$$P_{x,k}(p) = E[\delta x_k(p)\delta x_k^T(p)], \quad P_{u,k}(p) = E[\delta u_k(p)\delta u_k^T(p)]$$

and $e_{\omega k}(p) = r(p) - y_{\omega k}(p)$. Suppose also that $y_{\text{ref}}(p)$, $x_k(0)$, $u_0(p)$, $P_{x,k}(0)$, $P_{u,0}(t)$, and k_{fin} are given. Then the ILC law in [233] can be applied by using the following procedure:

(i) Set $k = 0$

(ii) $Q_\mu = E[\mu_k(p)\mu_k^T(p)]$ and $R_\omega = E[\omega_k(p)\omega_k^T(p)]$ and compute for $p = [1, N-1]$

$$x_k(p+1) = Ax_k(p) + Bu_k(p) + \mu_k(p)$$

$$y_{\omega,k}(p) = Cx_k(p) + \omega_k(p)$$

$$e_{\omega,k}(p) = y_r(p) - y_{\omega k}(p)$$

$$P_{x,k}(p+1) = AP_{x,k}(p)A^T + BP_{u,k}(p)B^T + S_\mu$$

$$K_k(p) = P_{uk}(p)(CB)^T[(CB)P_{u,k}(p)(CB)^T$$
$$+ (C - CA)P_{x,k}(p)(C - CA)^T + CQ_nC^T + 2S_\omega]^{-1} \qquad (12.122)$$

$$y_{\omega,k}(p+1) = Cx_k(p+1) + \omega_k(p+1)$$

$$e_{\omega,k}(p+1) = r(p+1) - y_{\omega,k}(p+1)$$

$$u_{k+1}(p) = u_k(p) + K_k(p)[e_{\omega,k}(p+1) - e_{\omega,k}(p)]$$

$$P_{u,k+1}(p) = [I - K_k(p)(CB)]P_{u,k}(p)$$

(iii) $k = k + 1$

(iv) If $k < k_{fin}$, then go to Step (i), else stop.

Figures 12.29 and 12.30 demonstrate the rate of trial-to-trial convergence of this ILC law is strongly dependent on the state initial conditions on each trial.

Consider the case when $P_{x,k}(0) = 10^3 \times \text{diag}[1\ 1\ 1]$, $P_{u,0}(t) = 10^3$, leads to Figures 12.31–12.34.

In this case, the reduction of $E(k)$ against the trial number is much slower (reduction by a factor of 5 takes approximately 500 trials).

The control law (12.122) uses the discretized version of the derivative of $e_{\omega,k}(p)$ and the unfiltered measurement signal. This results in noise amplification and the output $y(p) = Cx(p)$ is more

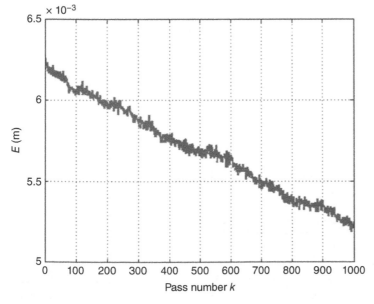

Figure 12.29 The progression of $E(k)$ against trial number for the design, $P_{x,k}(0) = \text{diag}[1\ 1\ 1]$, $P_{u,0}(t) = 1$. Source: Adapted from [232].

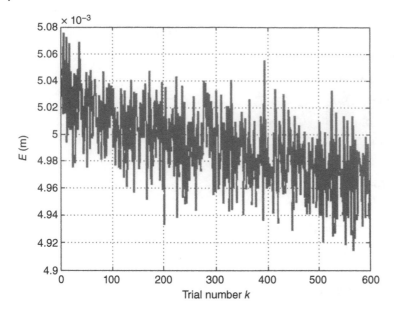

Figure 12.30 The progression of $E(k)$ against trial number for the design, $P_{x,k}(0) = \mathrm{diag}[1\ 1\ 1]$, $P_{u,0}(t) = 0.1$. Source: Adapted from [232].

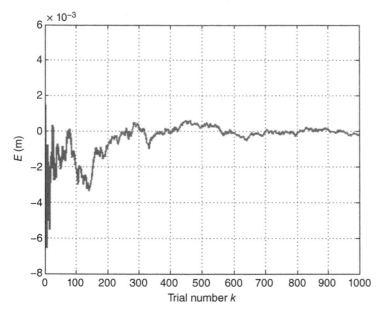

Figure 12.31 The progression of $E(k)$ against trial number for the design with $P_{x,k}(0) = 10^3 \times \mathrm{diag}[1\ 1\ 1]$, $P_{u,0}(t) = 10^3$. Source: Adapted from [232].

noise corrupted relative to the case when the new law (12.117) and (12.120) is used. Moreover, under (12.122), the rate of trial-to-trial error convergence critically depends on the initial values of covariance matrices $P_{x,k}(0)$ and $P_{u,0}(t)$, see Figures 12.29 and 12.30. Finally, (12.122) has a time-dependent gain and, hence, a more complicated implementation compared to (12.117).

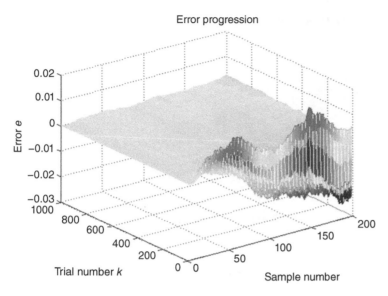

Figure 12.32 The progression of the error against trial number for the design with $P_{x,k}(0) = 10^3 \times \text{diag}[1\ 1\ 1]$, $P_{u,0}(t) = 10^3$. Source: Adapted from [232].

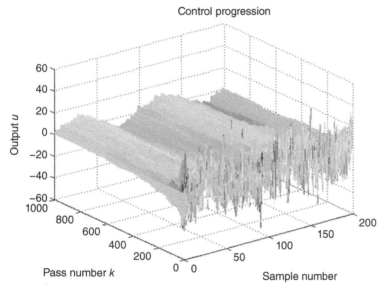

Figure 12.33 Control input progression against the design with $P_{x,k}(0) = 10^3 \cdot \text{diag}[1\ 1\ 1]$, $P_{u,0}(t) = 10^3$. Source: Adapted from [232].

12.5 Concluding Remarks

Stochastic ILC has seen relatively less development than deterministic systems. The results in this chapter give an overview of the results currently available. The experimental results with simple structure laws are very encouraging. Future efforts should include the case when the system dynamics are nonlinear. In this respect, the analysis of Section 8.1 for input saturation should immediately generalize to the stochastic case.

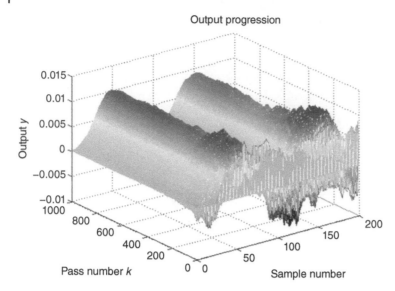

Figure 12.34 Output progression against trial number for the design, $P_{x,k}(0) = 10^3 \times \text{diag}[1\ 1\ 1]$, $P_{u,0}(t) = 10^3$. Source: Adapted from [232].

13

Some Emerging Topics in Iterative Learning Control

A particular feature of iterative learning control (ILC) analysis and design is the continual emergence of new application areas. Some of these can be addressed by existing designs, but others require some prior reformulations or developments of the underlying theory. Moreover, the emergence of data-driven control also has implications for ILC design. Similarly, the links with reinforcement learning, see, e.g. [247, 287] as a starting point for the literature, are not yet fully resolved. This chapter considers a representative cross-section of these active areas of ILC research.

13.1   ILC for Spatial Path Tracking

In most ILC applications, the desired tracking reference trajectory is prespecified over the trial length, i.e. it is the same for all trials. However, in possible applications, e.g. welding, laser cutting, and rehabilitation, the end-effector must follow a spatial path without any temporal information during the given time interval. In spatial tracking problems, substantial design freedom is available compared to the classical task description as the speed of travel along the path is not specified. Early research in [110, 169, 235] attempted to apply ILC to improve the spatial tracking performance in, e.g. two-axis gimbal systems, switched reluctance motors and microadditive manufacturing, where the solutions are application-specific.

In recent research, the norm optimal iterative learning control (NOILC) design (Chapter 5) has been extended to solve more general spatial path tracking problems [41–45]. The approach, based on [43], first introduces a spatial ILC framework with a reference trajectory, r, which specifies how fast the end-effector travels along the path. Then a spatial ILC algorithm with monotonic convergence property has been developed to update the input and reference trajectory, i.e. u_k, and r_k. The method is used to solve a minimum norm problem at the end of each trial.

Consider a discrete linear time-invariant system with state-space model in the ILC setting:

$$x_k(p + 1) = Ax_k(p) + Bu_k(p)$$
$$y_k(p) = Cx_k(p) \tag{13.1}$$

where on trial k, $x_k(p) \in \mathbb{R}^n$ is the state vector, $y_k(p) \in \mathbb{R}^m$ is the output vector and $u_k(p) \in \mathbb{R}^l$ is the control input vector. Or, in lifted form (assuming zero initial condition, see (1.37))

$$Y_k = GU_k \tag{13.2}$$

A spatial path is described as a subset of points in the output space, $\mathbb{R}^m$, defined by a continuous function, $\tilde{r}$, which maps each spatial position s in the interval, $[0, 1]$, to a point, $\tilde{r}(s)$, in the output

Iterative Learning Control Algorithms and Experimental Benchmarking, First Edition.
Eric Rogers, Bing Chu, Christopher Freeman and Paul Lewin.
© 2023 John Wiley & Sons Ltd. Published 2023 by John Wiley & Sons Ltd.

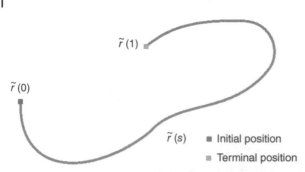

Figure 13.1 An example of a spatial path.

■ Initial position
■ Terminal position

space, $\mathbb{R}^m$, i.e. $\tilde{r} : s \in [0,1] \to \tilde{r}(s) \in \mathbb{R}^m$. An example of a spatial path is shown in Figure 13.1. In the spatial tracking problem, the spatial path profile, $\tilde{r}$, is known and hence, is independent of the temporal information.

The next step is to specify a speed profile, θ, to determine how fast the end-effector moves along the spatial path. This profile, θ, is defined as a function that maps each time index, p, in the interval, $[0, N]$, to a spatial position, s, in the interval, $[0, 1]$, i.e. $\theta : p \in [0, N] \to s \in [0, 1]$. Hence,

$$s_i = \theta(t_i), t_i = i, \quad i = 0, \dots, N, 0 = s_0 \leqslant s_1 \leqslant \cdots \leqslant s_N = 1 \tag{13.3}$$

where t_i denotes the ith sample time index and s_i denotes the spatial position corresponding to the time index, t_i. Moreover, the inequality constraint prevents the end-effector from going back along the path. Therefore, the initial and terminal positions of the end-effector are along the finite time interval $[0, N]$.

Define the spatial position allocation as follows:

$$\Pi = \begin{bmatrix} s_0 & s_1 & \cdots & s_N \end{bmatrix}^T \in \Theta \tag{13.4}$$

to represent the mapping of the speed profile in discrete formation, where Θ is the admissible set for the allocation, Π, defined as follows:

$$\Theta = \{\Pi \in \mathbb{R}^{N+1} : 0 = s_0 \leqslant s_1 \leqslant \cdots \leqslant s_N = 1\} \tag{13.5}$$

Using the definition of the path and speed profiles, the reference profile, R, is defined as follows:

$$R = \tilde{r}(\Pi) = \begin{bmatrix} r(0) & \cdots & r(N) \end{bmatrix}^T \in l_2^m[0, N] \tag{13.6}$$

where

$$r(i) = \tilde{r}(\theta(t_i)) = \tilde{r}(s_i), \quad i = 1, \dots, N \tag{13.7}$$

Since r is a variable for spatial position allocation, Π is not constant.

The design objective of the spatial tracking problem is: choose an input signal, U, such that the corresponding output trajectory, $Y = GU$, accurately tracks the reference profile, R, i.e.

$$R = Y = GU \tag{13.8}$$

Also, the spatial ILC tracking problem is: iteratively update the input signal, U_{k+1}, and reference profile, R_{k+1}, using a function of the previous trial's input signal, U_k, reference profile, R_k and output trajectory, Y_k, i.e.

$$(U_{k+1}, R_{k+1}) = \mathscr{F}(U_k, Y_k, R_k) \tag{13.9}$$

such that the spatial tracking error $E_k = R_k - Y_k$ converges to zero as $k \to \infty$, i.e.

$$\lim_{k \to \infty} E_k = 0 \qquad (13.10)$$

and the input signal also converges, i.e.

$$\lim_{k \to \infty} U_k = U^* \qquad (13.11)$$

where $R_k = \tilde{r}(\Pi_k)$ denotes the reference profile for trial k with respect to the spatial position allocation, Π_k.

Significant freedom in control design is available with the spatial tracking objective (13.8), but difficulties arise in design, e.g. the reference profile, R_k must also be updated at the end of each trial. Therefore, it is necessary to design a spatial ILC update law for (13.9) to achieve the spatial tracking objective (13.8).

A spatial ILC algorithm (Algorithm 13.1) is obtained by extending the constant reference profile, r, to a varying reference profile, R, depending on the spatial position allocation, Π. In this algorithm, u_0 and r_0 are suitable initial input signal and reference profile, respectively, and ϵ is a sufficiently small positive scalar specifying the practical tracking precision requirement.

Algorithm 13.1 Given system dynamics (13.1), spatial path profile, $\tilde{r}$, any initial input signal, $u_0 \in l_2^\ell[0, N]$, and any initial reference profile, $r_0 \in l_2^m[0, N]$, update the input sequence, $\{U_k\}_{k \geqslant 0}$, and reference profile sequence, $\{R_k\}_{k \geqslant 0}$, using the update law:

$$(U_{k+1}, \Pi_{k+1}) = \arg\min_{U,\Pi} \{J_k(U, R) \;:\; E = R - Y, Y = Gu, \quad R = \tilde{R}(\Pi), \Pi \in \Theta\} \qquad (13.12)$$

until convergence, i.e. $\|E_k\| < \epsilon$, where the cost function, J_k, is defined as follows:

$$J_k(U, R) = \|E\|_Q^2 + \|U - U_k\|_R^2 \qquad (13.13)$$

Remark 13.1 The ILC update law (13.12) is a more general version of the NOILC law in [4] (see also Chapter 5). In particular, if the reference trajectory does not change with k, the latter law is recovered.

The following theorem establishes that the algorithm above solves the spatial tracking problem with the desired convergence properties.

Theorem 13.1 *The spatial error sequence, $\{E_k\}_{k \geqslant 0}$, generated by (13.12) monotonically converges to zero, i.e.*

$$\|E_{k+1}\|_Q^2 \leqslant \gamma \|E_k\|_Q^2 \qquad (13.14)$$

where $0 < \gamma < 1$ is the largest eigenvalue of $(I + GG^*)^{-1}$.

This last theorem establishes that Algorithm 13.1 can iteratively solve the spatial ILC tracking problem with high spatial tracking accuracy. This appealing feature makes the practical application of this algorithm feasible. The following result establishes the detailed procedure to apply Algorithm 13.1 to improve the performance of applications relevant spatial tracking tasks. The ILC law (13.12) can be implemented by iteratively solving the optimization problem:

$$\min_{U,\Pi} \|R - GU\|_Q^2 + \|U - U_k\|_R^2 \;:\; Y = GU, \quad r(i) = \tilde{r}(s_i), \quad i = 0, \dots, N, \quad \Pi \in \Theta \qquad (13.15)$$

to update the reference profile, R_{k+1}, and the input signal, U_{k+1}, at the end of each trial.

The path profile function $\tilde{r}$ is, in general, nonlinear, the problem (13.15) is treated as a large scale nonconvex optimization problem, and there may exist many local optimal solutions. In addition, a heavy computation load can be incurred in solving this problem using standard solvers due to its nonconvex large-scale structure. A second-order cone programming (SOCP) problem based on linearized approximation of the nonlinear constraint is used to address this issue.

As the function, $\tilde{r}$, defines a spatial path, the constraint:

$$r(i) = \tilde{r}(s_i), \quad i = 0, \ldots, N \tag{13.16}$$

between the reference profile, R, and the spatial position allocation, Π, in (13.15) is generally nonlinear. Moreover, the nonlinear constraint (13.16) means that (13.15) is not convex. To eliminate the nonlinearity, the search range of the variable Π in (13.15) is reduced to a subset of Θ, by introducing the following assumption.

Assumption 13.1 The spatial position allocation Π in (13.15) is assumed to lie in the following set:

$$\Theta_k = \{\Pi \in \mathbb{R}^{N+1} : s_i \in [s_{i-1}^k, s_{i+1}^k], i = 1, \ldots, N-1\} \tag{13.17}$$

where s_i^k denotes the spatial position with respect to the ith time index on the previous trial k.

The last assumption applies an extra constraint (13.17) to problem (13.15), such that for each spatial position, s_i, its value is searched over the local subinterval, $[s_{i-1}^k, s_{i+1}^k]$, instead of over the whole interval, $[0, 1]$. Linearized approximation of constraint (13.16) can be undertaken based on this assumption, as detailed next.

Suppose that the extra constraint (13.17) is added to Assumption 13.1, then the constraint (13.16) can be linearized as follows:

$$r(i) \approx r_k(i) + \frac{s_i - s_i^k}{s_{i+1}^k - s_{i-1}^k} \cdot (r_k(i+1) - r_k(i-1)) \tag{13.18}$$

for $i = 1, \ldots, N-1$, and $r(1) = \tilde{r}(0), r(N) = \tilde{r}(1)$.

Using the linearized approximation (13.18) of the constraint (13.16), the problem (13.15) is convex and, as the following result shows, can be reformulated as a (SOCP) problem.

Proposition 13.1 *Using the linearized approximation (13.18) based on Assumption 13.1, the optimization problem (13.15) can be equivalently written as follows:*

$$\min_{r,u,y,\Pi,d} \quad \sum_{i=0}^{N} d_i$$

$$subject\ to \quad Y = GU,$$

$$\left\| \begin{array}{c} 2 \cdot (r(i) - y(i)) \\ 2 \cdot (u(i) - u_k(i)) \\ 1 - d_i \end{array} \right\|_{\{Q,R,I\}} \leqslant 1 + d_i, \quad i = 0, \ldots, N \tag{13.19}$$

$$r(i) = r_k(i) + \frac{s_i - s_i^k}{s_{i+1}^k - s_{i-1}^k} \cdot (r_k(i+1) - r_k(i-1))$$

$$i = 1, \dots, N - 1, \quad r(1) = \tilde{r}(0), \quad r(N) = \tilde{r}(1)$$

$$\Pi \in \Theta \cap \Theta_k$$

which is a SOCP problem.

A SOCP problem (by the convexity property) such as (13.19) has a global optimal solution. Hence, existing standard parsers in MATLAB, e.g. YALMIP [149], can be employed. To test the algorithm, two axes, X and Z, of the gantry robot testbed of Section 2.1.1 are used to perform spatial path tracking tasks. A comparison with the NOILC design of Chapter 5 is also given.

The particular example considered is to follow a spatial path defined by

$$\tilde{r}(s) = \begin{bmatrix} 0.005 \cdot \cos(-0.5\pi \cdot s + \pi) + 0.005 \\ 0.005 \cdot \sin(-0.5\pi \cdot s + \pi) \end{bmatrix}, \quad s \in [0, 1] \tag{13.20}$$

The tracking time is $T = 2$ seconds, and the sampling time of the system is 0.01 seconds, i.e. $N = 200$. Also, the weighting matrices considered are of the form $Q = qI$, and $R = rI$, where q and r are positive scalars.

Algorithm 13.1 is applied to solve the spatial tracking task. The weighting parameters are chosen as $q = 30\,000$, $50\,000$, $80\,000$, $100\,000$, and $120\,000$, $r = 1$, respectively, the initial input signal is chosen as $U_0 = 0$, and the initial reference profile, R_0, is determined by the initial spatial position allocation:

$$\Pi_0 = \begin{bmatrix} 0 & 0.005 & 0.01 & 0.015 & \dots & 0.995 & 1 \end{bmatrix}^T \tag{13.21}$$

Figure 13.2 gives the norm of the spatial error on each trial for different choices of q. It is evident from this figure that the spatial tracking error monotonically decreases along the trial for all cases, which verifies the condition (13.14) in Theorem 13.1. Note that the convergence rate increases as

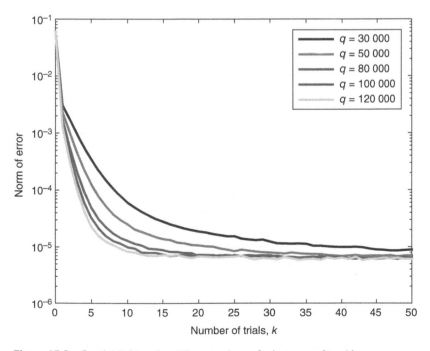

Figure 13.2 Spatial ILC law for different values of q in terms of tracking errors.

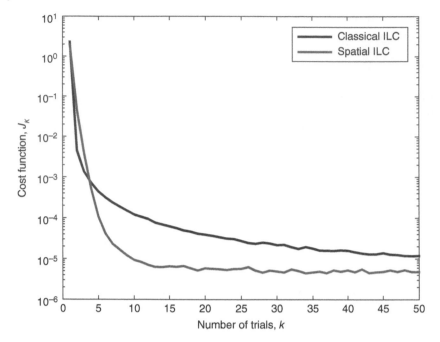

Figure 13.3 A comparison of spatial ILC and the NOILC design of Chapter 5 in terms of cost function values.

the value q increases, i.e. a larger weighting is applied to the norm of error in the cost function, $J_k(U, R)$. In addition, due to the measuring limit of the incremental encoder, all the tracking errors converge to approximately 10^{-5}. Therefore, it is concluded that Algorithm 13.1 can provide high-tracking accuracy while performing this spatial tracking task, i.e. the design requirement (13.10) for tracking error is satisfied.

As a comparison, the NOILC design of Chapter 5 has also been applied to the spatial path defined in (13.20) with $q = 120\,000$, $r = 1$, initial input signal, $u_0 = 0$, and fixed reference profile, r, determined by Π_0 in (13.21). The cost function of each trial is plotted in Figure 13.3 (dark gray curve). For comparison, the corresponding results using Algorithm 13.1 with the same q and r values are plotted in the same figure (light gray curve). Although the NOILC update provides a smaller cost function value for the first four trials, Algorithm 13.1 provides a much faster convergence rate in the longer term. Consequently, Algorithm 13.1 can deliver faster long-term convergence speed.

The monotonic convergence properties of the spatial ILC law considered are established. However, a rigorous robustness analysis is necessary to determine its convergence performance in the presence of model uncertainties. Also, alternative linearization methods should be considered within the convexification procedure used to obtain the spatial path. Moreover, an additional performance index, such as the control effort required, should be included in the cost function $J_k(U, R)$ and optimized over the trials. Experimental implementation is also required. The first results on some of these future research topics are in, e.g. [45].

13.2 ILC in Agriculture and Food Production

Agriculture and food production are a general area where operations arise with an inherent repetitive nature, e.g. batch processing and robotics-based systems for crop spraying and other tasks and, hence, the potential application of ILC laws. Previous results in this area include [34] for farm vehicle path following, and [161, 167, 229, 230] for automated irrigation and related issues in crop growing. (This area involves the use of distributed parameter systems models and hence, a potential further application area for the results of Chapter 9.)

Applications also exist for ILC in food production. The most obvious area is associated with the deployment of robotic systems to, e.g. collect items from a conveyor line and place them in packaging, another task which the gantry robot testbed of Section 2.1.1 provides the possibility of experimental validation of any new laws. This section considers recent research in food production, which provides a new application for terminal ILC (terminal ILC is also briefly discussed at the end of Chapter 5).

The food production application is the broiler, i.e. chicken meat, production process, and is based on [122]. Projections forecast that global demand for poultry meat will increase by 18% between 2015–2017 and 2027 to 139 billion kilogram [180, pp. 37], of which broiler (i.e. a chicken that is bred and raised specifically for meat production) meat will represent the majority. Industrial state-of-the-art broiler production typically has 30–40 000 broilers per batch, produces 2050 g broilers in 34 days from 42 g newly hatched broilers, and employs ad libitum feeding and drinking strategies, i.e. unrestricted access to feed and water. The optimization of the broiler feed conversion rate (FCR) reduces the amount of feed, water, and electricity required to produce a mature broiler.

Tight bounds on the production environment are needed to enable optimal growth, requiring a broiler application expert to tune each broiler house manually. Active feed control is not practically feasible in state-of-the-art broiler production as ad libitum feeding regimes, i.e. unlimited access to food, are used.

In this application area, temperature control is highly influential and practically feasible. Broiler production is mature in terms of data acquisition due to tight biosecurity and traceability requirements. These requirements, in turn, drive the need to automatically optimize performance in a data-driven framework by a suitably designed temperature controller. In what follows, a design based on combining ILC and a dynamic neural network (DNN) model of the data is described and evaluated in both simulation and implementation.

13.2.1 The Broiler Production Process

Consider the broiler growth model represented by the nonlinear discrete-time system

$$\begin{bmatrix} x_m\,[n+1] \\ x_f\,[n+1] \end{bmatrix} = \begin{bmatrix} x_m\,[n] \\ x_f\,[n] \end{bmatrix} + T_s \begin{bmatrix} G\left(u\,[n],\,x_m[n]\right) \\ R_f\left(x_m\,[n]\right) \end{bmatrix} \tag{13.22a}$$

$$\begin{bmatrix} y_w\,[n] \\ y_f\,[n] \end{bmatrix} = \begin{bmatrix} R_w\left(x_m\,[n]\right) \\ x_f\,[n] \end{bmatrix} + \begin{bmatrix} q_w\,[n] + q_{w,\text{bias}}\,[n] \\ q_f\,[n] \end{bmatrix} \tag{13.22b}$$

with initial conditions $x_m\,[N_s] = x_f\,[N_s] = 0$ and measured slaughter weight $\Gamma = z_w\,[N_e]$, where $x_m\,[n] \in \mathbb{R}_+$ is the broiler maturity in "effective growth days," $z_w\,[n] \in \mathbb{R}_+$ is the true broiler weight, $y_w\,[n] \in \mathbb{R}_+$ is the measured broiler weight, $x_f\,[n] \in \mathbb{R}_+$ is the cumulative feed consumption, $z_f\,[n] \in \mathbb{R}_+$ is the true cumulative feed consumption, $y_f\,[n] \in \mathbb{R}_+$ is the measured cumulative feed consumption, $u\,[n] \in \mathbb{R}$ is the temperature input, and $T_s \in \mathbb{R}_+$ is the sampling interval in days.

Under production conditions, the temperature input $u\,[n]$ is a reference trajectory for the climate control system, which, for simplicity, is assumed to achieve perfect tracking. In (13.22a), the smooth maturation function $G : \mathbb{R} \times \mathbb{R}_+ \mapsto [\beta, 1]$, where $\beta \in [0, 1]$ is the worst-case broiler growth rate,

$R_w : \mathbb{R}_+ \to \mathbb{R}_+$ and $R_f : \mathbb{R}_+ \to \mathbb{R}_+$ are smooth and strictly increasing functions mapping the broiler maturity $x_m[n]$ into broiler weight and feed consumption, $q_w[n] \in \mathbb{R}$ is the weight measurement noise, $q_{w,\text{bias}}[n] \in \mathbb{R}$ is the weight bias, and $q_f[n] \in \mathbb{R}$ is the feed measurement noise.

The growth and feed consumption of the widely used ROSS 308 fast-growing broiler strain are described by the manufacturer in [17, pp. 3] as follows:

$$R_w(t) = \frac{-18.3\,t^3 + 2.2551t^2 + 2.9118t + 54.739}{1000} \tag{13.23a}$$

$$R_f(t) = \frac{21.9 \times 10^{-6} t^4 - 4.232 \times 10^{-3} t^3 + 0.206t^2}{1000}$$

$$+ \frac{2.02t + 11.6}{1000} \tag{13.23b}$$

where $R_w(t) \in \mathbb{R}_+$ is the broiler weight reference in kg, $R_f(t) \in \mathbb{R}_+$ is the broiler feed uptake reference in kg/day, and $t \in [0, 59]$ days is the time in "effective growth days." Expressing broiler weight $R_w(x_m[n])$ and broiler feed uptake $R_f(x_m[n])$ in terms of the broiler maturity in "effective growth days" through $x_m[n]$ results in realistic weight and feed uptake behavior, as it captures the nonlinear nature of broiler growth. The polynomials result from statistical analysis by the manufacturer.

The maturation rate function $G : \mathbb{R} \times \mathbb{R}_+ \to [\beta, 1]$, where $\beta \in [0, 1)$ is a worst-case broiler growth rate, represents the influence of external stimuli u on the broilers' relative maturation. This function is impossible to construct from "first principles"; instead, a broiler application expert might heuristically specify the decreased growth rate for a specific temperature deviation from "optimal" growth conditions.

A modified normal distribution is chosen for G, as it has a unique maximum, and the standard deviation can easily be tuned to design how sensitive G is to temperature errors. Specifically,

$$G(u[n], x_m[n]) = \beta + (1 - \beta) \exp\left\{ \ln\left(\frac{\alpha + \beta - 1}{\beta - 1} \right) \left[\frac{u[n] - \bar{u}(x_m[n])}{\sigma_u} \right]^2 \right\} \tag{13.24}$$

where $\bar{u}(x_m[n])$ is the temperature maximizing G, $G(\bar{u}(x_m[n]), x_m[n]) = 1$, and $\sigma_u \in \mathbb{R}_+$ is the constant temperature sensitivity.

The temperature sensitivity is taken as the input error, $u[n] - \bar{u}(x_m[n])$, resulting in a decreased maturation rate of α – corresponding to $G(\bar{u}(x_m[n]) \pm \sigma_u, x_m[n]) = 1 - \alpha$ with $\alpha \in (0, 1 - \beta)$. Also, the parameters of the maturation rate function G are given in Figure 13.4.

The optimal temperature profile is unknown in this application, but typical temperature profiles for the ROSS 308 fast-growing broiler transition almost linearly between the initial temperature of $\bar{u}_s = 34\,°C$ at day $t_s = 0$ to $\bar{u}_e = 21\,°C$ at day $t_e = 34$. This corresponds to a temperature drop of $(\bar{u}_e - \bar{u}_s)$, which is modeled as proportional to the maturity $x_m[n]$, i.e.

$$\bar{u}(x_m[n]) = \bar{u}_s + \Delta T x_m[n], \quad \Delta T = \frac{\bar{u}_e - \bar{u}_s}{t_e - t_s} \tag{13.25}$$

Consequently, the optimal temperature at sample n depends on $x_m[n-1]$, which, in turn, depends on all prior inputs.

In [120], the role of a weight bias term $q_{w,\text{bias}}[n]$ was investigated and found to cause terminal weight measurement errors, $\tilde{y}_w - \tilde{z}_w$, with $-27.4\,g$ mean and $115.9\,g$ standard deviation through comparison with the accurately measured slaughter weight, see Figure 13.5a. Moreover, the weight bias onset was found to occur around day 15, which is heuristically assumed to increase linearly

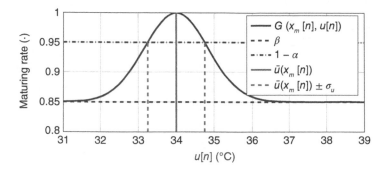

Figure 13.4 Visualization of the maturation rate function $G\left(x_m[n],\ u[n]\right)$ for $x_m[n] = 0$ with worst-case broiler growth rate $\beta = 0.85$, $\alpha = 0.05$, maximizing input $\bar{u}\left(x_m[n]\right) = 34\ °C$ and temperature error sensitivity $\sigma_u = 0.75\ °C$.

from zero at day 15 to $\mathcal{Q}_{\text{bias}} \sim \mathcal{N}\ (-27.4\ g,\ 115.9\ g)$ at the terminal sample, and hence,

$$z_{w,\text{bias}}[n] = \begin{cases} \frac{nT_s - 15}{N_e T_s - 15} \mathcal{Q}_{\text{bias}}, & 15 < nT_s \\ 0, & \text{otherwise} \end{cases} \tag{13.26}$$

In [120], it was found that using the measured slaughter weight, i.e. the terminal broiler weight, reduces the weight bias effect for broiler weight prediction on actual broiler production data.

The noise terms $q_w[n]$ and $q_f[n]$ result from analyzing the experimental test site's production data by constructing its frequency spectrum. As broiler weight is a smooth function of time, the "true" broiler weight is approximated by a second-order polynomial $\hat{y}_{w,\text{pol},2}$ between day 3 and 15, where the weight measurement y_w is expected to be the most reliable. The fit errors, $y_w - \hat{y}_{w,\text{pol},2}$, of 36 batches from the experimental test site are shown in the top plot of Figure 13.5b and are treated as measurement noise.

Subtracting the mean, concatenating all the fit errors, and computing the fast Fourier transform (FFT) produces the bottom magnitude plot. As this is not a standard distribution, random realizations of $q_w[n]$ with identical magnitude are obtained by randomly rotating the phases of the FFT and applying the inverse discrete Fourier transform. For more information on this approach, see [216]. Some realizations of $q_w[n]$ are shown in the top plot of Figure 13.5b. Similarly, the "true" cumulative feed uptake is approximated by a fourth-order polynomial $\hat{y}_{f,\text{pol},4}$ between day 3 and 30

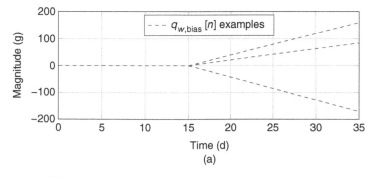

Figure 13.5 Measurement behavior for the heuristic broiler growth model. (a) Weight measurement bias $q_{w,\text{bias}}[n]$ samples using (13.26). (b) Visualization of the weight measurement noise $q_w[n]$. (c) Visualization of the feed uptake measurement noise $q_f[n]$.

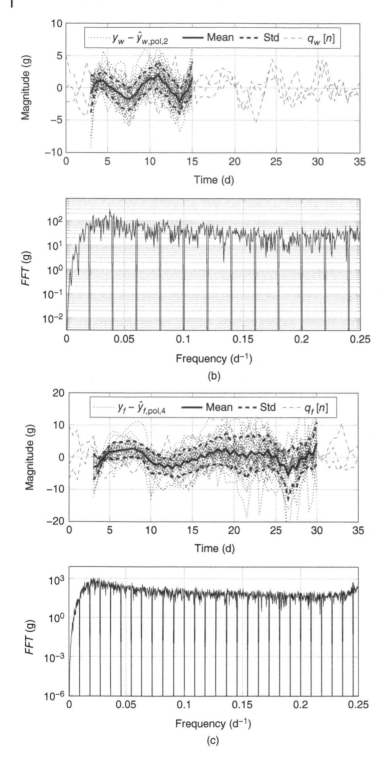

Figure 13.5 (*Continued*)

and shown in Figure 13.5c (using the same order of polynomial fit as proposed by the ROSS 308 manufacturer).

The control design objective includes the following considerations:

(i) **Weight maximization considerations:** Inspecting G shows that $x_m[n]$ is maximized by the unique input $\bar{u}\left(x_m[n]\right)$ that for all $u[n] \neq \bar{u}\left(x_m[n]\right)$ satisfies

$$G\left(u[n], x_m[n]\right) < G\left(\bar{u}\left(x_m[n]\right), x_m[n]\right) = 1$$

In the case when $\beta \leq G \leq 1$, the largest possible maturity $\bar{x}_m[n]$ is

$$\bar{x}_m[n] = \max\left\{x_m[n]\right\} = T_s \sum_{i=1}^{n} \max\left\{G\left(u[i], x_m[i]\right)\right\}$$

$$= nT_s$$

As R_w is strictly increasing, the largest possible accurate broiler weight $\bar{z}[n]$ is

$$\bar{z}_w[n] = \max\left\{z_w[n]\right\} = \max\left\{R_w\left(x_m[n]\right)\right\}$$

$$= R_w\left(\max\left\{x_m[n]\right\}\right) = R_w\left(nT_s\right)$$

This last step ensures that suboptimal control results in suboptimal weight, as expected in actual broiler production, where either a too low or too high-temperature results in decreased broiler growth.

In Figure 13.6, the behavior of the broiler model is shown for different temperature inputs. In particular, if the temperature is positively suboptimal, e.g. $u[n] = \bar{u}(\bar{x}_m[n]) + 1 \,°C$, then G converges to 1 for $n \to \infty$, and if the temperature is negatively suboptimal, e.g. $u[n] = \bar{u}(\bar{x}_m[n]) - 1 \,°C$, then G converges to β for $n \to \infty$. This is caused by the decreasing $\bar{u}\left(x_m[n]\right)$ for increasing $x_m[n]$. The fact that broiler farmers tend to use positively suboptimal temperatures is similar to the first state-of-the-art strategy.

(ii) **Feed minimization considerations:** if $\beta \leq G \leq 1$, the lowest maturation rate is governed by

$$\underline{x}_m[n] = \min\left\{x_m[n]\right\} = T_s \sum_{i=1}^{n} \min\left\{G\left(u[i], x_m[i]\right)\right\}$$

$$= T_s \beta n \tag{13.27}$$

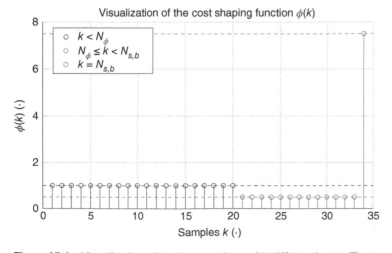

Figure 13.6 Visualization of broiler growth y_m with different inputs. The top plot depicts the maturation rate function $G\left(x_m[n], u[n]\right)$ as a function of the input $u[n]$ and the bottom plot depicts the output $y_m[n]$. The model settings equal that of Figure 13.4 with $T_s = 1$ day.

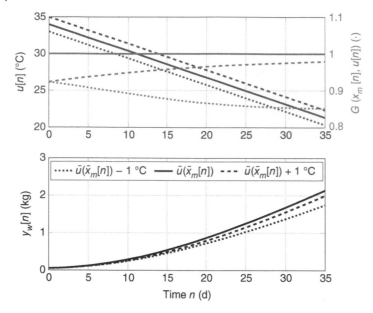

Figure 13.7 Visualization of different optimization strategies with $N_e = 34$, days, $T_s = 0.5$ day, and $\beta = 0.85$. A FCR difference of 1.14×10^{-3}, equivalent of 1.1%, exists between *Growth maximization* and *FCR minimization*, which potentially makes *FCR minimization* a better strategy.

As R_f is strictly increasing, the lowest cumulative feed consumption is given by

$$\underline{x}_f[n] = \min \{x_f[n]\} = \min \left\{ T_s \sum_{i=1}^{n} R_f(x_m[i]) \right\}$$

$$= T_s \sum_{i=1}^{n} R_f(\min \{x_m[i]\}) = T_s \sum_{i=1}^{n} R_f(T_s \beta i) \tag{13.28}$$

This suggests that feed minimization and weight maximization are completely opposing goals.

(iii) **FCR minimization considerations:** the expression for FCR from the heuristic model is

$$z_{\text{FCR}}[n] = \frac{z_f[n]}{z_w[n]} = \frac{x_f[n]}{R_w(x_m[n])} = T_s \frac{\sum_{i=1}^{n} R_f(x_m[i])}{R_w(x_m[n])} \tag{13.29}$$

Contrary to maximizing the weight and minimizing the feed, developing an analytical expression for the lowest possible FCR is nontrivial. The reasons are due to these simultaneous and opposing objectives, which depend on the simulation duration N_e as shown in Figure 13.8.

In Figure 13.7, the proposed strategies are compared. This figure shows that minimizing FCR consists of an initial period of minimizing the feed followed by weight maximization – similar to the second state-of-the-art strategy. Feed minimization produces the highest FCR and is not a feasible solution. Moreover, weight maximization results in a 1.1% higher FCR than FCR minimization. This fact makes FCR minimization favorable despite another output's added complexity.

13.2.2 ILC for FCR Minimization

Consider a system described in lifted form. Terminal iterative learning control (TILC) is applicable to a repeating process to iteratively learn the input sequence $U_k \in \mathbb{R}^{N_u N_n}$ such that the terminal

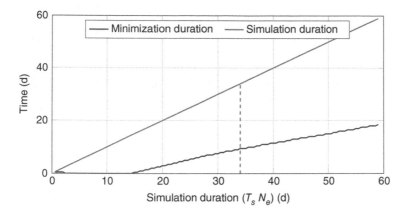

Figure 13.8 Minimization duration as a function of simulation duration ($N_e T_s$) with $\beta = 0.85$ and $T_s = 0.5$ days, which corresponds to the length of the initial period, where $G\left(x_m[n], u[n]\right) = \beta$. The dashed line indicates a simulation duration of 34 days with a minimization duration of 9.5 days, equivalent to Figure 13.7.

process output $\tilde{Y}_k\left(U_k\right) \in \mathbb{R}^{N_y}$ tracks the desired terminal reference vector $\tilde{R} \in \mathbb{R}^{N_y}$ denoted by

$$\lim_{k \to \infty} \tilde{Y}_k\left(U_k\right) = \tilde{R} \tag{13.30}$$

with the system dynamics written in the ILC setting as follows:

$$\tilde{Y}_k\left(U_k\right) = \tilde{P}U_k + \tilde{K} \tag{13.31}$$

where $\tilde{P} \in \mathbb{R}^{N_y \times N_u N_n}$ is the terminal system matrix, and $\tilde{K} \in \mathbb{R}^{N_y}$ represents terminal effects unrelated to the input vector $U \in \mathbb{R}^{N_u N_n}$.

This last problem can be solved using constrained Norm Optimal Point-To-Point ILC, which aims at tracking the output at specific samples. As TILC only aims at tracking the terminal output, TILC is a specialization of Point-To-Point ILC. Adapting the constrained Norm Optimal Point-To-Point ILC algorithm 1 in [51] to the particular case of the considered TILC problem is the solution mechanism. The intuition is to reduce the terminal tracking error by finding an input in the neighborhood of U_k that minimizes the cost function:

$$U_{k+1} = \arg\min_{U \in \Omega} \|\tilde{E}_k\|_{W_E}^2 + \|U_k\|_{W_{\Delta U}}^2 \tag{13.32}$$

subject to

$$\tilde{E}_k(U) = \tilde{R} - \tilde{Y}_k(U) \quad \text{and} \quad \tilde{Y}_k(U) = \tilde{P}U + \tilde{K} \tag{13.33}$$

where Ω denotes the set of valid inputs, $W_E \in \mathbb{R}^{N_y \times N_y} > 0$ is the tracking error weighting matrix, $W_{\Delta U} \in \mathbb{R}^{N_u N_n \times N_u N_n} > 0$ is the input weighting matrix, and $\tilde{E}(U_k)$ is the terminal tracking error given by the first equation in (13.33). Motivation for the choice of this cost function is to minimize the terminal tracking error on trial k by finding a minimizing input in the neighborhood of U_k.

The following results are Theorems 1 and 2, respectively, in [51].

Theorem 13.2 *If perfect tracking is feasible, i.e. $\exists U \in \Omega$ such that $\tilde{Y}_k\left(U\right) = \tilde{R}$; then (13.34) achieves monotonic convergence to zero-tracking error*

$$\left\|\tilde{E}_{k+1}\left(U_{k+1}\right)\right\|_{W_E} \leq \left\|\tilde{E}_k\left(U_k\right)\right\|_{W_E}, \quad k \geq 0 \tag{13.34}$$

and

$$\lim_{k\to\infty} \tilde{E}_k\left(U_k\right) = 0, \quad \lim_{k\to\infty} U_k = \bar{U} \tag{13.35}$$

Theorem 13.3 *If perfect tracking is not feasible, i.e. $\tilde{Y}_k(U) \neq \tilde{R} \quad \forall U \in \Omega$, then the input of (13.34) converges to*

$$\lim_{k\to\infty} U_{k+1} = \arg\min_{U\in\Omega} \left\| \tilde{R} - \tilde{P}U - \tilde{K} \right\|^2_{W_{\tilde{E}}} \tag{13.36}$$

This property is equivalent to the algorithm converging to the smallest possible tracking error. Moreover, this convergence is monotonic in the tracking error norm:

$$\left\| \tilde{E}_{k+1}\left(U_{k+1}\right) \right\|_{W_{\tilde{E}}} \leq \left\| \tilde{E}_k\left(U_k\right) \right\|_{W_{\tilde{E}}} \quad \forall k \in \mathbb{Z}_+ \tag{13.37}$$

The application of this algorithm requires a model of the process dynamics. One option is to use a heuristic model, but such a model will be less accurate and affect its performance. Next, the development of a data-driven model approach is described to enable control synthesis without a mathematical broiler FCR model by synthesizing $\tilde{P}$ and $\tilde{K}$ from (13.43) using past production data. A nonlinear discrete-time data-driven model is used to capture the broiler growth dynamic using data from the past N_b trials, $\left\{ \left\{ U_{k-N_b+1}, D_{k-N_b+1}, Y_{k-N_b+1} \right\}, \dots, \left\{ U_k, D_k, Y_k \right\} \right\}$, where D_k denotes the disturbance vector, and the N_b data indexes are represented by

$$\mathcal{B}_k = \left\{ k - N_b + 1, \dots, k \right\} \tag{13.38}$$

Data-driven model synthesis at trial k requires data from the trial indexes denoted by $\mathcal{B}_{k-1}$. Trial data prior to the first trial, $k < 1$, are denoted as *preliminary* trials, e.g. $\{U_{-2}, D_{-2}, Y_{-2}\}$. Hence, a total of N_b preliminary trials are required for model synthesis for the first trial, $k = 1$, denoted by the indices $\mathcal{B}_0 = \left\{ 1 - N_b, \dots, 0 \right\}$.

The data-driven representation chosen is a nonlinear autoregressive moving average model with exogenous input (NARMAX) type model implemented as a neural network with N_l input and output lags, a single hidden layer with N_N neurons, and a hyperbolic tangent activation function in the hidden layer:

$$\hat{y}_k\,[n+1 \mid \mathcal{W}, s] = W^o \tanh(\mathcal{X} + \theta^h) + \theta^o \tag{13.39}$$

with

$$\mathcal{X} = \sum_{i=0}^{N_l-1} W^h_{y,i}\hat{y}_k\,[n-i \mid \mathcal{W}, s] + W^h_{u,i}u_k[n-i] + W^h_{d,i}d_k[n-i]$$

where $W^o \in \mathbb{R}^{N_y \times N_N}$, $\mathcal{X} \in \mathbb{R}^{N_N}$, $\theta^o \in \mathbb{R}^{N_N}$, $W^h_{y,i} \in \mathbb{R}^{N_N \times N_y}$, $W^h_{u,i} \in \mathbb{R}^{N_N \times N_u}$, $W^h_{d,i} \in \mathbb{R}^{N_N \times N_d}$ and $\theta^h \in \mathbb{R}^{N_N}$ are model parameters stored in $\mathcal{W}$, $\hat{y}_k\,[n \mid \mathcal{W}, s]$ is the model output at sample n, initialized at sample s with model weights $\mathcal{W} \in \mathbb{R}^{N_w}$. The initialization, in this case, is,

$$\hat{y}_k\,[n \mid \mathcal{W}, s] = y_k[n] \quad \forall n \leq s \tag{13.40}$$

where n is implicitly lower bounded by the starting sample N_s, $N_s \leq n$, for both $y_k[n]$, $u_k[n]$ and $d_k[n]$.

The model weights were determined using the following training procedure:

$$\mathcal{W}(B) = \arg\min_{\mathcal{W}} \sum_{b\in B\setminus\min\{B\}} \frac{J_b(\mathcal{W})}{\#B - 1} \tag{13.41a}$$

with

$$J_b\left(\mathcal{W}\right) = \bar{\alpha}\left\|\mathcal{W}\right\|^2 + \sum_{i=1}^{N_S}\sum_{n=S_i}^{N_e}\frac{\mathcal{E}_b}{N_y\left(N_e - S_i + 1\right)} \tag{13.41b}$$

$$\mathcal{E}_b = \sum_{i=1}^{N_y}\begin{cases}\left\|\Gamma_b - \hat{y}_i\right\|_2^2\,\phi\left(k\right), & k = N_{s,b} \wedge i = i_w\\ \left\|y_i - \hat{y}_i\right\|_2^2\,\phi\left(k\right), & k \neq N_{s,b} \wedge i = i_w\\ \left\|y_i - \hat{y}_i\right\|_2^2, & \text{otherwise}\end{cases} \tag{13.41c}$$

$$\phi\left(k\right) = \begin{cases}1, & k < N_\phi\\ 1 + \left(N_{s,b} - N_\phi\right)\left(\gamma - 1\right), & k = N_{s,b}\\ \gamma, & \text{otherwise}\end{cases} \tag{13.41d}$$

where B is a set of batch-indices used for training, $S = \{S_1, \dots, S_{N_S}\}$ is the set of $N_S \in \mathbb{Z}_+$ initialization locations, which was found to speed up training as described in [121], Γ_b is the broiler slaughter weight of batch b, i.e. the true broiler weight prior to slaughter, $i_w \in \mathbb{Z}$ is the weight output index, $\phi : \mathbb{Z}_+ \to \mathbb{R}$ is the weight cost shaping function, $N_\phi \in \mathbb{Z}$ is the start weight cost-shaping sample number, and $\gamma \in (0, 1)$ is the weight cost-shaping parameter.

Automatic weighing pads are commonly used for weighing broilers and are known to be negatively biased onward from day 15, which is represented by (13.26) in the heuristic model. In [120], the weight cost-shaping function $\phi : \mathbb{Z}_+ \to \mathbb{R}$ in (13.41c) and (13.41d) was found to decrease the impact of this bias – one example of ϕ is shown in Figure 13.9. The slaughter weight is considered very accurate and is included by overriding the last measured local weight at sample $k = N_{s,b}$ of each batch. Extra emphasis is placed on the slaughter weight at sample $k = N_{s,b}$ in the cost function, while samples beyond $N_\phi \in \mathbb{Z}_+$ are less heavily weighted.

The cost function is minimized using the Levenberg–Marquardt algorithm with early stopping applied to the oldest batch index in B, which is denoted by $\min\{B\}$, in $J_{\min\{B\}}\left(\mathcal{W}\right)$, to prevent overtraining. The regularization constant $\bar{\alpha} \in \mathbb{R}_+$ is found iteratively through Bayesian regularization to prevent overfitting. Initialization of the model weights $\mathcal{W}$ is by the Nguyen–Widrow initialization scheme. For detailed information regarding the training, see, e.g. [120].

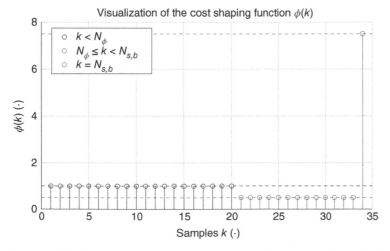

Figure 13.9 Visualization of the cost-shaping function $\phi(k)$ with $N_\phi = 20$, $N_{s,b} = 34$ and $\gamma = 0.5$. The dark gray, light gray (bottom), and light gray (top) values correspond to a separate case of (13.41d).

As (13.41a) is not a convex optimization problem, the weights $\mathcal{W}(B)$ are not guaranteed to be the global minimum. The ensemble mean of N_m models trained with different initial model weights is used to decrease the optimization terminating at a local minimum. The ensemble data-driven model simulated from sample N_s with data from batch b, $\{Y_b, D_b, U_b\}$, is

$$\hat{y}_{k,b}[n] = \frac{1}{N_m} \sum_{l=1}^{N_m} \hat{y}_b \left[n \mid \mathcal{W}_l \left(\mathcal{B}_k \setminus b \right), N_s \right] \tag{13.42}$$

where $\mathcal{W}_l \left(\mathcal{B}_k \setminus b \right)$ is the lth training of $\mathcal{W} \left(\mathcal{B}_k \setminus b \right)$ with the batch indexes $\mathcal{B}_k \setminus b$ to separate training and simulation data. The terminal super-vector ensemble data-driven model required for (13.34) is obtained by linearizing (13.42) along the trajectory of U_b (a past trial) using the first-order Taylor expansion:

$$\tilde{Y}_k(U) \approx \hat{\tilde{Y}}_{k,b} + \hat{\tilde{P}}_{k,b} \left(U - U_b \right) = \hat{\tilde{P}}_{k,b} U + \hat{\tilde{K}}_{k,b} \tag{13.43}$$

with

$$\hat{\tilde{P}}_{k,b} = \left. \frac{d\hat{\tilde{Y}}_{k,b}}{dU_b^T} \right|_{U_b} \quad \text{and} \quad \hat{\tilde{K}}_{k,b} = \hat{\tilde{Y}}_{k,b} \left(U_b \right) - \hat{\tilde{P}}_{k,b} U_b$$

where $U \in \mathbb{R}^{N_u N_n}$ is the super-vector input from above, and U_k is the super-vector representing the input for the current trial. The data-driven model is retrained for every k and b.

To use this model for FCR minimization requires an augmented data-driven model, denoted by $(\cdot)^*$. This model is given by

$$\tilde{Y}_k^*(U) = \frac{\tilde{Y}_{kf}(U)}{\tilde{Y}_{k,w}(U)} \tag{13.44}$$

where $\tilde{Y}_{k,w}(U) \in \mathbb{R}_+$ and the super-vectors $\tilde{Y}_{kf}(U) \in \mathbb{R}_+$, respectively, denote the weight and cumulative feed uptake – the equivalent of (13.22b). Linearized in U_b by a first-order Taylor expression similar to (13.43) results in

$$\tilde{Y}_k^*(U) \approx \hat{\tilde{Y}}_{k,b}^* \left(U_k \right) + \hat{\tilde{P}}_{k,b}^* \left(U - U_k \right) = \hat{\tilde{P}}_{k,b}^* U + K_{k,b}^* \tag{13.45}$$

with

$$\hat{\tilde{P}}_{k,b}^* = \frac{d\hat{\tilde{Y}}_{k,b}^*(U)}{d\hat{\tilde{Y}}_{k,b}^T(U)} \frac{d\hat{\tilde{Y}}_{k,b}(U)}{dU^T} = \frac{d\hat{\tilde{Y}}_{k,b}^*(U)}{d\hat{\tilde{Y}}_{k,b}^T(U)} \hat{\tilde{P}}_{k,b}$$

$$\hat{\tilde{K}}_{k,b}^* = \hat{\tilde{Y}}_{k,b}^* \left(U_k \right) - \hat{\tilde{P}}_{k,b}^* U_k = \frac{d\hat{\tilde{Y}}_{k,b}^*(U)}{d\hat{\tilde{Y}}_{k,b}^T(U)} \hat{\tilde{K}}_{k,b}$$

13.2.3 Design Validation

The design developed in this section has been verified both numerically and experimentally. Figure 13.10 shows relevant measured signals for $k = 1$, where the FCR@2.2 kg of trial $k = 1$ is approximately 6% smaller compared to $k = 0$. The terminal broiler weight is 200 g higher, and the terminal cumulative feed consumption is only 100 g higher, which is a disproportionate exchange rate. The initial input change is approximately 0.5 °C lower for days 0–4 and 9–15, and approximately 2 °C higher for day 27. The initial decrease in temperature decreased the broiler growth rate, as the operator reported mild signs of cold stress in the broilers on visual inspection.

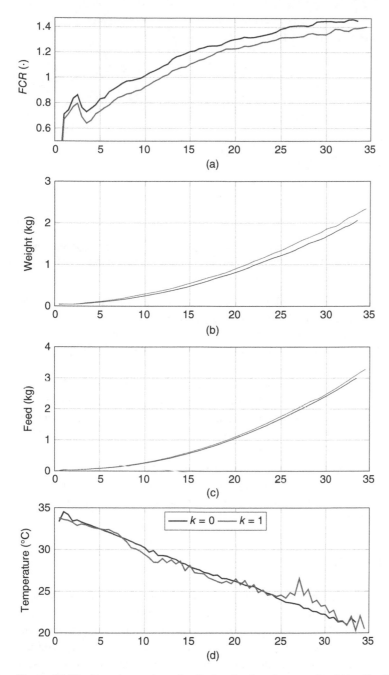

Figure 13.10 Experimental results for $k = 1$ using the new algorithm. The *FCR* (a), broiler weight (b), feed consumption (c), and measured temperature (d) is depicted for trial $k \in \{0, 1\}$.

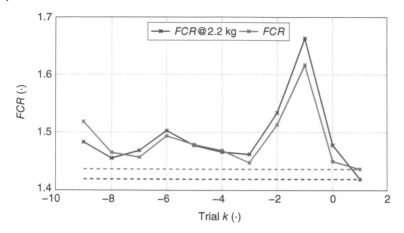

Figure 13.11 *FCR* and *FCR*@2.2 kg performance overview of the recent 10 trials $k \in \{-9, \ldots, 0\}$ and the current trial $k = 1$. Trial $k \in \{-2, -3\}$ have unusually high *FCR* due to an unusually cold winter, rendering the temperature regulation unable to maintain the desired temperature.

Applying the new design results in a FCR@2.2 kg decrease of 5.9% and FCR decrease of 1.4% for trial $k = 1$ – calculated using the slaughter weight. In Figure 13.11, the historical performance of the house is given, which shows that trial $k = 1$ has a very promising historically low FCR. This result is very close to the trial-to-trial FCR decrease for the first trial of the simulation study with an FCR decrease of $\approx 1\%$.

These experimental results demonstrate the basic feasibility of the new algorithm and provide a basis for onward development. One obvious area is to verify the algorithm's long-term results, as sentient biological production can sometimes give misleading results. More generally, there is much activity in the research community on data-driven control. The results in this section add to the very few designs followed through to experimental testing.

The broiler house used for the test is among the best-performing broiler producers in Denmark. The potential FCR minimization for other producers could be expected to be even higher. Suggestions for future work include studying the long-term properties of this algorithm and decreasing the effects of the measurement weight bias.

13.3 ILC for Quantum Control

Recently, ILC has been applied to quantum control systems [278]. The high-precision implementation of universal quantum gates in practical quantum information processing is vital. Although the current control technology has met the minimum requirement for quantum error correction (e.g. the 0.6–1% error threshold for surface codes has been reached in superconducting circuits, ion traps, quantum dots, and nitrogen-vacancy centers in diamond), the achievable precision still needs to be improved to reduce the resource overhead required for scalable quantum computation. This section is based on [278].

Consider the following quantum control system governed by the Schrödinger equation:

$$\dot{U}(t) = -i \left[H_0 + \sum_{k=1}^{m} u_k(t) H_k \right] U(t)$$

where $U(t) \in C^{N \times N}$ represents the quantum gate operation on the states, with $U(0) = I$, and $u_k(t) \in R, k = 1, \ldots, m$ are the control fields imposed on the control system. The free Hamiltonian H_0 and the control Hamiltonians H_ks are Hermitian matrices that steer the unitary $U(t)$.

In practice, the above Hamiltonians are never precisely known. Thus, any numerical calculation must use a design model:

$$\dot{U}_D(t) = -i \left[H_{D,0} + \sum_{k=1}^{m} u_k(t) H_{D,k} \right] U_D(t)$$

In the design model, the control pulses $v_k(t)$ are often chosen as piecewise-constant to facilitate numerical simulation on a digital computer and in some experimental situations. Note that the (designed) control pulses $v_k(t)$ are usually not identical with the actual pulses $u_k(t)$ applied to the system because the control signal produced by an arbitrary waveform generator (AWG) can be distorted due to various factors, including electronic limitations and transmission through the control line to the qubit. Such distorted signals have rising and falling edges or other unanticipated features sometimes called quantum gate bleed through.

One model for the distortion is a linear filter of the form:

$$u_k(t) = \int_0^{\infty} h(t - \tau) v_k(\tau) d\tau$$

where $h(t)$ is the filter impulse response. The control pulse is distortion-free only when $h(t) = \delta(t)$ is the Dirac function. Next, the correction of these errors by learning from online data is described.

The goal of quantum gate tune-up is to develop a design to control the pulse sequences $v_k(t)$ such that the generated control $u_k(t)$ can steer the system propagator $U(T)$ as close as possible to a desired unitary matrix U_f. This requirement can be achieved by minimizing the infidelity function:

$$\mathcal{J} = \frac{1}{2N} \| U(T) - U_f \|^2$$

where the norm is defined as $\|X\|^2 = tr(X^{\dagger}X)$.

One interpretation of this problem is finding an input sequence $v_k(t)$ such that the terminal output tracks a given target as accurately as possible, using, as one choice, the gradient type ILC law:

$$v_k(t, l) = v_k(t, l) - \alpha(l) g_k(t, l)$$

where

$$g_k(t, l) = \frac{\Delta \mathcal{J}}{\delta v_k(t, l)} = \langle \hat{\Delta}(T, l), H_{D,k}(t, l) \rangle$$

and the error matrix $\hat{\Delta}(T, l)$ is found by estimation of $\Delta(T, l)$ through process tomography of $U(T)$ and $H_{D,k}(t, l)$ is calculated from the design model (see [278] for more details). In this way, the real data is used to deduce whether the learning algorithm converges to a correct solution such that $\Delta(T, l) = 0$. Moreover, its incorporation with $H_{D,k}(t, l)$ provides an approximate gradient whose deviation from the actual gradient depends on the accuracy of the design model. Figure 13.12 shows the overall control arrangement where the explicit use of the design model is the major difference with existing model-free learning control strategies in the literature. This algorithm is known as the data-driven gradient ascending pulse engineering (d-GRAPE).

The performance of the design can, as a first step, be examined by numerical simulations, where the following Hamiltonian is assumed:

$$H(t) = J\sigma_z^1 \otimes \sigma_z^2 + \sum_{i=1}^{2} [u_x^i(t)\sigma_x^i + u_y^i(t)\sigma_y^i]$$

where J is the coupling strength between the two qubits and the design model is

$$H_D(t) = (J + \delta J)\sigma_z^1 \otimes \sigma_z^2 + \sum_{i=1}^{2} \left[u_x^i(t)\sigma_x^i + u_y^i(t)\sigma_y^i \right]$$

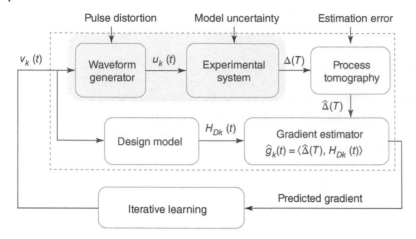

Figure 13.12 Schematic diagram of the data-driven grape (d-GRAPE) optimization procedure. The gradient is estimated from both the design model and the online data (for process tomography), which can in principle correct all deterministic errors such as the pulse distortion and the model uncertainty.

where δJ represents the identification error of J in the design model. Moreover, it is assumed that a linear filter distorts the control pulses:

$$u_{x,y}^i(t) = \int_0^t h(t - \tau)v_{x,y}^i(\tau)d\tau, \quad i = 1, 2,$$

with impulse response

$$h(t) = \frac{1}{t_r}e^{-t/t_r}$$

The time constant t_r characterizes the degree of pulse distortion by the steepness of the rising edge of distorted pulses, and the distortion is much heavier when t_r is long.

To demonstrate the ability of quantum gate tune-up by the d-GRAPE algorithm, the target of a controlled-not gate is taken as follows:

$$U_f = \begin{bmatrix} 1 & 0 & 0 & 0 \\ 0 & 1 & 0 & 0 \\ 0 & 0 & 0 & 1 \\ 0 & 0 & -1 & 0 \end{bmatrix}$$

In the simulation, the coupling constant is $J = 1$ and the final time as $T = 5$. The time interval is evenly divided into $M = 20$ subintervals, and hence, the duration of each subinterval is $\Delta t = T/M$.

Figure 13.13 shows simulation results for three cases with parametric error ΔJ in J and pulse distortion characterized by st_r. Each case includes results from 12 different initial random guesses. These fields are first optimized using the design model to obtain candidate pulses close to the optimal solution. Then, starting from these pulses, d-GRAPE based on the Broyden–Fletcher–Goldfarb–Shanno (BFGS) algorithm was applied. The estimation errors in the process tomography were simulated by injecting an additive random noise $\Delta U(l)$ (whose Frobenius norm is 2×10^5) to $U(T, l)$ in each trial. For comparison, the ideal d-GRAPE was also applied.

The simulation results show that the accuracy of the design model always limits the precision of the candidate pulses obtained from offline optimization (at the beginning of the optimization process shown in the plots). When the model error is relatively small (see Figure 13.13a), the

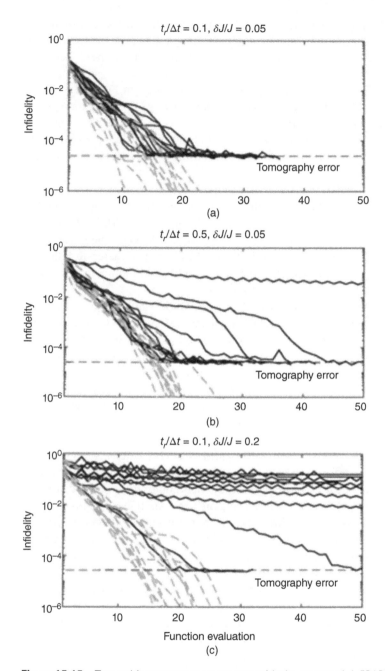

Figure 13.13 Two-qubit quantum gate tune up with the proposed d-GRAPE algorithm (solid black curves) for different model uncertainties and their comparison with an ideal GRAPE algorithm (light gray dashed curves). Each case include 12 runs from different initial guesses. (a) Small pulse distortion $t_r/\Delta t = 0.1$ and small parametric error $\Delta J//J = 0.05$; (b) large pulse distortion $t_r/\Delta t = 0.5$ and small parametric error $\Delta J//J = 0.05$; and (c) small pulse distortion $t_r/\Delta t = 0.1$ and large parametric error $\Delta J//J = 0.2$. The ultimate control precision is limited by the estimation error of the process tomography (indicated by the horizontal dashed line).

succeeding optimization based on the proposed d-GRAPE algorithm (solid curves) almost always converges to its optimal global solution that is limited by the tomography error. Compared with the ideal d-GRAPE optimization (see the light gray dash curves), its convergence speed is slightly reduced.

If the model error is not small enough (e.g. with severe pulse distortion in Figure 13.13b) or parameter deviation δJ in Figure 13.13c), fewer runs can quickly converge to the optimal global solution. Some runs still converge, but at the price of an increase in the number of iterations. The simulation results show that the accuracy of the design model always limits the precision of the candidate pulses obtained from offline optimization (at the beginning of the optimization process shown in the plots).

In cases where the model error is relatively small (see Figure 13.13a), the succeeding optimization based on the d-GRAPE algorithm (solid curves) almost always converges to its optimal global solution that is limited by the tomography error. Compared with the ideal GRAPE optimization (see the light gray dash curves), its convergence speed is slightly reduced. When the model error is not small enough (e.g. with severe pulse distortion in Figure 13.13b or parameter deviation δJ in Figure 13.13c) fewer runs can quickly converge to the optimal global solution. Some runs still converge, but at the price of an increase in the number of trials.

The model errors used in the simulations are relatively large (e.g. 20% in J and 50% in pulse distortion), and even under such a bad situation, the d-GRADE algorithm can still tune the gate to some extent. When the model error becomes more prominent, more and more optimizations become slower. There are even cases where the d-GRAPE algorithm gets lost and is trapped at a local false optimum solution. Thus, the d-GRAPE algorithm should not be applied with a very coarse model because of the potential traps and the increased experimental cost on process tomography. In practice, the precision of the design model should be improved as much as possible. Based on the high-precision model, the d-GRAPE algorithm can correct the error caused by the residue model imprecision within a few trials.

There is much room for improvement in the d-GRAPE algorithm. Several extensions of the algorithm are possible. First, extracting more knowledge from the offline model will improve online optimization. For example, the gradient can be enhanced by incorporating the pulse distortion function that can be offline identified from the waveform generator. Another option is to use a more sophisticated learning algorithm, e.g. a Newton algorithm because the Hessian matrix can be estimated using process tomography without increasing the number of experiments. Second, combined with adaptive tomography, it is possible to simultaneously improve the precision of the control and the process tomography, further accelerating the learning process.

The d-GRAPE algorithm can be extended to more general objectives, e.g. quantum state preparation problems, where the cost of state tomography is cheaper and hence can be more efficient. When the real quantum system undergoes open dynamics, open-system process matrices can replace the unitary propagators. However, the decoherence effects may limit the achievable precision. These are topics for possible future research.

13.4   ILC in the Utility Industries

These industries, e.g. water, gas, and electricity, have long been application areas for control systems, and there is increasing interest in applying ILC laws. As one example, some promising early research on ILC for pressure control in water distribution networks is considered, based on [118].

Scarcity of water is a global problem and also large amounts of water are lost due to leaks and other issues in distribution networks. Also, it is known that better pressure management is a cost-effective way of both lowering the amount of water lost to leaks and reducing the number of busts in a network. Considerable energy savings can be made by reducing the pumping effort required. In [118], the focus is on enabling the benefits of pressure management based on smart pumping station control.

Research in control and monitoring of water distribution systems has been focused on optimal pump scheduling in water networks with elevated reservoirs, e.g. [179], leakage detection [217], and contamination detection [98]. In the water industry, pressure control is achieved using various concepts [64]. In [118], an ILC law is used to satisfy pressure requirements at critical points in the network by controlling the pressures at the inlets.

In this application, see also [126], the ILC design is used in an architecture, where the critical point pressure inside the network is logged via data loggers, which transmit pressure measurements once a day. These logged data values are used to continuously calculate the optimal reference pressure for the pumping station's underlying pressure control. The ILC law makes use of the logged outputs directly instead of having to rely on an internal model, as in [126]. The application of ILC in this area depends on the fact that the consumption pattern of end-users in the system has a periodic profile, and next, the graph-based modeling is described.

Water distribution networks can be described by a directed and connected graph $\mathcal{G} = \{\mathcal{V}, \mathcal{E}\}$. The elements of the set $\mathcal{V} = \{v_1, \dots, v_n\}$ are denoted vertices and represent pipe connections with possible end-user water consumption. The ith vertex has three values associated to it: p_i, d_i, and h_i, where p_i and d_i are variables describing the absolute pressure and demand at the vertex, respectively.

The value h_i is the geodesic level at the vertex, and the elements of the set $\mathcal{E} = \{e_1, \dots, e_m\}$ are denoted edges and represent the pipes. Moreover, the pressure drop due to hydraulic resistance of the edge j is denoted by f_j and is a function of the flow q_j through the pipe (edge) j.

It is assumed that $f_i : \mathbb{R} \to \mathbb{R}$ has the following structure:

$$f_i(x) = \rho_i |x| x \qquad (13.46)$$

with $\rho_i > 0$ is a parameter of the pipe. Also, for the graph $\mathcal{G}$ the associated incidence matrix, H, is

$$H_{i,j} = \begin{cases} -1, & \text{if the } j\text{th edge is entering } i\text{th vertex} \\ 0, & \text{if the } j\text{th edge is not connected to the } i\text{th vertex} \\ 1, & \text{if the } j\text{th edge is leaving the } i\text{th vertex} \end{cases}$$

This matrix has dimension $n \times m$, where m denotes the number of edges and n is the number of vertices in the graph. Moreover, $m \geq n - 1$ for connected graphs.

Conservation of mass must hold for each vertex (also referred to as Kirchhoff's vertex law), i.e.

$$Hq = d \qquad (13.47)$$

where $q \in \mathbb{R}^m$ is the vector of flows in edges and $d \in \mathbb{R}^n$ is the vector of nodal demands, with $d_i > 0$ when demand flow is into vertex i. Also, by mass conservation in the network, there can only be $n - 1$ independent nodal demands and, hence,

$$\sum_{i=1}^{n} d_i = 0.$$

Let p be the vector of absolute pressures at the vertices and Δp be the vector of differential pressures across the edges. Then

$$\Delta p = H^T p = f(q) - H^T h \qquad (13.48)$$

where $p \in \mathbb{R}^n, f : \mathbb{R}^m \to \mathbb{R}^m, f(q) = (f_1(q_1), \dots, f_m(q_m))$, and f_i is strictly increasing. (This equation is the analogy of Ohm's law for electric circuits.)

The function f_i describes the flow-dependent pressure drop due to the hydraulic resistance and $H^T h$ is the pressure drop across the components due to differences in geodesic level between the ends of the components, and $h \in \mathbb{R}^n$ the vector of geodesic levels at each vertex expressed in units of potential (pressure).

To develop a reduced-order model that will prove beneficial in analysis, partition the n vertices of the underlying network graph into two sets, $\mathcal{V} = \bar{\mathcal{V}} \cup \hat{\mathcal{V}}$, where $\hat{\mathcal{V}} = \{\hat{v}_1, \dots, \hat{v}_c\}$ with $c \geq 1$ represents vertices in the graph corresponding to inlet vertices in the distribution network. The set $\bar{\mathcal{V}} = \{\bar{v}_1, \dots, \bar{v}_{n-c}\}$ represents the remaining vertices in the graph. Physically, the vertices in $\hat{\mathcal{V}}$ corresponds to network points where pumping stations supply the network.

Following on from above, $\hat{d}$ is the vector of supply flows to the network, and $\hat{p}$ is the vector of pressures at the supply points. The controller design will generate pressure set-points for the supply pumps, i.e. the vector $\hat{p}$ is the system input. Moreover, the vertices in $\bar{\mathcal{V}}$ correspond to end-users, hence, $\bar{d}$ is the vector of end-user water consumption.

The m edges of the network graph are likewise partitioned into two sets, $\mathcal{E} = \mathcal{E}_{\mathcal{T}} \cup \mathcal{E}_{\mathcal{C}}$. Here, $\mathcal{E}_{\mathcal{T}} = \{e_{\mathcal{T},1}, \dots, e_{\mathcal{T},n-c}\}$, $\mathcal{E}_{\mathcal{C}} = \{e_{\mathcal{C},1}, \dots, e_{\mathcal{C},m-n+c}\}$ and the partitioning is chosen such that the submatrix, say $\bar{H}_{\mathcal{T}}$, which maps edges in $\mathcal{E}_{\mathcal{T}}$ to vertices in $\bar{\mathcal{V}}$ is invertible.

Such a partitioning is always possible since any $(n-1) \times m$ submatrix of the incidence matrix of a connected graph has full rank $n-1$, [63]. Moreover, for $c = 1$ (single inlet), the graph $\mathcal{T} = \{\mathcal{V}, \mathcal{E}_{\mathcal{T}}\}$ is a spanning tree of the underlying network graph and the reduced model given in (13.52) below is the same as that derived in [126].

With the chosen partitioning, (13.47) and (13.48) can be rewritten as follows:

$$\bar{d} = \bar{H}_{\mathcal{T}} q_{\mathcal{T}} + \bar{H}_{\mathcal{C}} q_{\mathcal{C}}$$

$$\hat{d} = \hat{H}_{\mathcal{T}} q_{\mathcal{T}} + \hat{H}_{\mathcal{C}} q_{\mathcal{C}}$$

$$f_{\mathcal{T}}(q_{\mathcal{T}}) = \bar{H}_{\mathcal{T}}^T (\bar{p} + \bar{h}) + \hat{H}_{\mathcal{T}}^T (\hat{p} + \hat{h})$$

$$f_{\mathcal{C}}(q_{\mathcal{C}}) = \bar{H}_{\mathcal{C}}^T (\bar{p} + \bar{h}) + \hat{H}_{\mathcal{C}}^T (\hat{p} + \hat{h})$$

(13.49)

where $\bar{H}_{\mathcal{T}}$ $(\hat{H}_{\mathcal{T}})$ denotes the submatrix of H associated with edges $\mathcal{E}_{\mathcal{T}}$ and vertices $\bar{\mathcal{V}}$ $(\hat{\mathcal{V}})$; $\bar{H}_{\mathcal{C}}$ $(\hat{H}_{\mathcal{C}})$ denotes the submatrix of H associated with edges $\mathcal{E}_{\mathcal{C}}$ and vertices $\bar{\mathcal{V}}$ $(\hat{\mathcal{V}})$.

To obtain the reduced-order model, the following definitions of the vectors $a_{\mathcal{C}}$ and v are needed:

$$q_{\mathcal{C}} = a_{\mathcal{C}} \sigma, \quad \bar{d} = -v\sigma \tag{13.50}$$

where $\sigma = \sum_i \hat{d}_i > 0$ denotes the total inlet to the network and v is the vector that describes the distribution of the total inlet to the end-users and has the property that $\sum_i v_i = 1$. The following assumptions are required.

Assumption 13.2 At least one of the pressures at noninlet vertices is measured. Hence, there exists a vector $y \in \mathbb{R}^o$ (with $1 \leq o \leq n - c$ and $\{y_1, \dots, y_o\} \subset \{\bar{p}_1, \dots, \bar{p}_{n-c}\}$) of measured pressures at the noninlet vertices. Without loss of generality, the vertices are indexed such that $y_i = \bar{p}_i$ for $i \in \{1, 2, \dots, o\}$.

Assumption 13.3 The total head is the same at all inlets at any time, i.e.

$$\hat{p}(t) + \hat{h} = \kappa(t) I_c \tag{13.51}$$

for some $\kappa(t) \in \mathbb{R}$, and where $I_c \in \mathbb{R}^c$ denotes the vector with each entry equal to one.

Assumption 13.4   All noninlet vertices with nonzero demand have the same consumption profile, i.e. the vector v is constant.

The following result is Proposition 1 in [118].

Proposition 13.2   *Applying (13.46), Assumptions 13.3 and 13.4, the pressure at the ith noninlet vertex satisfies*

$$\bar{p}_i(t) = \alpha_i \sigma^2(t) + \kappa(t) + \gamma_i \tag{13.52}$$

with

$$
\begin{aligned}
\alpha_i &= (\bar{H}_{\mathscr{J}}^{-T})_i f_{\mathscr{J}} \, (-\bar{H}_{\mathscr{J}}^{-1} \bar{H}_{\mathscr{C}} a_{\mathscr{C}} + \bar{H}_{\mathscr{J}}^{-1} v) \\
\gamma_i &= -\bar{h}_i
\end{aligned}
\tag{13.53}
$$

where $(\bar{H}_{\mathscr{J}}^{-T})_i$ denotes the ith row of $\bar{H}_{\mathscr{J}}^{-T}$.

It can be shown that Assumption 13.3 is valid in at least some real-life cases. Moreover, for low consumption in the system, Assumption 13.3 must necessarily be satisfied to avoid reverse flow at the inlets. The details are in [118], which, in turn, cites the original work. Alternatively, a control law for the inlet pressures such that Assumption 13.3 is explicitly satisfied can be investigated.

In cases where Assumption 13.4 does not hold, the parameter α_i will be time-varying. Moreover, the demand profiles in a water distribution network exhibit a periodic behavior, i.e. $v(t + T) = v(t)$ and $\sigma(t + T) = \sigma(t)$ where T denotes the length of the period. Likewise, the parameter $\alpha_i(t)$ will exhibit periodic behavior in this case.

13.4.1 ILC Design

The control requirement is formulated as a minimum pressure requirement at the measured vertices, i.e.

$$y_i(t) \geq r_i \tag{13.54}$$

where $y_i(t) = \bar{p}_i(t)$ is the pressure at the ith measured vertex and r_i is the desired (constant) reference pressure for the ith vertex. Often, the pressure in the system is controlled using a pressure controller at one of the inlet vertices and the remaining inlet vertices are controlled to deliver the necessary flow to accommodate the demand. Consequently, the pressure at the controlled inlet is often set conservatively using feed-forward signals to ensure that (13.54) is satisfied while being robust toward events such as sensor dropouts, which would be critical in a feedback control structure.

The conservative control also means that pressure at the critical points may be unnecessarily high at times with low demand, increasing the risk of pipe bursts and higher-energy consumption. Using ILC in the current period, it is possible to adjust the inlet pressure to reduce the pipe stresses and overall energy consumption. Moreover, the control design gives pressure set points to all inlets in the network instead of flow set-points, thus reducing the need for flow measurements which are typically more expensive. Finally, the controller is not dependent on a system model and is therefore relatively easy to commission.

In the remainder of this section, the assumption is that the total demand $\sigma(t)$ is periodic with period T, that is, $\sigma(t + T) = \sigma(t)$, and the design parameter is the value $\kappa(t)$ in (13.52) which can subsequently be used to define the pressures at the inlet vertices using the relation (13.51) assuming that the vector $\hat{h}$ of geodesic levels at the inlet vertices are known.

Define the vector $\sigma_k \in \mathbb{R}^w$ for the kth period as

$$\sigma_k = (\sigma(kT), \sigma(kT + \tau), \sigma(kT + 2\tau), \ldots, \sigma(kT + (w-1)\tau))^T \tag{13.55}$$

where $\tau > 0$ is the sample time and w is the number of samples within a period. It is assumed that the sample time τ is chosen such that $w = T/\tau$ is integer. Hence, $\sigma(kT + w\tau) = \sigma((k+1)T) = \sigma(T)$ due to the periodic behavior of σ. Therefore, it follows that $\sigma_k = \sigma_{k-1} \equiv \sigma$ is constant across periods due to the periodic behavior. Lastly, the vectors $y_{i,k}$, $\bar{p}_{i,k}$ and κ_k are defined similarly to (13.55).

The error signal e_k for the kth period is now defined as follows:

$$
\begin{aligned}
e_k &= \max_i \{r_i - y_{i,k}\} \\
&= \max_i \{r_i - \bar{p}_{i,k}\} \\
&= \max_i \{r_i - \alpha_i \sigma^2 - \kappa_k - \gamma_i\} \\
&= \max_i \{r_i - \alpha_i \sigma^2 - \gamma_i\} - \kappa_k \\
&= c - \kappa_k
\end{aligned}
\tag{13.56}
$$

where $r_i = r_i\mathbf{I}$, $\gamma_i = \gamma_i\mathbf{I}$, and $\max\{\cdot\}$ and $(\cdot)^2$ are defined elementwise, where

$$c = \max_i \{r_i - \alpha_i \sigma^2 - \gamma_i\} \tag{13.57}$$

can be treated as a constant disturbance.

The ILC update for the control parameter vector κ_{k+1} on trial $k+1$ is

$$\kappa_{k+1} = \kappa_k + Ke_k \tag{13.58}$$

where $K \in \mathbb{R}$ is a control gain. Also, let $C(z)$ and $E(z)$ denote the z-transform of c and e, respectively, then the discrete closed-loop transfer-function from $C(z)$ to $E(z)$ is

$$
\begin{aligned}
E(z) &= \frac{z-1}{z + (K-1)} I_w C(z) \\
&= \frac{z-1}{z + (K-1)} C(z) \\
&\equiv G(z)C(z)
\end{aligned}
\tag{13.59}
$$

Using (13.59), $G(z)$ is stable if $0 < K < 2$. Furthermore, from (13.56) and (13.58), it follows that if $K = 1$, then $e_k = 0$ for all $k > 0$, since for $K = 1$

$$
\begin{aligned}
e_{k+1} &= c - \kappa_{k+1} \\
&= c - (\kappa_k + e_k) \\
&= c - \kappa_k - (c - \kappa_k) = 0
\end{aligned}
\tag{13.60}
$$

Since $G(z)$ is stable, the error vector e_k converges to zero. When $e_k = 0$, (13.56) gives that

$$
\begin{aligned}
0 &= \max_i \{r_i - \alpha_i \sigma^2 - \kappa_k - \gamma_i\} \Rightarrow \\
0 &\geq r_i - \alpha_i \sigma^2 - \kappa_k - \gamma_i \Rightarrow \\
r_i &\leq \alpha_i \sigma^2 + \kappa_k + \gamma_i = \bar{p}_{i,k} = y_{i,k}
\end{aligned}
\tag{13.61}
$$

and (13.54) is satisfied . Note also that for the index i that maximizes the expression in (13.56), $y_{i,k} = r_i$. Hence, the control will converge to the vector of minimum pressures κ such that (13.54) holds for all measurement vertices.

Consider the case where the vector σ_k of total demands during the kth period are not constant for all periods due to demand uncertainty. Consequently, define

$$\sigma_k = \sigma + \Delta_k^{\sigma} \qquad (13.62)$$

where σ is the (constant) vector of nominal demand over a period and

$$\Delta_k^{\sigma} = (\Delta^{\sigma}(kT), \Delta^{\sigma}(kT + \tau), \Delta^{\sigma}(kT + 2\tau), \dots, \Delta^{\sigma}(kT + (w-1)\tau))^T$$

where $\Delta^{\sigma}(t) \in [-\delta^{\sigma}, \delta^{\sigma}]$ is a bounded uncertainty. ("Constant" in this case means that, as before, σ is constant across periods, but that the demand can still vary within a period.)

The uncertainty in σ will, in turn, affect the disturbance c defined in (13.57). In this case, the disturbance c_k for the kth period can be separated as follows:

$$c_k = c^* + \Delta^{c_k} \qquad (13.63)$$

where c^* is a constant vector and $\Delta_k^c = (\Delta^c(kT), \Delta^c(kT + \tau), \Delta^c(kT + 2\tau), \dots, \Delta^c(kT + (w-1)\tau))^T$ and $\Delta^c(t) \in [-\delta^c, \delta^c]$ is again a bounded uncertainty. That $\Delta^c(t)$ is bounded follows from boundedness of $\Delta^{\sigma}(t)$ and continuity of the $\max\{\cdot\}$ function. Due to stability of $G(z)$ it follows that e_k converges to a bounded neighborhood of 0.

The results of a successful application of this ILC design are given in [118].

13.5 Concluding Remarks

This chapter has given an overview of four emerging areas of ILC research. These are representative of the continual emergence of new application areas for ILC design.

Appendix A

A.1 The Entries in the Transfer-Function Matrix (2.2)

$$G_{11}(s) = \frac{g_{11}(s)}{a_{11}(s)}$$

$$\begin{aligned}
g_{11}(s) = {} & 0.16s^9 + 14.51s^8 + 578.2s^7 + 1.3921 \times 10^4 \\
& + 2.26 \times 10^5 s^5 + 2.58 \times 10^6 s^4 + 2.09 \times 10^7 7s^3 \\
& + 1.17 \times 10^8 s^2 + 4.21 \times 10^8 s + 7.6 \times 10^8
\end{aligned}$$

$$\begin{aligned}
a_{11}(s) = {} & 5.25 \times 10^{-5} s^{12} + 0.01463s^{11} + 0.91s^{10} + 31.2s^9 + 714.1s^8 \\
& + 1.19 \times 10^4 s^7 + 1.45 \times 10^5 s^6 + 1.4 \times 10^6 s^5 + 1.01 \times 10^7 s^4 + 5.7 \times 10^7 s^3 \\
& + 2.3 \times 10^8 s^2 + 5.9 \times 10^8 s + 7.6 \times 10^8
\end{aligned}$$

$$G_{12}(s) = \frac{g_{12}(s)}{a_{12}(s)}$$

$$g_{12}(s) = -0.022s^7 - 3.24s^6 - 88.3s^5 - 1347s^4 - 1.06 \times 10^4 s^3 - 4.52 \times 10^4 s^2$$

$$\begin{aligned}
a_{12}(s) = {} & 5.25 \times 10^{-5} s^{10} + 0.014s^9 + 0.72s^8 + 20s^7 + 363s^6 + 4645s^5 \\
& + 4.3 \times 10^4 s^4 + 2.9 \times 005s^3 + 1.4 \times 10^6 s^2 + 4.18 \times 10^6 s + 6.323 \times 10^6
\end{aligned}$$

$$G_{21}(s) = \frac{g_{21}(s)}{a_{21}(s)}$$

$$g_{21}(s) = -0.16s^7 - 8.7s^6 - 194s^5 - 2498s^4 - 1.78 \times 10^4 s^3 - 6.64 \times 10^4 s^2$$

$$\begin{aligned}
a_{21}(s) = {} & 5.25 \times 10^{-5} s^{10} + 0.014s^9 + 0.67s^8 + 17.9s^7 + 316s^6 + 3963s^5 \\
& + 3.6 \times 10^4 s^4 + 2.42 \times 10^5 s^3 + 1.1 \times 10^6 s^2 + 3.5 \times 10^6 s + 5.3 \times 10^6
\end{aligned}$$

$$G_{22}(s) = \frac{g_{22}(s)}{a_{22}(s)}$$

Iterative Learning Control Algorithms and Experimental Benchmarking, First Edition.
Eric Rogers, Bing Chu, Christopher Freeman and Paul Lewin.
© 2023 John Wiley & Sons Ltd. Published 2023 by John Wiley & Sons Ltd.

$$g_{22}(s) = 0.027s^9 + 4.95s^8 + 264s^7 + 7394s^6 + 1.3 \times 10^5 s^5 + 1.69 \times 10^6 s^4$$
$$+ 1.5 \times 10^7 s^3 + 9.4 \times 10^7 s^2 + 3.8 \times 10^8 s + 7.6 \times 10^8$$

$$a_{22}(s) = 5.25 \times 10^{-5} s^{12} + 0.014s^{11} + 0.9s^{10} + 31s^9 + 714.1s^8 + 1.19 \times 10^4 s^7$$
$$+ 1.48 \times 10^5 s^6 + 1.4 \times 10^6 s^5 + 1.04 \times 10^7 s^4 + 5.7 \times 10^7 s^3 + 2.3 \times 10^8 s^2$$
$$+ 5.9 \times 10^8 s + 7.6 \times 10^8$$

A.2 Entries in the Transfer-Function Matrix (2.4)

The zero-interaction case ($c = 0$) has entries

$$G_{11}^0(z) = \frac{0.2024z^4 + 0.8463z^3 - 0.5635z^2 - 0.4583z - 0.01942}{z^5 - 2.538z^4 + 2.417z^3 - 1.097z^2 + 0.2397z - 0.0202}$$

$$G_{12}^0(z) = 0$$

$$G_{21}^0(z) = 0$$

$$G_{22}^0(z) = \frac{0.009843z^5 + 0.2231z^4 + 0.4622z^3 + 0.1756z^2 + 0.01227z + 7.857 \times 10^{-5}}{z^6 - 2.099z^5 + 1.782z^4 - 0.7816z^3 + 0.1868z^2 - 0.0231z + 0.001157}$$

the extremely low-interaction case ($c = 0.2$) has entries

$$G_{11}^{0.2}(z) = \frac{0.04094z^5 + 0.4376z^4 + 0.06153z^3 - 0.4429z^2 - 0.09037z - 0.001526}{z^6 - 2.864z^5 + 3.255z^4 - 1.91z^3 + 0.6158z^2 - 0.1041z + 0.007227}$$

$$G_{12}^{0.2}(z) = \frac{0.003084z^5 + 0.02531z^4 - 0.00146z^3 - 0.01818z^2 - 0.002509z - 2.83 \times 10^{-5}}{z^6 - 2.471z^5 + 2.304z^4 - 1.024z^3 + 0.2285z^2 - 0.02416z + 0.000926}$$

$$G_{21}^{0.2}(z) = \frac{0.07289z - 0.07261}{z^2 - 1.641z + 0.6435}$$

$$G_{22}^{0.2}(z) = \frac{0.00447z^5 + 0.1164z^4 + 0.2787z^3 + 0.1238z^2 + 0.01022z + 7.782e^{-05}}{z^6 - 2.419z^5 + 2.402z^4 - 1.254z^3 + 0.3629z^2 - 0.05519z + 0.003447}$$

the low-interaction case ($c = 0.4$) has entries

$$G_{11}^{0.4}(z) = \frac{0.04006z^5 + 0.4251z^4 + 0.05439z^3 - 0.4297z^2 - 0.08669z - 0.001449}{z^6 - 2.858z^5 + 3.237z^4 - 1.888z^3 + 0.6033z^2 - 0.1007z + 0.006891}$$

$$G_{12}^{0.4}(z) = \frac{6.657z^4 - 2.368z^3 - 10.88z^2 + 5.918z + 0.675}{z^5 - 2.135z^4 + 1.81z^3 - 0.7608z^2 + 0.1584z - 0.01305}$$

$$G_{21}^{0.4}(z) = \frac{0.05361z^2 - 0.03147z - 0.02167}{z^3 - 1.513z^2 + 0.5875z - 0.06804}$$

$$G_{22}^{0.4}(z) = \frac{0.004741z^5 + 0.1283z^4 + 0.3206z^3 + 0.1496z^2 + 0.01307z + 0.0001056}{z^6 - 2.476z^5 + 2.552z^4 - 1.401z^3 + 0.4325z^2 - 0.07113z + 0.004869}$$

the medium-interaction case ($c = 0.6$) has entries

$$G_{11}^{0.6}(z) = \frac{0.0188z^6 + 0.3547z^5 + 0.3027z^4 - 0.4516z^3 - 0.2097z^2 - 0.01343z - 7.471 \times 10^{-5}}{z^7 - 2.768z^6 + 3.045z^5 - 1.755z^4 + 0.5769z^3 - 0.1084z^2 + 0.01078z - 0.0004359}$$

$$G_{12}^{0.6}(z) = \frac{0.02691z^4 + 0.1916z^3 + 0.1263z^2 + 0.01223z + 0.0001116}{z^5 - 1.032z^4 + 0.4126z^3 - 0.07969z^2 + 0.007441z - 0.0002687}$$

$$G_{21}^{0.6}(z) = \frac{0.01502z^3 + 0.01894z^2 - 0.02858z - 0.004537}{z^4 - 2.363z^3 + 1.984z^2 - 0.7117z + 0.09282}$$

$$G_{22}^{0.6}(z) = \frac{0.00299z^5 + 0.08499z^4 + 0.2232z^3 + 0.1095z^2 + 0.01006z + 8.561 \times 10^{-5}}{z^6 - 2.635z^5 + 2.886z^4 - 1.681z^3 + 0.5497z^2 - 0.09563z + 0.006916}$$

the high-interaction case ($c = 0.8$) has entries

$$G_{11}^{0.8}(z) = \frac{0.02723z^6 + 0.4793z^5 + 0.3532z^4 - 0.5897z^3 - 0.2514z^2 - 0.01524z - 8.086 \times 10^{-5}}{z^7 - 2.58z^6 + 2.616z^5 - 1.393z^4 + 0.4268z^3 - 0.07582z^2 + 0.007265z - 0.0002904}$$

$$G_{12}^{0.8}(z) = \frac{0.01062z^5 + 0.1484z^4 + 0.1843z^3 + 0.03994z^2 + 0.001495z + 4.864 \times 10^{-6}}{z^6 - 1.257z^5 + 0.6206z^4 - 0.1528z^3 + 0.01938z^2 - 0.001142z + 2.102 \times 10^{-5}}$$

$$G_{21}^{0.8}(z) = \frac{0.001491z^6 + 0.02913z^5 + 0.02672z^4 - 0.03757z^3 - 0.01832z^2 - 0.001217z - 7.024 \times 10^{06}}{z^7 - 2.855z^6 + 3.254z^5 - 1.937z^4 + 0.6557z^3 - 0.127z^2 + 0.01307z - 0.0005537}$$

$$G_{22}^{0.8}(z) = \frac{0.007928z^5 + 0.1193z^4 + 0.09141z^3 - 0.1125z^2 - 0.03685z - 0.0008819}{z^6 - 3.532z^5 + 5.193z^4 - 4.068z^3 + 1.791z^2 - 0.4202z + 0.04104}$$

and the extremely high-interaction case ($c = 1$) has entries

$$G_{11}^{1}(z) = \frac{0.1717z^4 + 0.7177z^3 - 0.4803z^2 - 0.3911z - 0.01666}{z^5 - 2.531z^4 + 2.405z^3 - 1.093z^2 + 0.2402z - 0.02054}$$

$$G_{12}^{1}(z) = \frac{0.01476z^4 + 0.1868z^3 + 0.2275z^2 + 0.04229z + 0.0007591}{z^5 - 2.092z^4 + 1.734z^3 - 0.7122z^2 + 0.1448z - 0.01167}$$

$$G_{21}^{1}(z) = \frac{0.0007866z^6 + 0.02121z^5 + 0.03491z^4 - 0.02892z^3 - 0.02449z^2 - 0.002387z - 1.987 \times 10^{-5}}{z^7 - 3.612z^6 + 5.432z^5 - 4.424z^4 + 2.113z^3 - 0.593z^2 + 0.09063z - 0.005828}$$

$$G_{22}^{1}(z) = \frac{0.01502z^5 + 0.1607z^4 + 0.02365z^3 - 0.1598z^2 - 0.03216z - 0.0005313}{z^6 - 2.904z^5 + 3.338z^4 - 1.961z^3 + 0.6238z^2 - 0.1021z + 0.006716}$$

The transfer-function for the DC motor does not vary significantly with interaction and is hence given by

$$G_{23}(z) = \frac{0.1148z^4 + 0.07592z^3 - 0.433z^2 + 0.1949z + 0.04747}{z^5 - 3.638z^4 + 5.221z^3 - 3.694z^2 + 1.288z - 0.1769}$$

for all levels of interaction. Similarly, $G_{13}(z)$ is negligible at all interaction levels and will be assumed to be zero.

A.3 Matrices E_1, E_2, H_1, and H_2 for the Designs of (7.36) and (7.37)

$$E_1 = \begin{bmatrix} 0.061 & 0.024 & 0.092 & 0.019 & 0.0342 & 0.049 & 0.027 \\ 0.006 & 0.062 & 0.032 & 0.039 & 0.0794 & 0.095 & 0.043 \\ 0.006 & 0.031 & 0.042 & 0.099 & 0.0585 & 0.092 & 0.030 \\ 0.030 & 0.074 & 0.055 & 0.089 & 0.0304 & 0.019 & 0.036 \\ 0.041 & 0.100 & 0.028 & 0.088 & 0.0336 & 0.025 & 0.042 \\ 0.027 & 0.000 & 0.094 & 0.098 & 0.0946 & 0.035 & 0.085 \\ 0.015 & 0.062 & 0.082 & 0.006 & 0.0835 & 0.079 & 0.079 \end{bmatrix}, \quad E_2 = \begin{bmatrix} 0.024 \\ 0.093 \\ 0.039 \\ 0.060 \\ 0.014 \\ 0.050 \\ 0.060 \end{bmatrix}$$

$$H_1 = \begin{bmatrix} 0.0019 & 0.0042 & 0.0001 & 0.0016 & 0.0046 & 0.0003 & 0.0011 \\ 0.0038 & 0.0047 & 0.0045 & 0.0045 & 0.0005 & 0.0003 & 0.0017 \\ 0.0042 & 0.0008 & 0.0048 & 0.0047 & 0.0026 & 0.0033 & 0.0048 \\ 0.0016 & 0.0009 & 0.0001 & 0.0046 & 0.0018 & 0.0050 & 0.0021 \\ 0.0011 & 0.0010 & 0.0027 & 0.0006 & 0.0000 & 0.0010 & 0.0032 \\ 0.0050 & 0.0049 & 0.0041 & 0.0000 & 0.0018 & 0.0020 & 0.0013 \\ 0.0004 & 0.0005 & 0.0046 & 0.0010 & 0.0034 & 0.0047 & 0.0024 \end{bmatrix}$$

$$H_2 = \begin{bmatrix} 0.0050 & 0.0003 & 0.0021 & 0.0009 & 0.0037 & 0.0031 & 0.0032 \\ 0.0004 & 0.0021 & 0.0017 & 0.0010 & 0.0014 & 0.0001 & 0.0033 \\ 0.0047 & 0.0044 & 0.0017 & 0.0035 & 0.0020 & 0.0001 & 0.0028 \\ 0.0047 & 0.0043 & 0.0017 & 0.0035 & 0.0028 & 0.0019 & 0.0045 \\ 0.0035 & 0.0004 & 0.0040 & 0.0015 & 0.0018 & 0.0047 & 0.0039 \\ 0.0036 & 0.0046 & 0.0010 & 0.0002 & 0.0030 & 0.0032 & 0.0047 \\ 0.0031 & 0.0010 & 0.0006 & 0.0027 & 0.0037 & 0.0039 & 0.0048 \end{bmatrix}$$

References

1 C. Adams, J. Potter, and W. Singhose. Input shaping and model-following control of a helicopter carrying a suspended load. *Journal of Guidance, Dynamics, and Control*, 38(1):94–105, 2015.

2 Hyo-Sung Ahn, YangQuan Chen, and K. L. Moore. Iterative learning control: brief survey and categorization. *IEEE Transactions on Systems, Man, and Cybernetics, Part C: Applications and Reviews*, 37(6):1099–1121, 2007.

3 A. Al-Gburi, M. C. French, and C. T. Freeman. Robustness analysis of nonlinear systems with feedback linearizing control. In *Proceedings of the 52nd IEEE Conference on Decision and Control*, pages 3055–3060, 2013.

4 N. Amann, D. H. Owens, and E. Rogers. Iterative learning control for discrete-time systems with exponential rate of convergence. *Proceedings of the Institution of Electrical Engineers on Control Theory and Applications*, 143(2):217–224, 1996.

5 N. Amann, D. H. Owens, and E. Rogers. Iterative learning control using optimal feedback and feedforward actions. *International Journal of Control*, 65(2):277–293, 1996.

6 N. Amann, D. H. Owens, and E. Rogers. Predictive optimal iterative learning control. *International Journal of Control*, 69(2):203–226, 1998.

7 N. Amann, D. H. Owens, E. Rogers, and A. Wahl. An H_∞ approach to linear iterative learning control design. *International Journal of Adaptive Control and Signal Processing*, 10:767–781, 1996.

8 P. B. Andersen, L. Henriksen, M. Gaunaa, C. Bak, and T. Buhl. Deformable trailing edge flaps for modern megawatt wind turbine controllers using strain gauge sensors. *Wind Energy*, 13:193–206, 2010.

9 J. D. Anderson. *Fundamentals of Aerodynamics, Fifth Edition in SI Units*. McGraw-Hill, New York, 2011.

10 B. D. O. Anderson and J. B. Moore. *Optimal Control — Linear Optimal Control*. Prentice-Hall, Englewood Cliffs, NJ, 1989.

11 S. Arimoto. Robustness of learning control for robotic manipulators. In *IEEE International Conference on Robotics and Automation*, pages 1528–1533, 1990.

12 S. Arimoto, S. Kawamura, and F. Miyazaki. Bettering operation of robots by learning. *Journal of Robotic Systems*, 2(1):123–140, 1984.

13 K. B. Ariyur and M. Kristic. *Real-time Optimization by Extremum Seeking Control*. Wiley Inter-science, New York, 2003.

14 H. Aschemann and J. Ritzke. Gain-scheduled tracking control for high-speed rack feeders. In *Proceedings of 1st Joint International Conference on Multibody System Dynamics (IMSD)*, 2010.

Iterative Learning Control Algorithms and Experimental Benchmarking, First Edition.
Eric Rogers, Bing Chu, Christopher Freeman and Paul Lewin.
© 2023 John Wiley & Sons Ltd. Published 2023 by John Wiley & Sons Ltd.

15 H. Aschemann and D. Schindele. Model predictive trajectory control for high-speed rack feeders. In *Model Predictive Control*, pages 183–198. InTech, 2010.

16 A. C. Atkinson, A. N. Donev, and R. D. Tobias. *Optimum Experimental Designs, with SAS*. Oxford University Press, Oxford, 2007.

17 Aviagen. *Ross 308 Broiler: Performance Objectives*, 2014.

18 K. E. Avrachenkov. Iterative learning control based on quasi-Newton methods. In *Proceedings of the 37th IEEE Conference on Decision and Control*, volume 1, pages 170–174, 1998.

19 G. Balas, J. C. Doyle, K. Glover, A. Packard, and R. Smith. *μ-Analysis and Synthesis Toolbox*. MUSYN and The MathWorks Inc., 1991.

20 H. T. Banks, R. C. Smith, and Y. Wang. *Smart material structures: modeling, estimation and control*. Wiley, Chicester, 1996.

21 R. Baratta and M. Solomonow. The dynamic response model of nine different skeletal muscles. *IEEE Transactions on Biomedical Engineering*, 37(3):243–251, 1990.

22 T. K. Barlas and G. A. M. van Kuik. Review of state of the art in smart rotor control for wind turbines. *Progress in Aerospace Sciences*, 46:1–27, 2010.

23 G. K. Batchelor. *An Introduction to Fluid Dynamics*. Cambridge Unversity Press, 1967.

24 M. Bergmann, L. Cordier, and J. Brancher. Optimal rotary control of cylinder wake using proper orthogonal decomposition reduced order model. *Physics of Fluids*, 17:97101–1–21, 2005.

25 M. Bergmann, L. Cordier, and J. Brancher. Optimal control of the cylinder wake in the laminar regime by trust-region methods and POD reduced-order models. *Journal of Computational Physics*, 227:7813–7840, 2008.

26 Z. Bien and K. M. Huh. Higher-order iterative learning control algorithm. *Proceedings of the Institution of Electrical Engineering on Control Theory and Applications*, 136(3):105–112, 1989.

27 M. Blackwell, O. R. Tutty, E. Rogers, and R. Sandberg. Iterative learning control applied to a non-linear vortex panel model for improved aerodynamic load performance of wind turbines with smart rotors. *International Journal of Control*, 89:55–68, 2016.

28 F. Boe and B. Hannaford. On-line improvement of speed and tracking performance on repetitive paths. *IEEE Transactions on Control Systems Technology*, 6(3):350–358, 1998.

29 J. Bolder. A flexible and robust iterative learning control framework with applications in high-tech printing systems. *PhD thesis*, Eindhoven University of Technology, The Netherlands, 2015.

30 J. Borggaard and A. Lattimer. POD models for positive fields in advection-diffusion-reaction equations. In *American Control Conference*, pages 3739–3802, 2017.

31 S. Boyd, L. Ghaoui, E. Feron, and V. Balakrishnan. *Linear Matrix Inequalities in System and Control Theory*. SIAM, Philadelphia, PA, 1994.

32 D. A. Bristow. Optimal iteration-varying iterative learning control for systems with stochastic disturbances. In *Proceedings of the American Control Conference*, pages 1296–1301, 2010.

33 D. A. Bristow, M. Tharayil, and A. G. Alleyne. A survey of iterative learning control. *IEEE Control Systems Magazine*, 26(3):96–114, 2006.

34 X. Bu, Z. Zhongsheng Hou, and R. Chi. Model free adaptive iterative learning control for farm vehicle path tracking. In *Proceedings of the 3rd IFAC International Conference on Intelligent Control and Automation Science*, pages 153–158, 2013.

35 R. S. Burington. On the use of conformal mapping in shaping wing profiles. *The American Mathematical Monthly*, 47:362–373, 1940.

36 M. Butcher, A. Karimi, and R. Longchamp. A statistical analysis of certain iterative learning control algorithms. *International Journal of Control*, 81(1):156–166, 2008.

37 Z. Cai, D. A. Bristow, E. Rogers, and C. T. Freeman. Experimental implementation of iterative learning control for processes with stochastic disturbances. In *Proceedings of the IEEE Symposium on Intelligent Control*, pages 406–411, 2011.

38 Z. Cai, C. T. Freeman, P. L. Lewin, and E. Rogers. Experimental comparison of stochastic iterative learning control algorithms. In *Proceedings of the American Control Conference*, pages 4548–4553, 2008.

39 Z. Cai, C. T. Freeman, P. L. Lewin, and E. Rogers. Iterative learning control for a non-minimum phase plant based on a reference shift algorithm. *Control Engineering Practice*, 16(6):633–643, 2008.

40 E. K. Chadwick, D. Blana, A. J. van den Bogert, and R. R. Kirsch. A real-time, 3-D musculoskeletal model for dynamic simulation of arm movements. *IEEE Transaction on Biomedical Engineering*, 56(4):941–948, 2009.

41 Y. Chen, B. Chu, and C. T. Freeman. Spatial path tracking using iterative learning control. In *55th IEEE Conference on Decision and Control*, pages 7189–7194, Las Vegas, US, 2016.

42 Y. Chen, B. Chu, and C. T. Freeman. Generalized norm optimal iterative learning control: constraint handling. In *The 20th World Congress of the International Federation of Automatic Control*, Toulouse, France, 2017.

43 Y. Chen, B. Chu, and C. T. Freeman. Norm optimal iterative learning control for general spatial path following problem. In *2018 IEEE Conference on Decision and Control (CDC)*, pages 4933–4938. IEEE, 2018.

44 Y. Chen, B. Chu, and C. T. Freeman. Point-to-point iterative learning control with optimal tracking time allocation. *IEEE Transactions on Control Systems Technology*, 26(5):1685–1698, 2018.

45 Y. Chen, B. Chu, and C. T. Freeman. Iterative learning control for minimum time path following. *IFAC-PapersOnLine*, 52(29):320–325, 2019.

46 C. Chen and J. Hwang. Iterative learning control for position tracking of a pneumatic actuated X-Y table. *Control Engineering Practice*, 13(12):1455–1461, 2005.

47 K. Chen and R. W. Longman. Stability issues using FIR filtering in repetitive control. *Advances in Astronautical Sciences*, 206:1321–1339, 2006.

48 Y. Q. Chen, K. L. Moore, J. Yu, and T. Zhang. Iterative learning control and repetitive control in the hard disk drive industry – a tutorial. *International Journal of Adaptive Control and Signal Processing*, 23:325–343, 2008.

49 C. T. Chen and S. T. Peng. Learning control of process systems with hard input constraints. *Journal of Process Control*, 9(2):151–160, 1999.

50 YangQuan Chen and C. Wen. Unified formulation of linear iterative learning control. In *Lecture Notes in Control and Information Sciences*, volume 248, Springer, London, UK, 1999.

51 B. Chu, C. T. Freeman, and D. H. Owens. A novel design framework for point-to-point ILC using successive projection. *IEEE Transactions on Control Systems Technology*, 23(3):1156–1163, 2015.

52 B. Chu and D. H. Owens. Iterative learning control for constrained linear systems. *International Journal of Control*, 83(7):1397–1413, 2009.

53 B. Chu, A. Rauh, H. Aschemann, and E. Rogers. Experimental validation of constrained ILC approaches for a high speed rack feeder. In *American Control Conference*, pages 3631–3636, 2015.

54 B. Chu, A. Rauh, H. Aschemann, E. Rogers, and D. H. Owens. Constrained iterative learning control for linear time-varying systems with experimental validation on a high-speed rack feeder. *IEEE Transactions on Control Systems Technology*, 30(5):1834–1846, 2022.

55 B. Cichy, K. Galkowski, E. Rogers, and A. Kummert. Control law design for discrete linear repetitive processes with non-local updating structures. *Multidimensional Systems and Signal Processing*, 24:707–726, 2013.

56 N. R. Clarke and O. R. Tutty. Construction and validation of a discrete vortex method for the two-dimensional incompressible navier-stokes equations. *Computers & Fluids*, 23:751–783, 1994.

57 P. Dabkowski, K. Galkowski, O. Bachelier, E. Rogers, and J. Lam. A new approach to strong practical stability and stabilization of discrete linear repetitive processes. In *Proceedings of the 19th International Symposium on Mathematical Theory of Networks and Systems (MTNS 2010)*, pages 311–317, 2010.

58 P. Dabkowski, K. Galklowski, E. Rogers, Z. Cai, C. T. Freeman, and P. L. Lewin. Iterative learning control based on relaxed 2-D systems stability theory. *IEEE Transactions on Control Systems Technology*, 21(3):1016–1023, 2013.

59 P. Dabkowski, K. Galkowski, E. Rogers, and A. Kummert. Strong practical stability and stabilization of discrete linear repetitive processes. *Multidimensional Systems and Signal Processing*, 20:311–331, 2009.

60 D. De Roover. Synthesis of a robust iterative learning controller using an H_∞ approach. In *Proceedings of the 35th IEEE Conference on Decision and Control*, pages 3044–3049, 1996.

61 D. De Roover and O. H. Bosgra. Synthesis of robust multivariable iterative learning controllers with application to a wafer stage motion system. *International Journal of Control*, 73(10):968–979, 2000.

62 L. G. Dekker, J. A. Marshall, and J. Larsson. Experiments in feedback linearized iterative learning-based path following for center-articulated industrial vehicles. *Journal of Field Robotics*, 36(5):955–972, 2019.

63 N. Deo. *Graph Theory with Applications to Engineering and Computer Science*. Prentice-Hall, 1st edition, 1974.

64 Deutsche Gesellschaft für Internationale Zusammenarbeit (GIZ). Guidelines for water loss reduction - a focus on pressure management, 2011. [Online].

65 V. Dinh Van. Design and experimental verification of iterative learning controllers on a multivariable test facility. *PhD thesis*, University of Southampton, UK, 2013.

66 V. Dinh Van, C. T. Freeman, and P. L. Lewin. Assessment of gradient-based iterative learning controllers using a multivariable test facility with varying interaction. *Control Engineering Practice*, 29:158–173, 2014.

67 H. Dou, K. K. Tan, T. H. Lee, and Z. Zhou. Iterative learning control of human limbs via functional elecrical stimulation. *Control Engineering Practice*, 7:315–325, 1999.

68 J. C. Doyle, Glover K., P. P. Khargonekar, and B. A. Francis. State-space solutions to standard H_2 and H_∞ control problems. *IEEE Transactions on Automatic Control*, AC-34(8):831–847, 1989.

69 M. Drela. Xfoil. *publically available from* http://web.mit.edu/drela/Public/web/xfoil/., 2012.

70 J. B. Edwards. Stability problems in the control of multipass processes. *Proceedings of the Institution of Electrical Engineers on Control Theory and Applications*, 121(11):1425–1432, 1974.

71 E. Fornasini and G. Marchesini. Two-dimensional linear doubly-indexed dynamical systems: state-space models and structural properties. *Mathematical Systems Theory*, 12:59–72, 1978.

72 B. A. Francis and W. M. Wonham. The internal model principle for linear multivariable regulators. *Journal of Applied Mathematics and Optimization*, 2:170–194, 1975.

73 C. T. Freeman. Newton-method based iterative learning control for robot-assisted rehabilitation using FES. *Mechatronics*, 24:934–943, 2014.

74 C. T. Freeman. Upper limb electrical stimulation using input-output linearization and iterative learning control. *IEEE Transactions on Control Systems Technology*, 23(4):1546–1554, 2015.

75 C. T. Freeman. Robust ILC with application to stroke rehabilitation. *Automatica*, 81:270–278, 2017.

76 C. T. Freeman, M. A. Alsubaie, Z. Cai, E. Rogers, and P. L. Lewin. *Preliminary results on the choice of the initial input for iterative learning control applications*, Research Report, School of Electronics and Computer Science, University of Southampton, UK, 2010.

77 C. T. Freeman, M. A. Alsubaie, Z. Cai, E. Rogers, and P. L. Lewin. Initial input selection for iterative learning control. *Journal of Dynamic Systems, Measurement, and Control*, 133(5):054504, 2011.

78 C. T. Freeman, M. A. Alsubaie, Z. Cai, E. Rogers, and P. L. Lewin. A common setting for the design of iterative learning and repetitive controllers with experimental verification. *International Journal of Adaptive Control and Signal Processing*, 27(3):230–249, 2012.

79 C. T. Freeman, A.-H. Hughes, J. H. Burridge, P. L. Chappell, P. H. Lewin, and E. Rogers. A model of the upper extremity using surface FES for stroke rehabilitation. *Journal of Biomechanical Engineering*, 131:031011 (12 pages, 2009.

80 C. T. Freeman, A.-M. Hughes, J. H. Burridge, P. H. Chappell, P. L. Lewin, and E. Rogers. Iterative learning control of FES applied to the upper extremity after stroke. *Control Engineering Practice*, 17(3):368–381, 2009.

81 C. T. Freeman, A.-M. Hughes, J. H. Burridge, P. H. Chappell, P. L. Lewin, and E. Rogers. A model of the upper extremity using fes for stroke rehabilitation. *ASME Journal of Biomechanical Engineering*, 131(3):031006–1–031006–10, 2009.

82 C. T. Freeman, P. L. Lewin, and E. Rogers. Experimental evaluation of iterative learning control algorithms for non-minimum phase plants. *International Journal of Control*, 78(11):826–846, 2005.

83 C. T. Freeman, P. L. Lewin, and E. Rogers. Further results on the experimental evaluation of iterative learning control algorithms for non-minimum phase plants. *International Journal of Control*, 80(4):569–582, 2007.

84 C. T. Freeman, E. Rogers, A.-M. Hughes, J. H. Burridge, and K. L. Meadmore. Iterative learning control in health care: electrical stimulation and robotic-assisted upper-limb stroke rehabilitation. *IEEE Control Systems Magazine*, 32(1):18–43, 2012.

85 M. French and E. Rogers. Non-linear iterative learning by an adaptive Lyapunov method. *International Journal of Control*, 73(10):840–850, 2000.

86 A. Friedlander, J. M. Martinez, and M. Raydan. A new method for large-scale box constrained convex quadratic minimization problems. *Optimisation Methods and Software*, 5(1):57–74, 1995.

87 K. K. Furuta and M. Yamakita. The design of a learning control system for multivariable systems. In *Proceedings of the IEEE International Symposium on Intelligent Control*, pages 371–376, 1987.

88 P. Gahinet and P. Apkarian. A linear matrix inequality approach to H_∞ control. *Internarional Journal of Robust and Nonlinear Control*, 4(0):421–448, 1994.

89 M. Garden. Learning control of actuators in control systems. *U.S. Patent 3555252*, 1971.

90 G. Gauthier and B. Boulet. Robust design of temrinal ILC with H_∞ mixed sesnitivty approach for a thermoforming oven. *Journal of Control Science and Engineering*, 2008:289391, 2008.

91 T. T. Georgiou and M. C. Smith. Robustness analysis of nonlinear feedback systems: an input-output approach. *IEEE Transactions on Automatic Control*, 42(9):1200–1221, 1997.

92 P. B. Goldsmith. On the equivalence of causal LTI iterative learning control and feedback control. *Automatica*, 38(4):703–708, 2004.

93 D. Gorinevsky. An approach to parametric nonlinear least square optimization and application to task-level learning control. *IEEE Transactions on Automatic Control*, 42(0):912–927, 1993.

94 D. Gorinevsky. Loop shaping for iterative control of batch processes. *IEEE Control Systems*, 22:55–65, 2002.

95 S. Grundelius and B. Bernhardsson. Control of liquid slosh in an industrial packaging machine. In *Proceedings of the 38th IEEE Conference on Decision and Control*, pages 1654–1659, 1999.

96 M. Guay, D. Dochain, and M. Perrier. Adaptive extremum seeking control of nonlinear dynamics with parametric uncertainties. *Automatica*, 39:1283–1293, 2003.

97 L. G. Gubin, B. T. Polyak, and E. V. Raik. The method of projections for finding the common point of convex sets. *USSR Computational Mathematics and Mathematical Physics*, 7:1–24, 1967.

98 M. Gugat. Contamination source determination in water distribution networks. *SIAM Journal of Applied Mathematics*, 72(6):1772–1791, 2012.

99 S. Gunnarsson and M. Norrlof. On the design of ILC algorithms using optimization. *Automatica*, 37(12):2011–2016, 2001.

100 S. Gunnarsson and M. Norrlof. On the disturbance properties of high order iterative learning control algorithms. *Automatica*, 42(0):2031–2034, 2006.

101 F. Gustafsson. Determining the initial states in forward-backward filtering. *IEEE Transactions on Signal Processing*, 44(4):988–992, 1996.

102 K. Hamamoto and T. Sugie. An iterative learning control algorithm with prescribed input-output space. *Automatica*, 37(11):1803–1809, 2001.

103 S. Hara, T. Yamamoto, Y. Omata, and M. Nakano. Repetitive control system: a new type servo system for periodic exogenous signals. *IEEE Transactions on Automatic Control*, 33(7):659–668, 1988.

104 T. C. Harte, J. Hatonen, and D. H. Owens. Discrete-time inverse model based iterative learning control: stability, monotonicity and robustness. *International Journal of Control*, 78(8):577–586, 2005.

105 H. Havlicsek and A. Alleyne. Nonlinear control of an electrohydraulic injection moulding machine via iterative adaptive learning. *IEEE/ASME Transactions on Mechatronics*, 4(3):312–323, 1999.

106 G. Heinzinger, D. Fenwick, B. Paden, and F. Miyaziki. Robust learning control. In *Proceedings of the 28th IEEE Conference on Decision and Control*, pages 436–440, 1989.

107 L. Hladowski, K. Galkowski, Z. Cai, E. Rogers, C. T. Freeman, and P. L. Lewin. Experimentally supported 2D systems based iterative learning control law design for error convergence and performance. *Control Engineering Practice*, 18(4):339–348, 2010.

108 L. Hladowski, K. Galkowski, Z. Cai, E. Rogers, C. T. Freeman, and P. L. Lewin. Output information based iterative learning control law design with experimental verification. *Journal of Dynamic Systems, Measurement, and Control*, 134(2):021012/1–021012/10, 2012.

109 L. Hladowski, K. Galkowski, N. Nowicka, and E. Rogers. Dynamical iterative learning control. *Control Engineering Practice*, 46:157–165, 2016.

110 D. J. Hoelzle and K. L. Barton. A new spatial iterative learning control approach for improved micro-additive manufacturing. In *2014 American Control Conference*, pages 1805–1810, Portland, Oregon, USA, 2014.

111 J. J. Hotonen, T. Harte, D. H. Owens, J. D. Ratcliffe, P. L. Lewin, and E. Rogers. Iterative learning control - what is it all about? In *Proceedings of IFAC Workshop on Adaption and Learning in Control and Signal Processing and the IFAC Workshop on Periodic Control Systems*, pages 547–553, Yokohama, Japan, 2004.

112 M. Hu, H. Du, S. Ling, Z. Zhou, and Y. Li. Motion control of an electrostrictive actuator. *Mechatronics*, 14:305–312, 2004.

113 A. M. Hughes, C. T. Freeman, J. H. Burridge, P. H. Chappell, P. L. Lewin, and E. Rogers. Preliminary results from a clinical trial for robotic-assisted stroke rehabilitation, Research Report, School of Electronics and Computer Science, University of Southampton, UK, 2009.

114 A.-M. Hughes, C. T. Freeman, J. H. Burridge, P. H. Chappell, P. L. Lewin, and E. Rogers. Feasibility of iterative learning control mediated by functional electrical stimulation for reaching after stroke. *Neurorehabilitation and Neural Repair*, 23(6):559–568, 2009.

115 K. Inaba, C.-C. Wang, M. M. Tomizuka, and A. Packard. Design of iterative learning controller based on frequency domain linear matrix inequality. In *American Control Conference*, pages 246–251, 2009.

116 A. Isidori. *Nonlinear Control Systems*, 2nd edition. Springer, New York, 1989.

117 T. Iwasaki and S. Hara. Generalized KYP lemma: unified frequency domain inequalities and design applications. *IEEE Transactions on Automatic Control*, 50(1):41–59, 2005.

118 T. N. Jensen, C. S. Kallesøe, J. D. Bendtsen, and R. Wisniewski. Iterative learning pressure control in water distribution networks. In *Proceedings of the 2018 IEEE Conference on Control Technology and Applications (CCTA)*, pages 583–588, 2018.

119 G.-M. Jeong and C.-H. Choi. Iterative learning control with advanced output data for non-minimum phase systems. In *Proceedings of the 2001 American Control Conference*, pages 890–895, Arlington, VA, 2001.

120 S. V. Johansen, J. D. Bendtsen, and J. Mogensen. Broiler slaughter weight forecasting using dynamic neural network models. In *Proceedings of the International Conference on Industrial Engineering and Applications*, 2019.

121 S. V. Johansen, J. D. Bendtsen, M. Riisgaard-Jensen, and J. Mogensen. Broiler weight forecasting using dynamic neural network models with input variable selection. *Journal of Computers and Electronics in Agriculture*, 159:97–109, 2019.

122 S. V. Johansen, M. R. Jenson, B. Chu, J. D. Bendtsen, J. Mogensen, and E. Rogers. Broiler FCR optimization using norm optimal terminal iterative learning control. *IEEE Transactions on Control Systems Technology*, 29(2):580–592, 2021.

123 S. J. Johnson, J. P. Baker, C. P. van Dam, and D. Berg. An overview of active load control techniques for wind turbines with an emphasis on microtabs. *Wind Energy*, 13:239–253, 2010.

124 S. J. Johnson, C. P. van Dam, and D. E. Berg. Active load control tecniques for wind turbines. *Sandia National Laboratories, SAND2008-4809*, 2008.

125 T. Kaczorek. *Two-Dimensional Linear Systems*, volume 68 of *Lecture Notes in Control and Information Sciences*. Springer-Verlag, Berlin, Germany, 1985.

126 C. S. Kallesøe, T. N. Jensen, and R. Wisniewski. Adaptive reference control for pressure management in water networks. In *Proceedings of the 14th European Control Conference*, pages 3268–3273, 2015.

127 E. W. Kamen. Foundations of linear time-varying systems. In W. S. Levine, editor, *CRC Controls Handbook*, pages 451–468. CRC Press, 1996.

128 T. Kavli. Frequency domain synthesis of trajectory learning control for robotic manipulators. *Journal of Robotic Systems*, 9(5):663–680, 1992.

129 S. Kawamura and N. Sakagami. Analysis on dynamics of underwater robot manipulators based on iterative learning control and time-scale transformation. In *Proceedings of the IEEE Intenrational Conference on Robotics and Automation*, pages 1088–1094, 2002.

130 P. P. Khargonekar, I. R. Peterson, and K. Zhou. Robust stabilization of uncertain linear systems: quadratic stabilizability and h_∞ control theory. *IEEE Transactions on Automatic Control*, 35(3):356–361, 1990.

131 S. Z. Khong, D. Nesic, and M. Kristic. Iterative learning control based on extremum seeking. *Automatica*, 66:138–245, 2016.

132 D.-I. Kim and S. Kim. An iterative learning control method with application for CNC machine tools. *IEEE Transactions on Industrial Applications*, 32(1):66–72, 1996.

133 D. Kowalow, M. Patan, W. Paszke, and A. Romanek. Sequential design for model calibration in iterative learning control of DC motor. In *20 International Conference on Methods and Models in Automation and Robotics (MMAR)*, pages 794–799, 2015.

134 E. Kreyszig. *Introductory Functional Analysis with Applications*. Wiley, 1978.

135 M. Kristic, I. Kannellakopoulos, and P. V. Kokotovic. *Nonlinear and Adaptive Control Design*, 1st edition. Wiley, Springer, New York, 1995.

136 J. E. Kurek and M. B. Zaremba. Iterative learning control synthesis based on 2-D system theory. *IEEE Transactions on Automatic Control*, 38(1):121–125, 1993.

137 M. A. Lackner and G. A. M. van Kuik. The performance of wind turbine smart rotor control approaches during extreme loads. *Journal of Solar Energy Engineering*, 132:011008, 2010.

138 J. Laks, L. Pao, and A. Alleyne. Comparison of wind turbine operating transitions through the use of iterative learning control. In *Proceedings of the American Control Conference*, pages 4312–4319, San Francisco, CA, 2011.

139 N. Lan, H. Q. Feng, and P. E. Crago. Neural network generation of muscle stimulation pattersn for control of arm movements. *IEEE Transactions on Rehabilitation Engineering*, 2(4):213–224, 1994.

140 B. Lautenschlager, S. Pfeiffer, C. Schmidt, and G. Lichtenberg. Real-time iterative learning control-two applications with time scales between years and nanoseconds. *International Journal of Adaptive Control and Signal Processing*, 33:424–444, 2019.

141 F. Le, I. Markovsky, C. T. Freeman, and E. Rogers. Identification of electrically stimulated muscle models of stroke patients. *Control Engineering Practice*, 18(4):396–407, 2010.

142 J. H. Lee, K. S. Lee, and W. C. Kim. Model-based iterative learning control with a quadratic criterion for time-varying linear systems. *Automatica*, 36:641–657, 2000.

143 Xiao-Dong Li, T. W. S. Chow, and J. K. L. Ho. 2-D system theory based iterative learning control for linear continuous systems with time delays. *IEEE Transactions on Circuits and Systems I: Regular Papers*, 52(7):1421–1430, 2005.

144 Y.-J. Liang and D. P. Looze. Performance and robustness issues in iterative learning control. In *Proceedings of the 32nd IEEE Conference on Decision and Control*, volume 1, pages 1990–1995, 1993.

145 T. Lin. *Newton method based iterative learning control for nonlinear systems*. PhD thesis, University of Sheffield, UK, 2004.

146 T. Lin and D. H. Owens. Monotonic Newton method based ILC with parameter optimization for non-linear systems. *International Journal of Control*, 80(8):1291–1298, 2007.

147 T. Lin, D. H. Owens, and J. Hatonen. Newton method based iterative learning control for discrete non-linear systems. *International Journal of Control*, 79(10):1263–1276, 2006.

148 L. Ljung. *System Identification - Theory for the user*. Prentice-Hall, New Jersey, USA, 1999.

149 J. Löfberg. YALMIP: A Toolbox for Modeling and Optimization in MATLAB. In *Proceedings of the CACSD Conference*, 2004.

150 R. W. Longman. Iterative learning control and repetitive control for engineering practice. *International Journal of Control*, 73(10):930–954, 2000.

151 D. G. Luenberger. *Optimization by Vector Space Methods*. Wiley, New York, 1969.

152 P. S. Lum, C. G. Burger, and P. C. Shor. Evidence for improved muscle activation pattern after retraining of reaching movements with the MIME robotic system in subjects with post-stroke hemiparesis. *IEEE Transaction on Neural Systems and Rehabilitation Engineering*, 12(2):186–194, 2009.

153 L. Magni, G. D. Nicolao, L. Magnani, and R. Scattolini. A stabilizing model-based predictive control algorithm for nonlinear systems. *Automatica*, 37(9):1351–1362, 2001.

154 S. Mandra, K. Galkowski, and H. Aschemann. Robust guaranteed cost ILC with dynamic feedforward and disturbance compensation for accurate PMSM position control. *Control Engineering Practice*, 65:36–47, 2017.

155 S. Mandra, K. Galkowski, H. Aschemann, and A. Rauh. Frequency domain design of a robust iterative learning control via convex optimization techniques. In *Proceedings of the 58th IEEE Conference on Decision and Control*, pages 5599–5604, 2019.

156 S. Mandra, K. Galkowski, A. Rauh, H. Aschemann, and E. Rogers. Iterative learning control for a class of multivariable distributed systems with experimental validation. *IEEE Transactions on Control Systems Technology*, 29(3):949–960, 2021.

157 S. Mandra, K. Galkowski, E. Rogers, A. Rauh, and H. Aschemann. Simplified implementation of repetitive process based iterative learning control designs, Research Report, School of Electronics and Computer Science, University of Southampton, UK, 2002.

158 R. Maniarski, K. Klimkowicz, W. Paszke, and E. Rogers. Design of iterative learning control schemes for spatially interconnected systems. In *58th IEEE Conference on Decision and Control*, pages 6518–6523, 2019.

159 J. A. Marshall, T. Barfoot, and J. Larsson. Autonomous underground tramming for center-articulated vehicles. *Journal of Field Robotics*, 25(6–7):400–421, 2008.

160 J. M. Martinez. Practical quasi-Newton methods for solving nonlinear systems. *Jpurnal of Computational Applied Mathemartics*, 124:17–24, 2000.

161 A. C. McCarthy, N. H. Hancock, and S. R. Raine. Development and simulation of sensor-based irrigation controlstrategies for cotton using the VARIwise simulation framework. *Computers and Electronics in Agriculture*, 101:148–162, 2014.

162 K. L. Meadmore, T. A. Excell, E. Hallewell, A.-M. Hughes, C. T. Freeman, M. Kutlu, V. Benson, E. Rogers, and J. H. Burridge. The application of precisely controlled functional electrical stimulation to the shoulder, elbow and wrist for upper limb stroke rehabilitation: a feasibility study. *Journal of Neuroengineering and Rehabilitation*, 11: 105, 2014.

163 K. L. Meadmore, A.-M. Hughes, C. T. Freeman, Z. Cai, D. Tong, J. H. Burridge, and E. Rogers. Functional electrical stimulation mediated by iterative learning control and 3D robotics reduces motor impairment in chronic stroke. *Journal of Neuroengineering and Rehabilitation*, 32, 9 2012.

164 R. Merry, R. van de Molengraft, and M. Steinbuch. The influence of distrubances in iterative learning control. In *Proceedings of IEEE Conference on Control Applications*, pages 974–979, 2005.

165 W. H. Moase, C. Manzie, and M. J. Brear. Newton-like extremum seeking for the control of thermoacoustic instability. *IEEE Transactions on Automatic Control*, 55(9):2094–2105, 2010.

166 K. L. Moore. *Iterative Learning Control for Dynamic Systems*. Advances in Industrial Control. Springer, New York, 1993.

167 K. L. Moore and Y. Q. Chen. Iterative learning control approach to a diffusion control problem in an irrigation application. In *Proceedings of the 2006 IEEE International Conference on Mechatronics and Automation*, pages 1329–1334, 2006.

168 K. L. Moore and O. El-Sharif. Internal model principle for discrete repetitive processes. In *Proceedings of the 48th IEEE Conference on Decision and Control*, pages 446–451, 2009.

169 K. L. Moore, M. Ghosh, and Y. Q. Chen. Spatial-based iterative learning control for motion control applications. *Meccanica*, 42:167–175, 2007.

170 K. L. Moore and A. Mathews. Iterative learning control with non-standard assumptions applied to the control of gas-metal arc welding. In Z. Bien and J.-X. Xu, editors, *Iterative Learning Control: Analysis, Design, Integration and Applications*, pages 335–349. Springer-Verlag, Berlin, Germany, 1998.

171 J. J. More and G. Toralodo. Algorithms for bound constrained quadratic-programming problems. *Numerische Mathematik*, 55(4):377–400, 1989.

172 B. R. Noack, K. Afanasiev, M. Morzynski, G. Tadmor, and F. Thiele. A hierarchy of low-dimensional models for the transient and post-transient cylinder wake. *Journal of Fluid Mechanics*, 497:355–363, 2003.

173 J. Nocedal and S. J. Wright. *Numerical Optimization*, 2nd edition. Springer, London, 2006.

174 M. Norrlof. Comparative study on first and second order ILC - frequency domain analysis and experiments. In *Proceedings of the 39th IEEE Conference on Decision and Control*, pages 3415–3420, 2000.

175 M. Norrlof and S. Gunnarsson. Distrubance aspects of iterative learning control. *Engineering Applications of Artifical Intelligence*, 14:87–94, 2001.

176 M. Norrlof and S. Gunnarsson. Time and frequency domain convergence properties in iterative learning control. *International Journal of Control*, 75(14):1114–1126, 2002.

177 M. Norrlof and S. Gunnarsson. Distrubance rejection using an ILC algorithm with iteration varying filter. *Asian Journal of Control*, 6:432–438, 2004.

178 W. N. Nowicka. Iterative learning control for load management in wind turbines with smart rotor blades. *PhD thesis*, University of Southampton, UK, 2020.

179 C. Ocampo-Martinez, V. Puig, G. Cembrano, and J. Quevedo. Application of predictive control strategies to the management of complex networks in the urban water cycles. *IEEE Control Systems Magazine*, 33(1):14–41, 2013.

180 OECD. *OECD-FAO Agricultural Outlook 2018–2027*. OECD Publishing, 2018.

181 M. C. D. Oliveira, J. Bernussou, and J. C. Geromel. Extended H_2 and H_∞ norm characterizations and controller parameterizations for discrete-time systems. *International Journal of Control*, 75(2):666–679, 2002.

182 C. L. Ooi, T. Asai, and H. Okajima. Extension of reference signals in iterative learning control for non-minimum phase systems. *SICE Journal of Control, Measurement, and System Integration*, 4(1):050–054, 2011.

183 J. M. Ortega and W. C. Rheinboldt. *Iterative Solution of Nonlinear Equations in Several Variables*, volume 6. Academic Press, London, 1970.

184 C. J. Ostafew, A. P. Schoellig, T. D. Barfoot, and J. Collier. Learning based nonlinear model predictive control to improve vision-based mobile robot path tracking. *Journal of Field Robotics*, 33(1):133–152, 2016.

185 D. H. Owens. Universal iterative learning control using adaptive high gain feedback. *International Journal of Adaptive Control and Signal Processing*, 7:383–388, 1993.

186 D. H. Owens. *Iterative Learning Control An Optimization Paradigm*. Springer, 2016.

187 D. H. Owens and B. Chu. Modelling of non-minimum phase effects in discrete-time norm optimal iterative learning control. *International Journal of Control*, 83(10):2012–2027, 2014.

188 D. H. Owens, E. Rogers, B. Chu, C. T. Freeman, and P. L. Lewin. Influence of nonminimum phase zeros on the performance of optimal continuous-time iterative learning control. *IEEE Transactions on Control Systems Technology*, 22(3):1151–1158, 2014.

189 D. H. Owens, J. J. Hatonen, and S. Daley. Robust monotone gradient-based discrete-time iterative learning control. *International Journal of Robust and Nonlinear Control*, 19(4):634–661, 2009.

190 D. H. Owens and R. P. Jones. Iterative solution of constrained differential/algebraic systems. *International Journal of Control*, 27:957–974, 1978.

191 D. H. Owens and G. Munde. Error convergence in an adaptive iterative learning controller. *International Journal of Control*, 73(10):851–857, 2000.

192 D. H. Owens and K. Peng. Parameter optimization in iterative learning control. *International Journal of Control*, 76(11):1059–1069, 2003.

193 D. H. Owens and E. Rogers. Comments on 'on the equivalence of causal LTI iterative learning control and feedback control'. *Automatica*, 40(5):895–898, 2004.

194 A. Packard and J. C. Doyle. The complex structured singular value. *Automatica*, 29(1):71–109, 1993.

195 F. Padieu and R. Su. An H_∞ approach to learning control systems. *International Journal of Adaptive Control and Signal Processing*, 4:465–474, 1990.

196 P. Pakshin, J. Emelianova, M. Emelianov, K. Galkowski, and E. Rogers. Dissipativity and stabilization of nonlinear repetitive processes. *Systems & Control Letters*, 91:14–20, 2016.

197 P. Pakshin, J. Emelianova, K. Galkowski, and E. Rogers. Iterative learning control for discrete linear systems. In *Proceedings of the European Control Conference*, pages 3766–3771, 2019.

198 P. Pakshin, J. Emelianova, E. Rogers, and K. Galkowski. Iterative learning control with input saturation. *IFAC PapersOnLine*, 52(29):338–343, 2019.

199 P. Pakshin, J. Emelianova, E. Rogers, and K. Galkowski. Repetitive process based stochastic iterative learning control design. *Systems & Control Letters*, 137:104625, 2020.

200 P. Pakshin, J. Emelianova, E. Rogers, and K. Galkowski. Iterative learning control for switched systems in the presence of input saturation. *IFAC-PapersOnLine*, 53(2):1444–1449, 2020.

201 P. Pakshin, S. Mandra, J. Emelianova, E. Rogers, K. Erwinski, and K. Galkowski. Experimentally valdiated vector Lyapunov function based iterative learning control deign under input saturation. *IEEE Transactions on Control Systems Technology*, page Submitted, 2022.

202 P. V. Pakshin, J. P. Emelianova, M. A. Emelianov, K. Gałkowski, and E. Rogers. Stochastic stability of some classes of nonlinear 2D systems. *Automation and Remote Control*, 79(1):89–102, 2018.

203 W. Paszke, E. Rogers, and K. Galkowski. Experimentally verified generalized KYP lemma based iterative learning control design. *Control Engineering Practice*, 53:57–67, 2016.

204 W. Paszke, E. Rogers, K. Galkowski, and Z. Cai. Robust finite frequency range iterative learning control design and experimental verification. *Control Engineering Practice*, 21(10):1310–1320, 2013.

205 M. Patan. A parallel sensor scheduling technique for fault detection in distributed parameter systems. In *Euro-Par 2008–Parallel Processing*, pages 833–843. Springer, 2008.

206 M. Patan. *Optimal Sensor Networks Scheduling in Identification of Distributed Parameter Systems*. Springer, New York, 2012.

207 M. Patan and D. Uciński. Time-constrained sensor scheduling for parameter estimation of distributed systems. In *49th IEEE Conference on Decision and Control*, pages 7–12, 2010.

208 M. Patan and D. Uciński. Cost-constrained d-optimum node activation for large-scale monitoring networks. In *American Control Conference*, pages 1643–1648, 2016.

209 M. Patan and D. Uciński. *d*-optimal spatio-temporal sampling design for identification of distributed parameter systems. In *55th IEEE Conference on Decision and Control*, pages 3985–3990, 2016.

210 M. Q. Phan, R. W. Longman, and K. L. Moore. Unified formulation of linear iterative learning control. *Advances in Astronautical Sciences*, 105:93–111, 2000.

211 A. M. Plotnik and R. W. Longman. Subtitles in the use of zero-phase low-pass filtering and cliff filtering in learning control. *Advances in the Astronautical Sciences*, 103:673–692, 1999.

212 A. D. Polyanin. *Linear Partial Differential Equations for Engineers and Scientists*. Chapman and Hall/CRC, New York, 2002.

213 D. Popovic and Popovic M. Tuning of a nonanalytical hierarchical control system for reaching with FES. *IEEE Transactions on Biomedical Engineering*, 45:203–212 1998.

214 Z. Povlej. Quasi-Newton's method for multiobjective optimization. *Journal of Computational and Applied Mathematics*, 255:765–477, 2014.

215 R. Prabel and H. Aschemann. Active oscillation damping for a truck drive train. In *9th International Conference on Methods and Models in Automation and Robotics (MMAR)*, pages 486–491, 2014.

216 D. Prichard and J. Theiler. Generating surrogate data for time series with several simultaneously measured variables. *Physical Review Letters*, 73(7):951–954, 1994.

217 R. R. Pérez, G. G. Sanz, V. V. Puig, J. J. Quevedo, M. A. M. A. Cugueró-Escofet, F. Nejjari, J. J. Meseguer, G. Cembrano, J. M. Mirats-Tur, and R. Sarrate. Leak localization in water networks – a model-based methodology using pressure sensors applied to a real network in Barcelona. *IEEE Control Systems Magazine*, 34(4):24–36, 2014.

218 J. D. Ratcliffe. Iterative learning control implemented on a multi-axis system. *PhD thesis*, University of Southampton, UK, 2004.

219 J. D. Ratcliffe, T. Harte, J. J. Hotonen, P. L. Lewin, E. Rogers, and D. H. Owens. Practical implementation of a model inverse iterative learning controller. In *Proceedings of IFAC Workshop on Adaption and Learning in Control and Signal Processing and the IFAC Workshop on Periodic Control Systems*, pages 687–692, 2004.

220 J. D. Ratcliffe, J. J. Hatonen, P. L. Lewin, E. Rogers, and D. H. Owens. Robustness analysis of an adjoint optimal iterative learning controller with experimental verification. *International Journal of Robust and Nonlinear Control*, 18(10):1089–1113, 2008.

221 J. D. Ratcliffe, P. L. Lewin, E. Rogers, J. J. Hatonen, and D. H. Owens. Norm-optimal iterative learning control applied to a gantry robots for automation applications. *IEEE Transactions on Robotics*, 22(6):103–107, 2006.

222 H. Rauh, L. Senkel, H. Aschemann, V. V. Saurin, and G. V. Kostin. An integrodifferential approach to modeling, control, state estimation and optimization for heat transfer systems. *International Journal of Applied Mathematics and Computer Science*, 26:15–30, 2016.

223 B. Ren, P. Frihauf, R. J. Rafac, and M. Kristic. Laser pulse shaping via extremum seeking. *Control Engineering Practice*, 20:674–683, 2012.

224 A. Rezaeizadeh, R. Kalt, T. Schilcher, and R. S. Smith. An iterative learning control approach for the radio frequency pulse compressor amplitude and phase modulation. *IEEE Transactions on Nuclear Science*, 63(2):842–848, 2016.

225 A. Rezaeizadeh, T. Schilcher, and R. S. Smith. Control of the Swiss free electron laser. *IEEE Control Systems Magazine*, 37(6):30–51, 2017.

226 A. Rezaeizadeh and R. S. Smith. Iterative learning control for the radio frequency subsystems of a free-electron laser. *IEEE Transactions on Control Systems Technology*, 26(5):1567–1577, 2019.

227 R. P. Roesser. A discrete state-space model for linear image processing. *IEEE Transactions on Automatic Control*, 20(1):1–10, 1975.

228 E. Rogers, K. Gałkowski, and D. H. Owens. *Control Systems Theory and Applications for Linear Repetitive Processes*, volume 349 of *Lecture Notes in Control and Information Sciences*. Springer-Verlag, Berlin, Germany, 2007.

229 E. Rogers, K. Galkowski, W. Paszke, K. L. Moore, P. H. Bauer, L. Hladowski, and P. Dabkowski. Multidimensional control systems: case studies in design and evaluation. *Multidimensional Systems and Signal Processing*, 26(4):895–939, 2015.

230 E. Rogers and K. Galkowski. *Control systems analysis for the Fornasini-Marchesini 2D systems model - progress after four decades*, Research Report, School of Electronics and Computer Science, University of Southampton, UK, 2018.

231 E. Rogers, D. H. Owens, H. Werner, C. T. Freeman, P. L. Lewin, S. Kichhoff, C. Schmidt, and G. Lichtenberg. Norm-optimal iterative learning control with application to problems in accelerator-based free electron lasers and rehabilitation robotics. *Europena Journal of Control*, 16(5):497–522, 2010.

232 S. S. Saab. A discrete time learning control algorithm for a class of linear time-invariant systems. *IEEE Transactions on Automatic Control*, 40(6):1138–1142, 1995.

233 S. S. Saab. A discrete-time stochastic learning control algorithm. *IEEE Transactions on Automatic Control*, 46(6): 877–887, 2001.

234 S. S. Saab. Stochastic P-type/D-type iterative learning control algorithms. *International Journal of Control*, 76(2):139–148, 2003.

235 S. K. Sahoo, S. K. Panda, and J. X. Xu. Application of spatial iterative learning control for direct torque control of switched reluctance motor drive. In *IEEE Power Engineering Society General Meeting*, pages 1–7, Tampa, FL, 2007.

236 R. M. Sanner and J. J. E. Slotine. Gaussian networks for direct adaptive control. *IEEE Transactions on Neural Networks*, 3(0):837–863, 1992.

237 L. C. Scales. *Introduction to Non-linear Optimization*. MacMillian, London, 1985.

238 G. Sebastian, Y. Tan, and D. Oetomo. Convergence analysis of feedback-based iterative learning control with input saturation. *Automatica*, 101:44–52, 2019.

239 T. Seel, T. Schauer, and J. Raisch. Monotonic convergence of iterative learning control systems with variable pass length. *International Journal of Control*, 90(3):393–406, 2017.

240 J. S. Shamma. Linearization and gain-scheduling. In W. S. Levine, editor, *CRC Controls Handbook*, pages 388–397. CRC Press, Boca Raton, FL, 1996.

241 J.-H. She, H. Kobayashi, Y. Ohyama, and X. Xin. Disturbance estimation and rejection - an equivalent input disturbance estimator approach. In *43rd IEEE Conference on Decision and Control*, pages 1736–1741, 2004.

242 G. Shue, P. E. Crago, and H. J. Chizeck. Muscle-joint models incorporating activitation dynamics, moment-angle, and moment-velocity properties. *IEEE Transactions on Biomedical Engineering*, 42:213–223, 1995.

243 S. Skogestad and I. Postlethwaite. *Multivariable Feedback Control: Analysis and Design*. Wiley, 2005.

244 M. Steinbuch. Repetitive control for systems with uncertain period-time. *Automatica*, 38(12):2103–2109, 2002.

245 A. Stoorvogel. The H_∞ control problem: a state-space approach. *PhD thesis*, Eindhoven University of Technology, 1990.

246 T. Sugie and F. Sakai. Projection based iterative learning control with its application to continuous-time identification. In *Proceedings of ICKS*, pages 95–102, 2007.

247 R. S. Sutton and A. G. Barto. *Reinforcement Learning: An introduction*. MIT Press, Cambridge, MA, 2018.

248 K. K. Tan, S. Zhao, and S. Huang. Iterative reference adjustment for high-precision and repetitive motion control applications. *IEEE Transactions on Control Systems Technology*, 13(1):85–97, 2005.

249 M. Tao, R. Koust, and A. Gurcan. Learning feedforward control. In *Proceedings of American Control Conference*, pages 2575–2579, 1994.

250 H.-F. Tao, W. Paszke, E. Rogers, K. Galkowski, and H.-Z. Yang. Modified Newton method based iterative learning control design for discrete nonlinear systems with constraints. *Systems & Control Letters*, 118:35–43, 2018.

251 A. Tayebi, S. Abdul, M. B. Zaremba, and Y. Ye. Robust iterative learning control design: application to a robot manipulator. *IEEE/ASME Transactions on Mechatronics*, 13(5):608–613, 2008.

252 A. Tayebi and M. B. Zaremba. Iterative learning control design is straightforward for uncertain LTI systems satisfying the robust performance condition. *IEEE Transactions on Automatic Control*, 48(1):101–106, 2003.

253 D. S. Tong, G. Pipeleers, and J. Swevers. Multi-objective iterative learning control using convex optimization. *European Journal of Control*, 33:35–42, 2017.

254 A. Trofino. Parameter-dependent Lyapunov functions for a class of uncertain linear systems: an LMI approach. In *Proceedings of the 38th IEEE Conference on Decision and Control*, volume 1, pages 2341–2346, 1999.

255 O. R. Tutty, M. Blackwell, E. Rogers, and R. Sandberg. Iterative learning control for improved aerodynamic load performance of wind turbines with smart rotors. *IEEE Transactions on Control Systems Technology*, 22:967–979, 2014.

256 M. Uchiyama. Formation of high-speed motion pattern of a mechanical arm by trial. *Society of Instrumentation and Control Engineers*, 14(6):706–712, 1978.

257 D. Ucinski. *Optimal Measurement Methods for Distributed Parameter System Identification*. CRC Press, New York, 2004.

258 D. Uciński and M. Patan. D-optimal design of a monitoring network for parameter estimation of distributed systems. *Journal of Global Optimization*, 39(2):291–322, 2007.

259 F. J. Valero-Cuevas. A mathematical approach to the mechanical capabilities of limbs and fingers. *Progress in Motor Control*, 629:619–633, 2009.

260 J. van de Wijdeven. Iterative learning control design for uncertain and time-windowed systems. *PhD thesis*, Eindhoven University of Technology, 2008.

261 J. van de Wijdeven and O. H. Bosgra. Using basis functions in iterative learning control: analysis and design theory. *International Journal of Control*, 84(4):661–675, 2010.

262 J. van de Wijdeven, T. Donkers, and O. Bosgra. Iterative learning control: Robust monotonic convergence analysis. *Automatica*, 45(10):2383–2391, 2009.

263 J. van de Wijdeven, T. Donkers, and O. Bosgra. Iterative learning control for uncertain systems: noncausal finite time interval robust control design. *International Journal of Robust and Nonlinear Control*, 21(14):1645–1666, 2011.

264 G. A. M. van Kuik and D. E. Berg (eds). Smart blades. *Wind Energy*, 13(2 & 3):101–274, 2010.

265 P. S. Veers, T. D. Ashwill, H. J. Sutherland, D. L. Laird, and D. W. Lobitz. Trends in the design, manufacture and evaluation of wind turbine blades. *Wind Energy*, 6:245–259, 2003.

266 M. Verwoerd. Iterative learning control: a critical review. *PhD thesis*, University of Twente, The Netherlands, 2004.

267 B. Wahlberg. System identification using Laguerre models. *IEEE Transactions on Automatic Control*, AC-36:551–562, 1991.

268 J. Wallen, M. Norrlof, and S. Gunnarsson. A framework for analysis of observer based ILC. *Asian Journal of Control*, 13:3–14, 2011.

269 D. Wang. Convergence and robustness of discrete time nonlinear systems with iterative learning control. *Automatica*, 34:1445–1448, 1998.

270 D. Wang. On D-type and P-type ILC designs and anticipatory approach. *International Journal of Control*, 73(10):890–901, 2000.

271 L. Wang. *Model Predictive Control Systems Design and Implementation using MATLAB*. Springer, 2009.

272 X. Wang. Repetitive process based higher-order iterative learning control law design. *PhD thesis*, University of Southampton, UK, 2017.

273 L. Wang, S. Chai, E. Rogers, and C. T. Freeman. Multivariable repetitive-predictive controllers using frequency decomposition. *IEEE Transactions on Control Systems Technology*, 20(6):1597–1604, 2012.

274 L. Wang, C. T. Freeman, S. Chai, and E. Rogers. Predictive-repetitive control with constraints: from design to implementation. *Journal of Process Control*, 23(7):956–967, 2013.

275 L. Wang, C. T. Freeman, and E. Rogers. Predictive iterative learning control with experimental validation. *Control Engineering Practice*, 53:24–34, 2016.

276 Y. Wang, F. Gao, and F. J. Doyle. Survey on iterative learning control, repetitive control, and run-to-run control. *Journal of Process Control*, 19(10):1589–1600, 2009.

277 Z.-B. Wei, Q. Quan, and K.-Y. Cai. Output feedback ILC for a class of nonminimum phase nonlinear systems with input saturation: an additive-state-decomposition-based method. *IEEE Transactions on Automatic Control*, 62(1):502–508, 2017.

278 Re-Bing Wu, B. Chu, D. H. Owens, and H. Rabitz. Data-driven gradient algorithm for high-precision quantum control. *Physical Review A*, 97(4):042122, 2018.

279 Z. Xiong and J. Zhang. Batch-to-batch optimal control of nonlinear batch processes based on incrementally updated models. *Proceedings of the Inatitution of Electrical Engineers, Control Theory and Applications*, 151:158–165, 2004.

280 J.-X. Xu. Analysis of iterative learning control for a class of nonlinear discrete-time systems. *Automatica*, 33:1905–1907, 1997.

281 W. Xu, B. Chu, and E. Rogers. Iterative learning control for robotic-ssisted upper limb stroke rehabilitation in the presence of muslce fatigue. *Control Engineering Practice*, 31:63–72, 2014.

282 J. Xu, M. Sun, and L. Yu. LMI-based robust iterative learning controller design for discrete linear uncertain systems. *Journal of Control Theory and Applications*, 3:259–265, 2005.

283 J.-X. Xu and Y. Tan. Robust optimal design and convergence properties analysis of iterative learning control approaches. *Automatica*, 38(10):1867–1880, 2002.

284 J.-X. Xu and Y. Tan. *Linear and Nonlinear Iterative Learning Control*. Springer-Verlag, Berlin, 2003.

285 J.-X. Xu, Y. Tan, and T.-H. Lee. Iterative learning control design based on composite energy function with input saturation. *Automatica*, 40:1371–1377, 2004.

286 N. Young. *An Introduction to Hilbert Space*. Cambridge University Press, Cambridge, UK, 2012.

287 Y. Zhang, B. Chu, and Z. Shu. A preliminary study on the relationship between iterative learning control and reinforcement learning. *IFAC Papers Online*, 52(29):314–319, 2019.

Index

Iterative Learning Control Algorithms and Experimental Benchmarking, First Edition.
Eric Rogers, Bing Chu, Christopher Freeman and Paul Lewin.
© 2023 John Wiley & Sons Ltd. Published 2023 by John Wiley & Sons Ltd.